A New Certificate Chemistry

A. HOLDERNESS, M.Sc., F.R.I.C.
*Formerly Senior Chemistry Master at
Archbishop Holgate's Grammar School, York*

JOHN LAMBERT, M.Sc.
*Formerly Senior Chemistry Master at
King Edward's School, Birmingham*

FIFTH EDITION

revised in collaboration with

J. J. THOMPSON, M.A., Ph.D., F.R.I.C.
*Head of Science Education,
Department of Educational Studies, University of Oxford*

Heinemann Educational Books
London

Heinemann Educational Books Ltd
22 Bedford Square, London WC1B 3HH

LONDON   EDINBURGH   MELBOURNE   AUCKLAND

HONG KONG   SINGAPORE   KUALA LUMPUR   NEW DELHI

IBADAN   NAIROBI   JOHANNESBURG

EXETER (NH)   KINGSTON   PORT OF SPAIN

ISBN 435 64424 6

First published 1961
Second Edition 1961
Reprinted 1962
Third Edition 1964
Reprinted 1965, 1966, 1968
Fourth Edition 1969
Reprinted 1970, 1971, 1973, 1974 (twice), 1975
Fifth Edition 1976
Reprinted 1977, 1978, 1979, 1980

Printed and bound in Great Britain by
Butler and Tanner Ltd, Frome, Somerset

Preface to the Fifth Edition

The developments in the teaching of school chemistry which have taken place during the past decade, and which have affected chemistry courses in every continent in the world, have prompted the major revision which has been made for this fifth edition of *New Certificate Chemistry*. These developments have involved not only a change in the *content* of the syllabuses but, more significantly, changes in the *methods* of teaching and examining the subject. One of the prominent aspects of the 'new chemistry' has been the increased emphasis on the place of practical work in the course, and as a consequence a major feature of this edition is the extended scope and presentation of the experimental work. It is therefore anticipated that students will find the book to be of value on the laboratory bench.

In addition to the revision of the text of the fourth edition, this book contains a substantial amount of new material, particularly in the chapters on states of matter, oxidation and reduction, chemical equilibrium, rates of reaction, and energy changes in chemical reactions.

The recommendations of the various Examination Boards and of the Association for Science Education have been taken into account in the system of units and nomenclature adopted in this edition. In particular the A.S.E. publications *Chemical Nomenclature, Symbols and Terminology* and *SI Units, Signs, Symbols and Abbreviations* have been consulted extensively in the preparation of the text. Some trivial names in common usage have been retained, and the *mole* is used exclusively throughout.

The overall objective has been to produce a book which is not only up to date in content and approach, but which also provides sufficient detail in both theoretical and experimental work for the student to feel confident in using it as a reference guide for his course.

The authors wish to record their debt to the practising teachers who advised them in the preparation of the first edition, more particularly Messrs A. B. Adamson, F. E. P. Alford, F. W. Ambler, H. G. Andrew, H. C. Cockroft, E. H. Coulson, J. B. Guy, I. G. Jones, P. N. Lawrence, E. W. Moore, T. A. Muir, J. Turpie, and E. Dickinson.

It is especially a pleasure to acknowledge the helpful comments received during the preparation of this new edition from science teachers throughout the world, and in particular the valuable contributions made by Mr Martyn Berry, Mr J. P. Chippendale, Mrs Freda Stevens, Mr I. O. Ikeobi, and Dr E. O. Arene, all of whom provided helpful advice on the content and structure, and read the manuscript and proofs, and Dr C. J. Devoy, who prepared the questions from recent examination papers. Thanks are especially due to Dr J. J. Thompson who undertook the major task of preparing the revised manuscript, including the writing of new material.

January 1976   J. L.

Examination Questions

Many of the questions at the end of each chapter are reproduced by kind permission of the following Examination Boards:

University of Cambridge Local Examinations Syndicate (C.)
Oxford and Cambridge Schools Examination Board (O. and C.)
Joint Matriculation Board (J.M.B.)
University of London (L.)
Associated Examining Board (A.E.B.)
Oxford Delegacy of Local Examinations (O.)
Welsh Joint Education Committee (W.)
Southern Universities Joint Board for School Examinations (S.)
Scottish Certificate of Education Examination Board (Scottish)
Northern Ireland Examination Board (N.I.)

A New Certificate Chemistry

BABAK·M·

Cover photographs

These show potassium reacting with cold water containing phenolphthalein indicator. Hydrogen is evolved and the heat of the reaction is such that the hydrogen ignites explosively. The photographs show a special demonstration of this effect carried out in carefully controlled conditions. In the school laboratory only a minute piece of potassium should be used and **under no circumstances must the quantity exceed a 3 mm cube.** *All safety precautions must be strictly observed (see Experiment 67, page 244).*

Contents

1 Physical and Chemical Change; Elements, Compounds, and Mixtures

Physical and chemical change

The science of Chemistry sets before itself, as its primary objects, the determination of the nature and properties of the non-living matter which surrounds us and the preparation of new substances, scientifically interesting or generally useful, from the materials which Nature has provided. In trying to determine the nature of substances, chemists have been greatly interested in the changes which these undergo when subject to conditions which they normally do not encounter: high temperature, high pressure, extreme cold, contact with other materials under varying conditions, and so on. It is largely from the changes which materials undergo when subject to these conditions that chemists have drawn conclusions about their nature.

But changes are many and various. Any one change may be viewed from many different angles. When iron rusts, a chemist is concerned with the different properties which the iron and rust possess. How does each react with acids, alkalis, and other reagents? He also tries to give an explanation of what occurred to the iron when it rusted. The physicist will want to know whether the density, the conductivity, and the specific heat of the rust are the same as that of iron, from which it has been made. An economist thinks of the huge cost which accompanies the change, for a vast amount of money is spent yearly in an endeavour to prevent iron from rusting. It may be that what the chemist and the physicist find out about the rusting of iron will help the economist (or, more directly, the manufacturer of iron articles). It is mainly as a result of chemical research that rustless and stainless steels have appeared, while during that research the physicist has been careful to ensure that the elasticity and tensile strength of the steel have not been impaired by the process which made them rustless. Clearly, we must attempt to define, if only roughly, the kind of changes in which the chemist is interested and to which the name 'chemical' can properly be applied. With the object of attaining some kind of definition of 'chemical changes', we will now examine a few changes in the hope that, from them, some conclusions may emerge.

Experiment 1
Heating metals in air

(a) Magnesium

Hold one end of a piece of magnesium ribbon in tongs and put the other end in a Bunsen flame. Note the intense brilliance of the flame of the burning magnesium and the nature of the residue – a white ash – which remains.

(b) Platinum (or Nickel)

Repeat the experiment with platinum (or nickel) by holding a loose coil of platinum (or nickel) wire in a Bunsen flame. Note the white-hot glow of the metal, but contrast its unchanged appearance, after cooling, with the white ash left by the magnesium.

Experiment 2
Dissolution of substances in water

(a) Sodium

Safety goggles should be worn for this experiment

Take a small piece of sodium in tongs from the oil under which it is kept, and, *never touching it with your fingers,* cut it into pieces about the size of a **very small** pea. Drop these pieces in turn on to the surface of a little distilled water in a small beaker. Note how the sodium melts into a ball, darts about the surface of the water, produces a hissing sound, and finally disappears with a small flash and explosion. Heat the resulting clear liquid in an evaporating basin until no more steam is given off. On cooling, a white solid is left. If added to water, this solid dissolves but does not show the same vigorous action as sodium because it is a new substance.

(b) Common salt

To distilled water in a beaker, add some common salt and stir the mixture. The common salt undergoes an obvious change; it gradually disappears forming a solution and being no longer visible as a white solid. Put the liquid into a porcelain dish and heat gently until all the water has evaporated off. The common salt reappears in its original white solid form.

Experiment 3
Heating of sulphur

(a) Strongly

Heat some roll sulphur on a deflagrating spoon. Note how the sulphur melts and later begins to burn with a blue flame. It gradually decreases in amount and finally the spoon will be left empty. The sulphur has not simply been annihilated. Its disappearance is due to its conversion into a new gaseous substance which is invisible, but whose presence in the air can be detected by its irritating smell or by burning the sulphur in a gas-jar and adding to the jar some blue litmus solution. The gas, sulphur dioxide, will turn it red.

(b) Gently

Powder some roll sulphur in a mortar, then heat it gently in a test-tube, shaking all the time. Notice how the sulphur melts to an amber-coloured liquid (other changes will occur if it is more strongly heated) and that this liquid, on cooling, returns to its original condition as a yellow solid.

The six changes we have considered above are not all the same in nature. They fall into two classes, which are distinguished by the following characteristics.

1. All the changes in Experiments (*b*) were **easily reversible**: the molten sulphur returned to the solid form when cooled; the platinum wire ceased to glow and regained its original appearance when removed from the flame; the common salt was recovered by evaporating off the water. Contrast these results with those

of the changes in Experiments (*a*). The white ash from magnesium was totally unlike the original magnesium, and it would be a difficult matter to obtain magnesium from it; on evaporation of water in Experiment 2(*a*) we recovered not sodium, but caustic soda, from which sodium cannot easily be obtained; the sulphur became part of a gas from which it would be difficult to recover sulphur. These changes are not reversible. The easily reversible type of change is called 'physical change'; the more permanent type 'chemical change'.

2. In none of the physical changes recorded was a new kind of matter formed; we began with platinum, common salt, and sulphur and, after the change, finished with just those materials. **In all the chemical changes recorded some new kind of matter was formed**; magnesium was converted to the white powdery ash, magnesium oxide, sodium to caustic soda, and solid sulphur to the gas, sulphur dioxide – a new kind of matter each time. This is characteristic of chemical change.

3. The physical changes were not accompanied by any marked external effects. The solution of common salt in water and the melting of sulphur were not violent changes. The action of sodium with water, however, produced enough heat to melt the sodium and was violent enough to be slightly explosive at the end; the burning of magnesium produced intense heat and light, and the burning of sulphur similar, but less intense, effects. The chemical changes were the more violent, and were accompanied by heat changes. This is commonly the case.

4. Although no weighings were taken during the experiments it can be shown that, in all three physical changes which were recorded, **no change of mass occurred**: the sulphur, platinum, and common salt weighed just as much before the changes as after them. In the three chemical changes, however, it can be shown that the white ash weighed more than the magnesium, the caustic soda more than the sodium, and the gaseous sulphur dioxide more than the sulphur. (These gains in mass are made at the expense of other materials which lose in mass correspondingly; the gains in the case of magnesium and sulphur were made at the expense of the air and, in the case of sodium, at the expense of the water.)

We thus distinguish two kinds of changes – chemical changes and physical changes.

Now consider the following suggestions about a few common changes and decide by comparison with those discussed above whether the changes are physical or chemical. The correct classification appears on page 4.

1. (*a*) Melting of ice; (*b*) conversion of water to steam.
   Are the changes easily reversed? Are there any noticeably violent external effects?

2. Burning of coal.
   Does the coal appear to weigh the same as the products after burning it? (Appearances here are deceptive, see page 23.) Can we easily obtain coal again from its products of combustion? Are there any noticeable external effects while coal in burning?

3. Rusting of iron.
   Can iron be easily recovered from the rust?

4. Magnetizing iron.
   Can the iron be readily de-magnetized? Are there any marked changes during magnetization?

5. A coal-gas explosion.
   Is this change violent? Is there considerable heat change?

6. Heating of the filament of an electric light globe by the current.
Is the filament readily cooled again? Does it appear changed when cooled?
7. The melting of candle-wax.
Is the liquid wax easily solidified again? Does it then appear the same as the original wax? Are there any marked heating effects as the melting occurs?

We may now summarize the characteristics of chemical and physical change in the following table.

Physical change	Chemical change
1. Produces no new kind of matter	1. Always produces a new kind of matter
2. Is generally reversible	2. Is generally not easily reversible
3. Is not accompanied by great heat change (except latent heat effects accompanying changes of state)	3. Is usually accompanied by considerable heat change
4. Produces no change of mass	4. Produces individual substances whose masses are different from those of the original individual substances
	Thus, if two substances, A and B, react chemically and are changed into substances C and D, the mass of C will be different from the mass of A or B, and the mass of D will be different from the mass of A or B
Examples	*Examples*
1. All cases of the melting of a solid to a liquid (or the reverse)	1. The burning of any substance in air
2. All cases of vaporization of a liquid (or the reverse)	2. The rusting of iron
3. Magnetization of iron	3. The slaking of lime
4. The heating of a metal wire by electricity	4. Explosion of coal-gas or hydrogen with air

Elements and compounds

We will begin our study of elements and compounds by considering two fairly simple chemical changes.

Experiment 4
Heating mercury(II) oxide

This experiment should only be carried out in a well-ventilated space.

Put a spatula full of red oxide of mercury (mercury(II) oxide) into a dry test-tube. Heat it, rotating the test-tube so that it does not become misshapen. A silvery mirror gradually appears on the upper part of the test-tube (where it is cool), and later silvery globules of mercury will be seen. When the mirror begins to appear, insert a glowing splint of wood into the test-tube. It is rekindled. This is because the invisible gas, oxygen, is coming off from the heated oxide.

It is clear that, under the action of heat, mercury(II) oxide has yielded two products – mercury and oxygen.

Experiment 5
Heating lead(II) nitrate

This experiment should only be carried out in a well-ventilated space, and safety goggles should be worn.

Repeat Experiment 4 using a small quantity of lead(II) nitrate. Brown fumes are given off in this case (they are called nitrogen dioxide), and, by the test given above, it can be shown that oxygen is also liberated. Finally a yellow solid will remain in the test-tube. This solid is known as 'massicot'.

These experiments show that both mercury (II) oxide and lead(II) nitrate must be fairly complex substances. This is obvious from the fact that mercury(II) oxide yielded, under the action of heat, two substances, mercury and oxygen, while lead(II) nitrate yielded three – massicot, nitrogen dioxide, and oxygen. The question now arises whether these products can themselves be split up further into still simpler substances. The answer to this question is that in two cases they can; from massicot we can, by suitable chemical means, obtain lead and oxygen, and from nitrogen dioxide, nitrogen and oxygen. This means that massicot and nitrogen dioxide are themselves complex substances. How much further can this process of splitting up into simpler products be carried? Can we obtain from the lead, nitrogen, oxygen, and mercury, into which we have resolved our original lead(II) nitrate and mercury(II) oxide, any substances which are simpler still? The answer now is that we cannot. By no chemical process whatever is it possible to obtain from lead, mercury, oxygen, or nitrogen any substance simpler than themselves. Clearly, these four simple substances are different from the more complex mercury(II) oxide and lead(II) nitrate. The number of substances which like lead, oxygen, mercury, and nitrogen, are incapable of being split up into simpler substances is small. There is very good reason to believe that about a hundred such substances may exist, and, of them, nearly all are actually known on the earth. To them the name 'elements' has been given. We may now define this term.

An element is a substance which cannot by any known chemical process be split up into two or more simpler substances.

A list of the elements is given on page 499. We may here mention a few of the commoner ones. All the metals – lead, zinc, iron, copper, tin, platinum, gold, silver, and the rest – are elements; so also are the oxygen and nitrogen of the air, together with carbon, sulphur, phosphorus, iodine, and others to the number of over one hundred. Remember the characteristic they all possess – they cannot by any known chemical process be made to yield substances simpler than themselves.

From this small band of elements, all other substances on the earth are made. The number of chemical substances known is approximately three million. All of these, except the elements themselves, are made up of two or more elements combined together. They are called 'compounds'. It is astounding to reflect that all compounds on the earth, from the simplest, which, like water, contain only two elements, to those complex materials of which our own bodily tissues are composed, are made from about one hundred simple, elementary materials. The elements are indeed a small select band.

There follows a short list of common compounds and the elements which compose them.

Compound	Elements contained
Water	Oxygen; hydrogen
Sugar	Oxygen; hydrogen; carbon
Common salt	Sodium; chlorine
Saltpetre	Potassium; nitrogen; oxygen
Marble	Calcium; oxygen; carbon
Copper(II) sulphate	Copper; sulphur; oxygen; hydrogen
Sulphuric acid	Hydrogen; sulphur; oxygen
Sand	Silicon; oxygen
Clay	Aluminium; silicon; oxygen; hydrogen

It is interesting to note how often oxygen appears in the above list of common compounds, and analysis of the earth's crust, the oceans, and the atmosphere reveals that oxygen is the most abundant element on earth, accounting for half the total mass.

The following table gives the approximate percentage composition by mass of the earth's crust, the oceans, and the atmosphere.

Percentage by mass of elements in the earth's crust		Percentage by mass of elements in the oceans		Percentage by mass of elements in the atmosphere	
Oxygen	47	Oxygen	86	Nitrogen	75.5
Silicon	28	Hydrogen	10.9	Oxygen	23
Aluminium	7.8	Chlorine	1.8	Argon	1.4
Iron	4.5	Sodium	1.0	Hydrogen	0.02
Calcium	3.5	Magnesium	0.1	Carbon	0.01
Sodium	2.5	Calcium	0.05	Others (total)	0.07
Potassium	2.5	Sulphur	0.05		
Magnesium	2.0	Potassium	0.04		
Titanium	0.5	Nitrogen	0.02		
Hydrogen	0.2	Bromine	0.01		
Carbon	0.2	Carbon	0.01		
Others (total)	1.3	Others (total)	0.02		

We may now define a compound.

A compound is a substance which contains two or more elements chemically combined together.

We have found it necessary to use the expression 'chemically combined'. The meaning of it is connected with the idea of chemical change which was discussed earlier. We must now try to obtain a clearer idea of the meaning of the expression and, to do this, we shall contrast the properties of mixtures and compounds in the work of the next section.

Mixtures and compounds

In the following experiments, we shall compare the properties of an original mixture of iron and sulphur with those of the black solid left after heating it.

Experiment 6
Effect of heat on a mixture of iron and sulphur

Weigh out 28 g of iron filings and 16 g of sulphur (any fractions of these masses will do equally well). Grind the two thoroughly in a mortar and put about half of the mixture into a dry test-tube. Heat the test-tube, at the bottom, with a small flame. The mixture will glow. When it does so, remove the flame, and hold the test-tube over a mortar as a precaution against breakage. The glow will then spread slowly through the mixture without further heating. Allow the test-tube to cool, then break it away from the mass of material left. A dark grey, almost black, solid will be found.

Now examine the original mixture and the solid left after heating as indicated below.

	Mixture before heating	Solid left after heating
(a) *Action with water*	Place enough of the mixture in a test-tube to fill about 2.5 cm of its depth. Half fill the test-tube with water, shake it well, then allow the test-tube to stand. The denser iron will settle more rapidly than the lighter sulphur and form a layer below it. The experiment separates the iron from the sulphur.	Carry out the same test. The solid settles as a single layer with no sign of separation of the iron from the sulphur.
(b) *Action of a magnet*	Rub one end of a bar magnet well into the mixture, raise it, and tap gently. The iron filings will have been attracted by the magnet and will adhere to it. The sulphur will not. They are separated.	Repeat with the other end of the magnet. A very little iron (left unattacked by the sulphur) may be attracted by the magnet but it will be very much less than before. The bulk of the iron is not attracted from the black solid and is not separated from the sulphur.
(c) *Action of dilute hydrochloric acid*	Add dilute hydrochloric acid to some of the mixture in a test-tube. Warm gently. There is rapid effervescence. Apply a lighted taper to the test-tube. The resulting slight explosion shows that the gas is hydrogen. The iron has reacted with the acid to produce this gas. The sulphur remains unchanged.	Repeat the experiment described opposite. Effervescence occurs again. Apply the following two tests to the gas: 1. Smell very cautiously. The rather disgusting smell is similar to that of rotten eggs. 2. Apply a lighted taper to the test-tube. The gas burns with a blue flame but without explosion. It is **hydrogen sulphide**.

It is clear that the solid left after heating the mixture of iron and sulphur differs greatly in properties from the original mixture in several important respects.

1. Before the heating, the iron could be separated from the sulphur by **physical** methods. For example, by shaking with water, we took advantage of the physical property of density to separate the denser iron from the lighter sulphur; we also separated the iron from the sulphur by using its physical magnetic properties. In none of these experiments was any chemical action involved, but these physical methods could not separate the sulphur from the iron after the mixture had been heated. Physical methods of separation were then useless.

2. Again, in the mixture before heating, the two elements clearly exercised

their own independent properties. The iron was attracted by the magnet just as it would have been if the sulphur had not been present. Similarly, during the action of dilute hydrochloric acid on the mixture, the iron reacted with the acid exactly as if no sulphur were present, while the sulphur itself remained unchanged. After the heating, however, the black solid left showed properties of its own. The iron was not attracted by a magnet, while the separate densities of the two elements were no longer available for use in their separation. The action of dilute hydrochloric acid gave an entirely different reaction, with evolution of hydrogen sulphide instead of hydrogen. So we see that during the heating the separate properties of the iron and sulphur were lost and the new properties of the black solid, iron(II) sulphide, appeared.

The reason for this difference is that before the heating the two elements were simply mixed together, while during the heating they underwent chemical combination, forming the compound *iron(II) sulphide*. As a result of this change, the elements were united by a chemical link or bond instead of being merely close together in space. The nature of this bond has been the object of much speculation, and it is now known to be electrical (see Chapter 9).

3. Another characteristic difference between physical mixing and chemical combination is apparent from this experiment. During the mixing of the iron and sulphur in the mortar, no change was observed except a kind of averaging of the colours of the two elements so that the mixture had a colour between the grey of iron and the yellow of sulphur. During the chemical combination, however, enough heat was given out to raise the whole mass to a bright red glow, once the action had been started by the external application of heat. Chemical combination is often accompanied by heat changes of this kind, but physical mixing is not.

4. A further and most important difference between mixtures and chemical compounds is that the composition of a compound by mass is fixed and unalterable, while that of a mixture may vary within wide limits. For example, pure iron(II) sulphide always contains the iron and sulphur in the proportion of 56 g of iron to 32 g of sulphur, and no variation from this proportion is ever found. Mixtures of iron and sulphur may, however, have any desired composition.

These differences between compounds and mixtures are summarized in the table below.

Mixtures	Compounds
(a) The constituents can be separated from one another by physical methods.	The constituent elements cannot be separated by physical methods; chemical reactions are necessary.
(b) Mixtures may vary widely in composition.	Compounds are fixed in their compositions by mass of elements present.
(c) Mixing is not usually accompanied by external effects such as explosion, evolution of heat, or volume change (for gases).	Chemical combination is usually accompanied by one or more of these effects.
(d) The properties of a mixture are the sum of the properties of the constituents of the mixture.	The properties of a compound are peculiar to itself and are usually quite different from those of its constituent elements.

Separation of mixtures

One of the distinctive characteristics of a mixture of substances is that it is usually possible to separate the constituents by physical means. There are a great many different physical methods used to separate a wide variety of mixtures, and the particular method employed for any given mixture depends upon the nature of its constituents. The following experiments illustrate some of the methods in wide use.

Experiment 7
The separation of a mixture of salt and sand

You will be provided with a mixture of common salt (sodium chloride) and sand, and approximately 10 g of this mixture should be placed in a 250 cm³ beaker. To the mixture add distilled water until the beaker is about one-third full. Place the beaker on a gauze, which itself rests on a tripod, and heat the water until it boils. Whilst the solution is still hot, pour it through a filter paper in a filter funnel, collecting the solution passing through (which is known as the *filtrate*) in another beaker (Figure 1). Note that the sand remains in the filter paper, and it may be washed and dried.

When filtration is complete transfer the filtrate to an evaporating basin which is then heated on a steam bath; a suitable form of steam bath is shown in Figure 2. Continue to heat the evaporating basin on the bath until all the water has evaporated from the basin. Examine the colourless crystals of salt which remain.

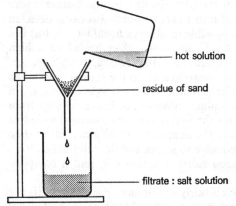

— hot solution

— residue of sand

— filtrate : salt solution

Figure 1

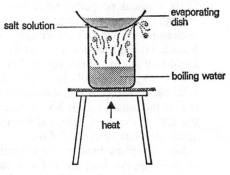

salt solution —

evaporating dish

— boiling water

heat

Figure 2

Experiment 7 represents a simplified laboratory version of what happens on an industrial scale when common salt is obtained from 'rock salt', a mixture of salt and sand. The process depends upon the fact that whereas salt is soluble in water, sand is not. Therefore when the mixture is heated in water, the salt dissolves to form a solution (substances tend to be more soluble in hot water than in cold water) whereas the sand remains as a solid. The process of *filtration* is designed to separate solids from the solution in which they are suspended, and in the above experiment the sand remains in the filter paper, which can be removed from the filter funnel and dried.

The second process involved in the experiment led to the recovery of the salt from the solution; this process is generally known as *crystallization* (see page 270). The purpose of the steam bath is to ensure slow crystallization of the salt; direct heating of the evaporating basin with the burner would result in the salt jumping out of the basin due to overheating at the stage when the water has almost completely evaporated – a phenomenon known as *decrepitation*.

This experiment has employed several techniques:

(*a*) *solution* and *filtration*, which can be used generally to separate an insoluble substance from a soluble one; and

(*b*) *crystallization*, which is used to obtain a soluble salt from its solution.

Experiment 8
The separation of a mixture of iodine and common salt

This experiment should be carried out in a fume cupboard or near the window.

Take a spatula measure of a mixture of iodine and common salt in a long test-tube, and heat it *gently* over a low flame. After a time the violet colour of iodine vapour will be seen, and careful examination of the top of the test-tube should reveal a black deposit of iodine. Continue the heating until the solid at the bottom of the tube becomes almost colourless.

In the case of the mixture of iodine and common salt, both of which are at least partially soluble in water, the process of solution and crystallization described in Experiment 7 is clearly unsuitable. It will, however, be obvious that the two substances in the mixture behave rather differently when heated, and it is this difference which forms the basis of the method of separation. Whereas the common salt remained in the solid form when heated by the burner under laboratory conditions, the iodine turned into a vapour which was easily detected by its violet colour. At no time was it possible to observe *liquid* iodine; this is a fairly unusual property, since most solid substances when heated to a high enough temperature change first of all to a liquid, and then to a vapour. Substances such as iodine, which pass straight from the solid state to the vapour state when heated, are said to *sublime*, and the phenomenon is known as *sublimation*. An important feature of this separation technique, however, is that the change from solid to vapour is easily reversible. Thus the iodine vapour condensed to iodine solid near the top of the test-tube where the temperature was much lower than that at the heated end. It is therefore possible to scrape out the solid iodine from the top of the tube before the common salt is emptied out, and a complete separation has been carried out.

Since the phenomenon of sublimation is fairly uncommon, the technique is not of wide application; some other substances which sublime when heated are anhydrous aluminium chloride, anhydrous iron(III) chloride, benzoic acid, and ammonium chloride.

Most inks consist of soluble dyes in water, and in order to obtain a sample of pure water from the mixture use is made of the fact that the temperature at which water boils is very much lower than that at which the dyestuff boils. Hence, if a mixture of dyestuff and water is heated, the water will boil first and will escape from the mixture as steam. This is what happened during the latter stages of the recovery of salt from brine (Experiment 7), a process known as *evaporation*. In that particular experiment the steam was allowed to escape freely into the atmosphere, since we were primarily interested in obtaining a sample of the salt. In Experiment 9, however, we wish to obtain a sample of the first liquid which boils from the ink, and therefore we use the apparatus illustrated in Figure 3, where the colourless liquid not only evaporates from the mixture, but is also subsequently condensed to the liquid state and collected. This process is known generally as *distillation*, and the liquid collected is called the *distillate*.

Experiment 9
The separation of pure water from ink

Into a 100 cm³ filter flask, fitted with a delivery tube to its side arm, place about 10 cm³ ink solution and insert a rubber bung carrying a thermometer (Figure 3). It is important that the bulb of the thermometer is situated opposite the side arm as shown. The delivery tube leads to an open test-tube clamped in a beaker of cold water.

With the very small flame, heat the ink solution carefully, making certain that the solution does not froth up to the top of the filter flask. When the solution is boiling, note the steady reading on the thermometer and look for the appearance of drops of colourless liquid in the test-tube. Continue to heat the solution until about 3 or 4 cm³ of liquid have been collected in the test-tube; do *not* continue to heat until all the liquid has been transferred from the flask to the test-tube.

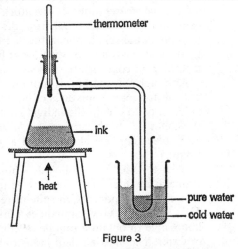

Figure 3

In the case of Experiment 9 the distillate was a colourless liquid, and the observation that the reading on the thermometer during the distillation process was approximately 100 °C (373 K) leads us to speculate that the liquid may be water. In order to confirm that the liquid is water it will be necessary to carry out further tests on it, such as determining its freezing point and density.

The process of distillation may be used to separate a liquid from a solution, or to separate two liquids whose boiling points differ by an appreciable temperature interval. Thus it is possible to separate a mixture of methylbenzene (toluene) and benzene, which are completely miscible (that is, form a clear solution when mixed in all proportions) and whose boiling points are 110 °C and 80 °C respectively.

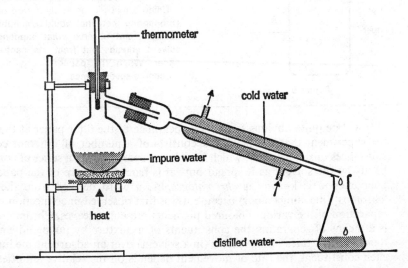

Figure 4

The apparatus usually employed to separate mixtures of liquids is different from that used in the simple distillation process in Experiment 9, and it is illustrated in Figure 4. Between the flask containing the liquid mixture and the vessel collecting the distillate there is a specially designed piece of apparatus, known as a *Liebig condenser* (after its inventor), the purpose of which is to bring about the condensation of the vapour from the flask in a more efficient manner than was possible in the simple apparatus of Figure 3. Cold water is continuously passed up the condenser against the flow of hot vapour from the flask, thereby creating a permanent cold surface on which the vapour can condense to liquid.

Although this method is more efficient than that of Experiment 9, it is still difficult to separate mixtures of liquids whose boiling points differ by only a few degrees, or to obtain the individual components of a mixture containing many substances. A further refinement in technique which enables such problems to be resolved consists of the use of a *fractionating column*, which is described in Chapter 23.

Even for mixtures of liquids whose boiling points differ by an appreciable temperature interval, it may not be possible to bring about a complete separation of the components, in spite of refinements such as a fractionating column. Thus, no matter how many times a mixture of ethanol and water is distilled, it is impossible to obtain a sample of ethanol (boiling point 78 °C) completely free of water. The most efficient industrial still can deliver a distillate which is 96% alcohol, but no more.

Experiment 10
The separation of the components of black ink

Lay a filter paper of fairly large diameter over the top of a beaker, as illustrated in Figure 5.

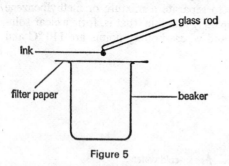

Figure 5

Using the end of a glass rod which has been drawn out to a fine tip, carefully transfer several drops of black ink from the ink bottle to the centre of the filter paper, making certain that the spot of ink does not grow too large. Allow the ink spot to dry between each application. About half a dozen transfers will be sufficient.

Using a teat pipette apply a drop of solvent (propanone or ethanol would be suitable) to the ink spot and observe what happens as the solvent spreads out from the centre of the paper. When the paper is dry, repeat the procedure several times.

It will be quite obvious from the appearance of the filter paper at the end of the experiment that the black ink consists of a number of different coloured substances mixed together, which have been separated into a series of concentric bands by the solvent as it spread outwards from the centre of the paper. This phenomenon is known as *chromatography*, a word derived from the Greek 'chromos' (meaning colour) because it was first observed in connection with the separation of the various coloured pigments present in leaves. Chromatography is a means of separating the constituents of a mixture by taking advantage of their different rates of movement (in a solvent) over an adsorbent medium. For each constituent, the rate of movement depends on the relative affinities of the constituent for the solvent and for the adsorbent medium. In the experiment above, the adsorbent medium is filter paper and the solvent is propanone or ethanol. The coloured band furthest from the centre of the filter paper contained the substance which was the least strongly adsorbed, and that remaining in the centre of the filter paper contained the substance most strongly adsorbed.

Chromatography can be used for mixtures other than those which are coloured, provided that the separated substances are capable of being identified in some convenient manner. For example, the paper may be examined in ultra-violet radiation which may cause fluorescence and so indicate the position of the substances, or the paper may be sprayed with a reagent which develops different colours according to the substances present (see Chapter 28). If desired, areas of paper on which separated substances have been detected may be cut out and subsequently stored, either as they are or dissolved in the solvent, as reference materials for future analysis. Two advantages of chromatography are that it can be performed with very small masses of material, and that the substances are not destroyed in the process.

In standard chromatography, rectangles of filter paper about 22 cm by 5 cm are usually used. A spot of solution containing the mixture under investigation is

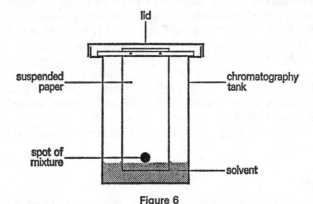

Figure 6

placed near the middle of one of the short ends of the filter paper and about 3 cm away from the end. Then the spot is allowed to dry. The paper is then suspended vertically with the spotted end immersed in the solvent but the spot well clear of the liquid level. The whole is enclosed by a covering vessel to prevent loss of solvent (Figure 6). The *chromatogram* will then develop as the solvent rises up the filter paper. Spots of the various constituents of the original mixture will collect, at different distances along a vertical line, above the original spot. When the ascending front of solvent is approaching the top of the paper (usually eight hours or more after the start), the paper is removed and dried. It can then be examined in various ways to locate the spots of material left and to identify their contents.

To secure better separations, the following device may be used. A square piece of filter paper is employed and a chromatogram is developed from a spot of material near one corner of the paper. This gives spots of separated materials along a line, as described above. The paper is then dried and a *second* chromatogram is developed, by using the same piece of paper at right angles to its former position and with a different solvent. Spots of material will then be found, after drying, much more widely spread over most of the surface of the paper.

Since the invention of chromatography as a separation technique at the beginning of the century, many applications have been found in a wide variety of fields, and, in one form or another, chromatography is one of the most widely used of all separation processes. Several of these applications will be described later in the book.

Pure substances

The separation techniques described above, together with a range of others which you will come across in the study of chemistry, are used primarily for the preparation of chemical substances in a pure state. It is not always a simple task, however, to find out whether or not a substance is pure; for example sea water *looks* like pure water, impure naphthalene has the same *smell* as pure naphthalene (the smell of 'moth balls'), and a piece of wire made of pure copper *feels* the same as one made of a copper alloy. It seems that the human senses are inadequate for deciding how pure a substance is, and in practice more scientific tests are applied. Since under the same conditions of temperature, pressure, and surroundings a pure substance always possesses a set of properties which are unique to that particular substance, and which may be reproduced at any time, certain characteristics of the substance are chosen by which the degree of purity of the sample may be determined. Among these characteristics are the melting point and density for a solid; boiling point, density, and refractive index for a liquid; and a variety of complex properties for gases. We have already noticed that the separation, or purification, techniques described above do not always lead to a completely pure substance as the end product, and industrial manufacturers spend a great deal of money each year in attempting to devise ways in which such techniques can be made more efficient. For this reason manufacturers of chemicals usually state the *degree of purity* of a substance on the label of the bottle in which it is contained. Examine the labels on some of the bottles in your laboratory.

Questions

1. What are considered to be the main distinctions between a chemical compound and a mixture? Explain why the liquid obtained by mixing sodium chloride with water is not regarded as a chemical compound. (O. and C.)

2. When a piece of sodium is placed in water it diminishes in size gradually, and finally disappears. In what way is the disappearance of the sodium different from the ordinary process of solution in water? Give experiments in support of your views. How could metallic sodium be recovered from the liquid? (J.M.B.)

3. Give *two* important differences between a mixture and a compound. Describe and explain experiments by which you could obtain: (a) nitrogen from a mixture of nitrogen and hydrogen; (b) nitric acid from a mixture of concentrated nitric and sulphuric acids; (c) sodium chloride from a mixture of sodium chloride, ammonium chloride, and sand. (A.E.B.)

4. Describe briefly how you would separate a pure sample of the first-named substance from the impurity in each of the following mixtures: (a) iron turnings contaminated with oil, (b) sodium chloride crystals contaminated with glass, (c) hydrogen sulphide contaminated with hydrogen chloride, (d) water contaminated with copper(II) sulphate (cupric sulphate), (e) copper powder contaminated with magnesium powder. (J.M.B.)

5. Outline how you would separate the first-named substance from the mixture in each of the following cases: (a) pure water from sea water, (b) the orange dye from the blue dye which together form black ink, (c) pure copper(II) sulphate (cupric sulphate) from a sample contaminated with a small quantity of sodium sulphate, (d) almost pure ethyl alcohol (ethanol) from whisky. (J.M.B.)

6. Describe, with the aid of a diagram, how you would determine the boiling point of a liquid found in a bottle whose label is too faded to read.
Explain how your procedure would enable you (a) to judge whether the liquid is pure,

(*b*) to identify the liquid by means of (i) a table of boiling points, (ii) a supply of known liquids whose boiling points are very close to the value you have determined.

What day-to-day changes in laboratory conditions would you have to take into account? (C.)

7. A mixture containing benzene, boiling point 80 °C, and toluene (methylbenzene), boiling point 111 °C, is amongst the products of the thermal decomposition of coal.

Explain, with the aid of a diagram, how you would obtain samples of benzene and toluene by fractional distillation of the mixture.

Name *one* crystalline solid which is soluble in benzene, and *one* crystalline solid which is insoluble in benzene. (S.)

8. It is required to separate a mixture of *three* solid dyes; one red, one yellow, and one blue. The following facts are known about the dyes. The blue and yellow dyes are soluble in cold water, while the red dye is insoluble. When an excess of aluminium oxide is added to a stirred, green aqueous solution of the mixed blue and yellow dyes and the aluminium oxide is filtered off and washed with water, it is found that the solid residue is yellow and the filtrate blue. When the yellow solid is stirred with ethyl alcohol (ethanol) and the mixture filtered, the solid residue is white and the filtrate yellow.

Describe how you would obtain dry samples of the three dyes. (J.M.B.)

9. What is the essential difference between a chemical and a physical change?

Indicate clearly the chemical and physical changes involved in the following processes, giving full reasons in each case: (*a*) the addition of metallic sodium to water; (*b*) the solution of sodium chloride in water; (*c*) the heating of magnesium in air; (*d*) the heating of ammonium chloride; (*e*) the addition of water to concentrated sulphuric acid. (L.)

10. Describe the experiments you would carry out in seeking to determine whether a given white powder is a pure substance or a mixture. If the substance is a pure chemical compound, how would you propose to ascertain whether it is (i) a salt, (ii) a basic oxide, or (iii) a peroxide? (J.M.B.)

11. Illustrate *three* differences between metallic and non-metallic elements by reference to the properties of iron and sulphur. Describe *three* tests by means of which you would prove that the compound, iron(II) sulphide, formed by heating a mixture of iron filings and sulphur, differs from the original mixture. (J.M.B.)

2 Atomic Theory

We have seen in the last chapter that there are about 100 kinds of simple matter called *elements*, and that all other kinds of matter have been formed by the chemical combination of two or more of these elements. The ancient Greek thinker, Democritus (about 400 B.C.), began speculations about the structure of matter, the Roman Lucretius (about 350 years later) took up the question, and, following these two ancient philosophers, there has appeared a succession of thinkers, European and Arabian, whose speculations culminated at the beginning of the nineteenth century in the ideas of an Englishman, John Dalton, a Manchester schoolteacher. His suggestions won universal acceptance for themselves and lasting scientific fame for their author.

Dalton

Dalton's lifetime abounded in famous names and exciting happenings – the French Revolution, Austerlitz, Trafalgar, Nelson, Napoleon, Wellington, Waterloo – but the thoughts of this Quaker, slowly maturing as he pursued an obscure and uneventful existence, have proved more potent in their influence on human modes of living in the succeeding century than all the wars and alarums of his day. His Atomic Theory is the foundation of modern chemistry.

Dalton's love of precision and truth is illustrated by the following story concerning him. Dalton had given a course of lectures, and at the end a student came to him with the request for a certificate of attendance. The great chemist looked up his records and found that the student had missed one lecture during the course. Dalton refused to sign the attendance certificate, but, after considering a few minutes, he said, 'If thou wilt come tomorrow I will go over the lecture thou hast missed.' Having quietened his conscience in this manner over the missed lecture, Dalton presumably signed the certificate.

Ideas about atoms

You will have seen, from the work of the last chapter, that, if ordinary powdered sulphur is mixed with water, it does not dissolve and can be removed by filtration. That is, the sulphur particles are too large to pass through the pores of the filter paper. If, however, we dissolve sulphur in carbon disulphide and filter the liquid, filtration does not separate the sulphur. That is, the sulphur particles are now small enough to pass through the pores of the filter paper.

If sulphur is *very* finely ground and mixed with water, filtration does not remove all the sulphur. Some is fine enough to remain permanently mixed with the water in what is called a *colloidal solution*. When this liquid is examined by the ultra-microscope, it can be seen that sulphur particles are in rapid random motion, which continues indefinitely at room temperature. This motion is now known to be caused by collisions between the minute sulphur particles and particles (molecules) of water. The movement is shown by all colloidal particles in water. It was first observed (1827) by the botanist, Brown, for particles of pollen in water and is known as the *Brownian movement*.

Experiment 11
Brownian movement in liquids

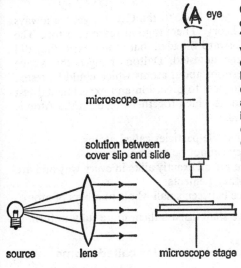

Figure 7

On to a microscope slide place a very small drop of colloidal graphite (sold commercially as 'Aquadag'), and very carefully add distilled water gradually from a fine teat pipette, stirring continuously until the solution is almost colourless. Place a cover slip over the solution and examine the slide through a microscope, illuminating the solution from the side as illustrated in Figure 7. By using a fairly high-power objective lens in the microscope it is possible to observe the irregular movement of the graphite particles.

Hypothesis and theory

The idea that elements are capable of being split up into very tiny particles was first evolved by Greek and Roman thinkers from another angle. They said that if we were to take a piece of, say, gold and cut it up into small pieces, and cut those pieces into smaller pieces, and those pieces into small pieces, and so on, a time would ultimately come when the dividing process would have to stop. The tiny particles of gold which we had then obtained would be incapable of being divided any further; they would be the smallest possible particles of gold which could ever be obtained. The Greeks gave them the name 'atoms'. As a temporary, though incomplete, definition of an atom we may say that it is the smallest, indivisible particle of an element.

The idea that elements are made up of atoms is called the **atomic theory.** This word 'theory' does not often mean very much to a beginner in Chemistry, so let us be quite clear about it. A scientific theory is a scientific idea which was thought of by somebody, suggested by him in a scientific book or journal, and accepted by other scientists after due consideration. 'So-and-so's theory' means 'So-and-so's accepted idea'. The process of getting an idea accepted may be a long one; there will be arguments, objections, testing by experiments, improvements of the idea, but, if it finally wins acceptance by scientists generally, it will be called a theory. When the idea is first put forward, and is still in the 'argument-and-objection' stage, it is called a **hypothesis**; later, if generally accepted, a **theory**.

Dalton's Atomic Theory

Dalton's Atomic Theory was first put forward in 1808. It soon gained general acceptance and then stood, virtually unchanged, for about a century. Discoveries made early in the twentieth century, however, showed that the theory must be modified. For this reason, the Atomic Theory will first be considered in its original, nineteenth century form; the changes made in it will be mentioned later, in Chapter 11.

The Atomic Theory, we have seen, goes back to the Greeks, yet we always speak today about *Dalton's* Atomic Theory. There is good reason for this. The reason is that, while the Greeks put forward the idea that atoms exist, they did no more. They left the idea vague and untested. Dalton changed this vague imagining into a set of concrete suggestions about atoms which could be tested by experiment. This change from vagueness to precision and experimental test justifies his claim to the theory. The ideas which together make up the Atomic Theory of Dalton (1808) are as follows.

1. Matter is made up of small, indivisible particles called atoms.
2. Atoms are indestructible and they cannot be created.
3. The atoms of a particular element are all exactly alike in every way and are different from the atoms of all other elements.
4. Chemical combination takes place between small whole numbers of atoms.

These ideas are so important that we shall discuss them more fully in turn.

Matter is made up of small, indivisible particles called atoms

Although the idea that all matter is composed of extremely small particles has been recognized for a very long time, it is quite a difficult task to verify this assertion since with even the most powerful microscope available we are unable to observe the particles directly. We therefore rely on *indirect* evidence; that is, the existence of such particles can be inferred from the results of certain experiments.

Experiment 12
The diffusion of coloured crystals in solution

Half fill a 250 cm³ conical flask with water and drop in a crystal of a deeply coloured salt (potassium permanganate or ammonium dichromate crystals are suitable). Over a period of ten to fifteen minutes observe closely what happens within the flask. After this period of time swirl the contents of the flask gently and note any significant changes.

The observation that the colour of the crystalline substance spreads throughout the whole of the solution can be explained only by supposing that the solid consists of extremely small particles, for if the crystal was thought to be composed of one continuous piece of material then the phenomenon of *diffusion* (or 'spreading out') would not be possible.

Diffusion also takes place between gases, as can readily be verified (by the class teacher) by placing a few drops of liquid bromine at the bottom of a gas-jar on top of which is placed a cover glass (Figure 8). After some time the colour of the bromine will be seen in the upper part of the gas-jar, in spite of the fact that bromine vapour is much denser than air. Again, only by assuming that liquid bromine is composed of particles can this observation be explained.

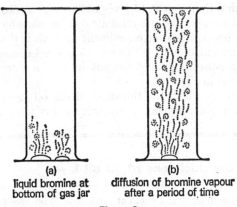

(a)
liquid bromine at
bottom of gas jar

(b)
diffusion of bromine vapour
after a period of time

Figure 8

It is clear that if matter is composed of particles then they must be extremely small. Although it is very difficult to visualize the size of an atom, some idea of how small it is may be gained through the following experiments.

Experiment 13
Dilution of a solution of potassium permanganate

You will be provided with a solution of potassium permanganate which contains 1 g of the salt in 1 dm^3 of solution. Using a measuring cylinder, take 10 cm^3 of the solution and place it in a 100 cm^3 beaker. What is the mass of potassium permanganate in the beaker? Now pour water into the beaker until it is full and stir or swirl. You will have diluted the potassium permanganate solution ten times, but the whole solution still contains the same mass of salt as did the 10 cm^3 of original solution.

Using the measuring cylinder once again remove 10 cm^3 of the diluted solution from the beaker and pour it into a second 100 cm^3 beaker. Then fill up the second beaker with water as before and note the decrease in the intensity of the colour. What is now the mass of potassium permanganate in the second beaker?

This procedure should be repeated (10 cm^3 of solution diluted to 100 cm^3 in a separate beaker) until the colour of the solution is so faint that it is only just visible.

Make a note of the number of times it is necessary to dilute the solution before the colour disappears, and on the basis that each time a dilution is performed one-tenth of the previous mass of potassium permanganate is removed, calculate the mass of salt in the final beaker.

Since the solution in the final beaker is coloured then it may be assumed that it contains at least one particle of potassium permanganate (in fact it still contains a large number!) and therefore the calculated mass represents the maximum for the potassium permanganate particle.

oil particle   oil layer

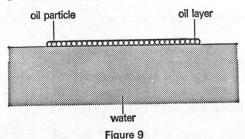

water

Figure 9

It is clear that Experiment 13 gives only a very approximate idea of the size of atoms, and a more accurate estimate can be made by the following method. The principle upon which the experiment depends is that when a drop of oil is allowed to come into contact with a water surface, because the two liquids do not mix (they are *immiscible*), the oil spreads out until it is one particle thick all over; this is illustrated in Figure 9.

Any method by which the thickness of the oil layer can be measured will therefore give a direct measure of the thickness of a molecule of oil.

Experiment 14
An estimate of the thickness of an oil layer

Fill a clean shallow tray with water to a depth of approximately 1 cm. Scatter a layer of talcum powder (flowers of sulphur or lycopodium powder from a pepper-pot can also be used) lightly over the surface of the water.

You will be supplied with a solution of an oil in petroleum ether (0.1 cm³ of the oil in 1 dm³ of solution) and, using a teat pipette, *one* drop of the oil solution should be placed on the water surface in the centre of the tray. The oil will spread out rapidly, and at the same time the petroleum ether will evaporate, leaving a monomolecular layer (that is, a layer one molecule thick) of the oil which has pushed the powder to one side, thus making the area over which the oil has spread clearly visible (Figure 10). Make an estimate of the area of the oil layer; this can most conveniently be done by assuming that it either corresponds approximately to a circle, a rectangle or a square. It is not necessary to measure the area exactly.

Before the experiment is concluded you should find out the volume of one drop of solution from the teat pipette. This can be carried out by counting fifty drops of solution from the pipette into a measuring cylinder and reading off the total volume.

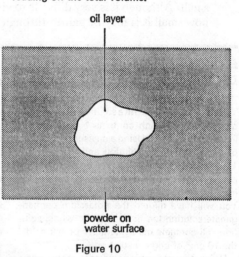

oil layer

powder on water surface

Figure 10

Specimen results
Volume of fifty drops of solution from teat pipette = 2.40 cm³
Approximate area of oil layer   = 16 cm²

Calculation
(a) *Volume of solution in one drop*
Volume of 50 drops of oil solution = 2.40 cm³

$$\text{Volume of } \textit{one} \text{ drop of oil solution} = \frac{2.40}{50} \text{ cm}^3$$

$$= 0.048 \text{ cm}^3$$

(b) *Volume of oil in one drop*
1 dm³ solution contains 0.1 cm³ oil

$$0.048 \text{ cm}^3 \text{ solution contains } 0.1 \times \frac{0.048}{1000} \text{ cm}^3 \text{ oil}$$

$$= 4.8 \times 10^{-6} \text{ cm}^3 \text{ oil}$$

(c) *Thickness of oil layer*

Approximate area of oil layer $= 16$ cm^2

If the thickness of the oil layer is t cm, then the volume of the layer will be $16 \times t$ cm^3

$$16 \times t = 4.8 \times 10^{-6}$$

$$t = \frac{4.8 \times 10^{-6}}{16} \text{ cm}$$

$$= 3 \times 10^{-7} \text{ cm}$$

Hence the thickness of the oil layer, and therefore the thickness of a molecule of oil, is approximately 3×10^{-7} cm.

Even though the result of your experiment will certainly indicate how very small atoms and molecules are, the actual size of the oil molecule which you measured is fairly large compared with many other molecules and atoms! The diameter of an atom is of the order of 10^{-8} cm (a unit previously called an Angström unit). It is difficult to have any conception of the size of atoms, but Figure 11 may help you to understand how small an atom really is.

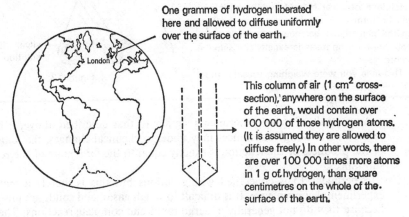

One gramme of hydrogen liberated here and allowed to diffuse uniformly over the surface of the earth.

London

This column of air (1 cm^2 cross-section), anywhere on the surface of the earth, would contain over 100 000 of those hydrogen atoms. (It is assumed they are allowed to diffuse freely.) In other words, there are over 100 000 times more atoms in 1 g of hydrogen, than square centimetres on the whole of the surface of the earth.

Figure 11

Atoms are indestructible and cannot be created

The important aspect of this idea is that, by chemical action, it is possible to alter only the state of combination of a number of atoms, not to reduce their number or add to it. If we start a chemical reaction with, say, a thousand million atoms of hydrogen, then we shall finish that reaction with exactly a thousand million, neither more nor less. They may have altered their state of combination – they may have become, for example, free hydrogen gas instead of being combined with oxygen in water – but the same number will be there. It follows that the total number of atoms present at the end of a chemical reaction must be the same as the number present at the beginning of it, though they will be differently combined. That is, the *total* mass of the products of a chemical action should be the same as the *total* mass of the starting materials. Experiment shows this to be true and the situation is expressed in the Law of Conservation of Mass.

The Law of Conservation of Mass

*The Law of Conservation of Mass (or Indestructibility of Matter)
states:* Matter is neither created nor destroyed in the course of chemical
action.

Experiment 15
An experimental test of the Law of Conservation of Mass*

Into a conical flask put some silver nitrate
solution and lower into it carefully by means of a
thread a small test-tube full of hydrochloric
acid. Insert the stopper (Figure 12). Place the
flask on the pan of the balance and weigh it.

Note that you have just weighed (besides
those portions of the apparatus which are
unchanged throughout) some *water* and *silver
nitrate* and *hydrogen chloride*.

Allow the two liquids to mix by tilting the
flask a little, and you will observe a white
precipitate of silver chloride whilst nitric acid
will be formed in solution. Replace the flask
on the pan of the balance and weigh again.
You will find the mass is exactly the same as
before.

This time you were weighing (besides those

portions of the apparatus which are unchanged
throughout) some *water* and *silver chloride*
and *hydrogen nitrate* (or nitric acid).

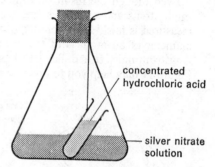

concentrated
hydrochloric acid

silver nitrate
solution

Figure 12

As a result of many experiments similar to that described above, it always
seems that although substances may undergo chemical changes, the *total* mass
of the products of the reaction is exactly equal to the *total* mass of the reacting
substances.

Experiment 15 is concerned with solutions but this is merely a matter of
experimental convenience. It is difficult to weigh gases, and solids are unsuitable
because they do not generally undergo rapid and complete reactions. The same
experimental results are obtained, however, when solids and gases are involved.

It must be noted, however, that a law of the kind we are discussing here is
valid only within the limits of experimental error involved. A very accurate set
of experiments carried out by Landolt about 1906 showed that the Law of Conservation of Mass is valid (for the cases investigated) to about one part in ten
million. For most purposes, errors of this order are negligible. It should be
noted also that, according to modern ideas, emission of energy during chemical
change involves loss of mass. The relation is expressed in Einstein's equation,
$E = mc^2$, where E is the energy, m the mass, and c the velocity of light (in
appropriate units). In ordinary chemical changes, the accompanying energy
changes are so small that their influence on mass cannot be detected by any
weighing apparatus in common use. Nevertheless, the true conservation is one
of mass-and-energy together.

* This experiment will work equally well with solutions of other substances, for instance
barium chloride and sodium sulphate, barium chloride and dilute sulphuric acid, lead nitrate
and potassium iodide, calcium nitrate and dilute sulphuric acid.

Apparent destruction of coal

Experiment 15 may not seem conclusive to you because in it there seems little possibility of loss. If you consider the burning of coal, where only a small ash is left, it seems much more likely that the matter of the coal has been destroyed. The only real difference is that some of the reactants and products are invisible

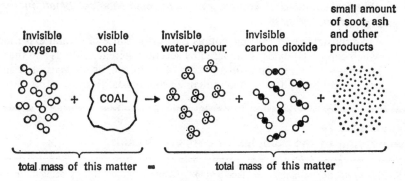

Figure 13

gases. Actually there is no loss of matter at all. If we could weigh all the oxygen which burns the coal, and all the ash, soot, water-vapour, and carbon dioxide into which the coal is changed by the burning, we should again find that the total mass of the materials before the reaction was the same as the mass of the total products after it (Figure 13).

The atoms of a particular element are all exactly alike in every way and are different from the atoms of all other elements

The most important point is that this statement includes in it the idea that all atoms of the same element are exactly alike in *mass*, but are different in mass from the atoms of any other element. The theory said that if we collected together, say, one thousand atoms of sulphur from all corners of the earth, every one of those atoms of sulphur would be exactly the same as every other. The same would be true of any number of atoms of copper. But the mass of each sulphur atom would be different from the mass of each copper atom.

Be sure you understand the universality of this idea. Consider, say, oxygen. Oxygen occurs in hundreds of thousands of compounds – water, sugar, massicot, copper(II) sulphate, alcohol, sulphuric acid, starch, and so on. If we were to collect one oxygen atom from each of these hundreds of thousands of compounds, every one of those oxygen atoms would, said the Atomic Theory, be absolutely and completely alike.

It is clear that we could not test the theory in this way because the atoms are so small that we could not examine them or weigh them even if we could separate them. However, it is obvious from their different properties that atoms of different elements differ from one another.

If atoms of the same element are all identical, it follows that all pure samples of the same chemical compound must be identical in composition by mass, however they are prepared. For example, samples of black copper(II) oxide, containing one atom of copper and one atom of oxygen in each 'molecule', should have the same composition by mass, irrespective of the sources of copper and oxygen. This can be tested by preparing copper(II) oxide in several different ways and analysing the various samples. Accurate experiment shows this identity of composition in all the samples. The facts are stated in the Law of Constant Composition (or Definite Proportions).

The Law of Definite Proportions (or Constant Composition)

The Law of Definite Proportions states: All pure samples of the same chemical compound contain the same elements combined in the same proportions by mass.

An illustration of this law entails the performance of several experiments, in which copper(II) oxide is prepared by several different methods and the samples are analysed by reduction in a stream of hydrogen, or coal-gas, and shown to contain copper and oxygen in the same proportions.

Experiment 16
Preparation of the samples of copper(II) oxide

Sample A. Starting from copper. Place a little clean, pure copper foil in a large crucible in a fume-chamber and carefully add concentrated nitric acid a little at a time until all the copper has dissolved. Brown fumes of nitrogen dioxide (poisonous) are seen and green copper(II) nitrate solution is formed. Evaporate the solution to dryness, then heat the green solid copper(II) nitrate until no more brown fumes of nitrogen dioxide are evolved. The black solid left is the first sample of copper(II) oxide. Store it in a desiccator to keep it dry.

Sample B. Starting from copper(II) sulphate. Put some copper(II) sulphate solution into a beaker and add excess of caustic soda solution. A blue gelatinous precipitate of copper(II) hydroxide appears. Heat the beaker and its contents on a tripod and gauze by means of a

Bunsen burner. The precipitate changes to black copper(II) oxide. Filter off the black solid, wash it several times with hot distilled water, and allow it to dry in a hot oven or on a porous plate.

Transfer the oxide to a crucible and heat it with a burner to drive off the last traces of water. Store the oxide in a desiccator.

Sample C. Starting from copper(II) carbonate. Place a little copper(II) carbonate in a dry crucible and warm it gently. It decomposes, turning from green to black, and carbon dioxide is given off. (Test – a drop of lime-water on the end of a glass rod is turned milky. The milkiness is caused by a precipitate of chalk.) Black copper(II) oxide is left. After heating it for some time, put the oxide into a desiccator to keep it dry.

Experiment 17
Analysis of the samples of copper(II) oxide by converting them to copper by heating in hydrogen

Weigh three clean, dry porcelain boats. Put into each boat 1–1½ grammes of one of the three samples of copper(II) oxide. Weigh all the boats again. Put them into a hard glass tube and connect up the apparatus as shown in Figure 14. The tube must slope so that the end C is the lower.

Turn on the supply of hydrogen, and pass the hydrogen for some time before lighting the jet D. Heat each boat in turn. All the samples of the oxide will glow and leave reddish-brown copper. The water formed by the combination of the oxygen of the oxide with the hydrogen condenses at C, where the tube is cooler. The end C is lower than the end A to prevent this

water from running back on to the hot part of the tube, which might be broken. When the action is complete, allow the tube to cool, keeping the gas stream passing so that air cannot enter and oxidize the copper again. When the boats are cool, weigh all the three boats again. Work out the results as shown below.

Within the limits of experimental error, the percentages of copper (and hence of oxygen) in all three samples of copper(II) oxide are the same. It is found that, however samples of a given compound are prepared, they always contain the same elements in the same proportions by mass.

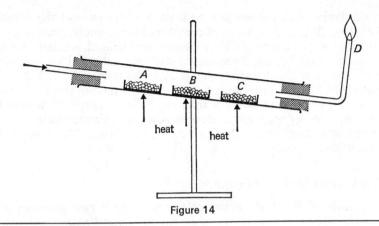

Figure 14

Specimen results

	Sample A	Sample B	Sample C
Mass of porcelain boat	3.01 g	2.50 g	2.70 g
Mass of porcelain boat and copper(II) oxide	4.26 g	3.65 g	4.14 g
Mass of porcelain boat and copper	4.02 g	3.42 g	3.85 g
Mass of copper	1.01 g	0.92 g	1.15 g
Mass of copper(II) oxide	1.25 g	1.15 g	1.44 g
Percentage of copper in copper(II) oxide	$\dfrac{1.01 \times 100}{1.25}$	$\dfrac{0.92 \times 100}{1.15}$	$\dfrac{1.15 \times 100}{1.44}$
	= 80.8	= 80.0	= 79.8

Chemical combination takes place between small whole numbers of atoms

It follows from the supposition that atoms are indivisible that they must combine in whole numbers. Dalton made the additional assertion that these whole numbers are small. By this he meant that atoms commonly combine in such numbers as 3 atoms of one element with 1 atom of another, or 2 atoms of one element with 5 atoms of another, or 1 atom of one element with 1 atom of another. Cases such as 67 atoms of one element combining with 125 atoms of another, or 322 atoms of one element with 27 atoms of another, were unknown. The numbers of atoms combining together are almost always small, though, of course, in any one laboratory experiment there will be millions of exactly similar combinations taking place.

We must now emphasize the fact that what has been stated above is a set of ideas. In science, ideas are treated with scant respect unless they can be backed up by experimental results. None of the statements made above can be tested by direct observation. We cannot line up a thousand oxygen atoms and inspect them to see if they are all alike, or count the number of hydrogen atoms which combine with one atom of oxygen to see if the number is small. We must deduce, from the Atomic Theory, some results which ought to follow from it, and which can be put to experimental test.

If atoms combine in small whole numbers, the compounds formed from elements with atoms A and B must be of the form: A_1B_1, A_1B_2, A_2B_1, A_2B_3, and so on. From this it follows that the mass A of one element may combine with the masses B, 2B, $\frac{1}{2}$B, $1\frac{1}{2}$B, etc. of the other element. But the mass A is a *constant* mass, because all atoms of a given element are identical. Similarly, the mass B is a *constant* mass. That is, the masses of B which combine with a *constant* mass of A in their various compounds should be in the ratio of $1 : 2 : 1/2 : 3/2$, which (multiplied by 2) is the ratio $2 : 4 : 1 : 3$. This is a ratio of small whole numbers. That is, if the atomic theory is true, the *different* masses of element B which combine with a *fixed* mass of element A are in a simple whole-number ratio. Experiment shows that these deductions are correct. The Law of Multiple Proportions expresses this.

The Law of Multiple Proportions

The Law of Multiple Proportions states: If two elements A and B combine together to form more than one compound, then the several masses of A, which separately combine with a fixed mass of B, are in a simple ratio.

In order to demonstrate the Law experimentally, we make use of the fact that copper and oxygen form two oxides, copper(II) oxide, CuO, and copper(I) oxide, Cu_2O. Pure samples of these two can be reduced in a current of hydrogen and the masses of copper which combine separately with, say, 100 g of oxygen in the two compounds, are calculated from the weighings.

Experiment 18
A test of the Law of Multiple Proportions

Weigh two clean dry porcelain boats and weigh them again containing samples of pure, dry, copper(I) oxide and copper(II) oxide respectively. Reduce the oxides to copper in a stream of dry hydrogen, as described in the last experiment. When the samples are cool, weigh each of them in the porcelain boats. Below is a table of specimen results and calculations.

Specimen results

	Copper(I) oxide	Copper(II) oxide
Mass of boat	6.90 g	7.30 g
Mass of boat and oxide	9.75 g	9.20 g
Mass of boat and copper	9.43 g	8.82 g
∴ Mass of copper	2.53 g	1.52 g
∴ Mass of oxygen	0.32 g	0.38 g

∴ 0.32 g oxygen is combined with 2.53 g copper. 100 g oxygen are combined with

$$\frac{2.53 \times 100}{0.32} \text{ g copper}$$

$$= 790 \text{ g}$$

∴ 0.38 g oxygen is combined with 1.52 g copper. 100 g oxygen are combined with

$$\frac{1.52 \times 100}{0.38} \text{ g copper}$$

$$= 400 \text{ g}$$

This ratio is 2 : 1 within limits of experimental error

Hence the masses of copper which have separately combined (*i.e.*, to form the two different oxides) with a *fixed* mass, 100 g, of oxygen are in the ratio 2 : 1. The law could have been illustrated just as easily by fixing the mass of copper.

If desired, the results of the above experiment can be expressed in the following way, given the relative atomic masses of oxygen and copper as 16 and 63.5 respectively. (For *mole*, see page 45.)

In copper(I) oxide

0.32 g of oxygen combine with 2.53 g of copper, so $\dfrac{0.32}{16}$ moles of oxygen atoms

combine with $\dfrac{2.53}{63.5}$ moles of copper atoms. That is, 0.02 moles of oxygen atoms

combine with 0.04 moles of copper atoms, *i.e.*, the ratio of moles is 1 : 2.

In copper(II) oxide

0.38 g of oxygen combine with 1.52 g of copper, so $\dfrac{0.38}{16}$ moles of oxygen atoms

combine with $\dfrac{1.52}{63.5}$ moles of copper atoms. That is, 0.024 moles of oxygen atoms

combine with 0.024 moles of copper atoms, *i.e.*, this ratio of moles is 1 : 1. From this, the numbers of moles of copper atoms which combine with one mole of oxygen atoms in the two compounds are in a simple ratio to one another (2 : 1). Since one mole of oxygen atoms is a *fixed* mass, this result agrees with the Law of Multiple Proportions. Also, the figures for mole ratios point to Cu_2O and CuO as the simplest formulae for *copper(I) oxide* and *copper(II) oxide* respectively. They were formerly known as cuprous oxide Cu_2O, the red oxide, and cupric oxide CuO, the black oxide.

The accuracy of the experiment described above depends on the purity of copper(I) oxide, a material extremely difficult to obtain in a pure state. Mercury(I) and mercury(II) chlorides can both be obtained in a high degree of purity (Analar quality is used), and the following experiment may be performed as a class illustration of the law.

Experiment 19
A further illustration of the Law of Multiple Proportions

Note: Mercury(II) chloride is poisonous and great care must be taken in its use.

Weigh a boiling-tube, add two or three grammes of mercury(I) chloride, and weigh again. Add an equal mass of sodium hypophosphite, half fill with water, immerse in a beaker of water, and warm. Repeat using mercury(II) chloride, taking care to distinguish between the two boiling-tubes. After about 20 minutes, globules of mercury are seen in each tube. Wash the mercury by decantation several times with water (pouring the water into a beaker so that if mercury is lost, it may be retrieved), then with methylated spirit, and finally with ether. (*Care! Extinguish all flames in the vicinity.*) Replace the tubes in the warm water for a minute to remove traces of ether, dry the outside, and weigh. Calculate the separate masses of chlorine associated with, say, one gramme of mercury in each of the two chlorides. These masses will be in the ratio 1 : 2.

Another law, the Law of Reciprocal Proportions, can also be shown to follow from the Atomic Theory we are considering.

The Law of Reciprocal Proportions

This is a fourth law which can be deduced from the Atomic Theory. It is expressed in the statement:

If an element A combines with several other elements, B, C, D, the masses of B, C, D, which combine with a fixed mass of A are the masses of B, C, and D which combine with each other, or simple multiples of those masses.

Like the Law of Multiple Proportions, this law can be derived from Dalton's assumption that combination takes place between small whole numbers of atoms, all the atoms of a particular element being identical.

We have now obtained, from the Atomic Theory, certain conclusions which can be made the subject of experimental test. It is the magnificent achievement of Dalton to have been the first to state the Atomic Theory, deduce from it these conclusions suitable for experimental check, and show that the experimental results support the Theory. It is a piece of work which stamps Dalton as a scientific genius of a very high order.

Molecules

We have seen that the smallest possible particle of an *element* is called an atom. It is obvious that the smallest possible particle of a *compound* must contain at least *two* atoms because a compound must contain at least *two* elements and cannot contain less than one atom of each. To the smallest possible particle of a compound is given the name 'molecule'.

The word molecule has, however, a wider meaning than this. We have seen that the smallest possible particle of an element is called one atom of it, but it does not follow, necessarily, that single atoms are the only particles normally existing in a mass of an element. Actually most elements usually exist as a mass of more complex particles, consisting of a number of atoms associated together and moving as a single particle. To this more complex particle is given the name 'molecule'. The distinction between the *atom* and the *molecule* of an element is that the *atom* is the smallest particle of it which can ever be obtained, and is the unit which is concerned in chemical reactions, while the *molecule* is the smallest particle of the element which is normally capable of a separate existence.

Try to get this distinction quite clear; the atom is the smallest particle which can participate in chemical reactions while the molecule is the smallest particle which can normally exist when the element is not concerned in chemical reaction.

The atom is the smallest, indivisible particle of an element which can take part in chemical change.

The molecule of an element or compound is the smallest particle of it which can normally exist separately.

Actually, most of the elementary gases consist of molecules each containing two atoms. The molecules of hydrogen, oxygen, nitrogen, and chlorine are all of this type; a proof of this statement cannot be given until we have dealt with the Molecular Theory more fully, but it will be found, applied to hydrogen, on page 109. This state of combination of the atoms is indicated by writing the molecules of these gases as H_2, O_2, N_2, and Cl_2, meaning a single unit consisting of two atoms of each of the gases. (2H or 2Cl would mean two separate atoms of each of these gases, a condition in which they do not normally exist.) The number of atoms in a molecule of an element is called its *atomicity*.

The atomicity of an element is the number of atoms in one molecule of it.

A molecule containing *one* atom is said to be **monatomic**, *e.g.*, He helium.
A molecule containing *two* atoms is said to be **diatomic**, *e.g.*, H_2, O_2, N_2, Cl_2.
A molecule containing *three* atoms is said to be **triatomic**, *e.g.*, O_3 ozone.
A molecule containing *four* atoms is said to be **tetratomic**, *e.g.*, P_4 phosphorus.
A molecule containing *many* atoms is said to be **polyatomic**, *e.g.*, S_8 sulphur.

molecules of hydrogen, diatomic, H_2

molecules of ozone, triatomic, O_3

molecules of phosphorus (in certain solvents), tetratomic, P_4

Figure 15

Relation between scientific law and theory

A scientific law is simply a generalized statement of observed facts. For example, in all cases examined, it has been found that the total mass of the products of a chemical action is equal to the total mass of the original reagents. This result is assumed to apply to *all* such cases and is generalized as the Law of Conservation of Mass. A scientific law is subject to two possibilities of error, first, it is not possible to examine every last case of the operation of the law. There may be millions of them. Second, the law can be accurate only to the limit of experimental error.

A theory is an idea put forward to explain the existence of one or more laws. In the above case, the law is explained by a part of Dalton's Atomic Theory, *i.e.*, the idea of the existence of atoms which can be neither created nor destroyed in chemical action. Once a scientific law has been definitely established it will usually stand for any relevant time, being well grounded in fact. A theory put forward to explain the law may, however, be replaced from time to time as knowledge increases.

Questions

1. State briefly Dalton's Atomic Theory. Explain why the theory is named after him in spite of the fact that he was not the first to bring forward the idea that matter consists of atoms.

2. Two plugs of cotton wool, one soaked in concentrated hydrochloric acid and the other in a concentrated solution of ammonia, are used to seal the ends of a horizontal long glass tube. After a time a white ring forms nearer to one end of the tube than to the other. Explain the formation and position of the white ring in terms of the movement of gaseous molecules. (J.M.B.)

3. Describe the change you would *see* on leaving for several days without stirring a beaker in which there is a layer of water on a layer of saturated copper(II) sulphate (cupric sulphate) solution.

Give the name of the phenomenon you have described. (J.M.B.)

4. What do you understand by diffusion of a gas? What causes a gas to diffuse?
Describe experiments (one in each case) to show: (i) that a dense gas or vapour can diffuse upwards; (ii) that hydrogen diffuses faster than air. (A.E.B.)

5. State (*a*) the Law of Definite (Constant) Proportions and (*b*) the Law of Conservation of Mass.
Describe experiments by which the Law of Definite Proportions can be illustrated.
What evidence do these laws provide for the atomic nature of matter? (A.E.B.)

6. State briefly what is meant by *Brownian movement* and explain its cause. (C.)

7. State the Law of Constant Composition (that is, Definite Proportions). You are required to verify this law by using the black oxide of copper, and for the purpose you are supplied with specimens of copper(II) nitrate and copper(II) carbonate. State clearly (*a*) how you would prepare specimens of copper(II) oxide; (*b*) how you would use them to verify the law. (J.M.B.)

8. A metal, X, forms two oxides, A and B. 3.000 g of A and B contain 0.720 g and 1.160 g of oxygen respectively. Calculate the masses of metal in grammes which combine with one gramme of oxygen in each case. What chemical law do these masses of metal illustrate? Explain briefly. If the oxide B has the formula XO, what is the formula of oxide A? (O = 16.)

9. Show that the following results obtained by the reduction of the two oxides of a metal are in agreement with the Law of Multiple Proportions.

	1st compound	2nd compound
Mass of boat	5.30 g	4.45 g
Mass of boat + oxide	13.85 g	13.05 g
Mass of boat + metal	12.12 g	12.08 g

10. State the Law of Multiple Proportions and show how it is explained by the Atomic Theory.

An element X forms two oxides containing 77.47 and 69.62 per cent of X respectively. If the first oxide has the formula XO, what is the formula of the second oxide? (C.)

11. State the Law of Multiple Proportions. Describe how you would attempt to verify this law experimentally, if you were given specimens of lead(II) oxide (PbO) and lead(IV) oxide (PbO_2). Name one other pair of substances which you could use to verify this law. (J.M.B.)

3 States of Matter

We already have evidence from the experiments described in Chapter 2 that matter is composed of tiny particles called atoms or molecules, and that these particles are in constant motion. We have not yet considered the reason why the same substance, say water, can exist in several different forms, for example as solid ice, liquid water, and gaseous steam. It is easily shown that in each of these *states* the substance is chemically the same, and therefore it can be concluded that the particles which make up the three states of matter for any given substance differ only in the manner in which they are arranged, and they are not different in kind. Most substances can be classified under one of the three states of matter – solid, liquid, or gas – under a given set of conditions, although it must be mentioned that there are a few substances for which it is difficult to give a definite classification into one of the three categories (among these are the so-called liquid-crystals and plasmas). The theory which deals with the way in which the arrangement of the particles of a substance determine the properties that substance will possess, and particularly the state in which it is likely to be found under a given set of conditions, is known as the *kinetic theory*.

Kinetic theory

The *kinetic theory* is so named because it deals with that property of the particles which is so crucial to an understanding of the three states of matter, namely the *motion* of the particles. Because the particles are in constant motion they possess kinetic energy, which tends to keep the particles well spaced out in any substance. Thus, in terms of the kinetic theory, the *gaseous* state is one in which the particles are moving independently of each other in all directions and at great speeds; a typical speed for a molecule of nitrogen in air at ordinary temperature and pressure has been found to be approximately 500 m s^{-1} (or, about 900 miles per hour). Because the particles are travelling so quickly, and because there are an enormous number of particles in even the smallest volume of gas under everyday conditions, each particle collides with another with extremely high frequency; again, it has been estimated that a nitrogen molecule makes 10^9 collisions each second! Thus a gas will rapidly spread out to fill any container into which it is placed, and it cannot therefore be said to have any shape of its own.

For a *liquid*, however, it is supposed that the particles are much closer together than in a gas, and they are held at such distances by forces of attraction which tend to bind them together. Such forces only become large enough to have the binding effect when the particles come close to one another, and therefore the theory postulates that the particles of a liquid are fairly randomly arranged (as in a gas) but consist of 'clusters' in which they are very close together. This makes a liquid have a definite *volume*, but since the particles are still fairly free to move it does not have any characteristic *shape*. Thus a liquid takes up the shape of the bottom of the container in which it is placed.

In the case of a *solid* the particles have approached each other so closely that the forces of attraction between them are very strong and free movement of the particles cannot take place. Thus in a solid the particles are arranged in a fixed

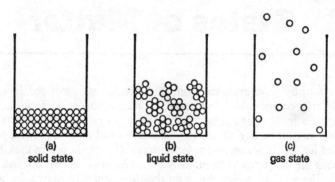

| (a) | (b) | (c) |
| solid state | liquid state | gas state |

Figure 16

pattern, and they form a *lattice* of vibrating masses. This makes a solid have a *fixed* shape, which can be altered only by applying strong forces.

The difference between the three states of matter so far as the arrangement of the constituent particles is concerned is illustrated in Figure 16.

Each of the three states of matter will now be considered in a little more detail, and the processes which take place when a substance in one state is converted into another state will be interpreted in terms of the kinetic theory.

Solid state

Substances in the solid state are made up of particles (atoms, molecules, or ions) which are packed so tightly together as to leave only slight unfilled space in the mass of solid. (Contrast the very wide separation of gas molecules.) The particles are bound together by forces strong enough to prevent movement of translation so that a solid has a definite shape which remains fixed (in constant conditions) unless sufficient force is supplied to shatter or distort the mass. Particles in a solid may however show a certain amount of movement of vibration.

Particles involved in the assembly of solid structures may be the following.

1. *Atoms*

Non-metals may form crystals by exercising *covalency* (page 81) to combine together very large numbers of their own atoms. An example of such a solid is *carbon* as *diamond*, forming what is known as a *giant covalent structure*. For a full account of this, and of the spatial extension of the covalency of carbon that makes it possible, see page 91.

Metals form solid masses in a different way. Their atoms can quite readily part with valency electrons from their outermost shells, as in the following examples.

$$Na \longrightarrow Na^+ + e^-; \quad Cu \longrightarrow Cu^{2+} + 2e^-$$

The resulting positively charged ions can then produce an ordered solid arrangement which depends on an equilibrium between their repulsive forces on each other and the binding effect of the electron cloud which moves, continually and at random, among them. See also *metallic bonding* (page 84).

2. *Molecules*

Many cases occur in which non-metallic elements form molecules by covalent combination, *e.g.*, iodine as I_2. (For the similar formation of a molecule Cl_2, see page 81.) In appropriate conditions of temperature and pressure, the atomic nuclei of one molecule and the electrons of another molecule attract each other sufficiently to bring about a close approach. As the molecules come together, the electrons of each begin to exert repulsive forces on each other. The forces of attraction and repulsion are balanced in the formation of a crystal. These *van der Waals forces* are, however, rather weak and the crystals tend to have low melting points. Simple molecules of compounds, *e.g.*, naphthalene, form crystalline solids by the operation of van der Waals forces in a similar way. For a further account of the iodine molecule see page 90.

3. *Ions*

When elements combine by exercising electrovalency (page 79), the resulting positively and negatively charged ions exercise powerful attractive forces on each other and can form a solid crystalline *lattice*. The arrangement of ions in such a lattice can usually be elucidated by X-ray diffraction. A well-known example is the lattice of solid sodium chloride formed from the ions Na^+ and Cl^-; for this see page 80. Imperfect lattices sometimes occur. For example, in iron(II) sulphide, some of the iron positions may be unoccupied and this will cause the compound to deviate slightly from the Law of Constant Composition.

Change of state – melting

If solids are heated, their constituent particles, atoms, molecules, or ions, acquire greater kinetic energy and vibrate more violently. Eventually a point may come at which vibration overcomes the binding forces and the particles become mobile. The crystalline structure then collapses and a *liquid* state is reached in which the particles are free to move. The temperature, t °C, at which this occurs is called the *melting point* of the solid and, at this temperature (and the prevailing pressure), the solid and liquid are in equilibrium. (Melting points are generally only slightly modified, *i.e.*, lowered, by moderate increase of pressure.) The energy which must be supplied to convert the solid at t °C to liquid at t °C is known as *latent heat of fusion*. Quantitatively, for example, 334 joules of heat energy convert one gramme of ice at 0 °C to one gramme of water at 0 °C (at standard pressure). If fairly simple molecules are only weakly bound together in crystals by van der Waals forces, the melting point of the crystals will be quite low, *e.g.*, naphthalene $C_{10}H_8$, 81 °C. In some cases, *e.g.*, iodine, some molecules may break away from the crystal directly into the vapour phase. This is known as *sublimation*. If binding forces are strong, *e.g.*, in lattices such as that of sodium chloride, or in some metallic crystals, melting points will be much higher, *e.g.*, for sodium chloride, 801 °C, and for copper, 1080 °C, though sodium metal melts at only 97 °C. When diamond is heated *in vacuo* (to prevent burning), it does not melt because the forces of covalency between the atoms are too great to be overcome by vibration with any ordinary source of heat. At very high temperature some carbon atoms will break away, but by sublimation into the gaseous state, not by melting.

Experiment 20
Determination of the melting point of a solid

The melting point of a solid (for example naphthalene or benzoic acid) may be determined using the apparatus shown in Figure 17.

A thin glass melting point tube about 4 cm long (which can be drawn out from ordinary glass tubing) is touched into a Bunsen flame to seal one end. When it has cooled, a little of the finely powdered solid is scooped into the open end and tapped down to the bottom of the tube to make a 2–3 mm layer. The tube will adhere to the thermometer by the surface tension effect of the liquid in the flask and the test material should be as close as possible to the thermometer bulb. The beaker containing glycerol is *gently* heated with continuous stirring and the melting point tube is carefully watched. At a certain point, the solid will be seen to melt and the corresponding temperature (melting point) is immediately read.

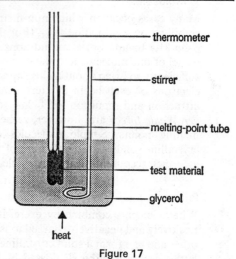

Figure 17

A *pure* solid will melt completely over a very narrow temperature range (less than 0.5 °C). Impurity depresses the melting point and usually causes gradual softening instead of sharp liquefaction. To melt sharply at its recognized melting point is good evidence of purity for a known solid.

Liquid state

It has already been mentioned that, when the particles which make up a solid acquire sufficient energy by heating, they become mobile, *i.e.*, the solid melts and forms a liquid. The particles remain close together by virtue of the attractive forces between them but are perpetually in random motion. A liquid spreads to fill a containing vessel as far as possible but leaves a level liquid surface if the capacity of the vessel exceeds the volume of liquid. This occurs because the great majority of particles (usually molecules) at the surface are held by attraction from the mass of particles below them; only relatively few particles of liquid (those with more than average energy) can escape.

Because of the perpetual random motion of particles, usually molecules, in a liquid, very small particles of insoluble solids are kept in similar random motion in liquids because they are continually and unevenly bombarded by molecules of the liquid. This is the cause of *Brownian movement* (see page 17).

An *ionic* liquid is obtained when a crystal lattice of an electrovalent solid (page 79) breaks down by the action of heat, but the temperature required is usually far above room temperature, *e.g.*, for K^+Cl^-, 768 °C. An ionic liquid is an electrolyte (page 149), *i.e.*, it conducts electric current by flow of ions to cathode and anode and is decomposed in the process, *e.g.*,

$$K^+ + e^- \rightarrow K \text{ (at cathode)}; \quad Cl^- \rightarrow \tfrac{1}{2}Cl_2 + e^- \text{ (at anode)}$$

A *molecular* liquid contains covalent molecules, *e.g.*, tetrachloromethane

CCl_4, trichloromethane $CHCl_3$, carbon disulphide CS_2. A great number of the simpler covalent compounds are liquid at room temperature and pressure, *i.e..* the melting points of the corresponding solids are below room temperature, Such liquids are non-electrolytes; they do not conduct electricity having no content of ions.

Vapour pressure

All molecules in a liquid are subject to attractive forces from neighbouring molecules. For molecules in the body of the liquid, these forces balance each other but, for molecules at the surface, there is a resultant attractive force acting downwards. In spite of this, some molecules of greater than average energy can escape from the surface of the liquid into the surrounding space and are lost. The liquid is said to *evaporate*. Escape of these particles of high energy lowers the average kinetic energy of those remaining; consequently, the temperature of the remaining liquid falls. This coldness can be felt if a small quantity of a liquid such as ether is allowed to evaporate on one's palm. Such a liquid, showing rapid evaporation at room temperature, is said to be *volatile* and has a low boiling point, *e.g.*, ether, 36 °C at standard pressure.

If a quantity of liquid is contained in an otherwise evacuated, sealed glass tube at a certain temperature, molecules will evaporate from the liquid surface into the enclosed space. Since these molecules are in rapid motion and cannot escape, some of them will collide with the liquid surface and re-enter it. It is found that, as a result of these opposite processes of evaporation and condensation, a state of equilibrium is finally reached, provided some liquid is left. This equilibrium is marked by the attainment of a *vapour pressure* which is *constant* (for the given temperature) and is called the *saturated vapour pressure* of the liquid at that temperature. This saturated vapour pressure is independent of the amount of liquid present (provided that it is not zero). For example, at 20 °C, the saturated vapour pressure of water is 17.5 mm of mercury.

Change of state – boiling

When a liquid is heated, its molecules acquire increased kinetic energy; therefore, the proportion of fast molecules increases, evaporation is more rapid, and the value of the saturated vapour pressure of the liquid rises. For example, for water at 40 °C, it is 55.3 mm and, at 95 °C, 634 mm. When a temperature is reached at which the vapour pressure of the liquid *equals* the prevailing atmospheric pressure, bubbles of vapour can form freely in the liquid and rise to the surface. The liquid is said to *boil*. That is, the boiling point of a given liquid at a pressure of P mm is the temperature at which the vapour pressure of the liquid is P mm. For pure water at 760 mm, the boiling point is 100 °C. At a higher pressure, water boils at a higher temperature because a higher vapour pressure must be reached to achieve free formation of vapour bubbles; conversely, at pressures below 760 mm, water has a boiling point lower than 100 °C, *e.g.*, 110 °C at 1075 mm, 90 °C at 526 mm.

Other liquids have different boiling points at standard pressure. For example, the binding forces between molecules of ethanol (alcohol) are such that the vapour pressure of this liquid reaches 760 mm at 78 °C, so this is its boiling point at standard pressure. The boiling point is higher than 78 °C at pressures above 760 mm.

Experiment 21
Determination of the boiling point of a liquid

The boiling point of a liquid is usually determined during the distillation of the liquid from a reaction mixture. However, it is sometimes required to check the purity of a liquid by taking its boiling point, and the following method may be used for low boiling point substances.

Because many of the low boiling liquids are flammable, it is advisable to avoid heating such liquids directly with a Bunsen flame. Therefore 5–10 cm³ of the liquid under test (for example propanone) is contained in a test-tube (150 mm × 25 mm), which is clamped in a beaker half full of water, as illustrated in Figure 18. A thermometer is held in the test-tube so that the bulb is situated 2–3 cm above the surface of the liquid. The water in the beaker should be heated gently – *with a fairly low Bunsen flame* – until the liquid begins to boil. If boiling occurs in too violent a manner it may be necessary to add a few pieces of porous pot to the liquid, which promotes smooth boiling. When the liquid is seen to condense on the thermometer bulb and run back into the bottom of the tube it is said to be *refluxing,* and the steady reading on the thermometer should be noted. Since boiling point varies with atmospheric pressure, the reading on the barometer should also be taken.

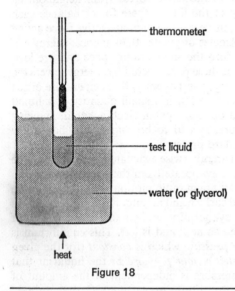

thermometer

test liquid

water (or glycerol)

heat

Figure 18

The above method is suitable only for those liquids with boiling points below 100 °C at standard pressure. If liquids of higher boiling points are used, then the water bath may be replaced by a beaker of glycerol (as in Experiment 20).

Experiment 22
Effect of dissolved solids on the boiling point of a solution

Put about 200 cm³ of water into a distilling flask (500 cm³) fitted with a sensitive thermometer, which should read to tenths of a degree. Add a few pieces of porous pot, clamp the flask at a suitable height, and heat the water to boiling point and notice at what temperature the water boils. Add a weighed amount of sodium chloride, say 5 g, and again determine the boiling point. Repeat this experiment, adding exactly 5 g of salt each time and noting the boiling point after each addition.

Specimen readings

Temperature of boiling water	99.7 °C
With 5 g of NaCl added	100.2 °C
With 10 g of NaCl added	100.7 °C
With 15 g of NaCl added	101.2 °C

From these figures it is clear that the *elevation* of the boiling point of water is directly proportional to the *concentration* of salt in solution, since for each 5 grammes of sodium chloride added the boiling point is raised by 0.5 °C. If a liquid contains non-volatile impurity in solution, this impurity will occupy some of the surface of the liquid, hindering vaporization. Consequently, the vapour pressure is lowered at any given temperature and the boiling point of the solution is *higher* than the boiling point of the pure liquid at the same pressure; correspondingly, the freezing point of the liquid is *depressed* by non-volatile impurity. For example, 5 g of sodium chloride raise the boiling point of 200 g of water by about 0.5 °C and depress the freezing point by about 2 °C at standard pressure. This is why sea water is much less readily frozen than fresh water in cold weather and why addition of common salt causes ice or snow to melt at temperatures not too far below 0 °C.

Gaseous state

Any gas consists of a collection of molecules of a particular kind which are in a state of rapid motion. The fact that the molecules are in motion is evident from the fact that if a small quantity of an odorous gas, such as hydrogen sulphide, is liberated at any point in a laboratory the smell of the gas soon pervades the whole room.

If the gas is confined in a closed vessel, some of the moving molecules strike the sides of the vessel and each impact exerts a small force upon the side. The number of molecules of gas inside such a vessel will normally be very large and, on the average, the same number of molecules will strike a given area on the sides of the vessel each second, so producing a steady pressure.

The kinetic theory makes the following assumptions about a 'perfect' gas.

1. Molecules of a gas move in straight lines at very great velocity until they collide with each other or with the wall of the containing vessel. Gas pressure is exerted in this vessel as the result of collisions between the gas molecules and the containing wall. The number of collisions in unit time being very great, the pressure appears constant (at constant temperature).

2. The total volume of the actual gaseous molecules is negligible relative to the capacity of the container.

3. Forces of attraction or repulsion between the molecules of the gas are negligible.

4. The average kinetic energy of the gas molecules measures the temperature of the gas.

These assumptions are approximately fulfilled by real gases at ordinary temperature and pressure.

Relation between pressure and volume of a gas

Suppose that one of the sides of a cylindrical vessel is a smooth piston and that there is a pressure, P, exerted on the piston just great enough to resist the pressure of the gas, of which the volume is V (Figure 19(a)). The piston will remain still.

Now suppose that the pressure on the piston is suddenly reduced to $\frac{1}{2}$P, without temperature change. The gas pressure is the greater and the piston will move up. As it does so the gas will fill the greater volume now available. The molecules will be more loosely packed in this larger space and so fewer will strike the sides of the vessel in a given time; that is, the pressure of the gas falls as the piston slides upwards. A stage will be reached when the gas occupies so large a volume that its pressure has also been reduced to $\frac{1}{2}$P and the piston will then stop (Figure 19(b)). This will happen when the volume of the gas has doubled; that is, a halving of the pressure causes the volume of the gas to be doubled. Similarly, it would be found that if the pressure on the piston was reduced to $\frac{1}{4}$P, it would come to a stop when the volume of the gas had increased to four times its original value.

Expressing the result generally, the pressure of a gas decreases in the same proportion as its volume increases.

From this, it is clear that if we multiply the varying volumes of a given mass of

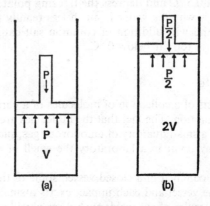

Figure 19

gas by the corresponding pressures, any decrease in the value of one of them will be exactly counterbalanced by the increase in the value of the other, and the result will always be the same.

Expressed mathematically, this may be stated in the form:

$$p_1 v_1 = p_2 v_2 \quad \text{(temperature constant)}$$

where p_1 and p_2 are two pressures and v_1 and v_2 the corresponding volumes of a given mass of gas.

This result is known by the name of its discoverer as Boyle's Law.

Boyle's Law: *The volume of a given mass of gas is inversely proportional* to its pressure, if the temperature remains constant.*

Boyle's Law is explained in terms of the kinetic theory as follows. If the volume of a given sample of gas is reduced at constant temperature, the average velocity of the gas molecules remains constant so they collide more frequently with the walls of the smaller containing vessel. The more frequent collisions cause higher pressure.

Effect of temperature change on the volume of a gas

It is common knowledge that a rise of temperature causes objects to expand and a fall of temperature causes contraction. The rule applies to gases, liquids, and solids, but the effect is much more marked in the case of gases than in the case of the other two. Charles found that, if pressure is constant, the volume of a gas increases or decreases by $\frac{1}{273}$ of its volume at 0 °C for every °C rise or fall of temperature; that is, if we take 273 cm³ of any gas at 0 °C, its volume will rise or fall by 1 cm³ for every °C rise or fall of temperature. Thus, at -1 °C, the volume will be 272 cm³; at -2 °C, 271 cm³; at -3 °C, 270 cm³ and so on. This leads to

---

* *Inversely proportional* is the mathematical expression of the fact that as the pressure *increases* the volume *decreases* in the same proportion.

the absurdity that, if the temperature falls to −273 °C, the volume of the gas will be 0 cm³ – the gas will have vanished! In actual practice, no substance can remain gaseous at such low temperatures; all become solids and Charles' Law does not then apply, but this temperature, at which the volume of a gas would theoretically be reduced to zero, gives us the lowest possible temperature that can ever be reached. It is called *absolute zero*.

A temperature scale is in use starting from this absolute zero as 0°, and using Centigrade (Celsius) degrees. This is called the *Kelvin (K) Scale* after Lord Kelvin who first suggested it. Measurements on this scale are stated in units, K, and a general kelvin scale temperature is represented by the capital letter symbol, T. Since absolute zero is the same as −273 °C, it is clear that the kelvin scale starts measuring temperature from a point 273 °C lower than the starting point of the Centigrade (Celsius) scale and, **to convert centigrade temperatures to kelvin temperatures, we must add 273°.** Thus:

−253 °C is the same as 20 K (−253 + 273)
0 °C is the same as 273 K (0 + 273)
15 °C is the same as 288 K (15 + 273)

Restating Charles' Law using kelvin temperature, we find that 273 cm³ of gas at 273 K (0 °C) will become 274 cm³ at 274 K (1 °C), 275 cm³ at 275 K (2 °C), 276 cm³ at 276 K (3 °C) and so on; for falling temperatures, the volume of gas will be 272 cm³ at 272 K, 271 cm³ at 271 K, 270 cm³ at 270 K and so on. This gives us the rule (known as Charles' Law) that the volume of a given mass of gas increases in the same proportion as its kelvin temperature, if pressure is constant.

Charles' Law: *The volume of a given mass of gas is directly proportional* to its kelvin (absolute) temperature if pressure is constant.*

From this it follows that if we divide the varying volumes of a given mass of gas by the corresponding kelvin temperatures, any increase in the volume will be exactly cancelled by the increase in the temperature, and the result will always be the same. This can be expressed in the form:

$$\frac{v_1}{T_1} = \frac{v_2}{T_2} \quad \text{(pressure constant)}$$

where v_1 and v_2 are the volumes of the gas at kelvin temperatures T_1 and T_2 respectively.

This relationship is accounted for in terms of the kinetic theory as follows: a fall of temperature represents a decrease in the average kinetic energy of the gas molecules; that is, average molecular velocity decreases (mass remaining constant). At constant pressure, this decreased velocity causes the sample of gas to occupy a smaller volume.

The Ideal Gas equation

By a combination of Boyle's Law,

$$p_1v_1 = p_2v_2$$

and Charles' Law,

$$\frac{v_1}{T_1} = \frac{v_2}{T_2}$$

* *Directly proportional* is the mathematical expression of the fact that the volume *increases* in the same proportion as the kelvin temperature *increases*.

we obtain the expression,

$$\frac{p_1 v_1}{T_1} = \frac{p_2 v_2}{T_2}$$

which is known as the *ideal gas equation*, since it strictly applies only to gases which behave in an 'ideal' manner (*i.e.*, those which fulfil the assumptions stated on page 37). However, as already mentioned, real gases obey these laws to a high degree at ordinary temperatures and pressures. The importance of the ideal gas equation is that it can be used to find the volume, v_2, that a given mass of gas will occupy at any desired temperature and pressure (T_2 and p_2) from its volume, v_1, at a given temperature and pressure (T_1 and p_1).

EXAMPLE. *A certain mass of gas occupies* 211 cm³ *at* 18 °C *and* 740 *mm pressure. What volume will it occupy (still gaseous) at* −20 °C *and* 770 *mm pressure?*

$$18 \text{ °C} = (18 + 273) \text{ K} = 291 \text{ K}$$
$$-20 \text{ °C} = (-20 + 273) \text{ K} = 253 \text{ K}$$

Then

$$\frac{p_1 v_1}{T_1} = \frac{p_2 v_2}{T_2}$$

$$\frac{740 \times 211}{291} = \frac{770 \times v_2}{253}$$

$$v_2 = \frac{740 \times 211 \times 253}{770 \quad \times \quad 291} \text{ cm}^3$$

At this stage, inspect your fraction to see if it agrees with what common sense would lead you to expect.

Thus (*a*), the 211 of the numerator, is the original volume of the gas in cm³. The pressure is changing from 740 to 770 mm; that is, an *increase* of pressure. This should *decrease* the volume. The fraction $\frac{740}{770}$ is actually doing so and is therefore correct.

(*b*) The temperature is *falling* from 291 K to 253 K; that is, the volume of the gas should be *decreasing*. The fraction $\frac{253}{291}$ *decreases* the volume as required and is, therefore, correct.

Using logarithms, $v_2 = 176$ cm³.

Standard temperature and pressure

Since the volumes of gases change in such a marked manner with changes of temperature and pressure, it is necessary to choose a suitable value of each as standards to which gas volumes can be referred. The standards chosen are 0 °C and 760 mm pressure and these are known as *standard temperature and pressure*, usually contracted to *s.t.p.*

That is, **s.t.p. indicates standard temperature and pressure or 0 °C (273 K) and 760 mm pressure (1 atm).** In the past, the alternative, *n.t.p.* (normal temperature and pressure) was used, the values being the same.

EXAMPLE. *A certain mass of gas occupies* 146 cm³ *at* 18 °C *and* 738 *mm pressure. Calculate its volume at s.t.p.*

$$18\,°C = 291\,K \qquad \text{s.t.} = 273\,K \ (0\,°C)$$
$$\text{s.p.} = 760\,mm$$

$$\frac{p_1 v_1}{T_1} = \frac{p_2 v_2}{T_2}$$

$$\frac{738 \times 146}{291} = \frac{760 \times v_2}{273}$$

$$v_2 = \frac{738 \times 146 \times 273}{760 \quad \times \quad 291} \quad \text{(Inspect this fraction as described above)}$$

$$= 133\,cm^3$$

Dalton's Law of Partial Pressures

This law states that, in a mixture of gases which do not act chemically together, each gas exerts a *partial pressure* which is the pressure it would exert if it alone filled the containing vessel at the same temperature and pressure. Then the total gas pressure is the sum of the partial pressures of the constituent gases.

This law applies most commonly to the case of a gas collected over water. For example, suppose 100 cm³ of an insoluble gas are collected over water at a barometric pressure of 745 mm and at 15 °C. If it is saturated with water-vapour, the true pressure of the gas is (745 – 13) mm, since the vapour pressure of water at 15 °C is 13 mm.

Dry, at s.t.p., the gas would occupy $100 \times \dfrac{732}{760} \times \dfrac{273}{288}\,cm^3$

$$= 91.3\,cm^3$$

In terms of the kinetic theory this can be explained as follows. In a non-reactive mixture, each gas exerts a separate pressure on the container because of collisions of its molecules with the containing walls. The total pressure on the container is caused by the sum of all the collisions.

If gas pressure rises to very high values, molecules are crowded very much more closely together so that ultimately forces between them are no longer negligible and their volume becomes significant relative to the space occupied by the gas. Then the simple gas laws no longer apply.

Questions

	Initial volume	Initial temperature and pressure	Final temperature and pressure
(a)	273 cm³	0 °C and 760 mm	14 °C and 861 mm
(b)	1638 cm³	0 °C and 819 mm	15 °C and 864 mm
(c)	1000 cm³	−23 °C and 750 mm	23 °C and 800 mm
(d)	500 cm³	17 °C and 870 mm	48 °C and 750 mm
(e)	1000 cm³	182 °C and 722 mm	s.t.p.
(f)	760 cm³	27 °C and 700 mm	s.t.p.
	The examples below need the use of logarithms		
(g)	700 cm³	17 °C and 740 mm	s.t.p.
(h)	133 cm³	14 °C and 745 mm	17 °C and 750 mm
(i)	55 cm³	14 °C and 744 mm	s.t.p.
(j)	574 cm³	s.t.p.	15 °C and 735 mm
(k)	70 cm³	s.t.p.	18 °C and 745 mm
(l)	121 cm³	150 °C and 780 mm	120 °C and 742 mm
(m)	534 cm³	s.t.p.	−15 °C and 740 mm

1. Calculate the volumes which will be occupied at the given final temperatures and pressures by the gases whose initial volumes, temperatures, and pressures are given on page 41.

2. The apparatus partially shown in Figure 20 was used by a student to find the melting point of a solid. List three errors that he had made in setting up the apparatus. (N.I.)

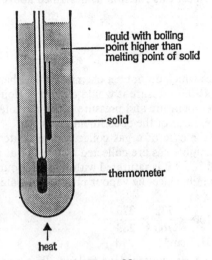

liquid with boiling point higher than melting point of solid

solid

thermometer

↑
heat

Figure 20

3. A strong, round-bottomed flask was half-filled with water which was then boiled vigorously for five minutes. Heating was discontinued and the flask was closed with a well-fitting rubber stopper. The flask was inverted and cold water was poured on the outside.

(i) What would be seen in the flask?
(ii) Give the reasons for your expected observations. (J.M.B.)

4. If a small quantity of ether is placed on one's skin a cold sensation is felt. Explain this effect. (S.)

5. The pressure inside a sealed can of 400 cm³ capacity at 27 °C is 600 mm of mercury. What would be the pressure at 127 °C if the size of the can did not alter?

If the original volume, temperature, and pressure had been 800 cm³, 27 °C, and 600 mm of mercury, what would the final pressure be at 127 °C? (J.M.B.)

6. By what factors must you multiply the volume of a gas measured at 20 °C and 750 mm of mercury pressure in order to find what the volume would be at s.t.p.? (S.)

4 Relative Atomic Mass

Introduction of the chemical balance

As soon as the existence of atoms was recognized, it was obviously desirable to try to obtain as much information as possible about the atoms of different elements, and, more particularly, to show how their masses compared with one another. This was necessary to enable quantitative experiments of various kinds to be made, to test doubtful points about the Atomic Theory, and also to enable chemists to make calculations of the quantities of materials involved in their experiments. Indeed, until the work of chemists could be made the subject of accurate weighings, and calculations could be tested by experiment, little progress could be expected. This was recognized more particularly by Black, a Scottish scientist working in Glasgow, and his name will always be honoured among chemists for his persistent and pioneer use of the chemical balance for checking his ideas. He taught scientists the importance of obtaining definite quantitative results instead of the vague qualitative statements with which they had been satisfied. The question for chemists became not only 'What happens?' but also 'What mass of each material is involved when it happens?'

Chaos of 1820–1850

Finding out how the masses of different atoms compared with each other proved extremely difficult to carry out. For about forty years after the Atomic Theory was suggested by Dalton, chemistry was in a state of chaos. It is difficult for us now to read with understanding any book on chemistry written between about 1820 and 1850, because chemists were simply groping about trying to solve the problem of comparing the masses of atoms, and they were making little progress. Let us try to understand where the difficulty lay and to follow the stages by which full knowledge was finally achieved.

In the first place, we must understand that chemists were not attempting to obtain the masses of *individual* atoms. It was clearly recognized that atoms were very small indeed and that there was no hope whatever at that time of obtaining the actual mass of a single atom of any element. The question was, rather: 'How do the masses of the atoms of different elements compare with one another? Is, for example, a sodium atom heavier than an atom of oxygen, and, if so, by how much? Is a silver atom heavier than an atom of gold, and, if so, by how much? We can now find the real mass in grammes of a single atom of any element; we have accomplished what the chemists of 1830 regarded as a vain aspiration, but the process has taken over 100 years, and when you read that the mass of a hydrogen atom is 0.000 000 000 000 000 000 000 001 7 g, you will not be surprised that science has consumed a great deal of time, and the patient efforts of many, to reach a stage at which such a minute quantity can be reasonably accurately measured.

Returning to the simpler problem of comparing the masses of atoms, it became necessary first to fix some standard of mass with which all the atoms could be compared. We were measuring potatoes in stones, jam in pounds, and masses of steel in tons. In what mass-units could we express the masses of atoms? To compare masses of atoms by using grammes would, in fact, have been far less sensible than trying to express the mass of grains of sand in tons. Chemists recognized this and decided to compare the masses of all other atoms with the mass of a hydrogen atom. The hydrogen atom was chosen because it is the lightest of all the atoms and would give numbers greater than unity for the comparative masses of all the other heavier atoms. This is more convenient than working in fractions or decimals, which the choice of any other atom as standard would necessitate.

Thus the first definition of the **relative atomic mass of an element** (formerly known as **atomic weight**) involved the direct comparison of the mass of one atom of the element with the mass of a single atom of hydrogen.

First definition. The relative atomic mass of an element is the mass of one atom of the element compared with the mass of one atom of hydrogen.

You must understand clearly what this most important characteristic of an atom is. If you look in a 'Table of Relative Atomic Masses', you will see some statement such as $O = 16$ or $Na = 23$ or $P = 31$. This is the chemist's shorthand way of saying 'The relative atomic mass of oxygen is 16, of sodium 23, and of phosphorus 31'; or, more fully, 'Every atom of oxygen has a mass 16 times greater than an atom of hydrogen'; or 'Every atom of sodium has a mass 23 times greater than an atom of hydrogen'; or 'Every atom of phosphorus has a mass 31 times greater than an atom of hydrogen'. All this is conveyed by $O = 16$, $Na = 23$, $P = 31$. You do not yet know how these figures have been obtained; take them for granted for the moment. It will be obvious that if the masses of all atoms are compared with the mass of a hydrogen atom in this way, the masses of all atoms are also compared with each other. When we say, for example, that the relative atomic mass of oxygen is 16 and that of sodium is 23, we also state that the sodium atom is heavier than the oxygen atom in the proportion of 23 to 16.

The fact that we have chosen the mass of a hydrogen atom as our unit of atomic mass leads to the statement that the atomic mass of hydrogen is 1 ($H = 1$). The most modern choice of standard for the unit of atomic mass is that of an isotope of carbon, $^{12}_{6}C$ (see page 75).

A table of the relative atomic masses of the elements is given on page 499.

Relative molecular mass

The masses of molecules of elements and compounds are also defined in terms of the mass of an atom of $^{12}_{6}C$ (see page 75 for an explanation of this symbol).

Definition. The relative molecular mass of an element or compound is the mass of one molecule of the element or compound compared with the mass of an atom of $^{12}_{6}C$ which is arbitrarily assigned as 12.000.

The relative molecular mass was formerly known as the *molecular weight*.

Atomic mass measurement. Mass spectrometer

A direct mode of determination of atomic mass has been available since 1920 through the mass spectrometer introduced by Aston. It must be remembered, however, that, by that date, the atomic masses of about 90 elements were fairly accurately known, and, at first, the results from Aston's work had to be checked against known atomic masses previously obtained from chemical experimentation. Since 1920, however, the mass spectrometer has been greatly improved and it can now give reliable, very accurate results. To a large extent, it has rendered chemical determination of atomic masses superfluous.

In principle the mass spectrometer, schematically illustrated in Figure 21, first produces positive ions of the element under investigation by causing its

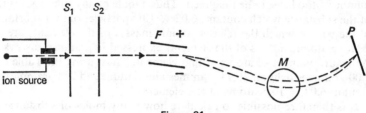

Figure 21

atoms to lose electrons. This may be done, for example, by very strong electrical heating of a trace of a salt of the element on a tungsten filament, or by subjecting a sample of the element in gaseous form to electron bombardment. For a univalent element, X,

$$X \rightarrow X^+ + e^-$$

The stream of ions, X^+, is selected by slits, S_1 and S_2, into a narrow beam which is deflected and dispersed by the electrical field between metallic plates at F. An electromagnetic field, M, at right angles to the field, F, then focuses the beam on to a photographic plate, P, so that all particles of the same charge/mass ratio fall on to a line resembling an optical spectrum line This line is an image of the selecting slits. The whole process occurs at very low pressure. By adding oxygen, $O = 16$, to the element under test, early experimenters could supply known lines at mass-change ratios 8, 16, and 32 (representing ions O^{2+}, O^+, and O_2^+) and use these as measuring standards for the lighter elements. Modern instruments can be made to record a graph on a moving paper roll. The graph shows a peak corresponding to each ion beam, from which the mass of the corresponding atom can be calculated, and the height of the peak indicates the intensity of the beam.

Aston's work of 1920 was notable not so much for its measurement of atomic masses as for its confirmation of the existence of isotopes (see page 79). For example, chlorine (chemical atomic weight 35.5) showed two kinds of atoms (isotopes) to be present in it by producing two mass spectrum lines at 35 and 37 with intensities in the proportion of about 3:1 (see Chapter 8).

The mole

It should be obvious that the relative atomic mass has *no units*; since it is essentially a ratio it is simply a *number* which states the relative mass of the atom concerned on a scale which gives the standard atom, $^{12}_{6}C$, a mass of 12.000 units. If, however, the *gramme unit* is attached to the relative atomic mass, then a specific *quantity* of the element concerned is being referred to, and a particular *number* of atoms of that element is implied. Thus the number of atoms in 12.000 grammes of $^{12}_{6}C$ is a special number, for it is the chemists' unit for the *amount of substance*, and it is known as *the mole*.

Definition. *The mole is the amount of substance which contains as many elementary entities as there are carbon atoms in 12.000 grammes of the $^{12}_{6}C$ isotope.*

By 'elementary entities' in the above definition is meant any type of particle that may represent the chemical constitution of the substance under investigation, and it clearly includes atoms, molecules, ions, electrons, protons, and any other particles. The actual number of atoms of carbon in 12.000 g of $^{12}_{6}C$ has been determined in a variety of ways and it is known as the *Avogadro number* or *Avogadro constant*, N_A (in honour of a great Italian scientist). The accepted value for the Avogadro number is 6.02×10^{23}, which is clearly a very large number – too large to be imagined! Thus a mole of any substance is that amount of the substance which contains 6.02×10^{23} particles of that substance. Because of the way in which the relative atomic masses of the elements are defined, the relative atomic masses of the elements expressed in grammes always contain the Avogadro number of atoms of that element. Thus, using the Table on page 499, 4.00 grammes helium, 26.98 grammes aluminium, and 238 grammes uranium all contain 6.02×10^{23} atoms of the elements.

It is therefore possible to calculate how many moles of substance correspond to a given mass of that substance, and also how many actual particles are present.

EXAMPLE. *A sample of aluminium has a mass of 5.4 g. Calculate (a) the number of moles of aluminium present, and (b) the number of atoms of aluminium in the sample.*

(a) The relative atomic mass of aluminium is 27. Thus a sample of aluminium of mass 27 g corresponds to 1 mole of aluminium.

$\therefore$ 5.4 g aluminium correspond to $\dfrac{5.4}{27}$ moles

$$= 0.2 \text{ mole aluminium}$$

(b) 1 mole of any substance contains 6×10^{23} particles.
$\therefore$ 0.2 mole aluminium contains $0.2 \times 6 \times 10^{23}$ atoms of aluminium

$$= 1.2 \times 10^{23} \text{ atoms}$$

Questions

1. Copy and complete the following table.

Element	Mass/g	Number of moles	Number of particles
Sodium	9.2		
Gold		2×10^{-3}	
Iron			2×10^{21}
Uranium	0.119		
Tin			10^{22}
Silver		5.5	
Copper	2.54		

Element	Mass /g	Number of moles	Number of particles
Helium			3×10^{24}
Carbon		6×10^{-2}	
Lead	18.63		
Potassium			4×10^{23}
Xenon		0.4	
Mercury			5×10^{20}
Hydrogen, H_2	0.2		

2. What mass of mercury contains ten times as many atoms as 4 g of carbon?

3. Explain the difference between the terms *relative atomic mass* and *relative molecular mass*.

4. Explain the meaning of the term *mole* and indicate its importance. (S.)

5. Given that 5 g of neon contain X atoms, state in terms of X the number of atoms in (a) 0.05 g of neon, (b) 5.6 g of silicon, (c) 1.1 g of manganese.

(Relative atomic masses: Ne = 20, Si = 28, Mn = 55.)

5 Symbols and Formulae

Symbols

Having reached the conclusion that elements are made up of atoms, scientists needed some means of denoting *atoms*. Dalton invented a number of symbols for the atoms of elements, a few of which are illustrated in Figure 22.

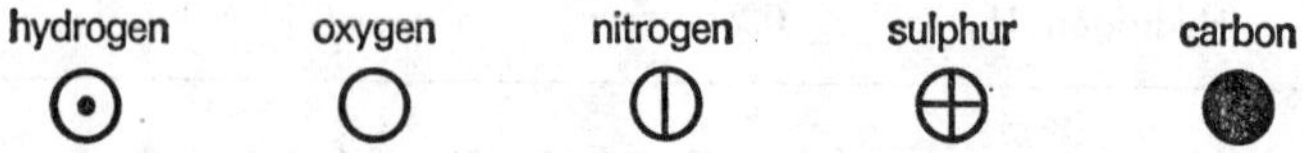

hydrogen   oxygen   nitrogen   sulphur   carbon

Figure 22

He then indicated the formulae of compounds by combining the necessary numbers of these symbols, writing each one separately. Using modern knowledge of these compounds, his formulae would be those given in Figure 23.

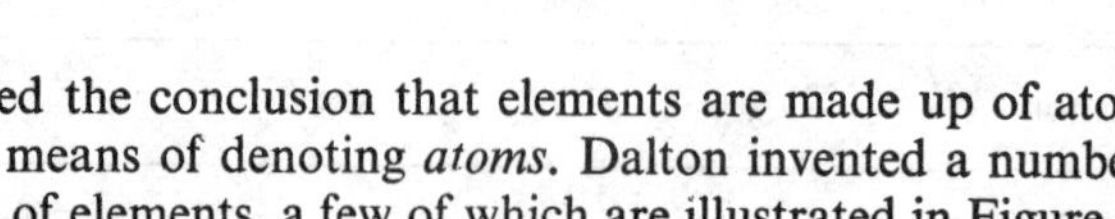

water   methane   sulphuric acid   ammonia

H_2O CH_4 H_2SO_4 NH_3

Figure 23

This very laborious system is quite unsuited to the representation of complex molecules. Think, for example, of the task of representing on this system the formula of cane sugar $C_{12}H_{22}O_{11}$. It would entail the drawing of 12 separate symbols for carbon, 22 for hydrogen and 11 for oxygen, and the result would be an unwieldy and confusing jumble of 45 separate signs. The system was soon abandoned for the modern simple system of representing atoms suggested by Berzelius, which consists generally of using the initial letter of the name of the element to stand for one atom of it, *e.g.*, one atom of hydrogen is denoted by H, one atom of oxygen by O, one atom of nitrogen by N. This rule cannot be universally applied because 100 elements have to share 26 letters. The difficulty has been readily overcome by using, for some of the elements, a symbol consisting of the initial letter, as a printed capital, together with one small letter from its name; for example, one atom of each of the elements carbon, chlorine, cerium, calcium, and caesium is denoted by the symbol C, Cl, Ce, Ca, and Cs respectively. In the case of the metals, the Latin names have sometimes been used as the source of the symbol. Some examples are shown in the table on page 49.

The last two metals in the table were unknown to the Romans, but a kind of pseudo-Latin name has been bestowed upon each, and from this its symbol is taken. Recently isolated *metals*, however, have been given names ending

Metal	Latin name	Symbol
copper	cuprum	Cu
iron	ferrum	Fe
lead	plumbum	Pb
silver	argentum	Ag
gold	aurum	Au
mercury	hydrargyrum	Hg
sodium	(natrium)	Na
potassium	(kalium)	K

in *-ium*, or *-um*, *e.g.*, radium, platinum, osmium, aluminium, while recently named *non-metals* have been given names ending in *-on*, *e.g.*, argon, xenon.

A list of the symbols of the known elements is given on page 499.

Formulae of elements and compounds

From Berzelius' system of symbols is derived a simple method of denoting molecules of a compound or element.

Anticipating a little, we have already seen that a molecule of an element is denoted by writing the symbol of the element and, to the right and below it, a number expressing the number of atoms in the molecule; for example,

H_2 denotes one molecule of hydrogen containing two atoms in combination.
P_4 denotes one molecule of phosphorus containing four atoms in combination.
S_8 denotes one molecule of sulphur containing eight atoms in combination.

The same device is adopted in representing the formulae of compounds, though here, of course, at least two symbols must appear because at least two elements must be present. Again, the small figure, to the right of a symbol and below it, expresses the number of atoms of the element present, the figure 1 being omitted. A few examples will make the idea clear.

H_2O denotes the formula of water, which contains two atoms of hydrogen and one atom of oxygen in combination;

H_2SO_4 denotes the formula of sulphuric acid, which contains two atoms of hydrogen, one atom of sulphur, and four atoms of oxygen in combination;

$CaCO_3$ denotes the formula of calcium carbonate (chalk), which contains one atom of calcium. one atom of carbon, and three atoms of oxygen in combination.

The close proximity of the symbols indicates that the atoms of the elements represented in the formulae of compounds are in chemical combination, and they therefore will behave quite differently in a chemical sense than when they are present in the separate elements. It is very important to realize this.

When a group of symbols is common to a class of compounds, it is frequently written as a bracketed group in their formulae, together with a number to indicate the number of groups present. For example, all metallic nitrates are derived from *nitric acid* HNO_3, and they all contain the nitrate group or **radical**, NO_3, in their formulae. When formulae of nitrates are written, this group is preserved intact, and, if two or more are needed, the number is indicated by enclosing the NO_3 group in a bracket and writing the number needed below and to the right. This arrangement is convenient because it emphasizes the relation of the nitrates to nitric acid. For example, $Ca(NO_3)_2$ means the same as CaN_2O_6, because the 2 multiplies everything inside the bracket, but $Ca(NO_3)_2$ indicates the relation of calcium nitrate to nitric acid HNO_3, more clearly than does CaN_2O_6. Similarly the formula of aluminium nitrate is written $Al(NO_3)_3$ rather than AlN_3O_9.

The sulphate **radical**, SO_4, which is common to sulphuric acid H_2SO_4, and to all sulphates, is similarly treated. Thus the formula of aluminium sulphate, written as $Al_2(SO_4)_3$, indicates the derivation of this compound from sulphuric acid H_2SO_4, more clearly than if written as $Al_2S_3O_{12}$. The hydroxyl group, OH, is also preserved in formulae to emphasize the relation of hydroxides to water H.OH, which is regarded as hydrogen hydroxide. Thus the formula of iron(III) hydroxide is written $Fe(OH)_3$, not FeO_3H_3.

If it is necessary to indicate a number of molecules of a compound, this is done by writing the appropriate number before the formula of the compound; for example,

$2H_2SO_4$ means two molecules of sulphuric acid.
$8HNO_3$ means eight molecules of nitric acid.
$4HCl$ means four molecules of hydrogen chloride.
$10H_2O$ means ten molecules of water.

It is important to notice carefully that the figure in front of the formula multiplies the *whole* of it. $2H_2SO_4$, for example, denotes two molecules of sulphuric acid each containing two atoms of hydrogen, one atom of sulphur, and four atoms of oxygen, or four atoms of hydrogen, two atoms of sulphur, and eight atoms of oxygen in all.

It is a common mistake of beginners to think that the figure multiplies only the symbol which immediately follows it; for example, that in $2H_2SO_4$ the 2 multiplies only the H_2 and not the SO_4. This is quite wrong. The 2 multiplies the *whole* of the formula H_2SO_4.

Nomenclature of compounds

The word 'nomenclature' means 'system of naming', and a fairly straightforward scheme has been designed which will enable you to state the composition of most chemical substances directly from their names.

Binary compounds

The name ending -ide is given to compounds containing only two elements, and the nature of the elements is indicated in the two words of the name, *e.g.*, copper(II) oxide CuO; hydrogen sulphide H_2S.

The chief exceptions to this are water H_2O, ammonia NH_3, methane CH_4, and other non-metal hydrides. Hydroxides are an exception to this rule but in these compounds the three elements present are indicated in the name. Another exception is an acid salt of hydrogen sulphide, *e.g.*, sodium hydrogensulphide NaHS, but there again the name is self-explanatory. In salts like ammonium chloride NH_4Cl, the NH_4 group has been treated as if it were an element.

Acids

A great many acids contain hydrogen, oxygen, and a third element, *e.g.*, H_2SO_4, HNO_3, H_3PO_4. The commonest and most stable of such acids is usually highly oxidized and to it is given a name which ends in *-ic* and is derived from the element it contains in addition to hydrogen and oxygen, *e.g.*, sulphuric acid H_2SO_4, and nitric acid HNO_3. An acid containing the same elements but *less oxygen* has the name-ending changed to *-ous*, *e.g.*, sulphurous acid H_2SO_3, and nitrous acid HNO_2.

Acid radicals

Acid radicals are present in salts which are derived from particular acids, and the naming of the acid radicals depends to a large extent on the acid from which they are derived. Thus those acid radicals derived from *-ic* acids take an ending *-ate* (*e.g.*, sulphate SO_4 from sulphuric acid, nitrate NO_3 from nitric acid, phosphate PO_4 from phosphoric acid). Those acid radicals derived from *-ous* acids take an ending *-ite* (*e.g.*, nitrite NO_2 from nitrous acid, sulphite SO_3 from sulphurous acid).

When metal atoms are incorporated in the acid radical the ending *-ate* is again used, *e.g.*, zincate ZnO_2, plumbate PbO_3, stannate SnO_3, manganate MnO_4.

Metal compounds

For metals which form only one series of salts with acids (have a fixed 'valency' – see page 101), the name of the metal is followed by the name of the acid radical, thus

$NaNO_3$	sodium nitrate,
$Al_2(SO_4)_3$	aluminium sulphate, and
$MgCl_2$	magnesium chloride.

For metals of variable valency Roman numerals are included in the name, to indicate the valency of the metal, thus

$FeCl_2$	iron(II) chloride, and
$FeCl_3$	iron(III) chloride.

Organic compounds

Organic compounds are those derived from carbon; they usually contain carbon, hydrogen, and oxygen and may also contain other elements. Since these compounds are in a special class of their own, a different system of naming them has been devised, and this will be explained in Chapter 28.

Historical note

Recent recommendations of the Chemical Society, the International Union of Pure and Applied Chemistry, and the Association for Science Education have made certain chemical names 'trivial' and they are to go out of use. Examples are names in *-ic* and *-ous* derived from names of *metals*, *e.g.*, *ferric* and *ferrous*, *cupric* and *cuprous*. The recommended replacement takes the form of the ordinary name of the metal with its operative valency stated in brackets in Roman numerals. Where necessary, the majority of the names in the present edition have been changed to conform to this notation. Some cases of the old usages have been left because they occur in chemical literature before about 1960 and reagent bottles in many laboratories still retain them, and therefore students should be aware that differences still exist. Examples are the following:

Old	*New*
Cuprous oxide	Copper(I) oxide Cu_2O
Cupric oxide	Copper(II) oxide CuO
Ferrous sulphate	Iron(II) sulphate $FeSO_4$
Ferric sulphate	Iron(III) sulphate $Fe_2(SO_4)_3$
Plumbous chloride	Lead(II) chloride $PbCl_2$
Plumbic chloride	Lead(IV) chloride $PbCl_4$
Manganese dioxide	Manganese(IV) oxide MnO_2

Determination of the formulae of compounds

The only way to be certain about the formula of an unknown compound is to perform experiments by which the compound is broken down directly into its elements, or into familiar compounds which can be shown unambiguously to have been obtained from the elements in the original compound under investigation – a process known as *analysis*. Alternatively, an attempt may be made to prepare the unknown compound from its elements either directly or indirectly, and compare the properties of the prepared compound with that of the unknown – a process known as *synthesis*.

The following experiments illustrate the way in which the formula of the black oxide of copper can be determined, either by analysing it by reduction in hydrogen (with heat) or by synthesizing it by oxidation of the pure metal, both processes being quantitative.

Experiment 23
Formula of copper(II) oxide by reduction

Black copper(II) oxide contains copper and oxygen chemically combined. By the removal of oxygen in a stream of hydrogen the masses of copper and oxygen which were in combination can be determined (Figure 24).

Then light the hydrogen at the jet (see Experiment 17, page 24) and warm the copper(II) oxide. Soon a glow spreads through the oxide (an indication that chemical change is taking place). A reddish-brown powder, copper, is

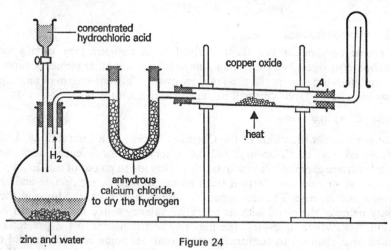

Figure 24

The tube must slope downwards towards *A*, otherwise water condensed during the experiment might run back on to the heated part and crack the tube.

Weigh a porcelain boat and weigh again with some pure dry copper(II) oxide in it. Place the boat inside a hard glass tube. Generate hydrogen as in Figure 24 and pass it through a calcium chloride tube to dry it. Allow the hydrogen to pass over the oxide until, when collected in a test-tube as shown, it burns quietly on exposure to a flame. This shows that all the oxygen in the apparatus has been expelled.

left, and drops of water collect at *A*.

Allow the copper to cool in a current of hydrogen so that air cannot enter and so convert the red metallic copper into the oxide again.

Weigh the boat and copper when cool. (The boat and contents should, to ensure complete reaction, be heated to constant mass.)

Note that the copper(II) oxide has been **reduced** by the hydrogen to red metallic copper, whereas the hydrogen has been **oxidized** to water.

copper(II) oxide + hydrogen ⟶ copper + water

Specimen analysis

$$\text{Mass of boat} = 4.32 \text{ g}$$
$$\text{Mass of boat} + \text{oxide} = 5.61 \text{ g}$$
$$\text{Mass of boat} + \text{copper} = 5.35 \text{ g}$$
$$\therefore \text{ Mass of oxygen} = 0.26 \text{ g}$$
$$\therefore \text{ Mass of copper} = 1.03 \text{ g}$$

0.26 g of oxygen has combined with 1.03 g of copper

Taking the relative atomic masses of oxygen and copper to be 16 and 63.5 respectively, the number of moles of oxygen atoms and copper atoms which combined together are:

$$\frac{0.26}{16} \text{ and } \frac{1.03}{63.5} \text{ or 0.016 (approx.) in both cases}$$

This means that, in black copper(II) oxide, copper and oxygen atoms combine together in *equal* numbers and the simplest formula of the oxide is CuO.

For class use, the following method is satisfactory.

Blow a small hole in a test-tube and weigh the test-tube. Weigh again with some pure dry copper(II) oxide in it. Push the rubber tube from the Bunsen

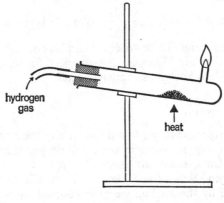

hydrogen gas

heat

Figure 25

burner into the test-tube, pass hydrogen through it and light it as shown in Figure 25.

Conduct the experiment in a similar manner to that described in the previous experiment and work out the result.

Experiment 24
Formula of copper(II) oxide by oxidation of copper

Weigh a small evaporating dish and clock-glass (7.5 cm diameter is suitable), and weigh again having added one or two small pieces of copper (not more than 0.5 g). Remove the clock-glass, add about 10 cm³ of bench dilute nitric acid (approx. 4M), replace the clock-glass, and heat *gently* on a tripod and gauze in a fume-chamber (Figure 26). There is a vigorous effervescence, brown fumes of nitrogen dioxide are seen, and the copper finally dissolves, giving a blue solution of copper(II) nitrate.

Continue to heat the solution (increasing the size of the flame to maintain a steady but not too vigorous evolution of vapour) until the whole of the excess nitric acid has been driven off and the copper(II) nitrate converted to black copper oxide.

Heat the dish very strongly for a few minutes

to decompose any copper nitrate on the sides of the dish or on the clock-glass, allow to cool (in a desiccator if possible) and weigh the dish and clock-glass. Repeat the heating to constant mass. (The clock-glass minimizes loss of liquid by spurting but does not materially reduce the rate of evaporation.)

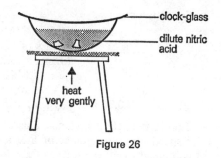

Figure 26

Specimen weighings

$$\begin{aligned}
\text{Mass of evaporating dish and clock-glass} &= 20.210 \text{ g} \\
\text{Mass of the above} + \text{copper} &= 20.634 \text{ g} \\
\text{Mass of the above} + \text{copper(II) oxide} &= 20.741 \text{ g} \\
\therefore \text{ Mass of copper used} &= 0.424 \text{ g} \\
\therefore \text{ Mass of oxygen combined} &= 0.107 \text{ g}
\end{aligned}$$

From these figures, 0.107 g of oxygen combined with 0.424 g of copper. Given that $O = 16$ and $Cu = 63.5$, $\dfrac{0.107}{16}$ moles of oxygen atoms combine with $\dfrac{0.424}{63.5}$ moles of copper atoms. These are equal numbers of moles of atoms at the value of 0.0067 approximately. That is, the simplest formula of the black oxide is CuO. This oxide is known as copper(II) oxide.

Empirical formula

In the above experiments we found the *ratio* of the number of copper atoms to the number of oxygen atoms in the black copper oxide, and inferred that the formula was CuO, since one atom of copper was in combination with one atom of oxygen. Strictly speaking we have still not determined the actual formula, since by the method we have adopted we would not be able to distinguish between oxides of copper with the formulae CuO, Cu_2O_2, Cu_3O_3, etc. It is necessary to carry out further experimental work in order to distinguish between these possibilities, and the manner in which this can be done will be described later.

That formula which indicates the *ratio* of the different atoms present in a compound is known as the *empirical formula* (sometimes, the *simplest formula*), and it is further discussed in Chapters 7 and 11.

Questions

1. State the meaning of the following formulae: KNO_3, $CuSO_4.5H_2O$, $PbCl_2$, $Na_2CO_3.10H_2O$ (washing soda), *e.g.*, the formula H_2O (water) means that one molecule of water contains two atoms of hydrogen and one atom of oxygen.

2. How many atoms of the various elements are indicated by the following formulae:

$$2H_2O, \ 5HCl, \ 7HNO_3,$$
$$20PbSO_4, \ 11Cu(NO_3)_2?$$

3. What do you understand by the term 'atomicity'? What is the atomicity usually assigned to the elements helium, chlorine, phosphorus?

4. Why is the formula for lead nitrate written $Pb(NO_3)_2$? (It could also be written PbN_2O_6.) Give three similar examples.

5. Give the names and formulae of *three* oxides of lead. Describe in detail how you would determine the percentage of lead in

any one of them, paying particular attention to the precautions necessary to obtain an accurate result. Name and state the law that could be verified by determining the percentage of lead in *two* different oxides. Starting from one of the oxides of lead, how would you prepare a pure dry sample of lead nitrate crystals? (J.M.B.)

6. 420 g of a compound of iron with sulphur only contained 224 g of sulphur. Calculate the empirical formula of this compound. (O.)

7. You are required to determine the percentage of copper in a sample of copper(II) oxide.

(i) Give a labelled diagram of the apparatus you would use.

(ii) State *three* precautions you would take when carrying out this determination and explain why they are necessary.

(iii) Describe the changes which would be observed during the reaction which occurs.

(iv) State the weighings which you would record, and outline how you would calculate the required result.

On reduction, 12.00 g of a metallic oxide, A, gave initially 11.20 g of a lower oxide, B, and ultimately 10.40 g of the metal, X.

(i) Calculate the mass of X which is combined with 16 g of oxygen in each oxide.

(ii) State the chemical law illustrated by these results.

(iii) The atomic mass of X is 208. Deduce the simplest formula of the oxide A. (A.E.B.)

8. If 20 g each of copper(II) oxide (cupric oxide), copper(II) sulphate (cupric sulphate), copper(II) nitrate (cupric nitrate), and copper(II) chloride (cupric chloride) are reduced to copper, which compound will give the highest mass of copper and why? Calculate this mass of copper. (A.E.B.)

9. Lead (Pb) is a metal low down in the activity series (electrochemical series). This metal forms several oxides, one of which, A, when heated decomposes to another, B, and oxygen is liberated. Devise and describe experiments by which you could find the formulae of the oxides A and B. (You should use a method different in principle for each oxide, though you may use the results of one in the calculation of the other.)

Suggest a set of readings that you might obtain in your experiments if the two oxides actually have the formulae PbO_2 and PbO, and show how these readings would lead to the answer.

When sulphur dioxide (SO_2) is passed over 2.39 g of the warm oxide PbO_2, the latter glows red and a white residue weighing 3.03 g is obtained. What do you think has happened? (L.)

10. Sketch and label an apparatus suitable for heating an oxide of copper in a current of dry *hydrogen*. State two precautions (with reasons for them) which are necessary for the safe conduct of the experiment. Describe what happens to the oxide of copper when heating begins. If 0.429 g of oxide were reduced to 0.381 g of copper in an experiment, which oxide of copper (CuO or Cu_2O) was used? ($Cu = 63.5$; $O = 16$.)

11. When suitably heated in oxygen, lead can produce a red compound said to be Pb_3O_4. (This red compound can be entirely reduced to lead at red heat in a current of dry hydrogen.) Describe, in detail, an experiment by which (using this reduction) you would test the accuracy of the quoted formula. Outline any required calculation. ($Pb = 207$; $O = 16$.)

12. Two oxides of lead, A and B, were heated in dry hydrogen to reduce them to metallic lead. In case A, 0.446 g of oxide left 0.414 g of lead; in case B, 0.717 g of oxide left 0.621 g of lead. Show that these figures are in accordance with the Law of Multiple Proportions and, given $Pb = 207$ and $O = 16$, calculate the simplest formulae of A and B.

6 Chemical Equations

Representation of reactions

We have already seen in Chapter 5 that it is possible to represent the chemical constitution of a compound in its formula – a type of chemical 'shorthand'. Having established the formulae of many compounds it would be convenient to be able to represent chemical **reactions** themselves in a shorthand manner. This is the purpose of the chemical equation.

Quite clearly any method for representing a chemical reaction will need to satisfy certain basic requirements:

(i) the chemical nature of the substances combining together (called the **reactants**) must be clear;

(ii) similarly, the chemical nature of the substances formed as a result of the reaction (called the **products**) must also be clear;

(iii) the mole proportions in which the reactants combine, and the mole proportions in which the products are formed, must be deducible;

(iv) the **direction** of the reaction (i.e. which substances are reactants and which are products) must be established.

We shall see later how it is possible to express other aspects of a chemical reaction through the equation, particularly the state of matter in which the reactants and products are present (solid, liquid, gas, or aqueous solution), and also how much energy is liberated or absorbed during the reaction.

Setting up an equation

The setting up of a chemical equation involves satisfying the requirements mentioned above, and the only way in which these requirements can ultimately be met is by conducting a series of experiments. Thus, an equation represents the results of an *empirical* investigation (an investigation through experiment) either on the reaction itself, or by the determination of the nature and formulae of the products (the nature and formulae of the reactants are normally assumed).

The first step in setting up the equation therefore, involves the establishment of the formulae of all the substances taking part in the reaction, both as reactants and products. The determination of chemical formulae has already been described in Chapter 5.

The second step is to write down the formulae of the reactants and products in some agreed manner, and conventionally the reactants are placed on the left-hand side of the page and the products on the right-hand side. An arrow from left to right indicates that the reaction proceeds from reactants to products as written. Thus, if we wish to represent the fact that one mole of copper(II) oxide reacts with one mole of sulphuric acid to produce one mole of copper(II) sulphate and one mole of water, we can do so in the form:

$$CuO + H_2SO_4 \rightarrow CuSO_4 + H_2O$$

This kind of statement is called 'a chemical equation'.

The + sign on the left of the equation means 'reacts with', but on the right it means simply 'and', while the arrow means 'producing'.

Take another simple chemical equation.

$$Zn + H_2SO_4 \rightarrow ZnSO_4 + H_2$$

This means: one mole of zinc reacts with one mole of sulphuric acid producing one mole of zinc sulphate and one mole of hydrogen.

It is important to notice that there must be the same number of each kind of atom on the right of a chemical equation as on the left, otherwise the equation would imply that atoms had been created or destroyed, which is impossible. Hence, whenever we write a chemical equation we are assuming the Law of Conservation of Mass, established in Experiment 15, Chapter 2. Hence it is important that whenever a chemical equation is written we must be certain that the number of atoms of any given element on one side of an equation is exactly the same as the number of atoms of the same element on the other side of the equation. This process of equalization is known as 'balancing' the equation.

Balancing an equation

It has already been explained that, by a balanced equation, we mean one which has the same number of each kind of atom on the right of the equation as on the left. An unbalanced equation implies that atoms have been created or destroyed; it is therefore wrong, and calculations based on it are certainly unreliable. It is absolutely essential, then, to obtain a balanced equation.

Facility in producing balanced equations is attainable only with practice, but this absolutely inviolable rule must be remembered:

The formula of a compound is absolutely fixed and unalterable and an equation must be balanced by taking appropriate numbers of molecules of the substances concerned, not by attempting alteration of their formulae.

The following brief account will illustrate the process of balancing.

To obtain a balanced equation for the action of hydrogen sulphide, H_2S, on sulphur dioxide, SO_2, producing water, H_2O, and sulphur, S.

The skeleton, but unbalanced and incorrect, 'equation' will be:

$$H_2S + SO_2 \rightarrow H_2O + S \quad \text{(UNBALANCED)}$$

We cannot alter any of these formulae or the symbols.

Now, both the oxygen atoms of the SO_2 molecule form water, therefore $2H_2O$ must be obtained. This gives:

$$H_2S + SO_2 \rightarrow 2H_2O + S \quad \text{(UNBALANCED)}$$

The $2H_2O$ on the right now requires $2H_2$, which must be provided by taking $2H_2S$. This gives:

$$2H_2S + SO_2 \rightarrow 2H_2O + S \quad \text{(UNBALANCED)}$$

We now have, on the left, 2S from the $2H_2S$ (remember the 2 multiplies *all* the H_2S) and S from the SO_2, therefore we must have 3S on the right and the balancing is complete.

$$2H_2S + SO_2 \rightarrow 2H_2O + 3S \quad \textbf{(balanced equation)}$$

Some such balancing process is necessary for all equations, but many of them become so familiar with frequent use that they can be set down correctly at once.

Deriving equations from experimental work

If it is important to establish the nature of the substances involved in a chemical reaction it is also important that the mole proportions in which those substances react is determined. The mole ratio in which the reactants combine, together with the mole ratio in which the products are formed, is known as the **stoichiometry** of the reaction. The following experiments are designed to establish the stoichiometry of a few chemical reactions.

Experiment 25
Stoichiometry of the reaction between magnesium and hydrochloric acid

Many metals displace hydrogen from dilute acids. For experimental purposes, magnesium is convenient because it reacts rapidly with either dilute hydrochloric or dilute sulphuric acid at room temperature. The purpose of the present experiment is to determine (given Mg = 24 and H = 1.008) what numbers of atoms of magnesium and hydrogen displace one another in the reaction. Magnesium is weighed normally on a chemical balance but hydrogen (being the lightest known gas) is more conveniently measured by volume and converted to mass from the fact that one dm³ (litre) of hydrogen at s.t.p. weighs 0.09 g.

Weigh a small watch-glass, then put on to it about 0.25 g of cleaned magnesium ribbon and weigh again. Obtain the mass of the metal by difference. Without the conical flask in position (Figure 27), fill the siphon tube with water by blowing gently into the short tube attached to the aspirator. Close the clip and place the delivery tube in the measuring cylinder. (This procedure is necessary because at the end of the experiment the siphon tube will be filled with water and, if it were left empty at the start, the volume of water delivered would be too low.) Using a funnel (to keep acid away from the upper sides) pour about 30 cm³ of dilute hydrochloric acid into a conical flask, carefully place the weighed magnesium as shown and connect the conical flask to the aspirator (supporting it suitably).

Open the clip. If the apparatus is airtight, water will flow for a second or two, then stop. (If it continues longer a leak exists and must be remedied.) Shake the magnesium into the acid. When all the metal has dissolved, allow time for the apparatus to cool. Then adjust the levels of water in the measuring cylinder and aspirator to equality (by raising or lowering one of the two as required). This gives atmospheric pressure in the aspirator. Read the volume of water in the measuring cylinder, after closing the clip and removing the cylinder. This volume is also the volume of hydrogen liberated. Room temperature and the barometer reading are required.

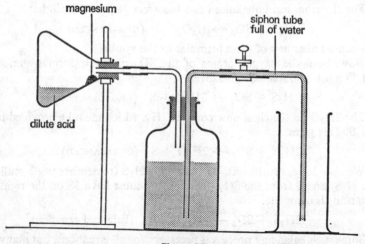

magnesium

siphon tube
full of water

dilute acid

Figure 27

Note: This experiment may be alternatively carried out by a 'syringe' method. See Experiment 28.

Specimen readings

Mass of magnesium $= 0.260$ g
Volume of hydrogen $= 252$ cm^3
Room temperature $= 10$ °C
Barometer reading $= 752$ mm (corrected for vapour
pressure of water)

1. *Reduce the volume of hydrogen to s.t.p.* (see Chapters 3 and 12)

$$\frac{p_1 v_1}{T_1} = \frac{p_2 v_2}{T_2} \quad \text{or} \quad \frac{752 \times 252}{283} = \frac{760 \times v_2}{273}$$

From this, $\qquad v_2 = \dfrac{752 \times 252 \times 273}{283 \times 760} = 240.5 \text{ cm}^3$

2. *Find the mass in grammes of this volume of hydrogen*

1000 cm^3 of hydrogen at s.t.p. weigh 0.09 g
so $\qquad 1$ cm^3 of hydrogen at s.t.p. weighs 0.00009 g
and $\quad 240.5$ cm^3 of hydrogen at s.t.p. weigh 0.00009×240.5 g
or $\quad 0.02165$ g

Approximating slightly on the experimental results: 0.0217 g of hydrogen are displaced by 0.260 g of magnesium. Assuming the relative atomic masses of hydrogen and magnesium to be 1.008 and 24 respectively, the number of moles of hydrogen and magnesium which replace one another are:

$$\frac{0.0217}{1.008} \quad \text{and} \quad \frac{0.260}{24}, \quad \text{or } 0.0215 \text{ and } 0.0108.$$

This is an almost exact ratio of $2 : 1$ in moles, *i.e.*, one mole of magnesium has displaced two moles of hydrogen atoms from the acid. Since the hydrogen molecule is diatomic, this corresponds to the equation:

$$Mg + H_2SO_4 \rightarrow MgSO_4 + H_2$$

Experiment 26
Stoichiometry of the reaction between zinc and copper(II) sulphate solution

Crush a quantity of copper(II) sulphate crystals (about 5 g) in a mortar and dissolve them in water in a beaker, warming to hasten the solution. Weigh a small watch-glass, add a few strips of pure zinc foil (mass about 1 g) and weigh again. The mass of the zinc can be obtained by difference. Put the zinc into the copper(II) sulphate solution. Immediately the zinc becomes coated with a red film of copper and, on stirring, the copper falls to the bottom of the beaker. After a while the whole of the zinc will have disappeared and there will be a layer of red metallic copper on the bottom of the beaker. (More copper(II) sulphate crystals can be added if the colour indicates that the solution is dilute.)

Filter off the copper using a weighed filter paper and wash the small particles of copper adhering to the beaker into the filter paper by means of a jet of water from a wash-bottle. Wash the copper several times with hot distilled water, and finally two or three times with methylated spirit (care being taken to extinguish any burners likely to set fire to the spirit). Allow the copper to dry and weigh it, together with the filter paper.

Specimen calculation

$$\text{Mass of zinc used} = 0.920 \text{ g}$$
$$\text{Mass of copper displaced} = 0.890 \text{ g}$$

Given Zn = 65 and Cu = 63.5, $\dfrac{0.920}{65}$ moles of zinc atoms displaced $\dfrac{0.890}{63.5}$ moles

of copper atoms. That is, the number of moles of zinc and copper atoms concerned is *equal* at 0.014 approximately. This is in accordance with the equation:

$$\text{Zn} + \text{CuSO}_4 \rightarrow \text{ZnSO}_4 + \text{Cu}$$

in which one atom of zinc replaces one atom of copper.

Experiment 27
Stoichiometry of the reaction between barium chloride solution and potassium chromate solution

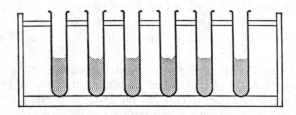

Figure 28

When barium chloride solution and potassium chromate solution are mixed together an instant precipitate is formed; the precipitate is barium chromate. The purpose of the experiment is to determine the mole proportions in which the two reactants combine together.

Because the reaction takes place in solution it is necessary to know what the concentrations of the two solutions are. You will be provided with molar solutions of each reactant (which means that 1 dm³ of each solution contains one mole of substance; see Chapter 15).

Arrange a series of test-tubes (six would be a convenient number) in a test-tube rack, as indicated in Figure 28.

The test-tubes should be selected for uniform internal diameter, and should preferably be of size 125 × 16 mm. Into each test-tube run *exactly* 5.0 cm³ molar barium chloride solution from a burette. If possible the test-tubes should be heated in a beaker of hot water immediately prior to the addition of the solution. From a second burette containing molar potassium chromate, run into the first test-tube 2.0 cm³ of solution and stir with a glass rod. Into the remaining test-tubes run successively volumes of 3.0 cm³, 4.0 cm³, 5.0 cm³, 6.0 cm³, and 7.0 cm³ of molar potassium chromate and stir each tube after the addition. Allow the precipitate of barium chromate to settle and inspect the test-tubes as follows.

(1) Observe the tube which contains *excess* potassium chromate solution, as revealed by the yellow colour of the solution above the precipitate in the tube. Since potassium chromate is in excess, it is in the *previous* test-tube that the two solutions are reacting together in the proportion demanded by the equation. Thus if in the last test-tube no yellow solution above the precipitate occurs on addition of 5.0 cm³ potassium chromate to 5.0 cm³ barium chloride, then, since the solutions are both *molar* solutions, the two substances must be reacting together in *equimolar* proportions.

(2) After the precipitates in the test-tubes have been allowed to settle (which

may take some time), measure the height of the precipitate from the bottom of the tube (Figure 29). It will be seen that the height of the precipitate increases with increasing volumes of potassium chromate solution added up to a certain volume, after which the addition of more potassium chromate solution does not cause any further increase in the quantity of precipitate. The point at which the two substances are therefore reacting in their stoichiometric proportions is given by the test-tube in which the height of precipitate first reaches the maximum value – tube number 4 in Figure 29.

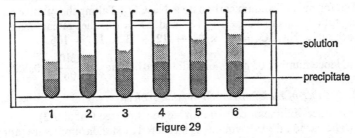

Figure 29

Before the equation may be written for this reaction it should strictly be necessary to identify all the products by chemical methods. This is difficult to do at this stage, and so you should assume that the reaction taking place is:

barium chloride + potassium chromate

$\longrightarrow$ barium chromate + potassium chloride
(precipitate)

You should now be able to write the full chemical equation.

Experiment 28
To write an equation for the reaction between sodium carbonate and hydrochloric acid

In this reaction it is accepted that the gas evolved when the sodium carbonate reacts with dilute hydrochloric acid is carbon dioxide, and that sodium chloride is formed along with water. The problem remaining to be solved is that of establishing the amount of carbon dioxide that is evolved from a given amount of sodium carbonate.

Weigh accurately a sample of anhydrous sodium carbonate of approximately 0.2 g and place it in a small test-tube. Carefully lower the small test-tube into a much larger tube containing approximately 20 cm³ 5M hydrochloric acid (this acid will probably have already had a small quantity of sodium carbonate added to it by the teacher, so that the solubility of the carbon dioxide evolved in the reaction will be very small). The large test-tube is connected to a gas syringe, of capacity 100 cm³, by a right-angled glass tube as shown in Figure 30. Note the reading on the gas syringe. Tip the large test-tube so that the acid mixes thoroughly with the sodium carbonate, and carbon dioxide is

evolved which pushes back the piston in the syringe. When evolution of the gas ceases, read the syringe, and obtain the volume of carbon dioxide evolved by subtraction of the original reading.

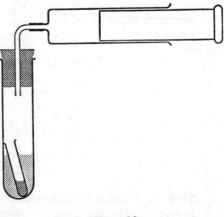

Figure 30

Specimen results

$$\text{Mass of anhydrous sodium carbonate} = 0.210 \text{ g}$$
$$\text{Volume of carbon dioxide evolved} = 47 \text{ cm}^3$$
$$\text{Room temperature} = 10 \text{ °C}$$

Calculation

1. *Number of moles of sodium carbonate*

The formula of sodium carbonate is Na_2CO_3. Taking the relative atomic masses as follows: $Na = 23$, $C = 12$, and $O = 16$, the molar mass of sodium carbonate, $Na_2CO_3 = (2 \times 23) + 12 + (3 \times 16) = 106$.

Hence number of moles of sodium carbonate $= \dfrac{0.210}{106} = 0.002$ mole

2. *Number of moles of carbon dioxide evolved*

Reduce volume of carbon dioxide to s.t.p. (see Chapter 3).

The carbon dioxide volume is measured at standard pressure (approximately) and therefore the major correction to be made is that due to temperature.

The volume of carbon dioxide at 0 °C would be

$$\frac{47 \times 273}{283} = 45.4 \text{ cm}^3$$

Now 22.4 dm³ of any gas at s.t.p. contains one mole of gas.

∴ 45.4 cm³ of carbon dioxide at s.t.p. contains

$$\frac{45.4}{22.4 \times 10^3} = 0.002 \text{ mole}$$

Hence one mole of sodium carbonate produces one mole of carbon dioxide. It is now possible for the equation to be written.

Note: Experiment 25 may be carried out by a syringe method, as described above.

State symbols in equations

It was mentioned earlier (page 56) that it is possible to indicate through the equation the states of matter in which the substances are present. This is usually done by using so-called *state symbols* which are placed after the formula of the substance in the equation, and are usually enclosed in parentheses. The table below gives the principal state symbols.

State of the substance	State symbol
Solid	s
Liquid	l
Gas	g
Aqueous solution	aq

Thus, the reaction between sodium chloride solution and lead nitrate solution to give a precipitate of lead(II) chloride may be written:

$$2NaCl(aq) + Pb(NO_3)_2(aq) \rightarrow PbCl_2(s) + 2NaNO_3(aq)$$

and the equation for the burning of ethanol (C_2H_5OH) in air may also be written:

$$C_2H_5OH(l) + \tfrac{7}{2}O_2(g) \rightarrow 2CO_2(g) + 3H_2O(g).$$

It is not conventional to write equations with 'halves' in front of species such as the oxygen molecule in the above equation, since it is said that 'half a molecule' of oxygen cannot exist. In such cases the equation is often rewritten as:

$$2C_2H_5OH(l) + 7O_2(g) \rightarrow 4CO_2(g) + 6H_2O(g).$$

Note that if the products of the burning of alcohol are collected in a cooled receiver the water may be in the liquid form, and the equation would then read:

$$2C_2H_5OH(l) + 7O_2(g) \rightarrow 4CO_2(g) + 6H_2O(l)$$

which obviously conveys more information than the simple equation without the state symbols.

In order to draw the readers' attention to the fact that either a precipitate is formed in a reaction or a gas is evolved, an arrow has been used in the vertical position by some authors, the arrow pointing down for a precipitate and up for a gas, and placed immediately after the formula for the substance. Two examples are given below.

1. A precipitate of copper(II) sulphide is obtained by passing hydrogen sulphide gas through an aqueous solution of copper(II) sulphate.

$$CuSO_4 + H_2S \rightarrow CuS{\downarrow} + H_2SO_4$$

This is clearly not so informative as

$$CuSO_4(aq) + H_2S(g) \rightarrow CuS(s) + H_2SO_4(aq)$$

2. Carbon dioxide is evolved when dilute hydrochloric acid reacts with calcium carbonate.

$$CaCO_3 + 2HCl \rightarrow CaCl_2 + CO_2{\uparrow} + H_2O$$

which, in turn, is not so useful as

$$CaCO_3(s) + 2HCl(aq) \rightarrow CaCl_2(aq) + CO_2(g) + H_2O(l)$$

The interpretation of chemical equations in a quantitative manner is a most important feature of modern chemistry, and it is the subject of Chapter 7.

Questions

A variety of questions on quantitative aspects of chemical equations will be found at the end of Chapter 7.

1. Write down balanced equations for the following reactions: (*a*) the action of heat on red lead (Pb_3O_4); (*b*) the action of warm concentrated hydrochloric acid on manganese(IV) oxide; (*c*) the action of warm concentrated sulphuric acid on copper. (S)

2. Write equations for the action of heat on (*a*) iron(II) sulphate; (*b*) a mixture of sodium ethanoate and sodium hydroxide in the form of soda lime; (*c*) ammonium nitrate; (*d*) concentrated nitric acid. (S.)

3. 1.00 g of the metal manganese (symbol Mn) was added to 100 cm^3 of molar hydrochloric acid (an excess of acid). The manganese dissolved and 436 cm^3 of hydrogen, measured at room temperature and 1 atm pressure, were produced.

(*a*) Use the information given to write an equation for the reaction between manganese and dilute hydrochloric acid.

(*b*) Some hydrochloric acid will remain when all the manganese has dissolved. What volume of molar sodium hydroxide will be required to neutralize the acid remaining?

(*c*) After the solution of manganese in dilute hydrochloric acid has been neutralized with sodium hydroxide, it is possible to deposit the manganese by electrolysis. What is the minimum time for which a current of 1 ampere must be passed through the solution to deposit all the manganese present?

(One mole of gas at room temperature and one atmosphere pressure occupies approximately 24 dm^3.) (L.)

4. On heating, 8.5 g of a white crystalline compound gave 5.4 g of silver, 2.3 g of nitrogen dioxide, and 560 cm^3 of oxygen (measured at s.t.p.) as the only products.

(*a*) (i) Calculate the mass of the white compound needed to produce, on heating, 22.4 dm³ of oxygen at s.t.p.
(ii) Calculate the masses of nitrogen dioxide and silver formed at the same time.
(iii) How many moles of silver, nitrogen dioxide (as NO_2), and oxygen are produced in (a) (i) and (a) (ii)?
(*b*) Suggest a formula for the white compound and write the equation for its decomposition.
(*c*) The boiling points of nitrogen dioxide and oxygen are 22 °C and −183 °C respectively. Draw a labelled diagram of the apparatus you would use to obtain separate samples of nitrogen dioxide and oxygen from the white compound.
(*d*) What would happen if the gases formed on heating the white compound were (i) passed through a long tube filled with heated copper turnings, (ii) passed into a solution of sodium hydroxide? (C.)

5. When copper(II) sulphate crystals are heated, the water of crystallization is evolved.

$$CuSO_4.5H_2O \rightarrow CuSO_4 + 5H_2O$$

Calculate the mass of anhydrous copper(II) sulphate which can be obtained from 25.0 g of crystals. (Relative atomic masses: H = 1; O = 16; S = 32; Cu = 64.)
When 16.0 g of anhydrous copper(II) sulphate are heated strongly for about 30 minutes mass is lost until only 8.0 g of solid remain. Name the remaining solid.
Write a possible equation for the reaction. (J.M.B.)

6. This question is about a set of experiments in which magnesium was added to 2 molar hydrochloric acid. A fixed mass of magnesium (0.6 g) was used in each experiment but a different volume of acid was chosen each time. A gas was formed and its volume was measured at room temperature and pressure. The results of the experiments are given in the table below.

Volume of 2M HCl used in cm³	Volume of gas formed in cm³
5	120
15	360
25	600
35	600
45	600

(*a*) Name the gas formed in this experiment and mention one property which could be used to identify it.
(*b*) Plot a graph of volume of gas formed against volume of acid used. Volume of gas formed should be plotted along the vertical axis.
(*c*) Use your graph to predict: (i) the volume of gas produced if 50 cm³ of 2M HCl is added to 0.6 g of magnesium, (ii) the volume of 2M HCl which must be added to 0.6 g of magnesium to produce 480 cm³ of gas.
(*d*) Use your graph to find the volume of 2M HCl which would just dissolve 0.6 g of magnesium. You must explain clearly how you have used the graph.
(*e*) Use the experimental results to predict: (i) the volume of 2M HCl which would just dissolve 1 mole of magnesium atoms, (ii) volume of gas which would be formed when 1 mole of magnesium atoms dissolves in hydrochloric acid.
(*f*) Use your answers in part (*e*) to write an equation for the reaction between magnesium and hydrochloric acid. You *must* show clearly how you have used the information from part (*e*). (L.)

7. Explain the use of a chemical equation and the information which it conveys.
Give equations representing three chemical reactions and state the exact meaning of each equation. (O. and C.)

7 Calculations involving Masses

We have seen in Chapter 6 that every definite chemical reaction can be represented by means of an equation. In Chapter 4 the question of relative atomic masses was considered. In the present chapter we shall combine the knowledge obtained in both and show how a quantitative meaning can be assigned to an equation in terms of the commonly used mass units – more particularly the scientific unit, the gramme.

To calculate the formula mass of a compound from its molecular formula

We have noted that the formula of a compound indicates the kind of atoms present in the substance and their number; thus the formula H_2SO_4, for sulphuric acid, indicates that one molecule of the acid contains 2 atoms of hydrogen, 1 atom of sulphur, and 4 atoms of oxygen. The relative atomic mass of each of these elements is known and can be obtained from tables. The formula mass of sulphuric acid can now be calculated by allowing the appropriate number of mass units for each element present and adding to obtain the total. ($H = 1$; $S = 32$; $O = 16$.)

Thus

$$\begin{array}{ccc} H_2 & S & O_4 \end{array}$$
$$(2 \times 1) + 32 + (4 \times 16)$$
$$= 2 + 32 + 64$$
$$= 98 = \text{formula mass of sulphuric acid}$$

EXAMPLE. *Calculate the formula mass of red lead oxide, Pb_3O_4.* ($Pb = 207$; $O = 16$.)

$$\begin{array}{cc} Pb_3 & O_4 \end{array}$$
$$(3 \times 207) + (4 \times 16)$$
$$= 621 + 64$$
$$= 685 = \text{formula mass of red lead oxide}$$

If the formula of the compound contains bracketed acid radicals, it will be simpler and more accurate for you to remove the brackets first and then proceed as above.

EXAMPLE. *Calculate the formula mass of calcium nitrate, $Ca(NO_3)_2$.* ($Ca = 40$; $N = 14$; $O = 16$.)

$$Ca(NO_3)_2$$
$$\text{or } CaN_2O_6$$
$$40 + (2 \times 14) + (6 \times 16)$$
$$= 40 + 28 + 96$$
$$= 164 = \text{formula mass of calcium nitrate}$$

D

Later, with practice, you will be able to carry out the removal of the bracket mentally.

It is also possible to calculate the mass of each element present in a given mass of compound from its formula. This information is usually stated as the percentage composition of the compound.

To calculate the percentage by mass of each element present in a compound from its formula

EXAMPLE. *Calculate the percentage by mass of each element in calcium sulphate, $CaSO_4$. (Ca = 40; S = 32; O = 16.)*

First calculate the molar mass of calcium sulphate.

$$CaSO_4$$
$$40 + 32 + (4 \times 16)$$
$$= 40 + 32 + 64$$
$$= 136.$$

40 of these 136 units of mass are calcium, that is, the fractional mass of calcium in calcium sulphate is 40/136. Then the percentage mass is $\frac{40}{136} \times 100$ or 29.4.

Similarly the fractional mass of sulphur is $\frac{32}{136}$ and the percentage is $\frac{32}{136} \times 100$ or 23.5.

The percentage mass of the third element, oxygen, need not be calculated as it is given by the expression 100 − (% of calcium + % of sulphur),

$$\text{or } 100 - (29.4 + 23.5)$$
$$= 47.1$$

These calculations can be set out completely as below.

EXAMPLE. *Calculate the percentage composition of calcium hydroxide, $Ca(OH)_2$.* (Ca = 40; O = 16; H = 1.)

$$Ca(OH)_2$$
$$\text{or } CaO_2H_2$$
$$40 + (2 \times 16) + (2 \times 1)$$
$$= 40 + 32 + 2$$
$$= 74.$$

Calcium.   Fractional mass $\frac{40}{74}$. Percentage mass $\frac{40}{74} \times 100 = 54.1$

Oxygen.   Fractional mass $\frac{32}{74}$. Percentage mass $\frac{32}{74} \times 100 = 43.2$

Hydrogen.   Percentage mass $= 100 - (54.1 + 43.2) = 2.7$

It is also possible to calculate the formula of a compound from its composition by mass.

To calculate the simplest formula of a compound from its composition by mass

This calculation is illustrated by the following worked example.

EXAMPLE. *Calculate the formula of a compound which has the following per-centage composition: sodium 43.4, carbon 11.3, oxygen 45.3.* (Na = 23; C = 12; O = 16.)

The fact that the relative atomic mass of sodium is 23 means that every 23 parts by mass of sodium in the compound represent one atom of sodium.

Thus 43.4 parts by mass of sodium represent $\frac{43.4}{23}$ atoms of sodium. Similarly for the other elements present:

11.3 parts by mass of carbon represent $\frac{11.3}{12}$ atoms of carbon.

45.3 parts by mass of oxygen represent $\frac{45.3}{16}$ atoms of oxygen.

∴ Number of atoms represented is:

	sodium	carbon	oxygen
	$\frac{43.4}{23}$	$\frac{11.3}{12}$	$\frac{45.3}{16}$
or	1.89	0.94	2.83

These cannot be the actual numbers of atoms present because fractions of atoms are impossible. We have to find the **whole numbers** which are in the ratio 1.89 : 0.94 : 2.83. To do this, divide all these figures by the lowest or, if this does not result in a whole number ratio, by the smallest difference. Then the number of atoms of each element is:

	sodium	carbon	oxygen
	$\frac{1.89}{0.94}$	$\frac{0.94}{0.94}$	$\frac{2.83}{0.94}$
or	2	1	3

That is, the formula is Na_2CO_3.

The calculation is set out compactly in the following example:

EXAMPLE. *Calculate the formula of a compound which has the composition: magnesium 9.8%, sulphur 13%, oxygen 26%, water of crystallization 51.2%.* (Mg = 24; S = 32; O = 16; H_2O = 18.)

	Magnesium	Sulphur	Oxygen	Water
% by mass	9.8	13	26	51.2
Ratio of atoms or molecules	$\frac{9.8}{24} = 0.408$	$\frac{13}{32} = 0.406$	$\frac{26}{16} = 1.63$	$\frac{51.2}{18} = 2.84$
Divide by smallest (or smallest difference)	$\frac{0.408}{0.406}$	$\frac{0.406}{0.406}$	$\frac{1.63}{0.406}$	$\frac{2.84}{0.406}$
	1	1	4	7

∴ the formula is $MgSO_4.7H_2O$

Empirical and Molecular Formulae are discussed on page 113.

Note: Experimental error nearly always results in a small deviation from whole numbers when calculations such as those outlined above are carried out. It is accepted practice to 'round off' these values to the nearest whole number.

Calculations from equations

We have seen already that the equation:

$$CuO \ + \ H_2SO_4 \ \rightarrow \ CuSO_4 \ + \ H_2O$$
$$(63.5 + 16) \quad (2 + 32 + 64) \quad (63.5 + 32 + 64) \quad (2 + 16)$$
$$79.5 \qquad\qquad 98 \qquad\qquad\quad 159.5 \qquad\qquad 18$$

means: 'one mole of copper(II) oxide reacts with one mole of sulphuric acid producing one mole of copper(II) sulphate and one mole of water'.

The appropriate formula masses having been inserted, as above, it also means that 79.5 parts by mass of copper(II) oxide react with 98 parts by mass of sulphuric acid, producing 159.5 parts by mass of copper(II) sulphate and 18 parts by mass of water. These 'parts by mass' may be any desired mass units – tonnes, ounces, pounds, kilogrammes – provided that the same unit is used throughout.

Obviously, the figures given by the equation can be used to calculate any required information about the masses of the four substances concerned.

EXAMPLE. *What mass of copper(II) sulphate could be obtained by starting with 10 g of copper(II) oxide?*

From the equation, 79.5 g copper(II) oxide yield 159.5 g copper sulphate.

$$\therefore \ 10 \text{ g copper(II) oxide yield } 159.5 \times \frac{10}{79.5} \text{ g copper(II) sulphate}$$

$$= 20.1 \text{ g copper(II) sulphate}$$

EXAMPLE. *What mass of pure sulphuric acid would be needed to react with 15 tonnes of copper(II) oxide?*

From the equation, 79.5 tonnes of copper oxide need 98 tonnes of sulphuric acid.

$$\therefore \ 15 \text{ tonnes of copper(II) oxide need } 98 \times \frac{15}{79.5} \text{ tonnes of sulphuric acid}$$

$$= 18.5 \text{ tonnes sulphuric acid}$$

This means that equations have now been given a quantitative meaning in terms of ordinary mass units, instead of simply in terms of atoms and molecules. This makes them extraordinarily useful for making calculations of the masses of materials needed for chemical reactions and the masses of products obtainable. Chemical manufacturers base all their calculations of masses of materials on equations.

Insertion of formula masses into the equation

Here, it is very desirable to remember that it is unnecessary to insert the formula masses of any materials unless they are actually concerned in the calculation you are performing.

Consider, for example, this problem.

EXAMPLE. *Calculate the mass of calcium nitrate which would be formed by treating 148 grammes of calcium hydroxide, $Ca(OH)_2$, with excess of dilute nitric acid. (Ca = 40; O = 16; H = 1.)*

Here the balanced equation is:

$$\underset{\substack{\text{calcium}\\\text{hydroxide}}}{Ca(OH)_2} + \underset{\substack{\text{nitric}\\\text{acid}}}{2HNO_3} \rightarrow \underset{\substack{\text{calcium}\\\text{nitrate}}}{Ca(NO_3)_2} + \underset{\text{water}}{2H_2O}$$

In the problem, the only two substances mentioned quantitatively are calcium nitrate, of which a mass is to be calculated, and calcium hydroxide, of which the mass is given. (Nitric acid is only mentioned as 'excess'.) The only formula masses we need insert are, therefore, those of calcium nitrate and calcium hydroxide. $2HNO_3$ and $2H_2O$ may be ignored once the equation has been balanced, because the problem is not concerned with them.

So we get,

$$Ca(OH)_2 + 2HNO_3 \rightarrow Ca(NO_3)_2 + 2H_2O$$
$$40 + 32 + 2 \qquad\qquad 40 + 28 + 96$$
$$74 \qquad\qquad\qquad 164$$

From the equation, 74 g calcium hydroxide yield 164 g calcium nitrate.

$$\therefore \text{ 148 g calcium hydroxide yield } 164 \times \frac{148}{74} \text{ g calcium nitrate}$$

$$= 328 \text{ g calcium nitrate}$$

EXAMPLE. *Calculate the mass of lead which would be obtained by heating 34.25 g of red lead oxide in a stream of hydrogen and the mass of water formed at the same time.* (Pb = 207; H = 1; O = 16.)

Writing the balanced equation and inserting the formula masses of the materials concerned in the calculation, we have:

$$Pb_3O_4 + 4H_2 \rightarrow 3Pb + 4H_2O$$
red lead
oxide
$$(621 + 64) \qquad\qquad 4(2 + 16)$$
$$685 \qquad\qquad 621 \qquad 72$$

(i) From the equation, 685 g red lead oxide yield 621 g lead.

$$\therefore \text{ 34.25 g red lead oxide yield } 621 \times \frac{34.25}{685} \text{ g lead}$$

$$= 31.05 \text{ g lead}$$

(ii) From the equation 685 g red lead oxide yield 72 g water.

$$\therefore \text{ 34.25 g red lead oxide yield } 72 \times \frac{34.25}{685} \text{ g water}$$

$$= 3.6 \text{ g water}$$

Questions

1. How many tonnes of copper could be obtained by displacing copper from copper(II) sulphate solution by 16.25 tonnes of zinc?

2. What mass of sodium oxide, Na_2O, could be made from 1.15 g sodium?

3. Find the empirical formulae of the following compounds from their compositions by mass:

(a) Zn 47.8%; Cl 52.2%
(b) Na 39.3%; Cl 60.7%
(c) Cu 39.5%; S 20.3%; O 40.2%
(d) Pb 62.5%; N 8.45%; O 29.05%

4. Calculate the percentage by mass of each element in the following compounds:

(a) Sodium hydrogencarbonate $NaHCO_3$
(b) Calcium chloride $CaCl_2$
(c) Ammonium sulphate $(NH_4)_2SO_4$
(d) Sodium thiosulphate $Na_2S_2O_3$

5. What mass of dilute nitric acid (containing 10% of the pure acid) will be required to dissolve 5 g of chalk, calcium carbonate?

6. How many grammes of hydrogen sulphide would be necessary to precipitate 7.5 grammes of copper sulphide from a copper(II) sulphate solution?

7. 76.5 g of sodium hydrogencarbonate were heated strongly. What mass of carbon dioxide was obtained? If a dilute acid had been added, what mass of carbon dioxide would have been obtained in this case?

8. What mass of nitrogen dioxide could be obtained by heating 11.1 g of lead nitrate?

9. 50 g of ammonium chloride were heated with 40 g of calcium hydroxide. What mass of ammonia gas would be evolved? Which of the reagents is in excess and by how much?

10. How many grammes of hydrochloric acid, containing 20% by mass of hydrogen chloride, would be required to dissolve 13 g of zinc? (J.M.B.)

11. Give a brief account of the chemical reactions involved in the extraction of iron from its ores. 0.1867 g of a sample of iron containing carbon as an impurity was dissolved in dilute sulphuric acid, filtered, and the filtrate heated with a slight excess of concentrated nitric acid. An excess of ammonium hydroxide solution was then added to the solution and the resulting precipitate was filtered off, washed, dried, and finally heated to redness until the mass was constant. The mass of the product was 0.2600 g. Give the reactions involved in this process and calculate the percentage of iron in the original sample. (Fe = 56.) (L.)

12. Find the simplest formula of a substance whose percentage composition is hydrogen 5, nitrogen 35, oxygen 60. (S.)

13. Describe the action of water (or steam) upon (i) calcium oxide, (ii) sodium, (iii) magnesium.

When 3.18 g of copper(II) oxide were carefully heated in a stream of dry hydrogen, 2.54 g of copper were formed and 0.72 g of water was collected. From these figures calculate the mass of hydrogen which combines with one mole (16.0 g) of oxygen atoms. (O.)

14. When 3.56 g of a hydrate of copper(II) sulphate were heated to constant mass, 3.20 g of the anhydrous salt remained. The hydrate has a formula $CuSO_4.xH_2O$. Calculate the value of x. (Relative atomic masses: H = 1.0, O = 16.0, S = 32.0, Cu = 64.0.) (J.M.B.)

15. 1.195 grammes of an oxide of lead were reduced to 1.035 grammes of the metal. Calculate the formula of the oxide. (S.)

16. 6.2 grammes of an element X burn in oxygen to produce 14.2 grammes of oxide. Write the equation for the reaction. (Relative atomic mass of X = 31.) (S.)

17. Calculate the percentage by mass of nitrogen in ammonium sulphate. (S.)

18. Give the name of *one* important ore of iron, and state briefly how the metal may be obtained from it. Describe the reaction of iron with (a) chlorine, (b) copper(II) sulphate solution.

Calculate the mass of sulphur which combines with 3.5 grammes of iron, and the volume of hydrogen sulphide (measured at s.t.p.) which would be evolved if the whole of the product was treated with excess of hydrochloric acid.

19. An oxide of lead was weighed in a porcelain boat and it was then reduced to lead by heating it in a stream of hydrogen. The boat with the lead in it was then allowed to cool with the hydrogen still passing over it, and it was then weighed. It was reheated in hydrogen, recooled, and reweighed until a constant mass was attained for the boat and the lead.

The following weighings were obtained:

Mass of boat = 10.20 g
Mass of boat + lead oxide = 17.37 g
Final mass of boat + lead = 16.41 g

(Relative atomic masses: Pb = 207.0, O = 16.0.)

(a) Name a drying agent which could be used to dry the hydrogen used in the experiment.

(b) Why was (i) the boat cooled with the hydrogen still passing over it, (ii) the experiment repeated until a constant mass was attained?

(c) (i) What mass of lead was produced in the experiment?
    (ii) What mass of oxygen was originally combined with this mass of lead?
    (iii) Calculate the mass of oxygen which combines with 1 mole of lead.
    (iv) How many moles of oxygen combine with 1 mole of lead?
    (v) Write the formula and the name of the oxide of lead used in the experiment.

(d) In a second experiment 4.14 g of lead was obtained from 4.46 g of another oxide of lead.
    (i) Calculate how many moles of oxygen combine with 1 mole of lead to form this oxide.
    (ii) Give the name and formula of this oxide. (J.M.B.)

20. How many grammes of sodium carbonate can be obtained by heating 4.2 grammes of sodium hydrogencarbonate until no further change in mass occurs? (S.)

8 Atomic Structure

It seems fairly certain that for most of the nineteenth century atoms were regarded as very small spherical particles like very minute lead shot. It was believed that no smaller particle could exist, and that atoms were solid and homogeneous. This state of affairs was very greatly changed in the first half of this century, mainly by the pioneer work of Lord Rutherford.

It is now believed that atoms are themselves built up from many smaller particles, three of which are of direct interest to the chemist. These are the proton, the electron, and the neutron. The **proton** is a positively charged particle of mass about equal to that of a hydrogen atom. The **electron** is negatively charged, its charge being equal but opposite to the charge on a proton. It has a very small mass, about 1/1850 of the mass of the proton. The **neutron** has no charge, and its mass is about equal to the mass of a proton.

	Charge	Mass ($^{12}_{6}C = 12.000$)
proton	$+1$	1
electron	-1	1/1850
neutron	nil	1

Discovery of these particles

Electrons. In the late nineteenth century, a great deal of work was done on the effects of electrical discharge at very high voltage (by induction coil) through elementary gases at very low pressure. This led to the discovery of *cathode rays*. These emerge at right angles to the cathode and travel in straight lines. If passed through an electrostatic field, they are deflected away from the negative plate, *i.e.*, they are negatively charged. The rays can exert mechanical pressure and convey substantial amounts of kinetic energy so that a metallic object on which they impinge will be heated and may even disintegrate. From these facts, it was concluded that cathode rays consist of a stream of negatively charged particles in rapid motion (about 10^7 m s^{-1}). These particles were given the name of *electrons*. They are produced by all known gases and have been shown to have a mass of 9.1×10^{-28} g and a charge of 1.6×10^{-19} coulomb.

At the same time, positively charged particles leave the area of the anode, their nature depending on the identity of the gas used. Hydrogen, for example, yields hydrogen ions, H$^+$. The effect of the discharge is to ionize hydrogen atoms into *electrons* and *hydrogen ions*.

Protons. Rutherford carried out experiments early this century in which hydrogen was bombarded by fast alpha-particles (helium ions, He^{2+}) from a radioactive source. He found that very penetrating particles were produced, of approximate mass 1 and carrying an electrical charge equal to that of the electron, but *positive*. The particles were named *protons* and are, in fact, the same as the positive particles produced in a hydrogen discharge tube (above). Alpha-particles ionized hydrogen atoms by knocking out electrons from them.

$$H \rightarrow H^+ + e^-$$

Neutrons were discovered (1932) as a very penetrating radiation knocked out of boron nuclei when this element was subjected to the action of alpha-particles from the radioactive element, polonium. A neutron has very nearly the same mass as a proton but no electrical charge. This lack of electrical properties explains why neutrons were detected so much later than electrons and protons.

Arrangement of these particles in the atom. Nuclear Theory

In 1906, Rutherford noticed that if a stream of alpha-particles was passed, as a very thin pencil, through gold leaf (of thickness about one-millionth of a centimetre) and then on to a photographic plate, a certain scattering of the alpha-particles was evident. Later (1909) Geiger and Marsden studied this scattering more accurately and extended the observations by using a zinc sulphide screen on a rotating arm (Figure 31). The screen scintillates when an

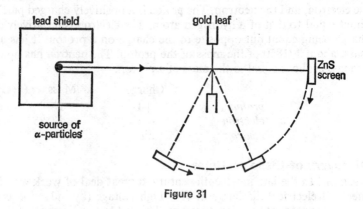

Figure 31

alpha-particle strikes it. It was found that the straight-line path of the great majority of the alpha-particles was little affected by the gold leaf but about one in 8000 of them was deflected through an angle of 90° or more from its original direction. Rutherford commented later that, considering the thinness of metal used and the high velocity and mass of the alpha-particles, these large deflections were 'about as credible as if you had fired a fifteen-inch shell at a piece of tissue paper and it came back and hit you'. Rutherford deduced that these few large deflections of alpha-particles required atoms of gold to contain a region (the *nucleus*) which:

1. is positively charged because it repels the positively charged alpha-particles;
2. is relatively massive and highly charged because the deflections are large;
3. must occupy a very small space because so very few alpha-particles are materially affected.

According to this nuclear theory of Rutherford, the protons and neutrons of an atom are concentrated into the nucleus which is, therefore, positively charged and relatively massive but minute. The electrons, equal in number to the protons, and so making the atom electrically neutral, are outside the nucleus. The whole bulk of the atom, defined by the outermost electron ring, is very great compared with that of the nucleus.

Arrangement of electrons in the atom

Bohr (1913) put forward a theory of electron positioning which is still generally accepted for chemical purposes. It was developed originally in connection with the hydrogen atom, which contains only one proton (as nucleus) and one electron. Bohr suggested the existence of certain circular orbits (or shells) at definite distances from the nucleus in which the electron may rotate. These orbits are associated with definite energy content of the electron, increasing outwards from the nucleus. While the electron adheres to any one orbit (usually the innermost, of lowest energy), the atom radiates no energy. If, however, the atom absorbs energy, the electron may jump to one of the outer orbits of greater energy. If it later falls back to the inner orbit of lower energy, energy will be radiated as light of a definite colour (or frequency). Thus, if several outer, higher energy orbits are involved, an optical spectrum should be obtained, showing several lines at definite frequencies. Bohr was able to calculate the theoretical frequencies for such a spectrum and show that they accorded well with observed hydrogen spectra. It is also known that spectra of other elements indicate similar electronic orbits in their atoms.

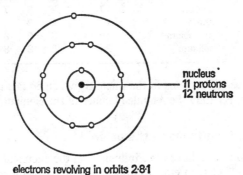

nucleus
11 protons
12 neutrons

electrons revolving in orbits 2·8·1

Figure 32

It was known that many cases occurred in which pairs or triplets of spectral lines occurred close together, *e.g.*, the two yellow lines of sodium. To explain these slightly varying energy levels in the same electron shell, it was suggested that some orbits are circular and some elliptical, but this feature can usually be ignored for chemical purposes, at any rate at the present level.

This, and much subsequent work, leads to the following conclusions:

1. Several groups of electrons may occur in an atom and each group is known as an *electron shell*. Shells are numbered 1, 2, 3, etc. outwards from the nucleus. All electrons in a given shell have approximately equal energy. This energy increases in successive shells outwards from the nucleus.

2. The *maximum* possible number of electrons in a shell numbered n is $2n^2$, i.e., in successive shells, 2, 8, 18, 32, ... electrons.

3. In the outermost shell of any atom, the maximum number of electrons possible is 8.

A diagrammatic representation of a typical atom would be as in Figure 32.

The simplest atom is that of hydrogen. It consists of a single proton with one electron rotating round it. In order of complexity, the simpler atoms are made up as in the table on page 74, neutrons being omitted.

Element	Protons	Electrons in each shell			
hydrogen	1	1			
helium	2	2			
lithium	3	2	1		
beryllium	4	2	2		
boron	5	2	3		
carbon	6	2	4		
nitrogen	7	2	5		
oxygen	8	2	6		
fluorine	9	2	7		
neon	10	2	8		
sodium	11	2	8	1	
magnesium	12	2	8	2	
aluminium	13	2	8	3	
silicon	14	2	8	4	
phosphorus	15	2	8	5	
sulphur	16	2	8	6	
chlorine	17	2	8	7	
argon	18	2	8	8	
potassium	19	2	8	8	1
calcium	20	2	8	8	2

The number of protons in the nucleus (which equals the number of electrons in the shells) is called the **Atomic Number** of the element.

Function of neutrons in the atom

Neutrons appear to have little influence on the chemical properties of the atom. Their chief effect is in contributing to the mass of the atom. For example, the sodium atom with 11 protons and 12 neutrons has a total mass of 23 units. The 11 electrons are relatively negligible in mass.

Many cases occur, however, in which two atoms contain the same number of protons but differing numbers of neutrons. Having equal numbers of protons, these atoms must also have equal numbers of electrons. These are arranged in the same way and give the atoms identical chemical properties. But the differing numbers of neutrons cause the atoms to have different masses. An element showing these characteristic properties – that is possessing atoms of similar chemical properties but different masses – is said to show **isotopy** and the varieties of the atom are called **isotopes** of the element.

A well-known example of isotopy occurs in chlorine.

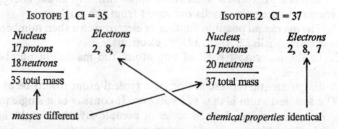

ISOTOPE 1 Cl = 35

Nucleus
17 *protons*
18 *neutrons*
_____
35 total mass

Electrons
2, 8, 7

ISOTOPE 2 Cl = 37

Nucleus
17 *protons*
20 *neutrons*
_____
37 total mass

Electrons
2, 8, 7

masses different

chemical properties identical

Dalton believed that all atoms of the same element were exactly alike. The existence of isotopes has proved him wrong. But, except in a few cases, elements contain isotopes **in almost constant proportions** and so appear to act as if all their atoms are equal in mass. Chlorine, for example, is found to have its isotopes mixed in such proportions that its average relative atomic mass appears constant. The lighter isotope predominates, giving a value of 35.5.

It may also be mentioned that, in spite of the development of nuclear fission reactors and atomic bombs, the atom can still be regarded as an indivisible unit in **chemical** actions. Apart from modifications in the electron shells, an atom is conveyed as a whole unit in chemical actions.

Uranium has two principal isotopes. Both possess 92 protons and 92 electrons. One isotope has 146 neutrons, giving U = 238; the other has 143 neutrons, giving U = 235. The lighter isotope constitutes about 0.7% of natural uranium. If the atom U = 235 acquires one extra neutron in the nucleus, it becomes unstable and divides into two unequal parts. This 'fission' is accompanied by a small diminution of mass, the combined final products having a mass slightly smaller than that of the uranium. Consequently, there is a very great evolution of energy, principally as heat, according to the Einstein equation, $E = mc^2$, where E is the energy, m the mass lost, and c the velocity of light. At the same time, neutrons are emitted. Other atoms of uranium absorb them and undergo fission, so that, if the mass of uranium is large enough, a chain reaction is set up, causing an atomic explosion.

We have noted already that the number of protons in the nucleus of an atom is called the *Atomic Number* (Z) of the element concerned; the sum of the protons and neutrons in the nucleus is called the *Mass Number* (A) of the element and it also expresses the mass of the atom (on the scale of $^{12}_{6}C = 12$) to the nearest integer. If an element is isotopic, it has as many mass numbers as isotopes. The *relative atomic mass* of an element for practical purposes is the weighted average of the *accurate* masses of its isotopes (on the same scale as above) as they occur in the element in experimental practice. For a given element, atomic mass is always a constant, or very nearly so, except for products of radioactivity, such as lead (page 214). In some cases, recent atomic mass tables show variations of atomic mass caused by differences of isotopic distribution but, for ordinary purposes, they are very slight.

EXAMPLE

	Atomic number (Z) protons	*Mass number* (A) protons + neutrons	*Neutrons* (A − Z)	*Relative atomic mass*
Chlorine (2 isotopes)	17	$^{35}_{17}Cl$ 35	18	35.46
		$^{37}_{17}Cl$ 37	20	(3 atoms of 35 to 1 of 37 approx.)

Notation for isotopes

In order to distinguish between different isotopes of the same element in writing symbols and formulae a simple system is adopted. The isotope of any element X will have the symbol $^{A}_{Z}X$, where A is the *mass number* of the isotope and Z is the *atomic number* of any atom of X. Thus for all the isotopes of any one element Z is constant, and A varies because there are different numbers of neutrons in the different isotopes of the element. In the example above the two isotopes of chlorine were given the symbols $^{35}_{17}Cl$ and $^{37}_{17}Cl$. Since A represents the total number of neutrons and protons in the nucleus of the atom, and since Z is the number of protons (equals the atomic number), then the number of neutrons in the nucleus of a given isotope is given by:

$$\text{Number of neutrons in the nucleus} = (A - Z)$$

This is expressed in the example quoted above for the isotopes of chlorine.

Recent modifications of Dalton's Atomic Theory

1. *Elements are made up of small, indivisible particles called atoms*

The atomic nature of elements is not disputed. Atoms can, however, no longer be regarded as indivisible in the full meaning of the term. Radioactive elements are spontaneously dividing in the sense that the atomic nucleus is giving out particles and so producing two less complex atoms, *e.g.*, radium disintegrates to produce two noble gases, helium and radium emanation (radon).

2. *Atoms of a given element are all exactly alike*

This statement can no longer be accepted. The phenomenon of *isotopy* (page 74) contradicts it. Thus, potassium has isotopes $^{39}_{19}K$ and $^{41}_{19}K$. Both have 19 nuclear protons and 19 electrons, arranged 2, 8, 8, 1, so have the same atomic number and properties. But the 41-isotope has two extra neutrons on the nucleus and so is the heavier atom. Most elements exhibit isotopy in this way.

3. *Atoms cannot be created or destroyed*

This statement is still acceptable when applied to chemical reactions, in which, apart from electronic changes, atoms react as whole units. The changes associated with atomic fission, however, certainly destroy atoms of the element involved, in the sense that the nuclei are broken into smaller units which correspond to simpler atoms. For example, the nucleus of the uranium isotope, U = 235, can absorb a neutron and then break up into two unequal fragments with mass numbers of approximately 95 and 140.

4. *Atoms combine in small whole numbers*

This statement is still acceptable for most elements. Carbon, however, forms the very complex compounds of 'organic' chemistry (page 328) and the element silicon occurs in some very complex silicates.

Questions

1. Copy and complete the following table of atomic/electronic structure of some elements.

Element	Atomic number	Relative atomic mass	Number of protons	Number of neutrons	Number of electrons	Electronic configuration
X	7	14				
Y		23	11			
Z		56				2,8,14,2

(N.I.)

2. The particles of which atoms are composed include three of particular interest to a chemist – protons, neutrons, and electrons. What are the relative masses and charges of these particles? How many of each of these particles are present and how may they be represented in (i) a hydrogen atom, (ii) a hydrogen molecule?

The element chlorine (relative atomic mass $= 35.5$, atomic number $= 17$) is a mixture of two isotopes, ^{35}Cl and ^{37}Cl. The structure of an atom of ^{35}Cl can be represented as in Figure 33:

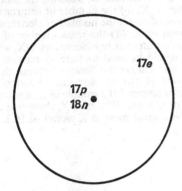

17e

17p
18n •

Figure 33

Draw a similar diagram to represent the structure of an atom of ^{37}Cl and suggest an explanation for each of the following observations:

(i) Atoms of ^{35}Cl and ^{37}Cl have the same chemical properties.
(ii) Atoms of ^{35}Cl differ from those of ^{37}Cl.
(iii) 1 mole of chlorine molecules has a mass of 71 grammes.

The relative atomic mass of fluorine is 19.00. Would you expect fluorine to be composed of several isotopes? Explain briefly why you have given this answer. (L.)

3. State briefly what you understand by the terms (*a*) proton, (*b*) neutron, (*c*) electron. State the numbers of each particle which occur in the potassium atom. (Relative atomic mass 39, atomic number 19.) (S.)

4. State *one* law of chemical combination, and show how it was explained by the atomic theory. How has Dalton's atomic theory had to be modified in more recent times?

Explain the terms nucleus, proton, electron, ion.

Sodium has a relative atomic mass of 23 and its atomic number is 11. Draw simple diagrams to show the particles which make up (*a*) a sodium atom, (*b*) a sodium ion, and give a brief explanation of the difference between them. (S.)

5. Silver is an element which exists naturally as a mixture of two isotopic forms. A and B represent atoms of these two isotopes. They occur in equal numbers.

A is $^{107}_{47}Ag$. B is $^{109}_{47}Ag$

In each case, the upper figure is the mass number and the lower figure is the atomic number.

State the number of (*a*) protons in atom A, (*b*) electrons in atom B, (*c*) neutrons in atom A, (*d*) neutrons in atom B.

What is the relative atomic mass of naturally occurring silver? (J.M.B.)

6. The numbers of protons, neutrons, and orbital electrons in particles A to F are given in the following table:

Particle	Protons	Neutrons	Electrons
A	3	4	2
B	9	10	10
C	12	12	12
D	17	18	17
E	17	20	17
F	18	22	18

(*a*) Choose from the table the letters that represent

(i) a neutral atom of a metal,
(ii) a neutral atom of a non-metal,
(iii) an atom of a noble gas (inert gas),
(iv) a pair of isotopes,
(v) a cation (positive ion),
(vi) an anion (negative ion)

(*b*) Give the formulae of the compounds you would expect to be formed between (i) C and D, (ii) E and hydrogen, (iii) C and oxygen. (You may use the letters C, D, and E as symbols for the elements.)

In each case, say whether the compound will be an acid, a base, or a salt.

(*c*) What will be the formula of a compound containing only particles A and B? Give *two* physical properties you would expect this compound to have. (C.)

7. The table shows the mass numbers and atomic numbers of atoms labelled T to Z.

	Mass number	*Atomic number*
T	2	1
V	3	1
W	3	2
X	6	3
Y	9	4
Z	11	5

(a) How many protons are there in an atom of Y?

(b) How many electrons are there in an atom of W?

(c) How many neutrons are there in an atom of Z?

(d) Which atoms are isotopes of the same element?

(e) Which atom would readily form an ion with a single positive charge?

(f) Which is an atom of a noble gas?

(J.M.B.)

8. Write out and complete the following table.

Particle	Mass number	Number of protons	Number of neutrons	Number of electrons
Li atom		3	4	
Li⁺ ion				
¹²C isotope	12			6
¹³C isotope	13	6		
N³⁻ ion	14	7		
Ne atom			10	10
S²⁻ ion	32			18

(a) Which particles, apart from the Ne atom, have the same total number of electrons as a noble gas?

(b) The electron configuration of the carbon atom can be written as C: 2. 4. Show in a similar way the configurations of (i) the Li atom, (ii) the S^{2-} ion.

(c) Write the formulae of (i) a compound containing the S^{2-} ion, (ii) a covalent compound of sulphur, (iii) a compound composed only of Li^+ and N^{3-} ions. (C.)

9. Taking the symbol $^{16}_{8}X$, to represent an atom of the element X, state (a) the atomic number of X, (b) the number of neutrons in an atom of X, (c) the number of electrons in an atom of X, (d) the mass number of X. If another atom is represented as $^{18}_{8}X$, what term would be used to state its relation to $^{16}_{8}X$ and what is the difference between them in terms of the number and situation of particles present? If a sample of X contained 90% of $^{16}_{8}X$ and 10% of $^{18}_{8}X$, show that the relative atomic mass of X would be 16.2.

9 Chemical Union and Structure

Types of chemical combination

We have already seen (Chapter 8) that neon and argon have *eight* electrons in their outermost electron layer, *i.e.*, an *octet* of electrons. This structure is very stable and extremely difficult to disturb. In consequence, these two gases are chemically inert and have great difficulty in forming compounds with other elements. They are self-satisfied. In the simpler noble gas, helium, the duplet of electrons is equally stable and functions like the octet.

The tendency of other elements is to try to attain this noble gas structure of a stable outer octet (or duplet) of electrons and their chemical behaviour is a reflection of this tendency. On this general principle, elements combine in two main forms of combination, known as the **electrovalent** (or ionic) and **covalent types.**

Electrovalent combination

In this type, an atom of a metallic element or group loses, from its outermost electron shell, a number of electrons equal to its **valency.** These electrons pass over to the outer electron shells of non-metallic atoms with which the metal is combining. By this means, an electron octet is *left behind* in the metal and *created* in the non-metal. Both elements now have the outer electron structure of a noble gas, but the metallic particles have a positive charge from the excess proton(s) left on the nucleus, while the non-metallic particles are negatively charged from the added electron(s). The particles are then known as **ions.**

Thus:

	Sodium atom		Chlorine atom	
	Protons	*Electrons*	*Protons*	*Electrons*
Before combination	11	2,8,1	17	2,8,7
After combination	11	2,8	17	2,8,8
		Sodium ion +		Chloride ion −

Both ions now possess stable outer electron octets, like a noble gas.

No molecules of sodium chloride are formed. Because of the attraction of the oppositely charged Na^+ and Cl^- ions for one another, the ions arrange themselves into a rigid, solid shape called a crystal, but they remain quite separate. The combination can be expressed only in ionic form as Na^+Cl^-, meaning an association of sodium and chloride ions in equal numbers. The manner in which the sodium and chloride ions are arranged in the crystal is shown in Figure 34.

Sodium chloride crystallizes as a *face-centred cube* (Figure 34). In an end face of the cube a Na^+ ion occupies the centre; then the four corners of the face are also occupied by Na^+ ions, with four Cl^- ions spaced equally between them. In the next face, the positions of Na^+ and Cl^- are reversed, and so on alternately. The attractive forces between the ions are relatively great. The only ionic motion is some vibration, consequently the solid appears rigid and has negligible electrical conductivity.

Calcium chloride and calcium oxide are further examples of electrovalent compounds. In the calcium ion, the two excess nuclear protons produce a double positive charge; in each chloride ion, the excess electron produces a single negative charge, *i.e.*, $Ca^{2+}2Cl^-$.

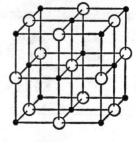

● = Na^+
○ = Cl^-

Figure 34

Calcium chloride:

	Calcium atom		Two chlorine atoms	
	Protons	*Electrons*	*Protons*	*Electrons*
Before combination	20	2,8,8,2	17 17	2,8,7 2,8,7

These valency electrons pass to the chlorine atoms

	Calcium ion + +		Two chloride ions −	
	Protons	*Electrons*	*Protons*	*Electrons*
After combination	20	2,8,8	17 17	2,8,8 2,8,8

Calcium oxide:

	Calcium atom		Oxygen atom	
	Protons	*Electrons*	*Protons*	*Electrons*
Before combination	20	2,8,8,2	8	2,6

These valency electrons pass to the oxygen atom

	Calcium ion + +		Oxide ion − −	
	Protons	*Electrons*	*Protons*	*Electrons*
After combination	20	2,8,8	8	2,8

As stated above, the calcium ion acquires a double positive charge; the two excess electrons produce a double negative charge on the oxide ion, *i.e.*, $Ca^{2+}O^{2-}$.

Note the presence of the outer *octet* of electrons in all the above ions.

Magnesium oxide is an electrovalent compound just like calcium oxide, and you should carry out the exercise for this compound as outlined above. The crystal structure of magnesium oxide is similar to that of sodium chloride with magnesium and oxide ions occupying the positions of the sodium and chloride ions respectively.

Characteristic properties of electrovalent (or ionic) compounds

(1) Ionic compounds do not contain molecules. They consist of aggregates of oppositely charged ions. In consequence, if they are melted, or dissolved in water, to make the ions mobile, they conduct electricity and are, therefore, electrolytes (Chapter 16).

(2) They are solids and do not vaporize easily.

(3) They will not usually dissolve in organic solvents such as toluene, ether, benzene, etc.

Salts, alkalis, and bases are electrovalent, and acids, when in solution in water, also show electrovalency.

Covalent combination

In this type of combination, electrons are not actually gained or lost by the atoms concerned. They pass into a 'shared' state.

Consider two chlorine atoms. Each has the electron structure 2,8,7. In covalency, the atoms contribute *one electron each* to a 'shared-pair'. In this way,

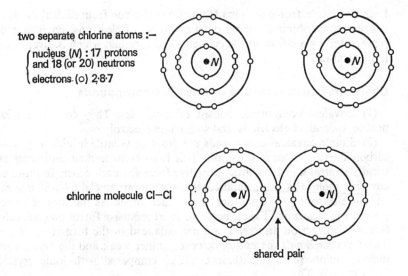

two separate chlorine atoms :—

{ nucleus (*N*) : 17 protons
and 18 (or 20) neutrons
{ electrons (o) 2·8·7

chlorine molecule Cl—Cl

shared pair

Figure 35

both obtain an approximation to the external octet by making fourteen electrons do the work of sixteen (Figure 35).

Here, actual molecules are produced, not ions. Each 'shared-pair' electron passes from an orbit controlled by the nucleus of *one* chlorine atom into an orbit controlled by the nuclei of *both* chlorine atoms. This joint control of the orbits constitutes the valency bond.

Other examples of molecules formed by covalency are shown in Figure 36.

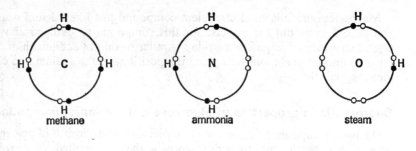

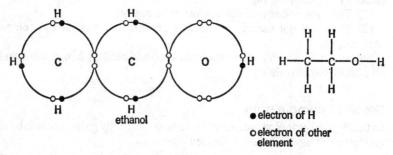

Figure 36

Each shared electron pair is made up of one electron from each atom concerned. All the atoms obtain a shared octet or duplet of electrons. Each shared-pair is represented by a stroke in the conventional formula, which is given for ethanol in illustration.

Characteristic properties of covalent compounds

(1) Covalent compounds consist of molecules. They contain no ions, are unable to conduct electricity and so are non-electrolytes.

(2) Simple covalent compounds are gases or volatile liquids, *e.g.*, ammonia, carbon dioxide, ethanol (alcohol). This is so because their molecules are electrically neutral and have little attractive force for each other. In more complex covalent molecules, *e.g.*, naphthalene, the atomic nuclei (+) of one molecule and the electrons (−) of another attract each other. As the molecules come together, the electrons of each begin to exert repulsive forces on each other. The forces of attraction and repulsion are balanced in the formation of a crystal. These *van der Waals forces* are, however, rather weak and the crystals have low melting points (*e.g.*, naphthalene 81 °C) compared with ionic crystals (*e.g.* Na^+Cl^-, 801 °C).

(3) Covalent compounds are usually soluble in covalent organic solvents, such as benzene or carbon disulphide.

Co-ordinate (dative) bond

Another variety of covalent bond (*i.e.*, electron sharing) has been given the name of **co-ordinate bond** (or **dative bond**). This bond is characterized by the fact that the two shared electrons are both supplied by one of the participating atoms (not one electron by each atom as in an ordinary covalent bond). A co-ordinate bond is formed (in simpler cases, at least) when one of the participants possesses a *lone pair* of electrons, *i.e.*, a pair not directly concerned in its existing valency bonds. This lone pair is donated to an atom needing them to build up, or complete, an electron octet or duplet of great stability. The ammonia molecule

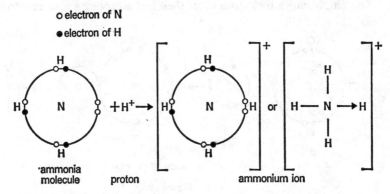

Figure 37

(Figure 37) possesses such a lone pair of electrons; it can be donated to a hydrogen ion (proton) from an acid to produce the ammonium ion, NH_4^+. This bonding supplies an electron duplet to the hydrogen nucleus while still maintaining (though with sharing) the electron octet of the nitrogen atom. The proton, combining with the NH_3 molecule, carries over its positive charge to give the ammonium ion, NH_4^+. The co-ordinate bond is often indicated by an arrow pointing from donor atom to acceptor atom as shown, though there is

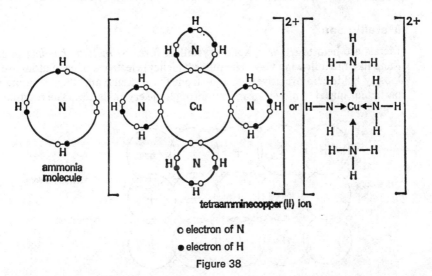

tetraamminecopper (II) ion

o electron of N

● electron of H

Figure 38

certainly internal equalization of the four bonds of NH_4^+. This ion is associated (electrovalently) with some anion as a *salt*, *e.g.*, $NH_4^+Cl^-$ or $NH_4^+NO_3^-$.

Again, the ammonia molecule participates by co-ordinate bonding in the formation of the tetraamminecopper(II) ion (also called cuprammonium ion) which gives the intense blue coloration when excess of ammonia is added to a solution of a copper(II) salt, such as copper(II) sulphate. In this copper(II) sulphate, copper is in the 2,8,17,2 electron state with the outermost 2 (valency) electrons transferred to the SO_4^{2-} ion and the outer electron shell vacant in the ion Cu^{2+}. Four NH_3 molecules form co-ordinate bonds with the Cu^{2+} ion by means of their lone pair electrons to create a shared electron octet in the vacant valency electron shell of the copper(II) ion (Figure 38).

Oxygen, having a 6-electron outer shell and accepting a lone pair to complete

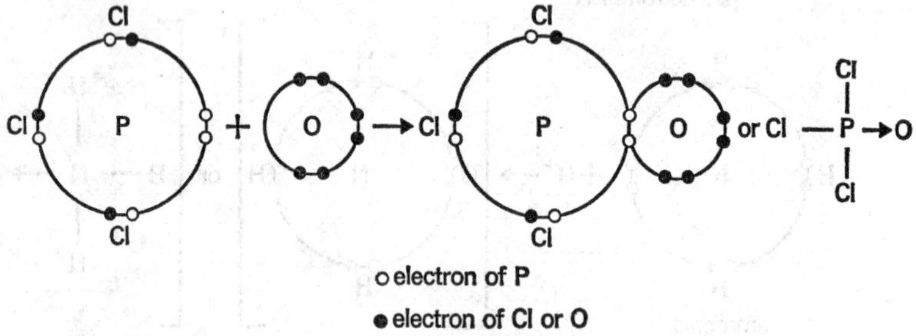

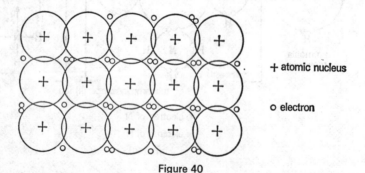

o electron of P

● electron of Cl or O

Figure 39

the shared octet, is a frequent participant in co-ordinate bonds (as acceptor atom). A simple example of this is seen in the relation of phosphorus trichloride to phosphorus oxychloride (Figure 39).

Compounds containing only co-ordinate bonds and covalent bonds are very similar in properties to purely covalent compounds. Both types are non-electrolytes (having no ions); the simpler examples are usually liquids in ordinary room conditions, but, with co-ordinate bonds present, tend to be less volatile.

Metallic bond

Metals are held together in solid crystalline form by *metallic bonding* in something like the following way. The outer (valency) electrons of each atom are only loosely held, being relatively distant from the nucleus, and they separate from particular nuclei to move at random through the crystal lattice. The residual ions,

+ atomic nucleus

o electron

Figure 40

now positively charged by loss of valency electrons, tend to repel each other but are held together by the moving electron cloud and some overlapping of residual electron orbits (Figure 40).

This type of bonding is very strong in some metals, *e.g.*, iron, which are difficult to shatter, but it is much weaker in, *e.g.*, sodium or potassium, which can be cut with a knife. If an electric potential difference is applied to the ends of a metallic rod, the free electrons lose their random motion and move towards the positive end of the rod, being steadily replaced by more from the source of potential difference. Hence metals are good electrical conductors. As freely moving electrons can convey heat energy, metals are also good conductors of heat.

Since the bonding agent in a metal is mainly a moving electron cloud, the ions of most metals will usually slide relative to one another, under stress, without shattering the lattice and produce a new position of stability. This accounts for the malleability and ductility of many metals, though special temperature conditions may sometimes be necessary.

The different ways in which atoms pack together in close contact is illustrated by the structures of metals. The most economical manner in which a layer of atoms can be arranged is that shown in Figure 41, where over 60% of the

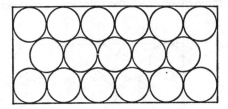

Figure 41

available space is occupied by the atoms. Each atom is in contact with six others in the same layer; these are known as *nearest neighbours*. In placing a second layer of atoms over the first, the atoms in the second layer will arrange themselves so that they fit into the hollows in the first layer. This is illustrated in Figure 42,

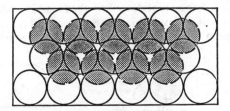

Figure 42

in which the second layer atoms are shown by broken lines and shaded in. A third layer may then be placed on top of the second in one of two different ways.

If the atoms of the third layer are placed in such a position that they coincide exactly with the atoms in the first layer, they will fit into the hollows marked with a dot in Figure 43. If this pattern is repeated throughout the three-dimensional structure of the metal crystal, it gives rise to what is known as *hexagonal close-packing*, since the centres of the six nearest neighbours surrounding each atom in each layer form a regular hexagon (see Figure 41). If the first layer is represented by A, and the second layer by B, hexagonal close-packing corresponds to the arrangement ABABAB etc.

However, if the third layer of atoms is placed in such a way that they fit into the hollows of the second layer marked with a cross in Figure 43 an arrangement which can be represented ABCABCABC etc. will result. Continued indefinitely

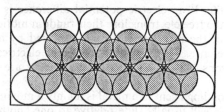

Figure 43

throughout the crystal structure in three dimensions, this gives rise to *cubic close-packing* (sometimes known as *face-centred cubic* since when the structure is examined in detail it is found that it consists of an atom in the centre of each face of a cube of atoms). Both of these structures are called close-packed because they give rise to the most economical manner of packing atoms in three dimensions, nearly three-quarters of the total available space being taken up by the atoms.

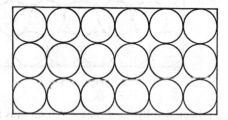

Figure 44

A completely different way of packing atoms is shown in Figure 44, where the atoms are arranged in columns and rows in each layer. This arrangement is not quite so space-filling as that described above, and in one layer the atoms take up only just over half the available space. A second layer can be added as shown in Figure 45, and a third will then be added to coincide with the first layer. Thus

Figure 45

an ABABAB etc. arrangement results once again, but in this case it is known as a *body-centred cubic* structure, since the three-dimensional lattice consists of a series of connected atoms arranged at the corners of a cube, with another atom at the centre of each cube. Each atom has eight nearest neighbours, whereas in the hexagonal and cubic-closed packed structures each atom has twelve nearest neighbours. The body-centred cubic arrangement fills 68% of the available space.

The structures of most metals fall into one of the three categories described above, as indicated in the table (Figure 46).

Li	Be
b	b

Na	Mg
b	h

h = hexagonal close-packed
c = cubic close-packed
b = body-centred cubic

K	Ca	Sc	Ti	V	Cr	Mn	Fe	Co	Ni	Cu	Zn
b	ch	ch	hb	b	b	—	cb	ch	ch	c	h

Rb	Sr	Y	Zr	Nb	Mo	Tc	Ru	Rh	Pd	Ag	Cd
b	c	h	hb	b	hc	h	ch	c	c	c	h

Cs	Ba	La	Hf	Ta	W	Re	Os	Ir	Pt	Au	Hg
b	b	ch	hb	b	b	h	ch	c	c	c	—

Figure 46

Shapes of some molecules

Carbon dioxide

In a carbon dioxide molecule, the carbon atom is bonded to each oxygen atom by two pairs of electrons in covalency. These electron groups exercise mutual

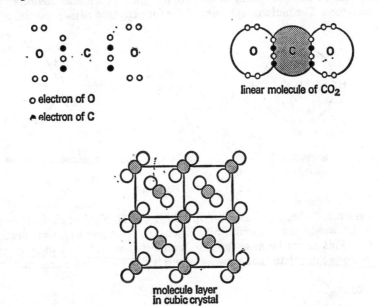

o electron of O

▲ electron of C

linear molecule of CO_2

molecule layer
in cubic crystal

Figure 47

repulsion and produce a linear molecule. In solid carbon dioxide, 'dry ice', these linear molecules take up a cubic formation (bonded by van de Waals forces). One layer of molecules in the cube is shown (Figure 47).

Methane

The four covalency bonds of the carbon atom in methane are distributed symmetrically in three dimensions, the angle between any pair of valency directions being $109\frac{1}{2}°$. Each bond represents a shared electron pair to which a carbon and hydrogen atom supply one electron each. The molecule of tetrachloromethane is similar with substitution of four chlorine atoms for the four hydrogen atoms of methane (Figure 48).

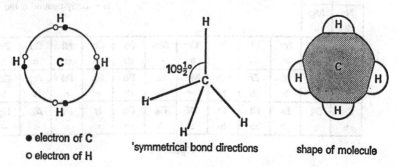

● electron of C
○ electron of H

'symmetrical bond directions

shape of molecule

Figure 48

Ammonia

An ammonia molecule consists of three hydrogen atoms bonded covalently, *i.e.*, by shared electron pairs, to a nitrogen atom, which also has one lone pair of electrons. The mutual repulsion of the four electron pairs produces a 'triangular

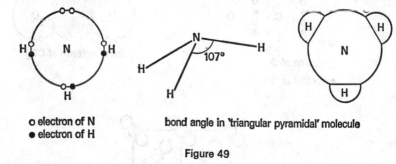

○ electron of N
● electron of H

bond angle in 'triangular pyramidal' molecule

Figure 49

pyramidal' shape of molecule in which the H–N–H angle is 107° (not 109½° which would be produced if all four electron pair repulsions were equal) (Figure 49). This arises because the repulsion between two lone pairs of electrons is much greater than that between any other electron pairs present.

Steam

In a molecule of steam, the two hydrogen atoms are bonded to the oxygen atom by shared electron pairs and the oxygen atom has two lone pairs of electrons not

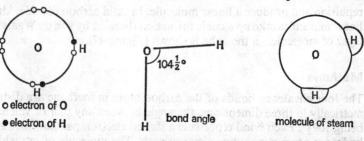

○ electron of O
● electron of H

bond angle

molecule of steam

Figure 50

concerned in valency bonding. The two lone pairs exercise a stronger mutual repulsion than that between other electron pairs (see above), and the resultant effect of this is to produce a 'bent' structure in which atomic nuclei all lie in the same plane with $104\frac{1}{2}°$ as the H–O–H angle (Figure 50).

Ice

The structure of ice is in marked contrast to that of many solids since close-packing of the molecules does not take place. The most striking feature of the structure is the 'open' nature of the arrangement of the molecules, giving rise to a low density for the solid. Each oxygen atom in the water molecule is surrounded by four other oxygens at the corners of a regular tetrahedron and the whole structure has hexagonal symmetry (Figure 51).

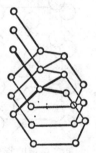

Figure 51

On melting, the whole structure begins to break up and to collapse, causing a decrease in volume which continues in the liquid state to 4 °C. Due to the close proximity of the water molecules in the solid state, there exists a weak interaction between them in which the hydrogen atom of one water molecule becomes associated with the oxygen atom of another (Figure 52). This is known as a **hydrogen bond,** which, although weak compared with the electrovalent and covalent bonds, can substantially affect the properties of those compounds in which it is present.

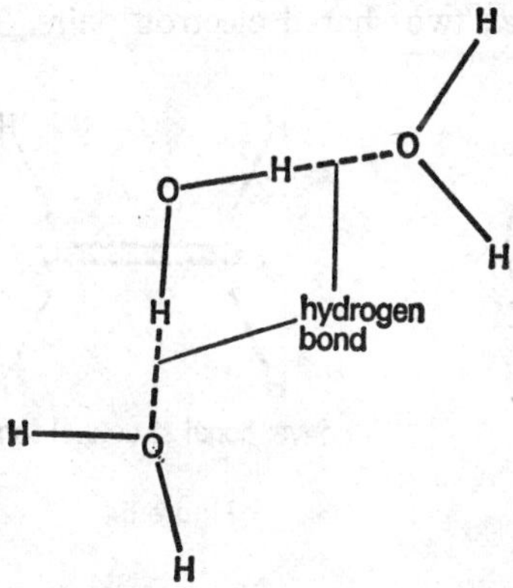

Figure 52

Iodine

Iodine produces a molecule, I_2, by formation of a covalent (shared-pair) bond to which each iodine atom contributes one electron (so completing, for each atom, a shared electron octet). By the operation of van der Waals forces, the

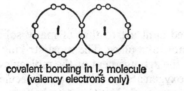

covalent bonding in I_2 molecule
(valency electrons only)

shape of molecule

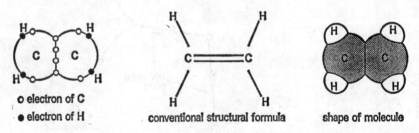

layer of I_2 molecules in crystal.
Next layer covers dotted lines

Figure 53

molecules then form crystals in which they lie in a herring-bone pattern which is reversed in successive layers. One layer of iodine molecules is shown in the square (Figure 53). In the next layer, molecules lie over the dotted lines.

Ethene (Ethylene)

In the ethene molecule C_2H_4, the two carbon atoms are combined by two covalent bonds, *i.e.*, two shared electron pairs. This indicates *unsaturation* in

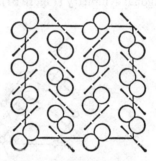

o electron of C
• electron of H

conventional structural formula

shape of molecule

Figure 54

the molecule. Each carbon atom is also combined covalently to two hydrogen atoms. The nuclei of all six atoms involved lie in the same plane, *i.e.*, the molecule is planar. The H–C–H angles are 117° (Figure 54).

Ethyne (Acetylene)

In the ethyne molecule C_2H_2, the two carbon atoms are covalently combined together by three shared electron pairs. This corresponds to a high degree of

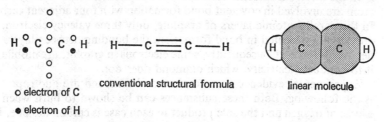

conventional structural formula

linear molecule

o electron of C

• electron of H

Figure 55

unsaturation in the molecule. Each carbon atom is also combined covalently with one hydrogen atom. The molecule is linear (Figure 55).

Allotropy (Polymorphism)

If an element can exist (without changing its state) in two or more different forms, the element is said to exhibit **allotropy**, or **polymorphism.**

The forms of the element are known as **allotropes** of it. They always exhibit different physical properties and may have different chemical properties also. Several elements exhibit this phenomenon, and the structures of some of their allotropes are given below.

Carbon

Carbon exists in two main allotropic modifications, diamond and graphite. Other allotropes have been described, and these are discussed on page 308. The crystal structures of diamond and graphite are as follows.

In diamond, covalency operates between the carbon atoms throughout and produces a crystal which is one single molecule which may become very large

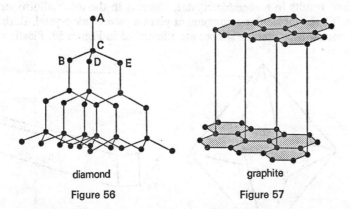

diamond

Figure 56

graphite

Figure 57

(a *giant molecule*). The crystal unit contains five atoms, ABCDE (Figure 56), and it is repeated indefinitely, forming interlacing buckled hexagons. The strength and uniformity of bonding make diamond very hard, non-volatile, and resistant to melting and to chemical attack. In *graphite*, carbon atoms are combined by covalency in hexagons in parallel planes. Between the planes, bonds are much weaker and are the result of *van der Waals forces* (page 33). These weak forces allow movement of the planes parallel to each other and this makes graphite very soft. The open structure (Figure 57) makes graphite more liable than diamond to chemical attack. (See also below.)

In the three-dimensional formation of diamond, all four valency electrons per atom are involved in covalent bond formation with four adjacent carbon atoms. In the parallel atomic layers of graphite, only three valency electrons per atom are definitely located in bond formation, the bonding between layers being van der Waals forces. Consequently, some electrons in graphite are mobile and allow it to conduct electricity, which diamond does not.

Experimental evidence that graphite and diamond are allotropes of carbon is the following. Both these substances can be shown to burn when heated in excess of oxygen and the sole product in each case is carbon dioxide. Further, if the evidence is made quantitative by absorbing the carbon dioxide in potassium hydroxide solution or soda lime, it can be shown that the mass of carbon dioxide obtained from one gramme of diamond is the same as from one gramme of graphite, *i.e.*, 3.67 g.

Sulphur

Sulphur has three allotropes in the solid state, two of which are well-defined crystalline solids, and the third is a 'plastic' form of the element. Experiments describing how the three forms can be prepared are given in Chapter 32 on pages 394 to 395. The two crystalline allotropes both consist of S_8 molecules, in which the sulphur atoms are arranged in the form of a 'puckered' ring (Figure 58). In one allotrope, stable up to a temperature of 96 °C, the molecular arrange-

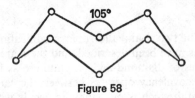

Figure 58

ment results in a *rhombic* crystal, whereas in the other allotrope, stable above 96 °C, the molecular arrangement gives a *monoclinic* crystal. Both rhombic and monoclinic crystalline shapes are illustrated in Figure 59. Plastic sulphur is not

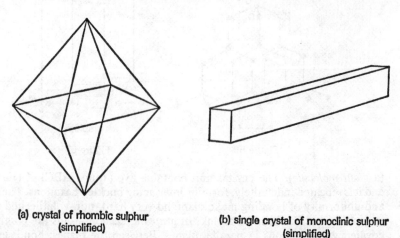

(a) crystal of rhombic sulphur (simplified)

(b) single crystal of monoclinic sulphur (simplified)

Figure 59

composed of S_8 rings at all, but is made up of *chains* of sulphur atoms which can be extremely long; over a period of time these chains revert to an S_8 ring structure, and rhombic sulphur is formed.

Phosphorus

Phosphorus exists in two main allotropic forms, both in the solid state. One allotrope, *white* phosphorus, is much less stable than the other, *red* phosphorus, and the white form changes spontaneously to the red form, albeit slowly at

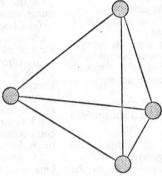

Figure 60

room temperature. White phosphorus is composed of individual molecules P_4, in which the atoms are at the corners of a regular tetrahedron (Figure 60); when it changes to red phosphorus these molecules link together to form an extended giant structure of phosphorus atoms, joined to each other in a random cross-linking manner.

Questions

1. The compounds named below are all covalent compounds. With the help of the table of electron-structures, give a diagram for a molecule of each of these compounds, showing the outermost electron shells only: (i) tetrachloromethane CCl_4; (ii) phosphorus trichloride PCl_3; (iii) silane SiH_4; (iv) trichloromethane $CHCl_3$; (v) phosphine PH_3; (vi) dichloromethane CH_2Cl_2.

2. The following compounds are electrovalent. With the help of the table of electron-structures, state what electronic changes take place when they are formed from their elements: (i) lithium oxide; (ii) potassium chloride; (iii) magnesium oxide; (iv) sodium sulphide. State briefly what kind of properties you would expect all these compounds to show by virtue of their electrovalent character.

3. Explain the meaning of the term: *metallic bond*. Explain why a typical metal is (*a*) a good conductor of electricity, (*b*) a good conductor of heat, (*c*) in some conditions at least, malleable and ductile.

4. Explain briefly in electronic terms why (1) the ammonia molecule, (2) the oxygen atom can participate readily in co-ordinate bonding. Give an example of the formation of an *ion* from ammonia by this means. Show, by electronic diagram, the formation of a co-ordinate linkage between phosphorus trichloride and oxygen. State *one* difference of chemical or physical behaviour you would expect between (i) phosphorus trichloride and its co-ordinate compound with oxygen, (ii) phosphorus trichloride and potassium chloride. Briefly explain your choice.

5. Graphite is called an 'allotropic form' of carbon. What do you understand by this statement? Give *one* other example of allotropy. How would you prove by a quantitative experiment that the statement is correct in the case of graphite and pure charcoal? (J.M.B.)

6. Discuss the chemical bonding in the following substances: magnesium oxide, hydrogen, dry hydrogen chloride, sodium chloride, paying particular attention to: (*a*) the types of particles involved; (*b*) the electronic changes occurring when the compounds are formed from their constituent elements; (*c*) the relationship between the type of bonding and the chemical and physical properties. (S.)

7. By means of electronic diagrams show how: (*a*) a calcium *ion* is formed from a calcium *atom*; (*b*) a chlorine *ion* is formed from a chlorine *atom*. (You may assume that (i) a calcium atom contains 20 electrons, (ii) a chlorine atom contains 17 electrons.) (S.)

8. The following table shows the symbols and atomic numbers of eight elements:

$_{11}$Na $_{12}$Mg $_{13}$Al $_{14}$Si $_{15}$P $_{16}$S $_{17}$Cl $_{18}$Ar

(*a*) Give the name and formula of a covalent hydride of *one* of the elements.

(*b*) Give the name and formula of an ionic chloride of *one* of these elements.

(*c*) From the list above name an element which does not readily form compounds.

(*d*) Give the name and formula of a compound containing *two* of these elements and oxygen.

(*e*) How does the electronic structure of a silicon atom differ from that of a phosphorus atom? (J.M.B.)

9. The atomic numbers of sodium and chlorine are 11 and 17 respectively.

(*a*) Give the electronic structures of the atoms of sodium and chlorine.

(*b*) Why do these elements show a valency of one in their ionic (electrovalent) compounds?

(*c*) Sketch the crystal lattice of sodium chloride.

(*d*) Explain how the crystal structure accounts for the high melting point of sodium chloride. (A.E.B.)

10. Explain the following statements:

(*a*) The movement of molecules in a gas differs from the movement of molecules in a solid.

(*b*) Sodium ($_{11}^{23}$Na) readily forms ionic (electrovalent) compounds.

(*c*) Ammonia adds to copper(II) (cupric) ions to give tetraamminecopper(II) ions (cuprammonium ions).

(*d*) The ethene (ethylene) molecule is planar while the methane molecule is not. (J.M.B.)

11. The atomic number and relative atomic mass of carbon are 6 and 12 respectively. (*a*) Give a labelled diagram showing the structure of a carbon atom. (*b*) Explain what is meant by a covalent bond. Give an electronic diagram for *one* named carbon compound.

Diamond, graphite, and the charcoal obtained from sugar are all pure forms of carbon. (*c*) How are the atoms arranged in graphite and in diamond? Show how this knowledge enables us to understand the physical characteristics of these substances. (*d*) Briefly state how it could be shown that graphite and the charcoal obtained from sugar are chemically identical. (Experimental details are *not* required.) (A.E.B.)

12. What do you understand by the term 'atomic number'?

The symbols of the elements of atomic number 6–17 inclusive are:

C, N, O, F, Ne, Na, Mg, Al, Si, P, S, Cl.

(*a*) Choose any *four* of these elements and give the formula of a chloride of each of them, indicating whether the bonding in the chloride is ionic or covalent in each case.

(*b*) Choose any *four* of these elements and give the name and formula of a covalent compound of each with hydrogen.

(*c*) Choose any *four* of these elements and write equations to show how an oxide of each reacts with water.

(*d*) Choose *two* of these elements, each of which can exist in two different crystalline forms. Indicate by a simple diagram the arrangement of the atoms in *one* form of *one* of these elements.

(*e*) Give the names and formulae of *two* compounds which each contain three of these elements.

(*f*) Atoms of some of these elements form ions which have identical electronic configurations. Atoms of another element in the list also have this electronic configuration. Name this other element, and write symbols for *three* of the ions in the manner Br⁻, Cu²⁺. (L.)

13. Name the particles present in the crystal lattices of (i) sodium; (ii) sodium chloride; (iii) diamond; (iv) carbon dioxide.

Explain the following in terms of the structures of the substances involved: (i) in the solid state sodium is a good conductor of electricity and is deformable whereas diamond is a non-conductor and is extremely hard; (ii) sodium chloride has a high melting point whereas carbon dioxide is a gas at room temperature and pressure. What do you understand by *allotropy*? Name *one* of the crystalline allotropes of sulphur. Describe its appearance and how you would prepare a sample of it in the laboratory. (A.E.B.)

14.

Element	Symbol	Atomic number
Lithium	Li	3
Beryllium	Be	4
Fluorine	F	9
Magnesium	Mg	12
Silicon	Si	14

f(*a*) The distribution of extra-nuclear electrons in the magnesium atom can be represented as

Mg: 2.8.2.

Show in a similar way the electron distribution in the atoms of lithium, beryllium, fluorine, and silicon.

(*b*) Write formulae for the oxides you would expect lithium, beryllium, and silicon to form. Which one of these oxides is most likely to be acidic?

(*c*) Write formulae for the simple ions formed by lithium, beryllium, fluorine, and magnesium.

(*d*) Fluorine forms compounds with all four of the other elements in the table. Which one of these fluorides is most likely to be covalent? Give the formula of this fluoride and explain the bonding in it in terms of electrons. (C.)

15. What is thought to happen when atoms are bonded (i) by an electrovalent (ionic) bond, (ii) by a covalent bond?

From the following list name *one* compound which is electrovalent and *one* which is covalent: ethanol, potassium chloride, calcium nitrate, starch, copper(II) sulphate. Describe *one* experiment applied to each of the chosen compounds which confirms your choice.

What is meant by the term *molecule*? Explain why the term has no meaning for an electrovalent compound. (Scottish)

16 What do you understand by the term *covalency*? Illustrate your answer by reference to the molecules of (*a*) chlorine, (*b*) oxygen.

An element M, relative atomic mass 40, atomic number 20, reacts with cold water to liberate hydrogen and forms a chloride whose relative molecular mass is 111. This chloride dissolves in water, and the solution conducts an electric current.

(*c*) Give simple diagrams to show and explain the difference between the M atom and the M ion. (*d*) State the type of valency which the chloride exemplifies. (*e*) Give the equation for the reaction of M with water. (*f*) Suggest a method by which you could prepare a small sample of M from the chloride. (S.)

17. A metal X (atomic number 11) burns in chlorine to produce a white, solid chloride Y. By means of diagrams illustrate the arrangement of electrons in X both before and after the reaction. Write an equation for the reaction.

Discuss the following properties of Y and account for them as far as possible: (*a*) melting point, (*b*) solubility, (*c*) electrical conductivity.

What type of bonding would you expect to find in a chloride which differed from Y in the above properties? State the name and formula of such a chloride. (S.)

10 Periodicity

The Periodic Table

Neglecting hydrogen (which has a uniquely simple atom – one proton and one electron) and helium, the lightest atoms have, in order of atomic number from left to right, the following electronic structures.

Group	I	II	III	IV	V	VI	VII	0
Period 2	Li 2,1	Be 2,2	B 2,3	C 2,4	N 2,5	O 2,6	F 2,7	Ne 2,8
Period 3	Na 2,8,1	Mg 2,8,2	Al 2,8,3	Si 2,8,4	P 2,8,5	S 2,8,6	Cl 2,8,7	Ar 2,8,8
Period 4	K 2,8,8,1	Ca 2,8,8,2						

It will be seen that, arranged in this way, *elements in the same vertical columns have the same number of valency electrons in the outermost shell of their atoms.* Because of this, the elements in each column tend to resemble each other closely in chemical behaviour. For example, the **noble gases**, He, Ne, and Ar, show a chemical inertness which is determined by the stable outer electron octet or duplet. These gases have not been found to form any compounds with other elements.

The occurrence of successive groups of elements showing strong chemical similarity in this way because of their similar outer electron shells is called *periodicity.* The Periodic Law has been developed fully to include all known elements and is expressed in a general Periodic Table, of which the above arrangement is the earliest part. The Groups of the table (written vertically) are numbered by Roman numerals, 0–VII. See the complete Periodic Table at the end of the book.

A brief description of the chemical properties of elements in the same group of the Periodic Table will be made for selected groups, to illustrate the similarity in chemical behaviour of such elements.

Group I

Sodium and **potassium**, each with one electron in the outer shell, resemble one another very closely. Both are univalent, ionizing by the loss of one electron per atom. Both are powerful reducing agents and good conductors of electricity.

$$Na \rightarrow Na^+ + e^- \text{ (K similar)}$$

Both are strongly electropositive, attacking cold water with liberation of hydrogen.

$$Na + H_2O \rightarrow Na^+OH^- + \tfrac{1}{2}H_2 \text{ (K similar)}$$

Both form strongly basic oxides, Na_2O and K_2O, and soluble hydroxides, NaOH and KOH, which are strong alkalis. The carbonates, Na_2CO_3 and K_2CO_3, are both soluble in water, giving mildly alkaline solutions, and are unaffected by ordinary heating. The nitrates of both metals liberate oxygen when heated and leave a nitrite.

$$2NO_3^- \rightarrow 2NO_2^- + O_2$$

Lithium resembles sodium and potassium in electropositive character and univalency, giving the ion, Li^+. It also gives a solid, electrovalent hydride, Li^+H^-, and attacks water liberating hydrogen, but less rapidly than for sodium and potassium. In this and in its almost insoluble carbonate and fluoride, it differs considerably from sodium and potassium and has some resemblance to the less electropositive *magnesium* of Group II.

Group V

Nitrogen and **phosphorus** show marked chemical similarity as non-metals. Both exercise a maximum valency of 5 (*i.e.*, the number of outer valency electrons) and also a lower valency of 3 (*i.e.*, the octet — 5). These valencies appear in N_2O_5 and P_4O_{10}, and in NH_3 and PH_3. Both the oxides are strongly acidic and combine with water to form acids, HNO_3 and HPO_3.

$$N_2O_5 + H_2O \rightarrow 2HNO_3 \text{ (P_4O_{10} similar)}$$

The chlorides, NCl_3 and PCl_3, are both covalent liquids, non-electrolytes, and rapidly hydrolysed by water. Both the hydrides, NH_3 and PH_3, form salts with hydrogen chloride, NH_4Cl and PH_4Cl, though the latter is much less stable, decomposing at about 35 °C.

Group VII

Group VII is the *halogen* group (see Chapter 30 for an explanation of 'halogen'); of its members, *fluorine* and *chlorine* have been shown in Periods 2 and 3, and *bromine* and *iodine* follow in later periods. This Group is important enough to be given rather special consideration. Its members show marked *general similarity* of properties and also, in many instances, a *consistent gradation* of behaviour. These elements and the electronic arrangements in their atoms are:

	F	Cl	Br	I
Electronic arrangement	2,7	2,8,7	2,8,18,7	2,8,18,18,7
Atomic number	9	17	35	53

The marked similarity of their chemical properties depends essentially on their possession of the same number (7) of electrons in the *valency* (highest energy) shell, but a pronounced gradation is imposed on the general similarity as the atoms increase in complexity from fluorine to iodine.

All the elements exhibit a **valency** of *one* in covalent combination with hydrogen and in electrovalent combination with metals. By formation of a shared electron pair with hydrogen in the first case, and capture of an electron from a metallic atom in the second, all these elements complete the external electron octet. Similarly, by formation of one shared electron pair, all produce diatomic molecules.

$$H_2 + X_2 \rightarrow 2HX; \quad X_2 + 2e^- \rightarrow 2X^-; \quad 2X \rightarrow X_2$$

where X is F, Cl, Br, or I.

Being electron acceptors, all the halogens are **oxidizing agents**. Examples are the following.

$$2Fe^{2+}(aq) + Cl_2(g) \rightarrow 2Fe^{3+}(aq) + 2Cl^-(aq) \text{ (iron(II) to iron(III)); } Br_2 \text{ similar}$$

$$S^{2-}(aq) + Br_2(l) \rightarrow 2Br^-(aq) + S(s) \text{ (}H_2S \text{ to sulphur); } Cl_2 \text{ and } I_2 \text{ similar}$$

As the number of electron shells increases, the atoms increase in size from the smallest, F, to the largest, I. Since each electron layer screens the outermost electrons from the attractive power of the nucleus, we should expect electrons to be most firmly held (or attracted) by fluorine atoms and least firmly held (or attracted) by iodine atoms. That is, fluorine should be the strongest, and iodine the weakest, in oxidizing behaviour. This is actually so. For example, in the order, F $\rightarrow$ Cl $\rightarrow$ Br $\rightarrow$ I, each halogen can oxidize the ions of those which follow it and liberate the free halogen. Examples are the following.

Chlorine liberates bromine from potassium bromide solution:

$$Cl_2(g) + 2Br^-(aq) \rightarrow 2Cl^-(aq) + Br_2(l)$$

Bromine liberates iodine from potassium iodide solution:

$$Br_2(l) + 2I^-(aq) \rightarrow 2Br^-(aq) + I_2(s)$$

There is a similar gradation from F to I in vigour of oxidation of hydrogen at room temperature.

Fluorine combines explosively with hydrogen even in the dark;
Chlorine combines slowly in daylight;
Bromine combines slowly in sunlight;
Iodine combines only when heated and then slowly and partially.

The same gradation is found in the stability of the hydrides towards heat and in the heat of formation of the sodium salts.

% dissociation at 720 °C		Heat of formation in kJ mol^{-1}	
HF	nil	NaF	568
HCl	0.0013	NaCl	410
HBr	0.2	NaBr	360
HI	28.0	NaI	288

Another illustration of the F $\rightarrow$ I gradation is given by the **boiling points** at standard pressure of the four halogen elements, which are:

F_2	Cl_2	Br_2	I_2
−188	−33.5	59	184 (all °C)

It must be allowed, however, that, in certain respects, halogen properties do show some variation. For example, while calcium fluoride is insoluble in water, the calcium salts of the other three halogens ($CaCl_2$, $CaBr_2$, CaI_2) are very soluble and deliquescent. Also, silver fluoride is quite soluble in water while the other three silver salts (AgCl, AgBr, AgI) have very low solubilities indeed. Also chlorine forms an oxide, Cl_2O_7, which is the characteristic oxide for a Group VII element.

None of the other halogens forms such an oxide. The highest oxide of iodine is I_2O_5 and of fluorine F_2O, while bromine forms no oxide stable at room temperature. On strict gradation, bromine would be expected to be intermediate between chlorine and iodine in solubility in water but, in fact, it is the most soluble of the three (with iodine the least soluble). Fluorine attacks water rapidly at room temperature, liberating a mixture of ozone, O_3, and oxygen. (For some further halogen comparisons relating to Cl_2, Br_2, and I_2 only, see page 375.)

Progression of properties in a Period

The progression of properties for elements in the same period can be illustrated from Period 3.

Period 3 (Groups in Roman numerals)

I	II	III	IV	V	VI	VII	0
Na	Mg	Al	Si	P	S	Cl	Ar
2,8,1	2,8,2	2,8,3	2,8,4	2,8,5	2,8,6	2,8,7	2,8,8

The figures represent the electron groupings in the various atoms, left to right being outwards from the nucleus.

The chemical inertness of argon, resulting from its very stable outer electron octet, has already been noticed (page 79).

Elements of Groups I to III (Na, Mg, Al)

Elements. The elements of Groups I, II, and III (Na, Mg, Al) all show marked *metallic* character by ionizing with electron loss but the metallic character weakens in the direction Na $\rightarrow$ Mg $\rightarrow$ Al. The valency of each element is equal to the Group number of the element and to the number of electrons in the outermost (valency) shell.

$$Na \rightarrow Na^+ + e^-; \quad Mg \rightarrow Mg^{2+} + 2e^-; \quad Al \rightarrow Al^{3+} + 3e^-$$

Sodium liberates hydrogen from cold water, showing its exceptionally electropositive nature; the other two metals liberate hydrogen from dilute acid although in the case of aluminium only from hot, concentrated hydrochloric acid. In all these cases, hydroxonium ion (page 268) is reduced by electron gain.

$$2H_3O^+(aq) + 2e^- \rightarrow 2H_2O(l) + H_2(g)$$

Chlorides. The chlorides of sodium and magnesium are electrolytes, Na^+Cl^- and $Mg^{2+}(Cl^-)_2$. Both are soluble in water; sodium chloride is chemically unaffected by water but magnesium chloride is hydrolysed slightly, showing the somewhat weaker electropositive (metallic) character of magnesium. Aluminium is still less characteristically metallic in its chloride. When anhydrous, the chloride is covalent (as Al_2Cl_6) and is much hydrolysed by water, *i.e.*, resembles the chlorides of non-metals, but it yields ions in aqueous solution, Al^{3+} (hydrated) and Cl^-, as do metallic chlorides.

Oxides. The oxides of sodium and magnesium are electrovalent compounds, $(Na^+)_2O^{2-}$ and $Mg^{2+}O^{2-}$. Both are soluble in water and yield *alkaline* solutions by forming hydroxide ions. The more electropositive sodium gives by far the stronger alkali.

$$O^{2-}(aq) + H_2O(l) \rightarrow 2OH^-(aq)$$

Aluminium oxide has basic properties, *e.g.*, in the reaction

$$Al_2O_3(s) + 6HCl(aq) \rightarrow 2AlCl_3(aq) + 3H_2O(l)$$

but it forms no alkali with water and so shows itself less electropositive than sodium or magnesium. Further, aluminium oxide shows slight acidic tendency (and is, therefore, *amphoteric*) by forming an aluminate with caustic alkali. This resembles the behaviour of a non-metal.

$$Al_2O_3(s) + 2NaOH(aq) + 3H_2O(l) \rightarrow 2NaAl(OH)_4(aq)$$

Hydride. Sodium (heated in dry hydrogen) forms a solid, electrovalent hydride. This occurs because the metal is so strongly electropositive as to reduce hydrogen by donating electrons to it.

$$2Na(s) + H_2(g) \rightarrow 2(Na^+H^-)(s)$$

This hydride yields hydrogen with cold water and is an electrolyte when molten, giving hydrogen at the *anode*. (Compare hydrides in Groups IV to VII later.)

$$H^-(s) + H_2O(l) \rightarrow OH^-(aq) + H_2(g)$$
$$2H^-(s) \rightarrow H_2(g) + 2e^- \text{ (to anode)}$$

Elements of Groups IV to VII (Si, P, S, Cl)

In Period 3, all the elements of Groups IV to VII exercise *covalency* numerically equal to their group number by using the valency electrons to form covalent pairs, *e.g.*, in oxides SiO_2, P_4O_{10}, SO_3, Cl_2O_7. Oxides with lower oxygen content also occur, *e.g.*, P_4O_6, SO_2, Cl_2O. In addition, each element exercises a covalency of (8 — Group number), *e.g.*, in hydrides SiH_4, PH_3, H_2S, and HCl. *Electrovalency* is shown by elements of Groups VI and VII only. This is exercised by acceptance of electrons to complete the outer octet and so form a negative ion. This electrovalency is numerically equal to (8 — Group number). Acceptance of more than two electrons per atom never occurs.

$$S + 2e^- \rightarrow S^{2-}; \ Cl + e^- \rightarrow Cl^-$$

This valency pattern is typical of non-metals.

Chlorides. The elements of Groups IV and V each produce a chloride by utilizing covalency numerically equal to the Group number, *i.e.*, $SiCl_4$, PCl_5. Both are rapidly hydrolysed by water with liberation of hydrogen chloride and, when pure, both are non-electrolytes.

$$SiCl_4(l) + 4H_2O(l) \rightarrow Si(OH)_4(aq) + 4HCl(aq);$$
$$PCl_5(l) + 4H_2O(l) \rightarrow H_3PO_4(aq) + 5HCl(aq)$$

The same elements also produce chlorides by exercising covalency of (8 — Group number), *i.e.*, $SiCl_4$ (as before) and PCl_3. This trichloride is a covalent liquid, a non-electrolyte, and rapidly hydrolysed by water.

$$PCl_3(l) + 3H_2O(l) \rightarrow H_3PO_3(aq) + 3HCl(aq)$$

Contrast the metallic type of chloride produced at the other end of the period in Groups I and II, where the elements have 1 or 2 valency electrons per atom and produce electrovalent chlorides, solids, electrolytes if molten or in aqueous solution, and either unaffected, or only slightly hydrolysed, by water.

Oxides. Elements of Groups IV to VII all form an oxide by exercising covalency equal to the Group number, *i.e.*, SiO_2, P_4O_{10}, SO_3, and Cl_2O_7. When combined with water, all these oxides produce *acids*. This is characteristic of non-metals.

$$SO_3(g) + H_2O(l) \rightarrow H_2SO_4(aq); \quad P_4O_{10}(s) + 2H_2O(l) \rightarrow 4HPO_3(aq)$$
$$Cl_2O_7(g) + H_2O(l) \rightarrow 2HClO_4(aq)$$

In addition, phosphorus and sulphur also form well-known oxides by exercising a lower covalency, *i.e.*, P_4O_6 and SO_2. These are also acidic oxides.

$$P_4O_6(s) + 6H_2O(l) \rightarrow 4H_3PO_3(aq); \quad SO_2(g) + H_2O(l) \rightleftharpoons H_2SO_3(aq)$$

Contrast the oxides with *basic* properties formed by the *metals* at the other end of the Period, Na_2O and MgO.

Hydrides. Elements of Groups IV to VII all form hydrides by utilizing covalency equal to (8 — Group number), *i.e.*, SiH_4, PH_3, H_2S, HCl. All these are typical of simple covalent compounds, *i.e.*, gaseous at ordinary temperature and pressure and non-electrolytes when water-free, having no chemical reaction with water (except ionization by H_2S and HCl). SiH_4 is spontaneously flammable in air. Contrast the *metallic* hydride formed by sodium at the other end of the Period — electrovalent (Na^+H^-), an electrolyte when molten (yielding hydrogen at the *anode*), attacked by water to yield hydrogen.

In physical properties, the Group I–III elements of Period 3 (Na, Mg, Al) are all *metallic*, *e.g.*, good conductors of heat and electricity. The Group V–VII elements of Period 3 are *non-metallic* in type, *e.g.*, very poor conductors of heat and electricity.

Notice the following patterns in Period 3 which are repeated in other periods.

	Group							
	I	II	III	IV	V	VI	VII	0
Valency towards Cl or H	1	2	3	4	3	2	1	—
Valency towards O (maximum)	1	2	3	4	5	6	7	—

Electropositive (metallic) character increasing ←

Electronegative (non-metallic) character increasing →

Chief compounds of the metals

It is important to know accurately the chemical formulae of the chief compounds of the common metals. They are listed below (by Groups of the Periodic Table) for study and reference.

Group of P.T.	I	II	III
	Li, Na, K	Mg, Ca, Ba	Al
Valency	1	2	3
Ion	Na^+	Ca^{2+}	Al^{3+}
Oxide	$(Na^+)_2O^{2-}$	$Ca^{2+}O^{2-}$	$(Al^{3+})_2(O^{2-})_3$
Hydroxide	Na^+OH^-	$Ca^{2+}(OH^-)_2$	$Al^{3+}(OH^-)_3$
Chloride	Na^+Cl^-	$Ca^{2+}(Cl^-)_2$	Al_2Cl_6 (covalent)
Sulphate	$(Na^+)_2SO_4^{2-}$	$Ca^{2+}SO_4^{2-}$	$(Al^{3+})_2(SO_4^{2-})_3$
Nitrate	$Na^+NO_3^-$	$Ca^{2+}(NO_3^-)_2$	$Al^{3+}(NO_3^-)_3$
Carbonate	$(Na^+)_2CO_3^{2-}$	$Ca^{2+}CO_3^{2-}$	none
Hydrogencarbonate	$Na^+HCO_3^-$	$Ca^{2+}(HCO_3^-)_2$	none
Sulphide	$(Na^+)_2S^{2-}$	$Ca^{2+}S^{2-}$	$(Al^{3+})_2(S^{2-})_3$ (decomposed by water)

The ionic formulae are shown in most cases.

Ammonium salts have formulae like those of sodium salts, with the ion, NH_4^+, instead of Na^+. There is no oxide.

Questions

1. (*a*) Copy the grid shown below which represents part of a blank periodic table, the numbers being the atomic numbers of the elements. In your grid, write,
 (i) V in a space which could be occupied by a noble gas,
 (ii) W in the space which the most active metal would occupy,
 (iii) X in the space which the most active non-metal would occupy,
 (iv) Y in a space which could be occupied by an element capable of forming a compound YX_2,

1							2
3	4	5	6	7	8	9	10
11	12	13	14	15	16	17	18

 (v) Z in a space which could be occupied by an element capable of forming a compound W_3Z.

(*b*) which of the following is the formula of the compound formed by Y and Z?

   YZ, Y_2Z_3, Y_3Z_2, Y_2Z, Y_3Z.
                                    (J.M.B.)

2. For each of the pairs of elements (*a*) sodium and potassium, (*b*) chlorine and bromine, give *three* properties that illustrate their chemical similarity.

Suggest an electronic explanation for this similarity. (O. and C.)

3. Carbon, atomic number 6, and silicon, atomic number 14, are elements in the same group of the Periodic Table. Give the electronic configuration for an atom of each element, and state in which group the elements occur.

Both carbon and silicon form dioxides. Give *one* similarity and *one* difference between these oxides. (J.M.B.)

4. The elements sodium, magnesium, aluminium, silicon, phosphorus, sulphur, chlorine, and argon form a series in the Periodic Table.

For each of *four* of these elements state (*a*) the electronic structure, (*b*) the formula for its hydride and the reaction, if any, between the hydride and water, (*c*) the formula of an oxide and whether this oxide is acidic, basic, amphoteric, or neutral. (O. and C.)

5. By reference to the properties of (*a*) the elements, (*b*) their oxides, (*c*) their chlorides, justify the inclusion of (i) sodium and potassium, (ii) nitrogen and phosphorus in the same group of the Periodic Table.

6. At the foot of the page is a section of the Periodic Table showing the first eighteen elements. From *these* choose elements to answer the questions which follow.

(*a*) Name *one* element which combines with other elements to form ionic compounds only. Give the formula of one such compound and show diagrammatically how the electron configurations of the elements change in forming the compound.

(*b*) Name *two* elements which combine together to form a covalent compound. Give the formula of this compound and show diagrammatically how the electron configurations of the elements change in forming the compound.

(*c*) Name *one* element whose atoms are converted into ions with three positive charges when the element is dissolved in dilute hydrochloric acid. State, with your reason, whether you consider the element is oxidized or reduced in this process.

1 H							2 He
3 Li	4 Be	5 B	6 C	7 N	8 O	9 F	10 Ne
11 Na	12 Mg	13 Al	14 Si	15 P	16 S	17 Cl	18 Ar

(*d*) Name *one* element which will dissolve in dilute hydrochloric acid so that 1 mole of atoms (1 gramme-atom) of the element will liberate 24 000 cm³ of hydrogen at room temperature and pressure from the acid. Write the equation for the reaction.

(*e*) Select *two* elements which have a large-scale use and indicate briefly what these uses are.

(*f*) Select *three* elements which form oxides such that

(i) the atomic ratio of element to oxygen is 1 : 1,
(ii) the atomic ratio of element to oxygen is 1 : 2, and
(iii) the atomic ratio of element to oxygen is 2 : 3.

Write the formulae of the three oxides. (L.)

7. Suppose that elements W, X, Y, Z are all in the same (early) period of the Periodic Table. Allot them to their correct Groups on the following evidence: an oxide, W_2O exists and is strongly basic; X forms a liquid, covalent chloride, XCl_3; the oxide of Y is Y_2O_3; Z produces an ion, Z^-. Answer the following questions:

(1) What would be the effect of adding W_2O to water and then spotting the liquid on to universal indicator paper? Explain. Write an ionic formula for the sulphate of W. Briefly characterize the hydride, WH.

(2) Write the formula of the highest oxide of X and an equation for its likely behaviour with water. If X forms a lower oxide, what is its likely formula? What is the formula of the simplest hydride of X and its likely physical state at s.t.p.? Relate this to its valency type.

(3) Write the empirical formula of the chloride of Y. Discuss (briefly) likely behaviour of this chloride with water. The oxide, Y_2O_3, is *amphoteric*. Explain this term.

(4) What is the likely behaviour of the element Z in the field of oxidation–reduction? Explain in electronic terms. Write an ionic formula for the compound formed between W and Z. Estimate qualitatively its melting point and describe the effect (if any) of an attempt to pass electric current through this compound (i) at room temperature, (ii) above its melting point.

8. For the main Groups of the Periodic Table, the metallic properties of the elements vary approximately with their positions as shown in the chart at the foot of the page. The direction of the arrows indicates an increase in metallic nature.

(*a*) Will the most metallic element be found at A, B, C, or D?

(*b*) Will the most non-metallic element be found at A, B, C, or D?

(*c*) The element indium has the symbol In. It is in the same Group as aluminium but, whereas aluminium is the second element in the Group, indium is the fourth. (i) What is the formula for indium oxide? (ii) Would you expect this oxide to be more basic or less basic than aluminium oxide?

(*d*) The names and symbols of the first five elements in Group II of the Periodic Table, reading from top to bottom, are beryllium, Be; magnesium, Mg; calcium, Ca; strontium, Sr; barium, Ba. (i) Which of these five elements will have the chloride of highest melting point? (ii) Write the formula for this chloride. (J.M.B.)

9. What feature of the atomic structure of the four halogen elements, fluorine, chlorine, bromine, iodine, justifies their common inclusion in Group VII of the Periodic Table? State and explain in electronic terms what valency behaviour they all share. Bromine is said to be intermediate in behaviour between chlorine and iodine. Justify this statement by reference to *two* examples of the chemical behaviour and *one* example of the physical behaviour of these three elements. Mention *one* respect in which bromine is not the intermediate element of the three and *two* features of the chemistry of fluorine in which it differs from the other three halogen elements. If astatine is another element of this Group (following iodine), give *one chemical* property you would expect it to possess and briefly justify your choice.

10. In the Periodic Table shown overleaf, lithium, carbon, oxygen, and neon have been placed in their correct positions. The positions of nine other elements have been represented by letters. These letters are not the real symbols for the elements concerned.

By reference to this table, answer the following questions.

(*a*) Give the letter of the most reactive metal.

I	II	III	IV	V	VI	VII	O
H							He
A						B	
C						D	

I	II	III	IV	V	VI	VII	0
Lithium			Carbon		Oxygen	L	Neon
X			J		G	Q	
Y						R	
Z						T	

(*b*) Give the letter of the most reactive non-metal.

(*c*) Name the 'family of elements' represented by L, Q, R, and T.

(*d*) Name one element in each case occurring in Groups II, III, and V.

(*e*) The element Q forms a compound with lithium and a compound with carbon. Suggest formulae for these two compounds (using Q as the symbol for the element), and compare (i) their solubilities in water, (ii) their relative melting points, (iii) their electrical conductivities when molten.

(*f*) Discuss briefly the bonding present in the compound formed between lithium and Q. Your answer should include a diagram to show what has happened to the electrons in the outer shell of atoms of each of these elements.

(*g*) Suggest a possible shape for a molecule of the compound formed between J and R. (L.)

11 Molecular Theory

The Gas Laws

We have already seen (Chapter 3) that the behaviour of gases when subject to temperature and pressure change, can be expressed by two simple laws, those of Boyle and Charles.

Boyle's Law. The volume of a given mass of gas is inversely proportional to its pressure, temperature remaining constant.

With the usual symbols, this is expressed mathematically as:

$$pv = \text{a constant } (T \text{ constant})$$

Charles' Law. The volume of a given mass of gas is directly proportional to its absolute temperature, pressure remaining constant.

With the usual symbols, this is expressed mathematically as:

$$\frac{v}{T} = \text{a constant } (p \text{ constant})$$

Gay-Lussac's Law of Gaseous Volumes

A third law, describing the behaviour of gases, when involved in chemical reactions, was stated by Gay-Lussac.

We can illustrate the Law of Gay-Lussac by quoting first some of the experimentally observed results of chemical reaction between gases, upon which the law is based. Temperature and pressure are to be considered constant throughout each statement.

1. *Ammonia*

   2 volumes of ammonia decompose to give 1 volume of nitrogen and 3 volumes of hydrogen.

2. *Steam*

   2 volumes of hydrogen combine with 1 volume of oxygen, giving 2 volumes of steam.

3. *Hydrogen chloride*

   1 volume of hydrogen combines with 1 volume of chlorine to give 2 volumes of hydrogen chloride.

4. *Nitrogen oxide*

   2 volumes of nitrogen oxide decompose to give 1 volume of nitrogen and 1 volume of oxygen.

Examining these experimental results (the methods by which they have been obtained are given in Chapter 13), we notice at once that all the volumes of the gases concerned are related to each other by simple whole-number ratios.

Whenever gases are concerned in chemical action, simple whole-number relations between their volumes are always found. This is the fact which was first noted by Gay-Lussac and expressed in his Law of Gaseous Volumes, which is now stated.

Gay-Lussac's Law of Gaseous Volumes. When gases react they do so in volumes which bear a simple ratio to one another, and to the volume of the product if gaseous, temperature and pressure remaining constant.

Simple behaviour of gases: an explanation required

These three Laws of Boyle, Charles, and Gay-Lussac express among them a highly interesting fact about gases – a curious similarity of behaviour. In chemical properties, and such physical properties as density and solubility in water, gases show marked variations. There are neutral gases such as nitrogen, oxygen, and hydrogen, acid-producing gases such as sulphur dioxide, hydrogen chloride, and nitrogen dioxide, alkali-producing gases such as ammonia; gases of very high solubility in water, such as hydrogen chloride (500 volumes of gas dissolve in 1 volume of water), gases of moderate solubility, such as hydrogen sulphide (3 volumes of gas dissolve in 1 volume of water), and gases of very low solubility, such as nitrogen (0.02 volumes dissolve in 1 volume of water); some gases are chemically very reactive, *e.g.*, chlorine, and some are entirely inert, *e.g.*, argon. **But, however great the variations in these properties may be, all the gases obey the Laws of Boyle, Charles, and Gay-Lussac quite closely.** There must be some explanation of this similarity. Note that it does not matter whether the gas is an element, *e.g.*, hydrogen, or a compound, *e.g.*, hydrogen chloride; each obeys the laws equally well.

Avogadro's Hypothesis

The explanation was put forward in 1811 by Avogadro, an Italian scientist, in the form known as Avogadro's Hypothesis. We have seen in Chapter 2 that the smallest particle of an element or compound which can exist separately is called a molecule of it. Avogadro's explanation of the simple behaviour of gases, especially as expressed in Gay-Lussac's Law, was that equal volumes of all gases, under the same temperature and pressure conditions, contain the same number of molecules. When this suggestion was put forward it was purely a hypothesis, that is, an idea which had occurred to Avogadro, which appeared to him sensible, but which still required to be tested further before it could be fully accepted. The truth of it has since become **experimentally** demonstrable, and it is frequently known, on that account, as Avogadro's Law. (See *Avogadro Constant*, page 46.)

Avogadro's Law. Equal volumes of all gases at the same temperature and pressure contain the same number of molecules.

This law has been of the greatest value in the development of chemistry since about 1860. It is a rather curious fact that its importance was at first unnoticed, and the full recognition of its implications, a few of which we shall now examine, is due to the work not of Avogadro himself, but of another Italian, Cannizzaro, some 47 years after the hypothesis had been first put forward, and after Avogadro himself was dead.

Why Avogadro's Law is important

The importance of the law lies in this fairly simple fact, that, since it asserts that equal volumes of gases contain equal numbers of molecules, it enables us to change over directly from a statement about *volumes* of gases to the same statement about *molecules* of gases. Every time we make a statement about *one volume* of any gas, we are also making a statement about a certain number of *molecules* of it, and that number, by Avogadro's law, is always the same, no matter what the gas may be. Consequently, we can change over at will, in any statement about gases, from volumes to molecules and vice versa.

This means that by applying the law to volume measurements of gases, we can probe right to the heart of a chemical reaction, to the actual molecules themselves. It is an enormous step to change directly from an experimental statement like:

2 volumes of hydrogen combine with 1 volume of oxygen giving
2 volumes of steam (temperature and pressure constant)

to

2 *molecules* of hydrogen combine with 1 *molecule* of oxygen giving
2 *molecules* of steam.

The second of these two statements goes right to the essentials of the reaction, to the very molecules themselves. The law is important because it gives us this power to reveal the molecules themselves at work in chemical reactions. **Note, however, that it applies only to gases.**

Other examples illustrating this important change from volume measurements to statements about molecules follow.

Ammonia

By experiment, 2 volumes of ammonia decompose to give 1 volume of nitrogen and 3 volumes of hydrogen.

Applying the Law,

2 molecules of ammonia contain 1 molecule of nitrogen and 3 molecules of hydrogen.

Hydrogen chloride

By experiment, 1 volume of hydrogen combines with 1 volume of chlorine to give 2 volumes of hydrogen chloride.

Applying the Law,

1 molecule of hydrogen combines with 1 molecule of chlorine to give 2 molecules of hydrogen chloride.

Nitrogen oxide

By experiment, 2 volumes of nitrogen oxide decompose to give 1 volume of nitrogen and 1 volume of oxygen.

Applying the Law,

2 molecules of nitrogen oxide contain 1 molecule of nitrogen and 1 molecule of oxygen.

(Temperature and pressure assumed constant throughout)

From these statements, it is only a further step to the deduction of the formulae of the gaseous compounds concerned. This step is given in Chapter 12 and the formulae of the common gases are considered in Chapter 13.

How Avogadro's Law explains Gay-Lussac's Law

Assume throughout the following paragraph that temperature and pressure are constant.

When gases react chemically, the reaction must take place between individual molecules of the gases. As Dalton suggested in the similar case of combination between atoms, the reactions will take place between small whole numbers of molecules of the reactants to produce small whole numbers of molecules of the products.

We have seen in the last section that, employing Avogadro's Law, we can change over directly from statements about molecules to statements about volumes, provided that gases only are concerned. Making this change, the last sentence of the last paragraph becomes: the reactions will take place between small whole numbers of *volumes* of the reactants to produce small whole numbers of *volumes* of the products (all being gases). This is what Gay-Lussac's Law states. Hence, Avogadro's Law has enabled us to deduce the experimentally observed Law of Gay-Lussac.

Avogadro's Law and the relative molecular masses of gases

We have seen, in Chapter 4, that relative molecular masses are expressed as the mass of one molecule of the substance compared with the mass of one *atom* of hydrogen. We have also seen, in this chapter, the very important relation which exists between the number of *volumes* of gases and the number of *molecules* of gases involved in chemical reaction. It is now necessary to find how the atom and the molecule of hydrogen are related to one another. This will lead us to a method of determining relative molecular masses.

The nature of the hydrogen molecule

By experimental work which is fully described later (page 123), it has been found that (at constant temperature and pressure):

1 volume of hydrogen combines with 1 volume of chlorine to give 2 volumes of hydrogen chloride.

Applying Avogadro's Law we can say at once:

1 molecule* of hydrogen combines with 1 molecule of chlorine to give 2 molecules of hydrogen chloride.

Now each of the two molecules of hydrogen chloride must contain some hydrogen. The least amount of hydrogen which can be contained in one molecule of hydrogen chloride is one atom, because the atom of hydrogen is indivisible. Consequently, the least amount of hydrogen there can be in two molecules of hydrogen chloride is 2 atoms of hydrogen. But these 2 atoms of hydrogen must have come from the 1 molecule of hydrogen marked *; therefore, a molecule of hydrogen must contain **at least** two atoms of hydrogen.

But hydrochloric acid forms with sodium hydroxide one salt only, sodium chloride. Two series of salts have never been obtained from hydrochloric acid as they have from, for example, sulphuric acid. Thus with sodium hydroxide, sulphuric acid can be made to form both **normal** sodium sulphate and **acid** sodium sulphate (see page 412). Since the hydrogen of hydrochloric acid cannot be replaced in two stages, there is only one hydrogen atom in the molecule. But the two hydrogen atoms necessary for two molecules of hydrogen chloride have been obtained from one molecule of hydrogen. Hence the molecule of hydrogen contains two atoms.

Try to visualize what this means. It means that, in ordinary gaseous hydrogen, no separate atoms of hydrogen exist. All the particles consist of two hydrogen atoms, locked in a chemical embrace, and moving always as a single unit, the molecule. It is as if the hydrogen atoms are paired off to run a perpetual three-legged race. The molecule, consisting of two hydrogen atoms, never breaks up, except for the purpose of engaging in chemical reactions. This fact is expressed by writing the hydrogen molecule as H_2, which means a single molecule of hydrogen containing two atoms. (2H would mean two separate hydrogen atoms.)

By a somewhat similar argument, it can be shown that the molecule of chlorine contains two atoms, and the reaction between hydrogen and chlorine may be diagrammatically expressed as in Figure 61.

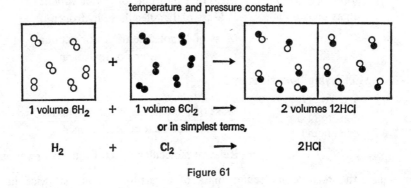

temperature and pressure constant

1 volume $6H_2$ +   1 volume $6Cl_2$ ⟶ 2 volumes 12HCl

or in simplest terms,

H_2 + Cl_2 ⟶   2HCl

Figure 61

Note that equal numbers of molecules of hydrogen, chlorine, and hydrogen chloride are contained in equal volumes.

It is known, from similar experimental evidence and argument, that nitrogen and oxygen also have two atoms per molecule and their molecules are written N_2 and O_2. This is expressed by saying that the molecules of hydrogen, chlorine, nitrogen, and oxygen are *diatomic* or that their *atomicity is 2*.

Atomicity. *The atomicity of an element is the number of atoms contained in one molecule of the element.*

Relation between vapour density and relative molecular mass

The relative densities of solids and liquids are expressed with reference to water, but it would be most inconvenient to deal with gases in this way because of the great difference between the densities of water and gases.

Definition. *The vapour density of a gas or vapour is expressed as the mass of a certain volume of the gas or vapour compared with the mass of the same volume of hydrogen at the same temperature and pressure.*

$$\text{Vapour density of a gas or vapour} = \frac{\text{Mass of 1 volume of gas or vapour}}{\text{Mass of 1 volume of hydrogen}}$$

(Temperature and pressure constant.)

Note that vapour density can be experimentally determined because it only involves weighing equal volumes of hydrogen and the vapour.

The relative molecular mass of a gas or vapour is expressed in the form:

Relative molecular mass of a gas or vapour $=$ $\dfrac{\text{Mass of 1 molecule of the gas or vapour}}{\text{Mass of 1 atom of hydrogen}}$

We shall now show that there is a simple relation between vapour density and relative molecular mass.

Vapour density of a gas or vapour $=$ $\dfrac{\text{Mass of 1 volume of gas or vapour}}{\text{Mass of 1 volume of hydrogen}}$

(Temperature and pressure constant.)

Applying Avogadro's Law, we can say directly:

Vapour density of a gas or vapour $=$ $\dfrac{\text{Mass of 1 molecule of gas or vapour}}{\text{Mass of 1 molecule of hydrogen}}$

$=$ $\dfrac{\text{Mass of 1 molecule of gas or vapour}}{\text{Mass of 2 atoms of hydrogen}}$

Multiplying both sides by 2:

2 × (Vapour density of a gas or vapour) $=$ $\dfrac{\text{Mass of 1 molecule of gas or vapour}}{\text{Mass of 1 atom of hydrogen}}$

$=$ Relative molecular mass of the gas or vapour

i.e., the relative molecular mass of a gas or vapour is twice its vapour density.

Relative molecular mass from vapour density. Regnault's Method

We have already noted that to find the vapour density of a gas or vapour it is only necessary to obtain the mass of a certain volume of the gas or vapour and the mass of an equal volume of hydrogen, both at the same temperature and pressure.

Unfortunately, direct weighing of hydrogen and other gases in this way is very difficult, partly because the actual masses of convenient volumes of the gases are small, and partly because changes of temperature, pressure, and humidity in the atmosphere introduce error during the course of the experiments.

In principle it is only necessary to evacuate a globe, weigh it, and fill it with hydrogen and weigh it again; then evacuate it again, fill with the gas and weigh again, temperature and pressure remaining constant. This is known as Regnault's Method. Then,

Vapour density of the gas $=$ $\dfrac{\text{(Mass of globe + gas) — (mass of globe)}}{\text{(Mass of globe + hydrogen) — (mass of globe)}}$

and relative molecular mass of the gas = 2 × vapour density.

A simpler method of finding the relative molecular mass, which can be applied to the particular case of oxygen, is described in Experiment 29.

Experiment 29
Relative molecular mass of oxygen

Weigh a hard glass test-tube containing approximately 4 g of dry red lead oxide. Attach

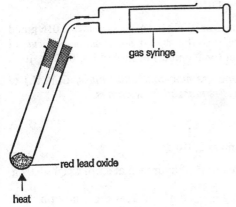

gas syringe

red lead oxide

heat

Figure 62

the hard glass test-tube and contents to a gas syringe, as illustrated in Figure 62, and heat the test-tube gently. After an initial increase in volume of the air in the test-tube on being heated, the red lead oxide will be decomposed to yield oxygen gas, which is collected in the gas syringe. (If no gas syringe is available the oxygen may be collected over water in a graduated cylinder in the usual manner.) When a convenient volume of oxygen has been collected (say approximately 80 cm³), cease heating the test-tube and allow the whole apparatus to cool. Note the final volume of the oxygen collected in the syringe (or in the graduated cylinder). Finally, remove the hard glass test-tube from the connection with the syringe and record the final mass of the tube plus contents. Note the temperature of the room, and the atmospheric pressure.

Specimen results

Mass of hard glass tube plus red lead oxide *before* heating = 17.397 g
Mass of hard glass tube plus contents *after* heating = 17.282 g
Volume of oxygen collected = 84 cm³
Temperature = 288 K
Atmospheric pressure = 750 mmHg

Calculation

Mass of oxygen $(17.397 - 17.282) = 0.115$ g

$$\text{Volume of oxygen at s.t.p.} = 84 \times \frac{273}{288} \times \frac{750}{760} \text{ cm}^3$$

$$= 81 \text{ cm}^3$$

$$\therefore 1000 \text{ cm}^3 \text{ oxygen weigh } \frac{0.115 \times 1000 \text{ g}}{81}$$

$$= 1.42 \text{ g}$$

But 1000 cm³ hydrogen at s.t.p. weigh 0.09 g

$$\therefore \text{Vapour density of oxygen} = \frac{1.42}{0.09}$$

$$= 15.8$$

$$\therefore \text{Relative molecular mass of oxygen} = 15.8 \times 2$$

$$= 31.6$$

If the oxygen is collected over water instead of in a gas syringe a pressure correction will be required before the above calculation is completed (see page 41).

Molar volume (or gramme-molecular volume) of hydrogen

It is shown on page 109 that the hydrogen molecule is diatomic and is written as H_2. Expressing this on the standard of $^{12}C = 12$ and $H = 1.008$, we have:

$$H_2$$
$$2 \times 1.008$$
$$\text{or } 2.016$$

If this is expressed in the scientific mass unit (gramme), it becomes 2.016 g and this is the mass of *one mole of molecules* of the gas (or *one gramme-molecule* of it). It contains the Avogadro Constant (or Number) of molecules, 6.02×10^{23}.

Definition. The molar mass (or one gramme-molecular mass, G.M.M.) of any gas is its relative molecular mass expressed in grammes.

We now have:

$$H_2$$
$$\text{Molar mass } 2.016 \text{ g}$$

By experiment, it has been found that 1 dm^3 of hydrogen at s.t.p. weighs 0.09 g.

Therefore, 2.016 g of hydrogen occupy $\dfrac{2.016}{0.09}$ dm^3, or 22.4 dm^3, at s.t.p. This

volume is called the **molar volume** (or **gramme-molecular volume**) of hydrogen.

We see, from this, that if we use grammes as our mass-unit, the formula, H_2, may denote either 2.016 g of hydrogen or 22.4 dm^3 of it at s.t.p. We have connected the molecular formula of hydrogen with a *volume*, rather a convenient result because hydrogen, a gas, is usually measured experimentally as a volume.

Molar volumes (or gramme-molecular volumes) of other gases

Consider the same volume, 22.4 dm^3 at s.t.p., of some other gas, say oxygen. Since we are considering the *same volume* of both oxygen and hydrogen in the same conditions, we know, by Avogadro's Law, that we must be considering the *same number of molecules* of the two gases. But we started from *one mole* of molecules of hydrogen; therefore, the 22.4 dm^3 at s.t.p. must represent the same number of molecules of oxygen, *i.e.*, one mole of molecules (or one gramme-molecule) of it. The same argument will apply to any other gas, which gives us this very important result:

One mole of molecules (or one gramme-molecule) of any gas occupies 22.4 dm^3 at s.t.p. This volume is called the molar volume (or gramme-molecular volume) of the gas.

Definition. The molar volume (or gramme-molecular volume, G.M.V.) of any gas is the volume occupied at s.t.p. by one mole of molecules (gramme-molecule) of the gas and is 22.4 dm³.

This is a most important and useful result because it means that if we write the molecular formula of any gas and refer it to grammes as mass-units, the volume of gas indicated is always 22.4 dm^3 at s.t.p.

It is often convenient to use the approximation that the molar volume of any gas at room temperature and pressure is 24 dm^3. Thus:

H_2	O_2	N_2	CO_2	H_2S	SO_2	Cl_2	
2.016	32	28	44	34	64	71	grammes
22.4	22.4	22.4	22.4	22.4	22.4	22.4	dm³ at s.t.p.

This means that the gases all have differing densities except where, by coincidence, their relative molecular masses are the same, *e.g.*, CO_2 and N_2O, 44.

Avogadro Constant and gas volumes

The *Avogadro Constant* has the value 6.02×10^{23} and was defined, on page 46, as the number of atoms in exactly 12 g of carbon-12. It is often called the *Avogadro number*. It is also the number of molecules contained in one mole of molecules of any gas. It is, for example, the number of molecules contained in 36.5 g of hydrogen chloride, 32 g of oxygen, or 64 g of sulphur dioxide, these figures being the relative molecular masses of the gases. The *volume* of gas which contains this number of molecules always occupies 22.4 dm³ at s.t.p.

Empirical and molecular formulae

A method was given on page 67 for the calculation of a formula for a compound from its percentage composition by mass. It is quite possible for two different compounds to have the same percentage composition by mass either because the compounds have different arrangements of the same atoms inside the molecule, or because the molecular formula of one is a multiple of that of the other. Considering the second of these possibilities, it is clear that ethyne C_2H_2, and benzene C_6H_6, both having 92.3% of carbon, will both appear to have the same formula, when the calculation of page 67 is applied. Thus:

	Carbon	Hydrogen
Per cent by mass	92.3	7.7
Number of atoms is represented by	$\dfrac{92.3}{12} = 7.7$	$\dfrac{7.7}{1} = 7.7$
Dividing by smallest	$\dfrac{7.7}{7.7} = 1$	$\dfrac{7.7}{7.7} = 1$

The formula appears to be CH for both. The reason is that this calculation always yields the *simplest* formula which expresses the composition of the substance by mass. Since the ratio of carbon atoms to hydrogen atoms is the same in CH, C_2H_2, and C_6H_6, the same composition by mass is expressed in all three. This simplest formula which expresses the composition of a compound by mass is called its **empirical formula.** Thus, the empirical formula of both benzene and ethyne is CH.

Clearly we must devise a method of finding the true or molecular formula of the compounds. This is simple enough. If the true formula is CH, the relative molecular mass is (12 + 1) or 13; if C_2H_2, 26; if C_6H_6, 78, and so on. Thus, a determination of relative molecular mass will at once decide the true formula, and, in practice, this means determining the vapour density (page 109) of the compound. The vapour density of ethyne is 13, and of benzene 39; that is, their relative molecular masses are 26 and 78 respectively. This gives a molecular formula C_2H_2 for ethyne and C_6H_6 for benzene.

Definition. *The empirical formula of a compound is the simplest formula which expresses its composition by mass, and which expresses the ratio of the numbers of the different atoms present in the molecule.*

Definition. *The molecular formula of a compound is one which expresses the actual number of each kind of atom present in its molecule.*

Another example will illustrate the point further.

EXAMPLE. *A gaseous compound of carbon and hydrogen contains 80% carbon by mass. One dm^3 of the compound at s.t.p. weighs 1.35 g. Find its molecular formula. ($C = 12$; $H = 1$. G.M.V. of any gas is 22.4 dm^3 at s.t.p.)*

	Carbon	Hydrogen
Per cent by mass	80	20
Number of atoms is represented by	$\frac{80}{12} = 6.7$	$\frac{20}{1} = 20$
Divide by smallest	$\frac{6.7}{6.7} = 1$	$\frac{20}{6.7} = 3$

The empirical formula is CH_3

∴ the molecular formula is C_nH_{3n}, where n is a whole number,
∴ the relative molecular mass is ($12n + 3n$).
From the problem, 1 dm^3 of the compound at s.t.p. weighs 1.35 g.

∴ 22.4 dm^3 of the compound at s.t.p. weigh $\dfrac{1.35 \times 22.4}{1} = 30.2$ g

$$12n + 3n = 30.2$$
$$15n = 30.2$$
$$n = 2$$

∴ Molecular formula is C_2H_6

Application of Avogadro's Law to determination of relative atomic masses

A development from Avogadro's Law, which we are now to consider, supplies a method of determination of the relative atomic mass of non-metals of a certain type. The method was first used by, and is named after, Cannizzaro.

The most convenient case for us to consider is that of carbon. Suppose we take the symbol X to denote the relative atomic mass of carbon. In a molecule of a carbon compound, there cannot be less than one atom of carbon, and there may be two, three, four, or any small whole number of carbon atoms. This means that, in the relative molecular mass of a carbon compound, there must be X, 2X, 3X, or nX units of mass of carbon (n is a small whole number).

The relative molecular mass of any carbon compound can be found by determining first its vapour density (by the method on page 110, provided that the compound is gaseous). The compound can then be analysed and the percentage by mass of carbon in it determined. The mass of carbon in the relative molecular mass is then given by the expression:

$$\frac{\text{Percentage of carbon}}{100} \times \frac{\text{Relative molecular mass of compound}}{1}$$

If this is applied to several carbon compounds, the results must represent the masses of the number of carbon atoms in the molecules. In the table on page 115, the figures are given for several compounds.

The figures in the last column correspond to the presence of one, two, three, or more carbon atoms. The lowest mass is 12 and the others are multiples of 12. Now it is obvious that, if we have included in our list any compound containing only one carbon atom per molecule, that compound will be the first, methane, or the last, methanal, because in these the mass of carbon is the least. If, therefore, the molecules of methane and methanal do actually contain only one carbon atom, the relative atomic mass of carbon is 12. This process has been applied to a very large number of carbon compounds, and the mass of carbon in the relative molecular mass has always been found to be 12, or a multiple of 12, but never less. From this we conclude that the least mass of carbon there can ever be in the relative molecular mass of one of its compounds is 12, that this mass corresponds to the presence of one carbon atom and that the relative atomic mass of carbon is 12.

The method can be applied to determine the relative atomic mass of any element forming a large number of gaseous or easily volatile compounds.

Compound	Vapour density by experiment	Relative molecular mass $(= 2 \times V.D.)$	% of carbon by mass (by experiment)	Mass of carbon in the relative molecular mass
Methane	8	16	75.0	$75 \times \dfrac{16}{100} = 12$
Ethane	15	30	80.0	$80 \times \dfrac{30}{100} = 24$
Propane	22	44	81.8	$81.8 \times \dfrac{44}{100} = 36$
Ethene	14	28	85.7	$85.7 \times \dfrac{28}{100} = 24$
Ethyne	13	26	92.3	$92.3 \times \dfrac{26}{100} = 24$
Methanal	15	30	40.0	$40 \times \dfrac{30}{100} = 12$

Application of Graham's Law to determination of relative molecular mass

One of the characteristic properties of substances in the gaseous state is that of *diffusion*, a process by which two gases mix freely when they come into contact. This phenomenon has been known for a very long time, and it has been used to postulate the existence of matter in a particulate form (see page 18). The *rate* at which gases mix together is, however, dependent on the nature of the two gases concerned, and it is possible to compare the rates of diffusion of two gases by allowing each in turn to diffuse, under carefully controlled conditions, through a small hole into a third gas. Provided that each of the gases under investigation is subject to the same physical conditions of temperature and pressure, it is found that the relative rates at which the gases escape through the small hole is entirely dependent on the vapour densities of the gases. The relationship is not a linear one but it is expressed in *Graham's Law*.

Graham's Law. The rate of diffusion of a gas is inversely proportional to the square root of its vapour density.

Expressed mathematically this is

$$\text{rate of diffusion } r \propto \frac{1}{\sqrt{d}}$$

where d is the vapour density of the gas. For any two gases, diffusing under the same conditions of temperature and pressure, we may write

$$r_1/r_2 = \sqrt{d_2/d_1}$$

Now since the vapour density of a gas is directly proportional to its relative molecular mass, M (page 110), we may also write

$$r_1/r_2 = \sqrt{M_2/M_1}$$

where M_1 and M_2 are the relative molecular masses of the two gases concerned.

This last expression allows a measurement of the relative molecular mass of an unknown gas to be made. By comparing the rate (r_1) at which the unknown gas (of relative molecular mass M_1) diffuses through a small hole in a container with the rate (r_2) at which a sample of gas of *known relative molecular mass* (M_2) diffuses through the same hole under the same conditions, the value of M_2 may be calculated.

Questions

Relative atomic masses will be found on page 499.

1. What is the mass of 22.4 dm³ (gramme-molecular volume) of the following gases at s.t.p. (*a*) ammonia; (*b*) hydrogen sulphide; (*c*) nitrogen; (*d*) chlorine; (*e*) dinitrogen oxide?

2. Calculate the relative molecular mass of the following gases from the statements:

(*a*) 0.8 g of oxygen occupied at s.t.p. a volume of 560 cm³.

(*b*) 1400 cm³ of sulphur dioxide measured at s.t.p. weighed 4 g.

(*c*) 1.12 dm³ of nitrogen oxide measured at s.t.p. weighed 1.5 g.

3. A hydride of carbon has a vapour density of 15, and it contains 20% by mass of hydrogen.

Calculate the formula of the hydrocarbon and write an equation to represent its complete combustion in oxygen. (J.M.B.)

4. State Gay-Lussac's law of combining volumes, and illustrate the law by referring to the reactions between (*a*) hydrogen and oxygen to form steam, the volumes of the gases being measured at 110 °C and atmospheric pressure, (*b*) ethane (C_2H_6) and oxygen to form carbon dioxide and water, gaseous volumes being measured at s.t.p. (J.M.B.)

5. An evacuated flask weighed 20.70 g. Filled with dry hydrogen, it was found to weigh 20.94 g. Filled with dry chlorine at the same temperature and pressure as the hydrogen, the flask weighed 29.22 g. Using these data alone, find the relative molecular mass of chlorine.

6. The following data were obtained in experiments with a gas.

Experiment 1
A flask was evacuated, stoppered, and weighed. It was then filled with the gas, stoppered, and reweighed.

Mass of the stoppered evacuated flask
$$= 45.340 \text{ g}$$

Mass of the stoppered flask and gas
$$= 45.547 \text{ g}$$

At the same temperature and pressure, the mass of hydrogen required to fill the flask was 0.009 g and the mass of an equal volume of air was 0.131 g.

Experiment 2
When the flask containing the gas was placed upside down in a beaker of water containing universal indicator, and the stopper removed, a rapid reaction took place, the flask quickly filled with water, and the green indicator turned red.

(*a*) Calculate the vapour density and hence the relative molecular mass of the gas. (*b*) Draw a diagram to show how this gas could be collected in a gas-jar. (*c*) Suggest a suitable agent which may be used for drying this gas. (J.M.B.)

7. Explain how Avogadro's Law can be used to establish (*a*) relative molecular masses; (*b*) relative atomic masses. Illustrate your answer by reference to oxygen and carbon. (O. and C.)

12 Calculations involving Gas Volumes

The formation of a gaseous substance is a common occurrence in chemical reactions, and therefore calculations involving volumes of gases are frequently required. The purpose of this chapter is to give a few examples of the types of calculations which are likely to be encountered in a certificate course, and to give practice at solving them.

It will be helpful if we summarize several points already dealt with in connection with the volumes of gases in previous chapters, for use will be made regularly of these in the examples which follow.

Summary of data on gaseous volumes

1. The volumes of all gases are subject to change by a variety of factors, especially by changes in temperature and pressure. It is therefore necessary to establish values of temperature and pressure which can serve as reference standards for purposes of comparing gas volumes. The standards chosen are 0 °C (273 K) and 760 mm mercury pressure (1 atmosphere), and when a given volume of gas is measured under these conditions of temperature and pressure it is said to be measured under standard conditions at s.t.p. (Chapter 3).

2. In the case of a mixture of gases being present in the same container, *each* gas has a volume equal to the total volume of the container (Chapter 3).

3. Equal volumes of all gases at the same temperature and pressure contain the same number of molecules (Avogadro's Law). The following statement is also true:

Equal numbers of molecules of different substances in the gaseous state occupy the same volume, provided all volume measurements take place under the same conditions of temperature and pressure.

These two statements enable us to change from statements concerning relative *volumes* of gases to statements concerning relative numbers of molecules of the substances concerned, and vice versa (Chapter 11).

4. The volume occupied by the molar mass of any gas is constant, and has a value of 22.4 dm^3 at s.t.p. It is known as the *molar volume* of the gas (or, sometimes, as the gramme-molecular volume of the gas) (Chapter 11).

These important points will be incorporated in the examples which follow.

Determination of the volume of a gas from the equation for the reaction

EXAMPLE 1. *Calculate the volume of oxygen at 12 °C and 745 mm pressure which could be obtained by heating 5 g of potassium chlorate. (K = 39; Cl = 35.5; O = 16. Molar, or gramme-molecular, volume of gases at s.t.p. is 22.4 dm^3.)*

Calculation

The first requirement is the equation:

$$2KClO_3(s) \rightarrow 2KCl(s) + 3O_2(g)$$

Note that, in the question, a **mass** of potassium chlorate is given and a **volume** of oxygen is wanted. Therefore, in the equation, we insert the **masses** appropriate to potassium chlorate and the **volume** appropriate to oxygen. It is quite unnecessary to insert a **mass** of oxygen. Using grammes as the mass units for potassium chlorate, O_2 (one mole, or gramme-mole, of oxygen) represents 22.4 dm^3 at s.t.p.; $3O_2$ represents, therefore, 3×22.4 dm^3, or 67.2 dm^3.

Inserting mass and volume, we have:

$$
\begin{array}{ccc}
2KClO_3 & \rightarrow & 2KCl \quad + \quad 3O_2 \\
2(39 + 35.5 + 48) \text{ g} & & 67.2 \text{ dm}^3 \\
245 \text{ g} & & \text{at s.t.p.}
\end{array}
$$

From the equation:

245 g of potassium chlorate yield 67.2 dm^3 of oxygen,

so,   5 g of potassium chlorate yield $\dfrac{67.2 \times 5}{245}$ dm^3 of oxygen,

$$= 1.37 \text{ dm}^3 \text{ of oxygen at s.t.p.}$$

Converting this volume to 12 °C and 745 mm pressure as the example requires:

$$\frac{p_1 v_1}{T_1} = \frac{p_2 v_2}{T_2}$$

$$\frac{760 \times 1.37}{273} = \frac{745 \times v_2}{285}$$

$$v_2 = \frac{760 \times 1.37 \times 285}{273 \times 745} \text{ dm}^3$$

$$= 1.46 \text{ dm}^3$$

∴ Volume of oxygen at 12 °C and 745 mm pressure = 1.46 dm^3.

EXAMPLE 2. *Calculate the volume of hydrogen sulphide at 14 °C and 770 mm pressure which will react with 10 g of lead(II) nitrate. (Pb = 207; N = 14; O = 16. Molar, or gramme-molecular, volume of gases is 22.4 dm^3 at s.t.p.)*

Calculation

We require first the equation; the nitrate is given as a **mass** so we insert **masses** under its formula, while the hydrogen sulphide is required as a **volume** so the **volume** appropriate to one mole of gas is put under its formula.

$$
\begin{array}{ccc}
H_2S & + & Pb(NO_3)_2 & \rightarrow PbS + 2HNO_3 \\
22.4 \text{ dm}^3 & & (207 + 28 + 96) \text{ g} \\
\text{at s.t.p.} & & 331 \text{ g}
\end{array}
$$

From the equation:

331 g of the nitrate react with 22.4 dm^3 of hydrogen sulphide,

so, 10 g of the nitrate react with $\dfrac{22.4 \times 10}{331}$ dm^3 of hydrogen sulphide at s.t.p.

$$= 0.677 \text{ dm}^3 \text{ at s.t.p.}$$

Converting this volume to 14 °C and 770 mm as the example requires:

$$\frac{p_1 v_1}{T_1} = \frac{p_2 v_2}{T_2}$$

$$\frac{760 \times 0.677}{273} = \frac{770 \times v_2}{287}$$

$$v_2 = \frac{760 \times 0.677 \times 287}{273 \times 770} \text{ dm}^3$$

$$= 0.703 \text{ dm}^3$$

∴ Volume of hydrogen sulphide at 14 °C and 770 mm pressure = 0.703 dm³.

Application of molar (gramme-molecular) volume of gases to determination of relative molecular masses

It follows from the above result that, to determine the relative molecular mass of a gas in grammes, we have simply to find the mass of the gas in grammes which occupies 22.4 dm³ at s.t.p.

EXAMPLE 3. *350 cm³ of a certain gas were found to weigh 1 g at s.t.p. conditions. What is the relative molecular mass of the gas?*

From the data given, the mass of 22.4 dm³ (or 22 400 cm³) of the gas at s.t.p. is $1 \times \dfrac{22\,400}{350}$ grammes or 64 grammes.

Therefore, the relative molecular mass of the gas is 64. This method of calculation is alternative to the vapour density method given on page 111.

Relative volumes of reacting gases

EXAMPLE 4. *Find the smallest volume of oxygen required to burn 1 dm³ of methane completely. What is the volume of carbon dioxide formed?*

Calculation

First of all we must write the equation for the reaction. Methane is a hydrocarbon of formula CH_4, and the products of the complete combustion of hydrocarbons are carbon dioxide and water.

$$2CH_4(g) + 3O_2(g) \rightarrow 2CO_2(g) + 2H_2O(g)$$

From the equation it can be seen that:

2 moles of methane require 3 moles of oxygen for combustion.

Therefore, by Avogadro's Law:

2 volumes of methane require 3 volumes of oxygen for combustion.

Hence:

2 dm³ methane require 3 dm³ of oxygen

∴ 1 dm³ methane requires $\frac{3}{2}$ dm³ of oxygen

∴ Volume of oxygen required = 1.5 dm³

To find the volume of carbon dioxide evolved we note from the equation that:

2 volumes of methane produce 2 volumes of carbon dioxide

Therefore:

1 dm³ methane produces 1 dm³ carbon dioxide.

Hence volume of carbon dioxide produced by complete combustion of 1 dm³ methane = 1 dm³.

EXAMPLE 5. *100 cm³ of hydrogen were sparked with 30 cm³ of oxygen, both gases at 110 °C and 760 mm. What is the total volume of gas left after cooling to the original temperature and pressure? What percentage of this gas by volume is steam?*

Again, the first step is to write the equation for the reaction.

$$2H_2(g) + O_2(g) \longrightarrow 2H_2O(g)$$

Since the final temperature is 110 °C, well above the boiling point of water, we are justified in writing $H_2O(g)$ rather than $H_2O(l)$ in the equation for the reaction.

From the equation it can be seen that 1 mole of oxygen requires 2 moles of hydrogen. Hence, by Avogadro's Law,

1 volume of oxygen requires 2 volumes of hydrogen

∴ 30 cm³ oxygen will require (2 × 30) cm³ hydrogen = 60 cm³ hydrogen

We may therefore conclude that of the original 100 cm³ hydrogen only 60 cm³ were actually used in the reaction. Therefore at the end of the reaction the vessel will contain (100 − 60) = 40 cm³ unused hydrogen. In addition the vessel will contain a volume of steam equivalent to 30 cm³ oxygen. Now, from the equation, one volume of oxygen gives two volumes of steam, and therefore:

30 cm³ oxygen produce (2 × 30) = 60 cm³ steam.

Hence finally in the vessel there will be:

40 cm³ unused hydrogen
and   60 cm³ steam

Hence total volume of gas left after cooling to the original temperature and pressure = (40 + 60) = *100 cm³*.

At the end of the reaction 100 cm³ of mixed gases remain, of which the volume of steam is 60 cm³.

Hence, *percentage of final gas mixture by volume which is steam*

$$= \frac{60}{100} \times 100\% = 60\%.$$

Other examples of the use of Gay-Lussac and Avogadro's Laws in calculations involving gas volumes will be found in Chapter 13, which deals with the establishment of the formulae of gaseous substances.

Questions

Relative atomic masses will be found on page 499.

1. What volume of carbon dioxide at s.t.p. could be obtained by dissolving 150 g of pure marble (calcium carbonate) in dilute nitric acid?

2. 1.16 g of magnesium was allowed to react with excess dilute sulphuric acid. What volume of hydrogen measured at s.t.p. was liberated?

3. An evacuated flask weighed 20.70 g. Filled with dry hydrogen, it was found to

weigh 20.94 g. Filled with dry chlorine at the same temperature and pressure as the hydrogen, the flask weighed 29.22 g. Using these data alone, find the relative molecular mass of chlorine.

4. 1 dm³ of ozone measured at 20 °C and 750 mm was converted into oxygen by heating. If the resulting oxygen was measured at 30 °C and 750 mm, what volume would it occupy?

5. Write the equation for the reaction between ethane and oxygen. Find the smallest volume of oxygen needed to burn 1 dm³ of ethane completely. What volume of carbon dioxide is formed? (All volumes measured at s.t.p.) (S.)

6. What volume of oxygen, measured at s.t.p., would be required to combust all the hydrogen produced by treating 6 grammes of magnesium with excess dilute sulphuric acid? (S.)

7. Explain carefully the term 'relative molecular mass'. Give a short account of the experiments and reasoning which lead to the conclusion that the relative molecular mass of steam is 18. (O. and C.)

8. Define the terms 'vapour density' and 'molecular weight' (relative molecular mass), and deduce from your definition the relation that exists between them, stating any assumption that you make in the deduction. A metallic chloride has a vapour density of 130 and contains 54.6% of chlorine by mass. How many atoms of chlorine does its molecule contain? ($Cl = 35.5$.) (C.)

9. What is the relation between relative molecular mass and vapour density, and how do you account for it?
At atmospheric pressure and 546° C, 50 cm³ of phosphorus vapour weigh 0.093 g. What is the relative molecular mass of phosphorus (1 dm³ of hydrogen at s.t.p. weighs 0.09 g)? (O. and C.)

10. Explain how Avogadro's Law can be used to establish (a) relative molecular masses; (b) relative atomic masses. Illustrate your answer by reference to oxygen and carbon. (O. and C.)

11. A flask of about 500 cm³ capacity, fitted with a rubber stopper, weighed 90.512 g when filled with air. Hydrogen chloride from a generator was passed in for a few minutes, the stopper was replaced, and the flask and its contents now weighed 91.230 g.
The flask was inverted in a trough of water, the stopper was removed under water, and eventually 456 cm³ of liquid entered when the levels had been adjusted. Given that the temperature of the gases was 13 °C and that the atmospheric pressure was 770 mm of mercury, find:

(a) The mass of a litre of hydrogen chloride at s.t.p.
(b) The relative molecular mass of hydrogen chloride from the given data. (One dm³ of hydrogen at s.t.p. weighs 0.09 g, one dm³ of air at s.t.p. weighs 1.293 g.)

12. Explain the use of a chemical equation and the information which it conveys.
Give equations representing three chemical reactions and state the exact meaning of each equation. (O. and C.)

13. Describe in detail the qualitative *and* quantitative experiments you would carry out to show that the equation

$$CuO + H_2 = H_2O + Cu$$

represents the action of hydrogen on red-hot copper(II) oxide. (O.)

14. The weights of equal volumes of certain gases were compared at the same temperature and pressure (N.B. This is *not* s.t.p.). The results are given in the table.

Gas	Mass of 1 dm³ of gas in grammes
Methane, CH_4	0.40
Carbon dioxide, CO_2	1.10
Hydrogen, H_2	0.05
Sulphur dioxide, SO_2	1.60

(a) For the four gases in the table, plot a graph of relative molecular mass against mass of 1 dm³. Molecular mass should be along the x axis (the horizontal axis) using a scale of 1 inch (or 2 cm) for a relative molecular mass of 10. Use a scale of 1 inch (or 2 cm) for a mass of 0.2 grammes along the y axis.

(b) The mass of 1 dm³ of a hydrocarbon, C_xH_y, is 1.45 grammes at the temperature and pressure of the experiment. Use the graph to estimate the relative molecular mass of this hydrocarbon. If y is 10, suggest a value of x which would fit the observations. Attempt to show how the atoms are arranged in a molecule of this compound (one isomer only).

(c) When ethanol vapour is passed over a heated catalyst, the following reaction can take place

$$C_2H_5OH(g) \longrightarrow C_2H_4(g) + H_2O(g)$$
ethanol　　　　ethene

In an investigation of this reaction, the water-vapour was removed from the gaseous product and a gas which was thought to be ethene was collected. If the gas really was ethene, what should the mass of 1 dm³ have been at the temperature and pressure at which the four gases in the table above were compared?

The mass of 1 dm³ of this gas was found in one experiment to be 0.30 grammes at the same temperature and pressure. This is considerably lower than might have been expected. Can you suggest any reason for this? (L.)

13 Formulae of Gases

In this chapter, we shall give the experimental evidence, and the reasoning from it, by which the formulae of a number of common gases have been established. It is most important that you should note carefully the way in which Avogadro's Law is continually employed in establishing these formulae, both directly in converting volume measurements into evidence of the numbers of molecules involved, and indirectly when vapour density measurements are employed. The importance of the Law in the following work cannot be over-estimated.

Because the experiments are difficult to set up and to carry out, or involve intricate safety precautions, it is not advisable for them to be carried out as class experiments. Nevertheless, fairly full experimental details are given, for it is important for the student to understand all that is involved in obtaining an accurate result.

Hydrogen chloride

All volume measurements made during the following experiment are under the same conditions of temperature and pressure.

Set up the apparatus of Figure 63 in diffused daylight, and allow the mixed

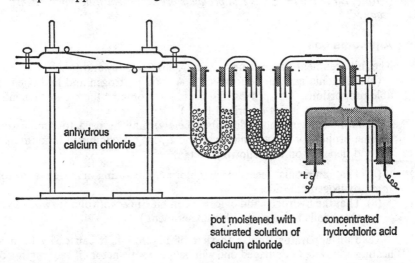

anhydrous
calcium chloride

pot moistened with
saturated solution of
calcium chloride

concentrated
hydrochloric acid

Figure 63

gases to pass for 5 to 10 minutes. Close both taps. Attach wires from a coil, and arrange a plate of thick glass between the tube and the observers. Close the circuit and a flash is seen but no sound is heard. Note that the green colour of chlorine has now disappeared. Fill the tube leading to one tap with mercury, and, holding the liquid in place with the finger, invert under mercury in a mortar.

Open the tap. No gas enters or leaves. Close the tap, replace the mercury in the tube by water, and open under the surface of water. The latter rises and almost fills the tube.

Note. The experiment on page 151 *(electrolysis of hydrochloric acid) shows that the gases which fill the eudiometer tube consist of equal volumes of hydrogen and chlorine.*

This proves that the whole of the gas in the tube was hydrogen chloride because any excess of hydrogen or chlorine would not have dissolved with this rapidity. This experiment proves that, starting with half a tubeful of hydrogen and half a tubeful of chlorine, we obtain, by their combination, a tubeful of hydrogen chloride (temperature and pressure constant) or:

1 volume of hydrogen combines with 1 volume of chlorine to give 2 volumes of hydrogen chloride (temperature and pressure constant).

Applying Avogadro's Law, we may substitute molecules for volumes, all the substances being gases.

∴ 1 molecule of hydrogen combines with 1 molecule of chlorine to give 2 molecules of hydrogen chloride.

But 1 molecule of hydrogen and 1 molecule of chlorine each contain 2 atoms.

∴ 2 molecules of hydrogen chloride contain 2 atoms of hydrogen and 2 atoms of chlorine.

∴ 1 molecule of hydrogen chloride contains 1 atom of hydrogen and 1 atom of chlorine.

∴ The formula of hydrogen chloride is HCl.

The equation for the above reaction is $H_2(g) + Cl_2(g) \rightarrow 2HCl(g)$.
Note. Since the reactants and products are all gaseous the vapour density is not required.

Ammonia gas

The full argument depends upon two experiments (*a*) and (*b*).

(*a*) Ammonia gas can be formed by sparking nitrogen and hydrogen in suitable proportions. This proves that ammonia contains these two elements only. (This experiment is not described here.)

(*b*) Hofmann's method (described below) may be used to demonstrate the volume proportions of the nitrogen and hydrogen combined in ammonia. The method depends on the following facts:

(i) That ammonia reacts with chlorine, liberating nitrogen and forming hydrogen chloride.

(ii) That the hydrogen and chlorine combine in *equal* volumes when hydrogen chloride is formed. (See last experiment.)

Take an apparatus similar to that of Figure 64. It can easily be made by heating a burette (a damaged one will be quite satisfactory if the length from the tap to the open end is 40 cm or more) and drawing out the heated portion until it forms a constriction of which the diameter is about that of ordinary glass tubing, and then making a file-scratch on the constriction and breaking *AB* separate from *C*. *AB* should be about 30 cm long and graduated into three equal portions. Attach *AB* to a chlorine generator in a fume-chamber and fill it with chlorine. Add the rubber tubing and clip, and the portion *C*, and clamp the whole apparatus in a vertical position. Put a little concentrated ammonia into

C, **cautiously** release the clip and allow a little of the ammonia to enter *AB*. There will be a flash of light and white fumes of ammonium chloride will appear. Carefully allow more ammonia to enter *AB* until there is no further reaction and a few drops of liquid have collected at *B*. Then put dilute sulphuric acid into *C*, colour it with a little litmus solution and run it into *AB* to neutralize the excess ammonia, taking care that *C* does not become empty of liquid or air will be drawn into *AB*. When the solution in *AB* is red (*i.e.*, when all the ammonia has been neutralized) remove *C* and place *AB* in a wide, deep vessel filled with water so that the tap *B* is well immersed. Open the tap. Water will enter. Push *AB* down into the vessel until the levels of water inside and outside are equal, so giving atmospheric pressure inside *AB*. It will then be found that the gas left occupies one-third of the volume of *AB*. The gas is nitrogen.

The original volume, *AB*, of chlorine combined with hydrogen from the ammonia to form hydrogen chloride. A volume, *AB*, of hydrogen from the

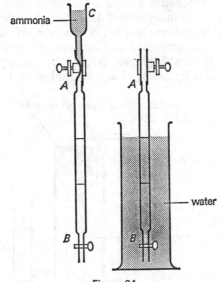

Figure 64

ammonia must have been used up to combine with this chlorine (see [ii] page 124). At the same time, one-third of a volume *AB* of nitrogen was liberated from the ammonia.

∴ 3 volumes of hydrogen were combined with 1 volume of nitrogen in ammonia gas (temperature and pressure constant).

Applying Avogadro's Law, we may replace 'volumes' by 'molecules'.

∴ 3 molecules of hydrogen combine with 1 molecule of nitrogen

∴ 6 atoms of hydrogen combine with 2 atoms of nitrogen or,

in the simplest terms,

3 atoms of hydrogen combine with 1 atom of nitrogen

∴ the simplest (empirical) formula for ammonia gas is NH_3 and its molecular formula is $(NH_3)n$ where *n* is a whole number

∴ the relative molecular mass of ammonia gas is $(14 + 3)n$ or $17n$.

We now need the relative molecular mass of ammonia gas to find the value of *n*.

The vapour density of ammonia gas is 8.5.

$$\therefore \text{ its relative molecular mass is } 17$$
$$\therefore 17n = 17$$
$$\therefore n = 1$$

∴ the molecular formula of ammonia gas is NH_3

Equations

$$2NH_3(g) + 3Cl_2(g) \rightarrow N_2(g) + 6HCl(g)$$

Then $6HCl(g) + 6NH_3(g) \rightarrow 6NH_4Cl(s)$

Adding $8NH_3(g) + 3Cl_2(g) \rightarrow N_2(g) + 6NH_4Cl(s).$

Steam

The apparatus (see Figure 65) consists of a stout eudiometer tube surrounded by a jacket which contains a vapour at 130 °C (pentanol boils at 130 °C at 760 mm pressure). The other limb of the eudiometer tube serves as a manometer, for mercury can be run into and out of this tube, so altering the pressure. One volume of oxygen is introduced and then two volumes of hydrogen. The open end of the tube is suitably plugged so that the mercury is not blown out. The mixture is exploded by means of an electrical spark and the plug is removed.

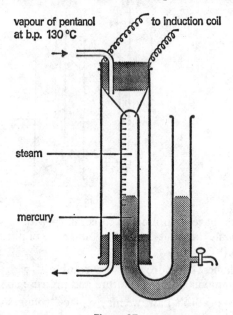

Figure 65

On allowing the gas to cool down to 130 °C and equalizing the mercury levels in the two tubes it is found that there are two volumes of steam left in the tube. (All the above measurements of volume are made at laboratory pressure and 130 °C.) On cooling the apparatus below 100 °C the steam condenses to water and the mercury rises to the top of the enclosed tube showing that all the oxygen and hydrogen have been used up since they would not condense to a liquid as does the steam.

From this experiment:

2 volumes of hydrogen and 1 volume of oxygen form 2 volumes of steam.

Applying Avogadro's Law we may substitute molecules for volumes, all the substances (at this temperature) being gaseous.

∴ 2 molecules of hydrogen and 1 molecule of oxygen form 2 molecules of steam

∴ 1 molecule of hydrogen and ½ molecule of oxygen form 1 molecule of steam.

But the molecule of hydrogen contains two atoms and the molecule of oxygen contains two atoms.

∴ Formula for steam is H_2O

Note. Since the reactants and the products are all gaseous under the conditions of the experiment the vapour density is not required.

Nitrogen oxide

A suitable volume of nitrogen oxide is measured at atmospheric pressure (levels *A* and *B* equal) in the hard glass tube (Figure 66). The spiral of iron wire is then electrically heated to red heat. The metal combines with the oxygen of the nitrogen oxide liberating nitrogen (the residual gas can be shown to be inactive).

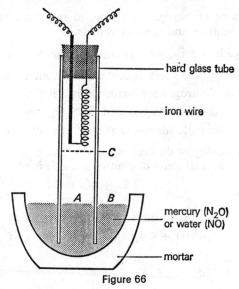

— hard glass tube

— iron wire

C

A *B*

— mercury (N_2O)
or water (NO)

— mortar

Figure 66

After about 20 minutes the electrical current is switched off and the tube allowed to cool. Water-level *A* rises towards *C*.

After the tube has been transferred to a deep vessel and lowered until the levels *C* and *B* are equal, the volume of nitrogen is measured and is found to be one-half of the original volume of nitrogen oxide. That is, 1 volume of nitrogen is contained in 2 volumes of nitrogen oxide (temperature and pressure constant).

Using Avogadro's Law, we may substitute 'molecules' for 'volumes'.

∴ 1 molecule of nitrogen is contained in 2 molecules of nitrogen oxide

∴ 2 atoms of nitrogen are contained in 2 molecules of nitrogen oxide

∴ 1 atom of nitrogen is contained in 1 molecule of nitrogen oxide

∴ the formula of nitrogen oxide is NO_x, where *x* is a whole number

∴ the relative molecular mass of nitrogen oxide is $(14 + 16x)$.

The vapour density of nitrogen oxide is 15.

∴ the relative molecular mass of nitrogen oxide is 30.

$$\therefore 14 + 16x = 30$$
$$x = 1$$

∴ formula of nitrogen oxide is NO

$$3Fe(s) + 4NO(g) \rightarrow Fe_3O_4(s) + 2N_2(g)$$

Note. Pure nitrogen oxide for this experiment is conveniently prepared by half-filling a small flask with iron(II) sulphate crystals, covering them with dilute sulphuric acid, warming, and dropping sodium nitrite solution into the mixture from a tap-funnel.

Dinitrogen oxide

Exactly the same experiment is performed as for nitrogen oxide except that the gas must be confined over mercury because dinitrogen oxide is fairly soluble in water. The hard glass tube should be only about half-filled with the gas at first. It is found that the volume of nitrogen left is equal to the volume of dinitrogen oxide taken.

∴ 1 volume of nitrogen is contained in 1 volume of dinitrogen oxide (temperature and pressure constant).

Using Avogadro's Law, we may substitute 'molecules' for 'volumes'.

∴ 1 molecule of nitrogen is contained in 1 molecule of dinitrogen oxide

∴ 2 atoms of nitrogen are contained in 1 molecule of dinitrogen oxide

∴ the formula of dinitrogen oxide is N_2O_x, where x is a whole number

∴ the relative molecular mass of dinitrogen oxide is $(28 + 16x)$.

The vapour density of dinitrogen oxide is 22.

∴ relative molecular mass of dinitrogen oxide is 44.

$$\therefore 28 + 16x = 44$$
$$\therefore x = 1$$

∴ the formula of dinitrogen oxide is N_2O

$$3Fe(s) + 4N_2O(g) \rightarrow Fe_3O_4(s) + 4N_2(g)$$

Hydrogen sulphide

A convenient volume of hydrogen sulphide is confined over mercury at atmospheric pressure (Figure 67). By means of an induction coil, electric sparks are passed between the platinum wires for some time. This decomposes the hydrogen sulphide into its elements. Solid sulphur (of negligible volume) is deposited and hydrogen is left. It is found that, when the tube has cooled, the volume of hydrogen left is exactly equal to the volume of hydrogen sulphide taken (temperature and pressure constant).

∴ 1 volume of hydrogen is contained in 1 volume of hydrogen sulphide (temperature and pressure constant).

Applying Avogadro's Law, we may substitute 'molecules' for 'volumes'.

∴ 1 molecule of hydrogen is contained in 1 molecule of hydrogen sulphide

∴ 2 atoms of hydrogen are contained in 1 molecule of hydrogen sulphide

∴ the formula of hydrogen sulphide is H_2S_x, where x is a whole number

∴ the relative molecular mass of the gas is $(2 + 32x)$.

The vapour density of hydrogen sulphide is 17.

∴ the relative molecular mass of hydrogen sulphide is 34.

$$\therefore 2 + 32x = 34$$
$$x = 1$$

∴ the formula of hydrogen sulphide is H_2S

$$H_2S(g) \rightarrow H_2(g) + S(s)$$

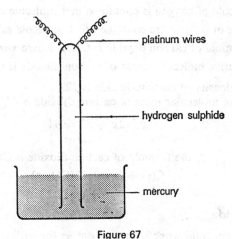

platinum wires

hydrogen sulphide

mercury

Figure 67

Carbon dioxide

The apparatus of Figure 68 is used. A little *dry* powdered charcoal (0.02 g is required for 40 cm³) is placed in the *dry* tube, and *dry* oxygen is passed through the whole apparatus for 3–4 minutes. The clip is closed and mercury poured into

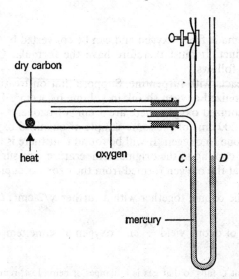

dry carbon

heat   oxygen   C   D

mercury

Figure 68

the manometer. The clip is momentarily released to make the pressure in the tube atmospheric. The graphite is heated with a small flame, and it burns (not always obviously) to carbon dioxide. Expansion causes the level at C to fall, and the level at D to rise,* but, when the bulk has cooled, the levels at C and D return to their original positions, that is, the volume of carbon dioxide formed is equal to the volume of oxygen used.

∴ 1 volume of oxygen is contained in 1 volume of carbon dioxide (temperature and pressure constant)

Applying Avogadro's Law, we may substitute 'molecules' for 'volumes'.

∴ 1 molecule of oxygen is contained in 1 molecule of carbon dioxide

∴ 2 atoms of oxygen are contained in 1 molecule of carbon dioxide

∴ the formula of carbon dioxide is C_xO_2, where x is a whole number

∴ the relative molecular mass of carbon dioxide is $(12x + 32)$.

The vapour density of carbon dioxide is 22.
∴ the relative molecular mass of carbon dioxide is 44.

$$\therefore 12x + 32 = 44$$
$$x = 1$$

∴ the formula of carbon dioxide is CO_2

$$C(s) + O_2(g) \rightarrow CO_2(g)$$

Sulphur dioxide

The same experimental work is carried out as for carbon dioxide, substituting sulphur for carbon. As in the case of the carbon dioxide the volume of sulphur dioxide formed is equal to the volume of oxygen used.

The reasoning to obtain the formula SO_2 is the same as for carbon dioxide, using the vapour density of sulphur dioxide, 32, and the relative atomic mass of sulphur, $S = 32$.

Ozone

Ozone can be made from oxygen and can be converted by heat into oxygen and no other product. It must therefore have the formula, O_n. The value of n is established as follows:

1. Ozone reacts with turpentine. Suppose that on treatment with turpentine, 200 cm^3 of ozonized oxygen shrink in volume by $x \text{ cm}^3$. This means that in the 200 cm^3 of ozonized oxygen there are $x \text{ cm}^3$ of ozone.

2. Another 200 cm^3 of the same sample of ozonized oxygen are heated, converting the ozone to oxygen. It will be found that there is an increase in volume of $x/2 \text{ cm}^3$ on cooling to the original temperature, pressure remaining constant. This means that the oxygen formed from the ozone occupies the $x \text{ cm}^3$ formerly occupied by the ozone, together with a further $x/2 \text{ cm}^3$, *i.e.*, $\dfrac{3x}{2} \text{ cm}^3$ in all.

∴ $x \text{ cm}^3$ of ozone yield $\dfrac{3x}{2} \text{ cm}^3$ oxygen at same temperature and pressure.

---

* If the level at C falls so that gas is in danger of being lost, remove the burner and close the mouth of tube D lightly with the finger.

Or 2 volumes of ozone yield 3 volumes oxygen at same temperature and pressure.

$\therefore$ By Avogadro's Law,

2 molecules of ozone yield 3 molecules of oxygen,

$$2O_n = 3O_2$$
$$\therefore n = 3$$

and the formula of ozone is O_3

Carbon monoxide

A measured volume of carbon monoxide, confined over mercury in a eudio-meter tube, is mixed with a measured volume of oxygen equal to several times its own volume. The mixture is exploded by a spark passed between platinum leads sealed through the glass. After cooling, the volume of residual gas is measured and the carbon dioxide is absorbed by allowing some concentrated caustic potash solution to rise above the mercury. (The potash solution is introduced at the bottom of the tube by means of a small pipette bent at the tip.) The diminution in volume caused by the potash represents the volume of carbon dioxide formed. The residual volume is the excess oxygen, and, by subtracting this from the original volume of oxygen taken, the volume of oxygen used up is obtained. (All measurements are taken at room temperature and atmospheric pressure.)

It will be found that:

2 volumes of carbon monoxide combine with 1 volume of oxygen to form 2 volumes of carbon dioxide.

Using Avogadro's Law, we may substitute 'molecules' for 'volumes'. Then, 2 molecules of carbon monoxide combine with 1 molecule of oxygen to form 2 molecules of carbon dioxide.

$\therefore$ 1 molecule of carbon monoxide contains $\frac{1}{2}$ molecule less oxygen than one molecule of carbon dioxide

$\therefore$ 1 molecule of carbon monoxide contains one atom of oxygen less than one molecule of carbon dioxide

But the formula of carbon dioxide is CO_2.

$\therefore$ the formula for carbon monoxide is CO.

Note. Since the reactants and products are all gaseous the vapour density is not required.

Questions

1. How would you show experimentally that two volumes of hydrogen combine with one volume of oxygen to yield two volumes of steam? What deductions can be drawn from these facts as to the number of atoms in the oxygen molecule? (C.)

2. Complete the following statements:

(a) Two volumes of hydrogen unite with —— volume(s) of oxygen to give —— volume(s) of steam, all the substances being at 100 °C and atmosphere pressure.

(b) Two volumes of hydrogen unite with —— volume(s) of nitrogen to give —— volume(s) of ammonia, all the gases being at the same temperature and pressure.

State the law which these statements illustrate, and give the name of its author.

Explain clearly how the formulae of steam and ammonia follow from the statements you have made. (J.M.B.)

3. Describe two experiments from the results of which the formula CO_2 for carbon

dioxide may be derived. Point out clearly what assumptions are made in the deduction of the formula.

4. Describe the preparation and principal properties of carbon monoxide. What is the evidence on which the formula CO is assigned to this gas? (L.)

5. Give an account of the experimental evidence on which the accepted formula for hydrogen chloride is based. 223 cm³ of hydrogen chloride and 250 cm³ of gaseous ammonia, both measured at 13 °C and 770 mm, were mixed. Calculate the mass of the solid product. (Cl = 35.5, N = 14.) (L.)

6. State Gay-Lussac's Law of Gaseous Combination, illustrating your answer by reference to *four* examples.

How would you show experimentally that a given volume of hydrogen sulphide contains twice as much hydrogen as an equal volume of hydrogen chloride? (L.)

7. (i) 15 cm³ of methane, CH₄, are exploded with 80 cm³ of oxygen, both gases being at room temperature and at atmospheric pressure. Calculate the volume which each of the residual gases would occupy after adjusting to the original conditions of temperature and pressure.

(ii) State an important commercial source of methane. (O. and C.)

8. Selenium (symbol Se) is a solid non-metallic element. It forms a gaseous compound with hydrogen called hydrogen selenide. This question is concerned with an investigation of hydrogen selenide.

(a) Analysis of hydrogen selenide shows that 2.47% by mass of the compound is hydrogen. What information does this give about the formula of hydrogen selenide?

(b) The mass of 96 cm³ of hydrogen selenide at room temperature and pressure is found to be 0.324 g. What information does this give about the formula of hydrogen selenide? (Assume that 1 mole of gas molecules occupy 24.0 dm³ at room temperature and pressure.)

(c) Hydrogen selenide can be decomposed completely into hydrogen and selenium by passing the gas through a heated tube. Write an equation for this decomposition.

(d) Calculate the volume of hydrogen which would be obtained by the complete decomposition of 100 cm³ of hydrogen selenide (all gaseous volumes being measured at room temperature and pressure).

(e) You have been asked to devise an experiment to find if your answer to part (d) of this question is correct. Sketch the apparatus you would set up, and describe briefly how you would use it. (L.)

9. Describe in detail how, in the laboratory, you would prepare and collect dry ammonia starting with a suitable ammonium salt.

Ammonia decomposes if sparked electrically. What would you expect to be the products of the decomposition? What percentage increase in volume would you expect to take place?

If the percentage increase in volume was in fact less than the value which you give for (b), what explanation would you offer? (L.)

10. 10 cm³ of a gaseous hydrocarbon were mixed with 30 cm³ of oxygen and the mixture was exploded. After the mixture had cooled to room temperature, 20 cm³ of gas remained. After shaking this gas with sodium hydroxide solution, its volume was reduced to 10 cm³. The remaining gas rekindled a glowing splint.

(a) Name the gas remaining at the end.

(b) What were the reacting volumes of the hydrocarbon and oxygen?

(c) Name the gas absorbed by the alkali.

(d) State the volume of the gas named in (c).

(e) Name the other product of the reaction.

(f) Work out the formula for the hydrocarbon. (J.M.B.)

11. In this question, all volumes were measured at room temperature and pressure.

(a) When 100 cm³ of a gaseous hydrocarbon, P, were heated in the presence of a catalyst with 100 cm³ of hydrogen, a single new gaseous compound, Q, was formed, there being no P and no hydrogen left over. Explain clearly what this information tells you about the gas P.

(b) When 100 cm³ of the gas Q were mixed with an excess of oxygen and an electric spark passed, there was an explosion; 350 cm³ of the oxygen had reacted, and 200 cm³ of carbon dioxide had been formed. The steam which was the only other product of the explosion had condensed to a drop of water of negligible volume. Explain clearly what this information tells you about the gas Q.

(c) Give the name and formula of the gas P.

(d) Describe a simple chemical test which will distinguish between P and Q.

(e) The gas P is an important raw material in industry. Name one substance that is manufactured from P, state briefly how the reaction is carried out, and give one important use of the substance manufactured. (L.)

12. Describe carefully any experiment which you have seen or read about, the aim of which is to determine the volume composition and the formula of a gas. (N.I.)

14 Volumetric Analysis

Units of volume measurement

This chapter is concerned with the measurement of volumes of aqueous solutions. The recent introduction of the so-called S.I. units (Système International d'Unités, or International System of Units) into scientific work has brought about a change in many of the units used for scientific measurements. In particular the units in which volumes of substances are measured have been affected, and it is now required that all volume measurements be based ultimately on the *metre* as a standard of linear measure. Thus the litre (l) and millilitre (ml) have been replaced by the *cubic decimetre* (dm^3) and the *cubic centimetre* (cm^3) respectively. In fact, it should be appreciated that the ml and the cm^3 are identical, but cm^3 is preferred and should therefore be used.

Standard and molar solutions

Volumetric analysis is a means of estimating quantities of certain materials (often acids or alkalis) by an analytical process which involves measurement of *volumes* of solutions, using pipettes, burettes, and (for approximate measurements) measuring cylinders. Weighings may also be involved.

Definition. A standard solution is a solution of known concentration.

For example, a solution known to contain, say, 12 g of sodium chloride in one dm^3 of solution is a standard solution.

The system now generally approved for volumetric work is based upon the *molar* (M) solution. Such a solution of a given compound contains the relative molecular mass in grammes (or one mole) of the compound in one cubic decimetre (dm^3) of the solution.

Definition. A molar (M) solution of a compound is one which contains one mole of the compound in one cubic decimetre (dm^3) of the solution.

For example, since $C_6H_{12}O_6 = 180$, a molar solution of glucose contains 180 g in 1 dm^3. In the case of strong electrolytes such as sodium chloride or sodium carbonate, which do not exist as molecules to any significant extent in dilute solution, the expression *molar mass* is often used as equivalent to one mole. For example, the molar mass of Na_2CO_3 is taken as 106 g ($Na_2CO_3 = 106$) and molar sodium carbonate solution contains 106 g of the anhydrous salt in one dm^3 of solution. In fact, the dissolved salt is almost entirely dissociated into ions (in dilute solution) in the proportion of $2Na^+$ to CO_3^{2-}. Also, with many salts, hydration of ions occurs in solution.

Molar solutions of some compounds commonly used in titration contain the following masses of the compounds in 1 dm^3 of solution:

sodium hydroxide NaOH	40 g
potassium hydroxide KOH	56 g

sulphuric acid H_2SO_4	98 g
hydrochloric acid HCl	36.5 g
sodium carbonate Na_2CO_3	106 g
sodium hydrogencarbonate $NaHCO_3$	84 g

The figures quoted are the molar masses of the respective compounds.

Derivative concentrations are also used, *e.g.*, 0.1M, 0.5M, 2M, and these contain one-tenth of, half of, and twice (respectively), the mass of solute present in an equal volume of the corresponding M solution.

Standard solutions in acid–base reactions

In the present chapter we shall be concerned entirely with acid–base reactions, as these are probably the most common types of reaction in volumetric analysis at an elementary level. However, it must be appreciated that it is possible to have a standard solution of many different types of substance, *e.g.*, a standard solution of an oxidizing agent or a standard solution of a reducing agent.

It is obvious that some accurately standard solution is required as a starting-point for volumetric estimations. Actually few compounds are suitable for the *direct* preparation of an accurately standard solution. Some compounds absorb water from the air and, for this reason, cannot be weighed out accurately without excessively difficult precautions, *e.g.*, sodium hydroxide, potassium hydroxide, and concentrated sulphuric acid. Others react with carbon dioxide of the air, *e.g.*,

$$2NaOH(s) \text{ (or } 2KOH) + CO_2(g) \rightarrow Na_2CO_3(s) \text{ (or } K_2CO_3) + H_2O(l)$$

Solutions may contain volatile constituents and be liable to change slowly in concentration during ordinary use, *e.g.*, concentrated hydrochloric acid and ammonia.

A compound which is commonly utilized for the direct preparation of an accurately standard solution is *anhydrous sodium carbonate*. It is best made from sodium hydrogencarbonate of high purity (which can be bought from chemical suppliers). This is done by heating the sodium hydrogencarbonate *to constant mass*, which ensures completion of the decomposition:

$$2NaHCO_3(s) \rightarrow Na_2CO_3(s) + H_2O(l) + CO_2(g)$$

The anhydrous sodium carbonate so formed is very pure and can be used in ordinary weighings with no appreciable change of composition.

Experiment 30
Preparation of a standard solution of sodium carbonate (0.1M concentration)

As stated above, molar sodium carbonate solution contains 106 g of solute in 1 dm³ (since $Na_2CO_3 = 106$). Consequently, a 0.1M sodium carbonate solution contains 10.6 g of solute in 1 dm³. The usual volume of solution made up in a laboratory experiment is 250 cm³ because this volume allows *three* titrations of 25 cm³ each, with a reserve for required washings of pipette and possible titration failures. For 250 cm³ of 0.1M solution, the mass of anhydrous sodium carbonate required is one-quarter of 10.6 g, or 2.65 g.

Put about 8 g of pure sodium hydrogen carbonate into a dry, clean evaporating dish and heat it with stirring for, say, 15 minutes. Place it in a desiccator to cool, then weigh it. *Without further stirring* (to prevent loss of solid), heat the dish again for five minutes or so, cool it in the desiccator and weigh, and so on until two consecutive weighings are the same. Store the dish and pure sodium carbonate in the desiccator.

Weigh a watch-glass (or, better, a stoppered weighing bottle) and weigh out, on to it exactly 2.65 g of anhydrous sodium carbonate. Tap the carbonate into a beaker (with lip) of about 400 cm³ capacity, containing about 50 cm³ of hot distilled water. Wash down the watch-glass (or weighing bottle) with a jet of hot distilled water from a wash-bottle and allow the washings to fall into the beaker (Figure 69). This procedure should wash *all* the sodium carbonate into the beaker. Stir with a glass rod till dissolution of the solid is complete, then cool the solution to room temperature. Leave the rod standing in the solution.

Place a very thin smear of Vaseline under the lip of the beaker and pour the solution down the glass rod into a measuring flask (250 cm³).

Wash the beaker out at least twice with jets of cold distilled water directed round the sides and pour the washings down the glass rod into the measuring flask (Figure 70). Shake the flask gently. Fill it up with cold distilled water almost to the mark, then add more distilled water *drop by drop* from a pipette till the lowest level of the meniscus is on the mark when at eye-level (to avoid parallax error). Stopper the measuring flask and shake well. The liquid should then be exactly 0.1M sodium carbonate solution.

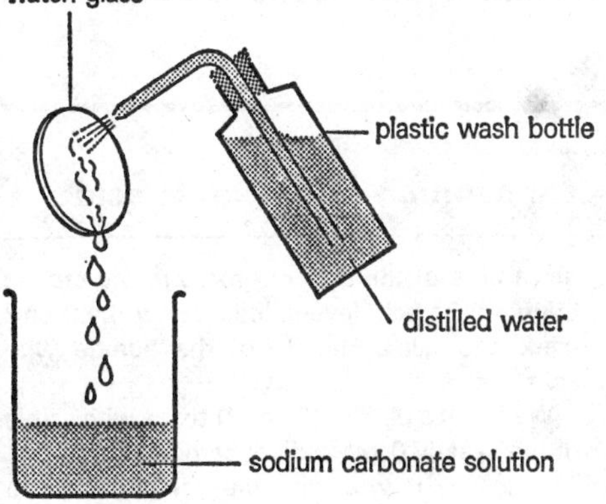

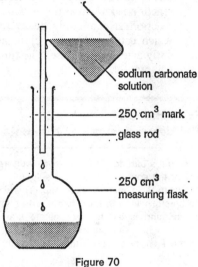

watch glass

plastic wash bottle

distilled water

sodium carbonate solution

Figure 69

sodium carbonate solution

250 cm³ mark

glass rod

250 cm³ measuring flask

Figure 70

The preparation of a standard solution of sulphuric acid cannot be carried out directly because concentrated sulphuric acid absorbs water rapidly from the air and is never reliably pure. A solution is prepared which is a little above 0.1M in concentration and it is then standardized and diluted with distilled water to exactly 0.1M.

A molar solution of sulphuric acid contains 98 g of pure acid in 1 dm³, so the 0.1M acid contains 9.8 g of the acid in 1 dm³. The concentrated acid has a density of about 1.8 g cm⁻³, so 9.8 g of it occupy about 5.5 cm³.

Thus, the preparation of a standard solution of sulphuric acid involves (*a*) diluting a concentrated solution of sulphuric acid to an approximate molarity, and (*b*) the estimation of the dilute acid solution using a previously standardized alkali solution.

Experiment 31
Dilution of concentrated sulphuric acid solution

Note: Safety goggles should be worn for this experiment.

With great care, because of the dangerously corrosive nature of the acid, take 5.5–6.0 cm³ of concentrated sulphuric acid in a small measuring cylinder; pour it, with stirring, into about 100 cm³ of cold distilled water in a beaker (with lip). Pour this solution into, say, 700 cm³ of cold distilled water in a measuring flask of capacity 1000 cm³, wash out the beaker with cold distilled water twice and add the washings to the measuring flask. Then add distilled water approximately to the mark on the measuring flask, stopper it, and shake well. This should give sulphuric acid of concentration a little above 0.1 M. It is now standardized with the 0.1 M sodium carbonate solution prepared above.

The estimation of the concentration of a solution of an acid by reaction with standard alkali solution is known as a *titration*, when carried out according to the process described in Experiment 32 which follows. The term titration also applies to the estimation of the concentration of a solution of an alkali by reaction with a standard acid solution. The point at which the acid has been added to the alkali in the mole proportions indicated by the equation for the reaction is known as the *end-point* of the reaction. The end-point of an acid–base reaction is commonly determined using a substance known as an *indicator*, which usually changes colour according to whether the solution is predominantly acidic or alkaline. The function of the indicator is more fully explained in Chapter 24.

Experiment 32
The determination of the molarity of a solution of sulphuric acid

Wash out a burette (50 cm³) twice with a few cm³ of the approximately 0.1 M sulphuric acid and run the acid out through the tap. (This leaves the burette wet with the liquid it is to contain and nothing is left to contaminate the acid solution when the burette is filled.) Fill the burette 1 cm or so above the 0 cm³ mark and run a little of the acid out *through the tap,* so bringing the acid level a little below the 0 cm³ mark and filling the tip of the burette with acid.

Wash out a pipette (25 cm³) twice with a little of the exactly 0.1 M sodium carbonate solution. This leaves it wet with the solution it is to measure. Draw this solution into the pipette above the mark and, with the mark at eye-level, allow the solution to run out very slowly till the lowest level of the meniscus is on the mark. Then allow the liquid to run from the pipette into a conical flask (Figure 71) which has been washed out with cold distilled water only and touch the tip of the pipette on to the surface of the liquid. (Do *not* blow out the last drop.) The flask will then contain exactly 25 cm³ of 0.1 M sodium carbonate solution (with a little distilled water which is of no account).

Add two drops of *methyl orange* solution (indicator) to the conical flask. This will turn the alkaline solution *yellow.* Read and note the level of acid in the burette. Run acid into the conical flask 2 or 3 cm³ at a time, with shaking, allowing a short drainage time between additions, until the liquid flashes pink. Then add the acid more slowly, eventually *drop by drop*, until the colour of the liquid is *orange* (*i.e.,* the neutral colour for methyl orange). This is the end-point of the titration. Slight excess of acid will turn it fully pink. Read and note the acid level. Repeat the titration to obtain at least two results which differ by not more than 0.1 cm³.

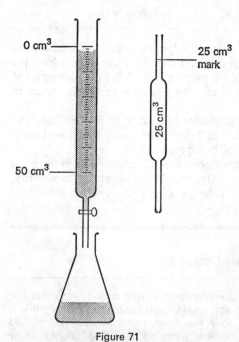

0 cm³

50 cm³

25 cm³ mark

25 cm³

Figure 71

Specimen readings

Volume 0.1M sodium carbonate solution used each time = 25.0 cm³.

Titration	1	2	3
2nd burette reading	24.0	23.9	24.0 cm³
1st burette reading	0.2	0.3	0.4 cm³
Volume of acid added	23.8	23.6	23.6 cm³

Since the first titration may be regarded as a 'trial run' it is usual in such experiments to neglect the reading and to average the subsequent titration results.

Therefore:

Volume of acid added at the end-point = 23.6 cm³

Calculation

The first step in calculating the molarity of any solution from the results of an acid–base titration is to write the equation for the reaction taking place.

$$Na_2CO_3(aq) + H_2SO_4(aq) \rightarrow Na_2SO_4(aq) + CO_2(g) + H_2O(l)$$

It can be seen from the equation that since the sodium carbonate and sulphuric acid react in *equi*molar proportions then the number of moles of sodium carbonate in 25.0 cm³ of solution is equal to the number of moles of sulphuric acid added at the end-point.

Hence, since the sodium carbonate solution is 0.1M:

1 dm³ sodium carbonate solution contains 0.1 mole sodium carbonate

$\therefore$ 25 cm³ sodium carbonate solution contains $\dfrac{0.1 \times 25}{1000}$ mole sodium carbonate

$$= 0.0025 \text{ mole}$$

Now, 23.6 cm³ sulphuric acid solution is required for completion of reaction, therefore:

23.6 cm³ sulphuric acid solution contain 0.0025 mole sulphuric acid

$\therefore$ 1 dm³ sulphuric acid solution contains $\dfrac{0.0025 \times 1000}{23.6}$ mole sulphuric acid

$$= 0.106 \text{ mole}$$

Hence, the concentration of the sulphuric acid solution is 0.106M.

To make the acid exactly equal in molar concentration to the sodium carbonate solution, 23.6 cm³ of the acid must be diluted to 25.0 cm³, *i.e.*, 1.4 cm³ of distilled water must be added to 23.6 cm³ of the acid. If 920 cm³ of the acid are left, it can be made exactly 0.1M by the addition of $920 \times \dfrac{1.4}{23.6}$ cm³ of distilled water, or 55 cm³. After the addition of this volume of distilled water, the acid should be shaken and should then be exactly 0.1M. This can be tested by titration of the 0.1M sodium carbonate solution, when 25.0 cm³ of the acid should be required to neutralize 25.0 dm³ of the carbonate solution.

These two standard alkaline and acidic solutions can be used to standardize other solutions, *e.g.*, sodium hydroxide, potassium hydroxide, hydrochloric acid, and, by their use, a wide range of acid–alkali estimations can be carried out.

Choice of indicators in acid–base titrations

When the technique of acid–base titrations is extended to a wide variety of acidic and alkaline solutions, care needs to be taken over the choice of indicator for any given reaction. The choice of an inappropriate indicator would lead to an incorrect result and it is therefore extremely important that the indicator is chosen carefully. The basis on which a choice of indicator is made concerns the *strength* of the acid or base involved in the reaction. (*Note:* the *strength* of an acid or base is not to be confused with the *concentration* of its solution; the exact meaning of the term 'strength' will be explained later.) Examples of 'strong' and 'weak' acids and bases are given in the table below.

ACID		BASE	
Strong	*Weak*	*Strong*	*Weak*
Hydrochloric HCl	Ethanoic (acetic) CH_3COOH	Sodium hydroxide $NaOH$	Ammonia NH_3
Nitric NHO_3	Methanoic (formic) $HCOOH$	Potassium hydroxide KOH	
Sulphuric H_2SO_4	Carbonic H_2CO_3	Sodium carbonate Na_2CO_3	

There are three common indicators which are used in titration experiments involving acids and bases; these are methyl orange, litmus, and phenolphthalein. Other indicators in less common use are malachite green, thymol blue, bromocresol green, and bromothymol blue. The table below shows the colours which each of these indicators takes up in acid or alkaline solution.

	Colour of indicator	
Indicator	*acid solution*	*alkaline solution*
Methyl orange	pink	yellow
Litmus	red	blue
Phenolphthalein	colourless	pink
Malachite green	yellow	blue/green
Thymol blue	red	yellow
Bromocresol green	yellow	blue
Bromothymol blue	yellow	blue

Indicators which are suitable for particular types of acid–base reactions are also given in the table below.

Acid–base titration	*Example*	*Choice of indicator*
Strong acid/strong base	H_2SO_4 and $NaOH$	Any indicator
Weak acid/strong base	CH_3COOH and KOH (ethanoic)	Phenolphthalein
Strong acid/weak base	HCl and NH_3	Methyl orange
Weak acid/weak base	CH_3COOH and NH_3	No satisfactory indicator available

Questions

1. Describe how you would transfer exactly 25.0 cm³ of a molar solution of sodium carbonate to a conical flask, using a pipette. How would you then set up and fill a wet burette with a given sample of dilute sulphuric acid and titrate the acid against the sodium carbonate solution? Mention throughout the precautions you would observe to secure accuracy and state the name and colour change of the indicator you would use.

2. It is required to determine the concentration of a sodium hydroxide solution by titration against 0.115M hydrochloric acid. A suitable indicator is provided, but it is one whose colour change you do not know. Describe carefully how you would investigate the colour change of the indicator, perform the titration and use the results to determine the concentration of the alkali solution.

Suppose that you found that 25.00 cm³ of alkali were neutralized by 20.50 cm³ of the hydrochloric acid. Calculate the concentration of the alkali solution. (J.M.B.)

3. Describe the experiments you would perform to confirm, as conclusively as possible, that a given liquid was correctly labelled 0.10M nitric acid.

Explain the reason for each of your experiments and the results you would expect to obtain from them. (N.I.)

4. You are provided with 2M sodium hydroxide solution and 1M sulphuric acid together with all equipment necessary for volumetric analysis. Describe fully how you would prepare crystalline samples of *two* different sodium salts of sulphuric acid. Describe carefully *two* ways by which you can prove that you have made two different substances. (C.)

5. Provided with sodium hydrogencarbonate (sodium bicarbonate), describe carefully how you could convert a suitable quantity completely into sodium carbonate and then make up 1 litre of 0.100M sodium carbonate solution.

You are given a burette that delivers a drop of volume 0.05 cm³. Calculate the percentage error that may arise due to the burette reading when you use the solution you have prepared to standardize a solution of sulphuric acid. The acid is known to be very nearly 0.100M and you use 20.0 cm³ portions of it. Name the indicator, state the colour change and assume that each drop of solution delivered by the burette near the end-point produces a detectable colour change. (C.)

6. 0.5 mole hydrogen chloride was dissolved in water and the solution made up to 1 dm³; 30.0 cm³ of this solution was found to neutralize exactly 25.0 cm³ of a solution of sodium carbonate. The equation for this reaction is

$$Na_2CO_3 + 2HCl \rightarrow 2NaCl + H_2O + CO_2.$$

Calculate the concentration of the alkali in moles per dm³ and the mass concentration in g Na_2CO_3 per dm³. (O.)

7. Give two chemical properties in each case (apart from their action on indicators) which are typical of (i) a dilute acid, (ii) a dilute aqueous alkali. Illustrate your answer by specific examples.

Describe carefully how you would determine the molarity of dilute aqueous sodium hydroxide, using hydrochloric acid of known molarity. (Only the experimental procedure should be described in this part of the question.)

If the alkali contained 1.50 g of sodium hydroxide in 250 cm³ of solution and the acid was 0.1 molar, what volume of acid would be needed to react with 20 cm³ of alkali? (L.)

8. What mass of sodium hydroxide would be needed to neutralize exactly 200 cm³ of a solution containing 49 grammes of sulphuric acid per dm³? (S.)

9. The following statement summarizes the results of a quantitative exercise involving an alkali and a standard acid solution:

0.12M aqueous nitric acid was titrated against 25.0 cm³ of aqueous sodium hydroxide contained in a conical flask. 22.5 cm³ of the acid was required to react completely with the base.

(a) (i) What is meant by a *standard* acid solution?
   (ii) Write a balanced equation for the reaction between the nitric acid and the sodium hydroxide.
(b) (i) Explain briefly why the volumes of the acid and the alkali are stated to one decimal place.
   (ii) Name a piece of apparatus which is suitable for measuring the nitric acid.
   (iii) In addition to making careful observations, how did the experimenter satisfy himself that the correct volume of acid was 22.5 cm³?
   (iv) Name one indicator, other than litmus or universal indicator in liquid or paper form, which would be suitable in the titration, and state the colour change.
(c) (i) Calculate the number of moles of nitric acid, HNO_3, involved in the complete reaction.

(ii) Calculate the molarity of the sodium hydroxide solution. (N.I.)

10. What volume of 0.100M nitric acid (*a*) contains 31.5 g of pure nitric acid, (*b*) neutralizes 25 cm³ of 0.125M sodium hydroxide solution, (*c*) neutralizes the solution formed by adding 25 cm³ of M sodium hydroxide solution to 20 cm³ of M hydrochloric acid? (J.M.B.).

15 Calculations involving Solutions

A high proportion of the chemistry investigated in school laboratories concerns reactions which take place in aqueous solution, and calculations involving the substances present in solution are often required. The purpose of this chapter is to present a few examples of the types of calculation which are common in chemistry courses, and to indicate the approach to the solving of them. Use will sometimes be made of the ideas already discussed in Chapters 7 and 12 dealing with calculations involving masses and gas volumes respectively, so the student must be prepared to turn back to those chapters where necessary.

As with the calculations concerning gas volumes, it is as well to bear a few ideas in mind in dealing with aqueous solutions of substances, and these are summarized below.

Summary of data on aqueous solutions

(1) The *concentration* of a substance in solution is a measure of the amount of substance dissolved in a known volume of solution, and is usually expressed in units of $g\ cm^3$ (grammes of substance per cubic centimetre of solution) or *mol* dm^{-3} (moles of substance per cubic decimetre of solution).

(2) A *standard* solution is one for which the concentration is known.

(3) A *molar* solution is one which contains one mole of substance in one dm^3 of *solution*.

Bearing these few simple ideas in mind we may now consider a variety of calculations which utilize them.

Conversion of a given mass of substance to a volume of solution of known molarity

EXAMPLE 1. *What volume of 0.25M solution of sodium hydroxide will contain 5 g of solute?*

We first of all need to calculate the molar mass of sodium hydroxide NaOH. Using the table of relative atomic masses on page 499,

$$\text{molar mass NaOH} = 23 + 16 + 1 = 40 \text{ g}$$

Hence 5 g of NaOH represent $\frac{5}{40} = 0.125$ mole.
Now the solution is 0.25M.

∴ 0.25 moles sodium hydroxide are contained in 1 dm^3 solution

Hence,

0.125 moles sodium hydroxide will be contained in $\dfrac{1000 \times 0.125}{0.25}$ cm^3 solution

$$= 500 \text{ cm}^3 \text{ solution}$$

Conversion of volumes of solution of known molarity to amounts of substances

EXAMPLE 2. *What mass of glucose, $C_6H_{12}O_6$, is contained in 40 cm^3 of 0.2M solution?*

We know that 1 dm^3 of glucose solution contains 0.2 mole glucose.

$$\therefore \text{40 cm}^3 \text{ solution contain } \frac{0.2 \times 40}{1000} \text{ mole glucose}$$

$$= 0.008 \text{ mole glucose}$$

Now the relative molecular mass of glucose, $C_6H_{12}O_6$, is

$$(6 \times 12) + (12 \times 1) + (6 \times 16) = 72 + 12 + 96 = 180.$$

Hence,

$$0.008 \text{ mole glucose corresponds to } 0.008 \times 180 \text{ g glucose}$$
$$= 1.44 \text{ g glucose}$$

Hence 40 cm^3 0.2M solution contains 1.44 g glucose

Conversion of volumes of solution to gas volumes

EXAMPLE 3. *What volume of dry HCl gas at s.t.p. will dissolve in 250 cm^3 of water to produce a 0.05M solution of hydrochloric acid? (Molar volume of gas at s.t.p. = 22.4 dm^3.)*

The final solution will have a concentration of 0.05M. Therefore,

$$1000 \text{ cm}^3 \text{ solution will contain 0.05 mole HCl}$$

$$\text{and} \qquad 250 \text{ cm}^3 \text{ solution will contain } \frac{0.05 \times 250}{1000} \text{ mole HCl}$$

$$= 0.0125 \text{ mole HCl}$$

We know that one mole of any gas at s.t.p. occupies a volume of 22.4 dm^3. Therefore, 0.0125 mole of hydrogen chloride gas at s.t.p. will occupy

$$22.4 \times 0.0125 \text{ dm}^3$$
$$= 22.4 \times 0.0125 \times 1000 \text{ cm}^3$$
$$= 280 \text{ cm}^3 \text{ at s.t.p.}$$

Volume of hydrogen chloride gas required at s.t.p. is 280 cm^3

EXAMPLE 4. *Calculate the maximum volume of ammonia gas at s.t.p. which could be obtained by boiling 5 dm^3 of a 0.04M solution of ammonia. (Molar volume of gas at s.t.p. = 22.4 dm^3.)*

We first of all need to calculate how many moles of ammonia are contained in solution. Since the solution is 0.04M,

$$1 \text{ dm}^3 \text{ solution contains 0.04 mole NH}_3$$
$$\therefore 5 \text{ dm}^3 \text{ solution will contain } 0.04 \times 5 = 0.2 \text{ mole NH}_3$$

If all the ammonia can be obtained from the solution by boiling, then 0.2 mole will be evolved.

Since 1 mole of NH_3 at s.t.p. occupies a volume of 22.4 dm^3,

0.2 moles NH_3 at s.t.p. will occupy a volume of $22.4 \times 0.2 = 4.48$ dm^3

> *Hence the maximum volume of ammonia obtained*
> *from the solution will be* 4480 *cm³ at s.t.p.*

In practice, the volume obtained may be rather less than this due to the great solubility of ammonia in water, even at quite high temperatures.

Conversion of gas volumes to solution volumes

EXAMPLE 5. *Calculate the volume of* 2M *hydrochloric acid which could be obtained by dissolving* 560 *cm³ hydrogen chloride gas (measured at s.t.p.) in water. (Molar gas volume* = 22.4 *dm³ at s.t.p.)*

22.4 dm³ hydrogen chloride gas correspond to one mole of hydrogen chloride at s.t.p.

$\therefore$ 560 cm³ hydrogen chloride gas correspond to $\dfrac{1 \times 560}{22\,400}$ mole HCl

$$= 0.025 \text{ mole HCl}$$

Now we require to produce a 2M solution of HCl, and since we only have 0.025 mole it is clear that the volume of water needed will be less than 1 dm³. Hence,

$$\text{volume of water} = \frac{1 \times 0.025}{2} \text{ dm}^3$$

$$= 0.0125 \text{ dm}^3$$

> *Thus* 560 *cm³ hydrogen chloride will produce*
> 12.5 *cm³* 2M *solution of hydrochloric acid*

Calculation of volumes of solution from the equation for the reaction

EXAMPLE 6. *Calculate the volume of* 0.1M *ammonia solution which could be obtained by heating* 2.675 g *ammonium chloride with excess sodium hydroxide, and absorbing all the ammonia evolved.*

First of all we require to write the equation for this reaction.

$$NH_4Cl(s) + NaOH(aq) \rightarrow NaCl(aq) + NH_3(g) + H_2O(l)$$

From the equation it can be seen that one mole of ammonium chloride will give rise to one mole of ammonia gas.

Hence we may write,

1 mole ammonium chloride, NH_4Cl, produces 22.4 dm³ NH_3 at s.t.p.

Since molar mass $NH_4Cl = 14 + 4 + 35.5 = 53.5$, we have $\dfrac{2.675}{53.5}$ mole

NH_4Cl present = 0.05 mole NH_4Cl.

$\therefore$ 0.05 mole NH_3 will be evolved.

In order to produce 0.1M solution we must dissolve the 0.05 mole NH_3 in $\dfrac{1000 \times 0.05}{0.1}$ cm³ solution.

$\therefore$ *Volume of* 0.1M *ammonia solution* = 500 *cm³*

EXAMPLE 7. *27.5 cm³ of a solution of sodium hydroxide neutralize 25.0 cm³ of M hydrochloric acid. Calculate (a) the molarity, (b) the concentration of the sodium hydroxide solution in g dm⁻³.*

The object of the calculation is to fix the molarity and concentration of the sodium hydroxide solution. We know the molarity of the hydrochloric acid; consequently we require to know neither the molar mass of the acid nor its concentration in g dm⁻³.

$$NaOH(aq) + HCl(aq) \rightarrow NaCl(aq) + H_2O(l)$$

1 dm³ of M   1 dm³ of M

40 g

The equation shows that M sodium hydroxide solution and M hydrochloric acid are equivalent to one another volume for volume. Because it requires the *smaller* volume in the titration, the acid used is the *more* concentrated (in molar terms) of the two reagents. That is, the sodium hydroxide solution is $\frac{25.0}{27.5}$M, or 0.909M.

Since the molar mass of sodium hydroxide is 40, the solution must contain 0.909×40 g dm⁻³ or 36.4 g dm⁻³.

EXAMPLE 8. *50 cm³ of M sulphuric acid are added to an excess of solid sodium hydrogencarbonate. Calculate (a) the mass of sodium sulphate produced, (b) the volume of carbon dioxide evolved at 15 °C and 770 mm pressure. ($Na_2SO_4 = 142$; molar volume of gas = 22.4 dm³ at s.t.p.)*

This calculation requires the equation for the reaction and the insertion into it of the quantities of the reagents *involved* in the *units in which they are expressed* in the data given. The sodium hydrogencarbonate is stated as in excess and can be ignored quantitatively. The acid is stated in terms of molarity and is so expressed in the equation. Sodium sulphate is required in grammes and the carbon dioxide in terms of volume. The required reaction statement then becomes:

$$2NaHCO_3(aq) + H_2SO_4(aq) \rightarrow Na_2SO_4(aq) + 2H_2O(l) + 2CO_2(g)$$

1 dm³ of M   142 g   2 × 22.4 dm³
                                    at s.t.p.

From the equation,

1000 cm³ of M H_2SO_4 produces 142 g of sodium sulphate

so   50 cm³ of M H_2SO_4 produces $142 \times \frac{50}{1000}$ g of sodium sulphate

∴ *Mass of sodium sulphate* = 7.1 g

From the equation,

1000 cm³ of M H_2SO_4 produces 44.8 dm³ of CO_2 at s.t.p.

so   50 cm³ of M H_2SO_4 produces $44.8 \times \frac{50}{1000}$ dm³ of CO_2, or

2.24 dm³ at s.t.p.

Converted to 15 °C and 770 mm pressure, this volume becomes:

$$2.24 \times \frac{760}{770} \times \frac{288}{273} \text{ dm}^3, \text{ or } 2.33 \text{ dm}^3$$

∴ *Volume of carbon dioxide at 15 °C and 770 mm pressure* = 2.33 dm³

EXAMPLE 9. *15.0 g of anhydrous sodium carbonate, containing some sodium chloride as impurity, were made up to 250 cm³ of solution and 25.0 cm³ of the solution were titrated by M HCl. 24.5 cm³ of M HCl were required to neutralize the 25.0 cm³ of solution. Calculate the percentage of sodium chloride in the solid.*

For this calculation, the equation is required, with the insertion of the quantity of hydrochloric acid in terms of molarity and of sodium carbonate in grammes. Sodium chloride does not react.

$$Na_2CO_3(aq) + 2HCl(aq) \rightarrow 2NaCl(aq) + H_2O(l) + CO_2(g)$$

106 g   2 dm³ of M

Mass of sodium carbonate in 25.0 cm³ of solution

$$= 106 \times \frac{24.5}{2000} \text{ g}$$

Mass of sodium carbonate in 250 cm³ of solution

$$= 106 \times \frac{24.5}{2000} \times \frac{250}{25.0} \text{ g} = 13.0 \text{ g}$$

Therefore, mass of sodium chloride in the mixture is:

$$(15.0 - 13.0) \text{ g}, \quad \text{or} \quad 2.0 \text{ g}$$

That is, percentage of sodium chloride $= \dfrac{2.0}{15.0} \times 100$

$$= 13.3\%$$

Questions

1. 12.5 cm³ of 0.5M sulphuric acid neutralize 50.0 cm³ of a given solution of sodium hydroxide. What is the molar concentration of the sodium hydroxide solution?

2. 25.0 cm³ of a solution of sulphuric acid required 32.0 cm³ of 0.1M sodium hydroxide solution for neutralization. Calculate the concentration of the acid in g dm⁻³.

3. 25.0 cm³ of 0.05M sulphuric acid neutralized 35.0 cm³ of a potassium hydroxide solution. What is the concentration of the alkali in terms of (a) molar concentration, (b) g dm⁻³?

4. 25.0 cm³ of 0.5M sulphuric acid are mixed with 30.0 cm³ of M sodium hydroxide solution. What volume of 0.1M sulphuric acid will just neutralize the excess of alkali?

5. What volume of a molar solution of sulphuric acid would exactly neutralize 25.0 cm³ of a sodium hydroxide solution containing 60 g of the alkali in 1000 cm³?

6. What volume of 0.1M hydrochloric acid would react exactly with 25.0 cm³ of a sodium carbonate solution containing 5.20 g of anhydrous salt in 1 dm³ of solution?

7. 25.0 cm³ of a solution containing 10.6 g of sodium carbonate in 1000 cm³ required 23.0 cm³ of hydrochloric acid for neutraliza-

tion. Calculate the molar concentration of the acid and its concentration in g dm⁻³.

8. 25.0 cm³ of a solution of sodium hydroxide containing 10.0 g in 1 dm³ required 24.0 cm³ of dilute sulphuric acid for neutralization. Calculate the molar concentration of the acid and its concentration in g dm⁻³.

9. 1.00 g of pure ammonium chloride was boiled with 20.0 cm³ of a solution of sodium hydroxide until the evolution of ammonia had ceased. If the resulting solution required 11.0 cm³ of 0.1M hydrochloric acid for neutralization, calculate (a) the volume of ammonia gas evolved at s.t.p., (b) the molar concentration of the sodium hydroxide solution.

10. 6.16 g of iron completely react with 100 cm³ of 2.2M hydrochloric acid. Write the equation for the reaction and calculate the relative atomic mass of iron. (C.)

11. It was found by titration that 20 cm³ of a 0.5 molar solution of hydrochloric acid reacted exactly with 25 cm³ of a solution of sodium carbonate, the equation for the reaction being

$$Na_2CO_3 + 2HCl = 2NaCl + H_2O + CO_2$$

Calculate the concentration of the carbonate expressing your result (i) as a molarity, (ii) in g Na_2CO_3 per dm³. (O.)

12. What do you understand by a **molar solution**? Describe carefully how you would find the molarity of a solution of sodium hydroxide, given a solution of sulphuric acid of known concentration.

25 cm³ of 0.12M sodium hydroxide were neutralized by 30.0 cm³ of a solution of a dibasic acid (H_2X) containing 6.30 g per dm³. Calculate: (i) the molarity of the acid solution; (ii) the relative molecular mass of the acid. (A.E.B.)

13. 20 cm³ of a hydrochloric acid solution exactly neutralized 25 cm³ of a molar solution of sodium hydroxide. Calculate (a) the molarity of the acid, and its concentration in g dm⁻³, (b) the volume of hydrogen, measured at s.t.p., evolved when 50 cm³ of the same acid react with an excess of magnesium. (J.M.B.)

14. Calculate how many cm³ of 0.5M sulphuric acid will be used in the titration of 25.0 cm³ of a solution which contains 70 g of sodium hydrogencarbonate in 1 dm³ of solution. State the molar concentration of the sodium hydrogencarbonate solution.

15. What is the molarity of a solution prepared by dissolving 3.15 g of nitric acid in water and diluting the solution to 250 cm³?

If 10 cm³ of M hydrochloric acid and 25 cm³ of M sodium carbonate solution were mixed, (i) what volume of carbon dioxide, measured at s.t.p., would be formed, (ii) what mass of sodium carbonate would remain in the final solution?

16. 100 g of an acid were analysed and found to have the following composition:

3.7 g of hydrogen,
37.8 g of phosphorus,
58.5 g of oxygen.

The mass of 1 mole of the acid is 82 g. Use this information to find the formula of the acid.

An aqueous solution of potassium hydroxide was prepared in order to investigate its reaction with this acid. The solution contained 112 g of potassium hydroxide KOH, in 1000 cm³ of solution. Calculate the molarity of the potassium hydroxide.

It was found that 100 cm³ of the potassium hydroxide solution were needed to neutralize completely 100 cm³ of a molar solution of the acid. Show how this information can be used to write an equation for the reaction

between potassium hydroxide and the acid. (L.)

17. Give *two* reasons why an aqueous solution of sodium hydroxide of accurately known concentration cannot be made directly by weighing a given amount of solid sodium hydroxide, dissolving in water, and making up to, say, 1000 cm³.

25.0 cm³ portions of an aqueous solution of sodium hydroxide were titrated with a solution of nitric acid containing 6.30 g of nitric acid in 1000 cm³ of solution. The following table gives the burette readings.

Titration number	Initial reading of burette, in cm³	Final reading of burette, in cm³
1	3.2	30.6
2	1.8	28.8
3	4.8	31.7
4	2.0	29.1

(a) Name an indicator that could be used in these titrations and state its colour change.

(b) Use the results in the table to determine, (i) the concentration of the sodium hydroxide solution in grammes per 1000 cm³ of solution, (ii) the volume of water that would have to be added to 500 cm³ of the solution to give an exactly 0.1M solution.

The equation for the reaction may be written as

$$NaOH + HNO_3 \rightarrow NaNO_3 + H_2O$$

or as $\quad OH^- + H^+ \rightarrow H_2O \qquad$ (C.)

18. (a) $\quad Al_2S_3 + 6H_2O \rightarrow$
$$2Al(OH)_3 + 3H_2S$$

Calculate the maximum volume of hydrogen sulphide, measured at s.t.p., which is evolved when an excess of water is added to 5.0 g of aluminium sulphide.

(b) $\quad Al(OH)_3 + 3HCl \rightarrow$
$$AlCl_3 + 3H_2O$$

What volume of 3M hydrochloric acid will just dissolve 3.12 g of aluminium hydroxide? (J.M.B.)

19. A typical sample of concentrated hydrochloric acid has a density of 1.16 g cm⁻³ and contains 32.0% of hydrogen chloride. What volume of this liquid would be needed to make 2 dm³ of M hydrochloric acid?

16 Electrolysis

We already know from the study of atomic structure, and from the manner in which atoms bond together, that electrons play a most important part in the chemical properties of substances. We should therefore expect that electricity, in the form of an electric current, will have some effect on chemical substances through which it is able to pass. Just as heat, which is a form of energy, has a variety of effects on a wide range of substances, so it is reasonable to expect that electricity, which is also a form of energy, will be as fundamental and as important in its effects on substances as heat has been seen to be. The aim of this chapter is to investigate the effects which electricity has on a range of substances, and to develop a satisfactory explanation of those effects in terms of our present knowledge of atomic structure.

Experiment 33
To investigate the conduction of electricity by a variety of substances in the solid and liquid states

Arrange the circuit shown in Figure 72, in which a supply of 6 volts d.c. is connected via a 6 volt

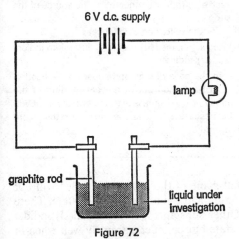

6 V d.c. supply

lamp

graphite rod

liquid under investigation

Figure 72

lamp bulb to a pair of crocodile clips. Use this arrangement to test the conductivity of the substances suggested in the list below. For the solid substances it is sufficient simply to attach the crocodile clips to the ends of the specimen; for the liquids it will be necessary to attach the crocodile clips to graphite rods (known as *electrodes*) which may then be inserted in the liquid as illustrated in Figure 72. In both cases the lighting of the bulb indicates that the circuit is complete, which means that the solid or liquid is a conductor of electricity.
  Try the following substances:

Solids

Small pieces of lead, copper, rubber, plastic, magnesium, paper, salt, sugar.

Liquids

Alcohol (ethanol), water, turpentine.

From the results of your experiments classify the substances you have tested as follows.
  1. If the substance is a solid, and conducts electricity it may be classified as a *conductor*. All metals will be included in this class.

2. If it is a solid, and does *not* conduct electricity then it is known as a *non-conductor* or *insulator*. Many non-metals are included in this class.

3. If the substance in the beaker is a liquid, but it does *not* conduct an electric current, and particularly, does not cause any chemical reaction to take place at the electrodes, it is known as a *non-electrolyte*.

4. If the liquid in the beaker conducts an electric current, so causing the lamp to light, and if the liquid undergoes considerable chemical decomposition around the electrodes, the substance is known as a *strong electrolyte*.

5. If the liquid in the beaker conducts an electric current only slightly, and the corresponding chemical decomposition at the electrodes is slight, the substance is known as a *weak electrolyte*.

Experiment 34
To investigate the conduction of electricity by molten substances

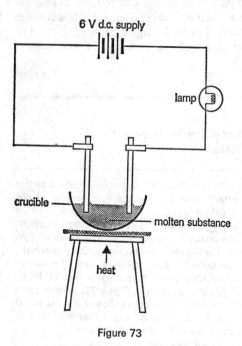

Figure 73

Set up the circuit shown in Figure 73, in which graphite electrodes are to be inserted into a crucible placed on top of a tripod and supported by a pipe-clay triangle. Place the solid substances into the crucible, filling it approximately two-thirds full. *Gently* heat the crucible until the substance just melts, and then arrange the height of the Bunsen flame so that the substance is kept in the molten state for the duration of the experiment. Note whether or not the substance conducts electricity, and pay particular attention to any changes which appear around the electrodes. In the case of the latter observation note also the terminal (positive or negative) to which each electrode is connected at the source of the supply.

Try the following substances:

Wax, sugar, lead(II) bromide, potassium iodide, lead(II) iodide

In the case of wax and sugar *gentle* heating must be carried out to prevent burning of the substance. In the case of the salts in the list it will be necessary to heat the crucible rather more strongly.

From this experiment we can see that some of the substances tested will not conduct electricity either in the solid state or in the molten (liquid) state. These substances are true non-conductors. Other substances, such as lead(II) iodide, do not conduct electricity in the solid state but conduct extremely well when in the molten form; because these substances also undergo considerable decomposition at the electrodes during the passage of the electric current they are therefore classified as strong electrolytes.

One class of substances as yet uninvestigated is that of aqueous solutions, *i.e.*, solutions of salts dissolved in distilled water. The next experiment enables the investigation to be carried out.

Experiment 35
To investigate the conduction of electricity by aqueous solutions

Set up the same circuit as for Experiment 33 (Figure 72) and place in the beaker a sample of distilled water. Note whether the lamp lights or not. Replace the beaker of distilled water by a beaker half full of one of the solutions listed below. Again, note whether electricity is conducted by the solution, and particularly whether any changes appear around the electrodes.

Before testing another solution wash the electrodes with distilled water and wipe them dry with absorbent paper. Record any changes you observe.

Try the following solutions:

Potassium bromide; potassium iodide; sugar; copper(II) chloride; urea; sodium chloride

---

As a result of Experiment 35 we can classify aqueous solutions of substances as non-conductors (for example, sugar) or as strong electrolytes (for example, sodium chloride). Although the water itself is an extremely poor conductor of electricity it appears that, when some substances which in the solid state are themselves non-conductors are dissolved in it, the whole solution becomes a very effective conductor. We shall see how the Ionic Theory accounts for this interesting observation.

Definitions

An **electrolyte** is a compound in solution or a molten compound which will conduct electric current with decomposition at the electrodes as it does so.

A **non-electrolyte** is a solution or a molten compound which cannot be decomposed by an electric current.

The **electrodes** are two poles of carbon or metal at which the current (as a flow of electrons) enters or leaves an electrolyte.

The **anode** is the positive electrode at which the electrons enter the external circuit.

The **cathode** is the negative electrode at which the electrons leave the external circuit.

The Ionic Theory

To account for the phenomena of electrolysis the Ionic Theory was put forward by Arrhenius about 1880.

Substances called **electrolytes** are believed to contain electrically charged particles called *ions*. Ions are derived from atoms (or groups of atoms) but differ from them by possessing electrical charges. These charges are *positive* for *hydrogen ions* and ions derived from *metals* (or metallic groups like NH_4), and *negative* for ions derived from *non-metals* or *acidic radicals*. The number of electrical charges carried by an ion is equal to the valency of the corresponding atom or group.

Some examples of ionization are:

COMPOUND	IONS	
Sulphuric acid	$2H^+$	SO_4^{2-}
Sodium chloride	Na^+	Cl^-
Sodium hydroxide	Na^+	OH^-
Copper(II) sulphate	Cu^{2+}	SO_4^{2-}
Lead(II) nitrate	Pb^{2+}	$2NO_3^-$
Hydrochloric acid	H^+	Cl^-

It is very important to notice that an ion is very different from the corresponding atom, as was explained in Chapter 9. A metallic ion is formed from the atom by *loss* of a number of electrons equal to the valency of the metal. Similarly, a non-metallic ion is formed from the corresponding atom by *gain* of a number of electrons equal to the valency of the atom, *e.g.*,

METAL	NON-METAL
$K - e^- \rightarrow K^+$ (univalent)	$\frac{1}{2}Cl_2 + e^- \rightarrow Cl^-$
$Ca - 2e^- \rightarrow Ca^{2+}$ (divalent)	$S + 2e^- \rightarrow S^{2-}$

These electronic changes give to the ions properties quite different from those of the corresponding (electrically neutral) atoms.

For example, by dissolving ordinary, electrically neutral, molecular chlorine in water, a solution is produced which is yellow in colour and is a vigorous bleaching agent, but a solution containing chlorine ions has neither of these properties. Similarly, ordinary metallic sodium, made up of neutral sodium atoms, attacks water liberating hydrogen, but sodium ions, Na^+, have no such action upon water. During the change,

$$Na(s) + \tfrac{1}{2}Cl_2(g) \rightarrow Na^+Cl^-(s)$$

the loss of an electron by Na and the gain of an electron by Cl causes the ionic particles, Na^+ and Cl^-, to assume a very stable, inert condition; hence their inactivity.

In **strong electrolytes**, the ionization is complete. Thus, there exist in a solution of common salt no molecules, NaCl, but only ions, Na^+ and Cl^-. All strong electrolytes, *i.e.*, *salts, the mineral acids*, and *the caustic alkalis*, are in this state of complete ionization in dilute solution.

In **weak electrolytes**, ionization is only slight and most of the electrolyte exists in solution in the form of unionized molecules; for example, in ordinary bench (2M) acetic acid, out of every 1000 molecules present, four are ionized and 996 are unionized.

$$CH_3COOH(aq) \rightleftharpoons CH_3COO^-(aq) + H^+(aq)$$

A solution of ammonia in water is also a weak electrolyte, containing a relatively small proportion of ammonium and hydroxyl ions.

$$NH_4OH(aq) \rightleftharpoons NH_4^+(aq) + OH^-(aq)$$

Most of the organic acids are weak electrolytes, *e.g.*, tartaric, citric, and carbonic. It is not possible to draw an absolutely sharp dividing line between strong and weak electrolytes, *e.g.*, trichloracetic acid is more highly ionized than acetic acid but is much less highly ionized that hydrochloric acid, and so lies between them in strength. For your present purpose, the strong electrolytes are the only group of considerable importance.

Non-electrolytes exist only in the form of molecules and are incapable of ionization, for example,

Trichloromethane	$CHCl_3$
Cane sugar	$C_{12}H_{22}O_{11}$
Alcohol (ethanol)	C_2H_5OH
Urea	CON_2H_4

Water as an electrolyte

Water is an electrolyte but is very weak.

$$H_2O(l) \rightleftharpoons H^+(aq) + OH^-(aq)$$

Exact measurement shows that in pure water, for every molecule of water ionized, furnishing one hydrogen ion and one hydroxyl ion, there are 600 000 000 molecules of water not ionized. The electrical conductivity of water, arising from these quantities of ions, is very small; but, even so, it must be clearly borne in mind that water is an electrolyte and has a small, but measurable, electrical conductivity.

Further, if by electrical or chemical action hydrogen or hydroxyl ions are removed, more water molecules can ionize. So, while at any moment H^+ and OH^- concentrations in water are very small, the water is potentially capable of yielding more of either ion as circumstances may demand.*

Mechanisms of electrolysis. Electrolysis of conc. hydrochloric acid

The apparatus of Figure 74 is suitable. The products are collected over calcium chloride solution because both are insoluble in it.

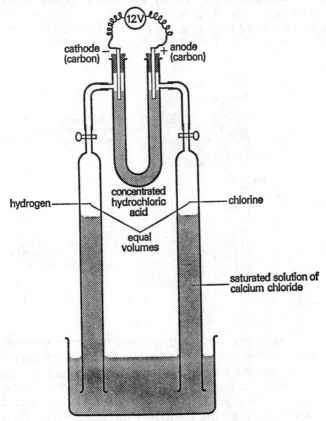

Figure 74

* The exact position is as follows:
  In pure water,

$$[H^+] = [OH^-] = 10^{-7} \text{ mol dm}^{-3} \text{ (at 25 °C)}$$
$$\therefore [H^+][OH^-] = 10^{-14} \text{ mol}^2 \text{ dm}^{-6} = K_w \text{ (a constant)}.$$

K_w is called the *ionic product of water* and is always maintained in aqueous liquids.

If H^+ or OH^- is withdrawn, water will ionize further to restore the value of K_w. When $[H^+] = [OH^-] = 10^{-7}$ mol dm^{-3}, the liquid is in the same condition as pure water, *i.e.*, neutral; if $[H^+]$ is greater than 10^{-7}, it is acidic; if $[H^+]$ is less than 10^{-7}, it is alkaline. But the product $[H^+] \times [OH^-]$ is always 10^{-14} mol^2 dm^{-6}.

When the current is switched on, gas will be found to collect in both tubes, which were full of saturated calcium chloride solution at the beginning. Equal volumes of gas will collect in the two tubes.

When sufficient gas has accumulated, disconnect the U-tube after switching off the current.

At the *cathode*, the gas will be colourless and can be tested by applying a light to it when it will burn in air. The gas is *hydrogen*.

At the *anode*, the gas will be pale yellowish-green. This gas can be tested by damp litmus paper, which will be bleached. It is *chlorine*.

Explanation of electrolysis by the Ionic Theory

So far we have seen that hydrochloric acid contains hydrogen ions and chlorine ions. The concentration of hydroxyl ion from water is so small that it plays no significant part in this electrolysis.

When no current is passing, the ions are wandering randomly about in the solution (Figure 75). The electrical circuit is closed and, immediately, the

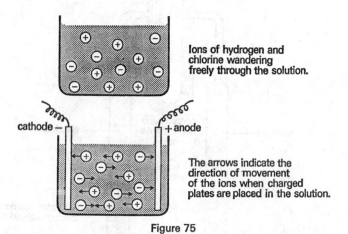

Figure 75

cathode becomes charged negatively, and the anode positively. The cathode attracts to itself the positive ions (that is, the hydrogen ions) whilst the anode attracts the negative ions (that is, the chlorine ions). A procession begins, hydrogen ions to the negative pole, chlorine ions to the positive pole.

The positive hydrogen ions strike the negative pole and acquire from it electrons which make them electrically neutral, and they become in this way atoms of hydrogen which link up in pairs into molecules and become ordinary gaseous hydrogen.

Note that it is the electron from the cathode which neutralizes the hydrogen ion and makes the latter become an atom instead of an ion.

Similarly, the chlorine ion, negatively charged, comes into contact with the anode, loses its electron and becomes an atom of chlorine. Pairs of these atoms combine and become molecules of ordinary gaseous chlorine, which comes off as a greenish gas. Thus the process can be summed up:

Hydrogen chloride yields ions H^+ and Cl^- on solution in water; no current is passing. When current passes and electrolysis begins,

$$H^+ + e^- \rightarrow H$$
ion   atom
              uncharged

$$H + H \rightarrow H_2$$
        molecule

At *cathode*
Hydrogen 1 volume

$$Cl^- - e^- \rightarrow Cl$$
ion   atom
              uncharged

$$Cl + Cl \rightarrow Cl_2$$
        molecule

At *anode*
Chlorine 1 volume

Experiment 36
To investigate the movement of ions during electrolysis

Cut a piece of filter paper approximately the size of a microscope slide, and place it on top of a microscope slide using crocodile clips to attach it in place (Figure 76). Moisten the filter paper with water (not *too* much, but enough to cover the whole surface of the paper by spreading), and place a small crystal of potassium permanganate midway between the points of attachment of the clips. Connect the slide and contents in a circuit with a 20 V d.c. supply, and observe the paper over a period of ten to fifteen minutes.

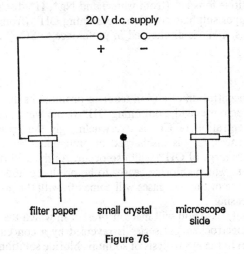

20 V d.c. supply

filter paper   small crystal   microscope slide

Figure 76

The movement of the colour towards one of the electrodes indicates that the electrovalent compound, potassium permanganate $K^+MnO_4^-$, is being 'pulled apart' by the migration of the positive ions towards the negative electrode (cathode) and the negative ions towards the positive electrode (anode). Can you decide which of the ions, K^+ or MnO_4^-, is the coloured one in the above experiment? Other coloured crystals may be used in a similar way.

Selective discharge of ions

When two or more ions of similar charge are present under similar conditions in a solution, *e.g.*, H^+ and Na^+, or OH^- and SO_4^{2-}, one is preferentially selected for discharge and the selection of the ion discharged depends on the following factors:

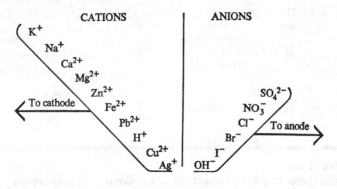

1. Position of the metal or group in the electrochemical series

Consider the arrangement above which is the same as that of the electro-chemical series. If all other factors (see below) are constant, any ion will be discharged from solution in preference to those above it, positive ions at the cathode and negative ions at the anode. For example, in caustic soda solution, containing positive ions H^+ (from water) and Na^+, H^+ discharges in preference to Na^+; in copper sulphate solution, containing OH^- (from water) and SO_4^{2-} as negative ions, OH^- is discharged in preference to SO_4^{2-}.

2. Concentration

Increase of concentration of an ion tends to promote its discharge, *e.g.*, in con-centrated hydrochloric acid, containing OH^- (from water) and Cl^- as negative ions, the concentration of Cl^- is overwhelmingly the greater of the two. In these circumstances, Cl^- is discharged in preference. But, if the acid is very dilute, some discharge of OH^- will also occur. As the acid is diluted, there will *not* be a point at which chlorine ceases to be produced and oxygen replaces it. Instead, a mixture of the two gases will come off, with the proportion of oxygen gradually increasing.

This is the only case you will meet at present in which the order of discharge stated by the electrochemical series is reversed by a concentration effect. The same case arises in the electrolysis of sodium chloride solution, because the same anions are involved.

3. Nature of the electrode

This factor may sometimes influence the choice of ion for discharge. The most important contrast is electrolysis of a solution of sodium chloride with mercury cathode, and with platinum cathode.

With platinum cathode, H^+ is discharged in accordance with the order of the electrochemical series, Na^+ being higher in the series. The cathode product is hydrogen gas (page 158).

If a mercury cathode is used, there is the possibility of discharging Na^+ to form sodium amalgam with the mercury. This requires less energy than the dis-charge of H^+ to hydrogen gas and so occurs in preference, and sodium amalgam is the product (see page 364).

The products of electrolysis of some important and typical solutions will now be considered.

Electrolysis of dilute sulphuric acid (so called electrolysis of water)

The apparatus used is shown in Figure 77 on page 156. Both electrodes are *platinum* foil. It is necessary to run the apparatus for some time, and then release any gases produced, in order to saturate the electrolyte with the gases so that when the experiment is repeated an accurate 2 : 1 volume ratio of hydrogen to oxygen is observed.

The following ions are present:

From sulphuric acid $H^+(aq)$ $SO_4^{2-}(aq)$
From water $H^+(aq)$ $OH^-(aq)$

CATHODE	ANODE
	$\boxed{SO_4^{2-}(aq) \text{ and } OH^-(aq)}$

migrates to the cathode, gains an electron and becomes a hydrogen atom.

$$H^+(aq) + e^- \rightarrow H(g)$$

Hydrogen atoms combine in pairs to give molecules.

$$H(g) + H(g) \rightarrow H_2(g)$$

Migration of SO_4^{2-} to the anode and discharge of H^+ are equivalent to decrease of concentration of sulphuric acid.

both migrate to the anode, where OH^-, being lower in the electrochemical series, is discharged in preference to SO_4^{2-}, in spite of the high concentration of the latter.

$$OH^-(aq) - e^- \rightarrow OH(g)$$

By interaction between the OH groups, water and oxygen are produced.

$$OH(g) + OH(g) \rightarrow H_2O(l) + O(g)$$
$$O(g) + O(g) \rightarrow O_2(g)$$

Discharge of OH^- disturbs the ionic equilibrium of water. More water ionizes to restore it.

$$H_2O(l) \rightleftharpoons H^+(aq) + OH^-(aq)$$

Excess H^+ so produced, with incoming SO_4^{2-}, is equivalent to increased concentration of sulphuric acid.

Summary
Hydrogen, 2 volumes
Acidity decreasing.

Summary
Oxygen, 1 volume
Acidity increasing.

The *total acidity* at anode and cathode together remains *constant*.

The final result, 2 volumes of hydrogen at the cathode and 1 volume of oxygen at the anode, is equivalent to the electrolysis of water. To see more clearly how the 2 : 1 volume ratio for the gases arises, consider the flow of four electrons through the solution.

At the cathode: $4H^+(aq) + 4e^- \rightarrow 2H_2(g)$
At the anode: $4OH^-(aq) \rightarrow 2H_2O(l) + O_2(g) + 4e^-$

Thus two moles of hydrogen molecules will be released for every one mole of oxygen molecules produced.

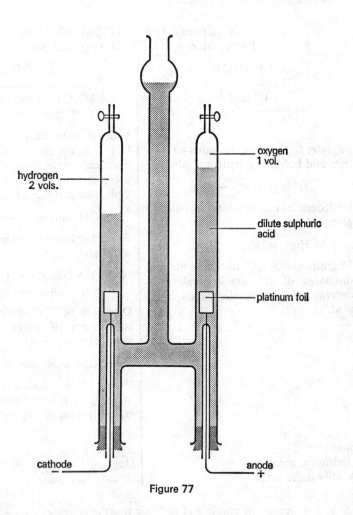

oxygen
1 vol.

hydrogen
2 vols.

dilute sulphuric
acid

platinum foil

cathode

anode
+

Figure 77

Electrolysis of caustic soda (sodium hydroxide) solution

The apparatus is the same as for the electrolysis of dilute sulphuric acid (page 156). The electrodes are again *platinum* foil.

The ions present are:

From sodium hydroxide	$Na^+(aq)$	$OH^-(aq)$
From water	$H^+(aq)$	$OH^-(aq)$

CATHODE   ANODE

both ions migrate to the cathode. H^+, being lower in the electrochemical series, is discharged in preference to Na^+, in spite of the high concentration of the latter.

$$H^+(aq) + e^- \rightarrow H(g)$$

The hydrogen atoms then combine in pairs to give molecules.

$$H(g) + H(g) \rightarrow H_2(g)$$

Discharge of H^+ disturbs the ionic equilibrium of water. More water ionizes to restore it.

$$H_2O(l) \rightleftharpoons H^+(aq) + OH^-(aq)$$

Excess OH^- so produced, with incoming Na^+, is equivalent to increase of concentration of sodium hydroxide.

ions migrate to the anode and discharge by loss of an electron.

$$OH^-(aq) - e^- \rightarrow OH(g)$$

By interaction between the OH groups, water and oxygen are produced.

$$OH(g) + OH(g) \rightarrow H_2O(l) + O(g)$$
$$O(g) + O(g) \rightarrow O_2(g)$$

Migration of Na^+ to the cathode and discharge of OH^- are equivalent to fall of concentration of sodium hydroxide.

Summary
Hydrogen, 2 volumes
Alkalinity increasing.

Summary
Oxygen, 1 volume
Alkalinity decreasing.

The total alkalinity at anode and cathode together is constant.

The process is equivalent to the electrolysis of water. The volume ratio of the gases is established by the same method as that used in the electrolysis of dilute sulphuric acid.

Electrolysis of sodium chloride solution

The apparatus is the same as for concentrated hydrochloric acid (page 151), or dilute sulphuric acid (page 156). The *cathode* may be *platinum* (or carbon), but the *anode* must be *carbon* to resist attack by chlorine.

The ions present are:

From sodium chloride	Na$^+$(aq)	Cl$^-$(aq)
From water	H$^+$(aq)	OH$^-$(aq)

<table>
<tr><td align="center">CATHODE</td><td align="center">ANODE</td></tr>
</table>

both migrate to cathode. H$^+$, being lower in the electrochemical series, discharges in preference to Na$^+$.

$$H^+(aq) + e^- \longrightarrow H(g)$$

Hydrogen molecules are then formed by combination of the atoms in pairs.

$$H(g) + H(g) \longrightarrow H_2(g)$$

Discharge of H$^+$ disturbs the ionic equilibrium of water. More water ionizes to restore it.

$$H_2O(l) \rightleftharpoons H^+(aq) + OH^-(aq)$$

Excess OH$^-$ so produced, with incoming Na$^+$, is equivalent to the presence of sodium hydroxide.

Summary
Hydrogen, 1 volume
Solution becomes alkaline by presence of sodium hydroxide.

both migrate to the anode. Cl$^-$ is discharged because it is present in much greater concentration than OH$^-$ (see page 154).

$$Cl^-(aq) - e^- \longrightarrow Cl(g)$$

The atoms then combine in pairs to give molecules.

$$Cl(g) + Cl(g) \longrightarrow Cl_2(g)$$

Summary
Chlorine, 1 volume
(N.B. In practice it will be less than one volume as some oxygen is *always* produced.)

Ideally, hydrogen and chlorine are produced in equal volumes. If the three-limbed voltmeter is used, chlorine will have to saturate the brine first.

Electrolysis of copper(II) sulphate solution

The ions present are:

From copper(II) sulphate	$Cu^{2+}(aq)$	$SO_4{}^{2-}(aq)$
From water	$H^+(aq)$	$OH^-(aq)$

CATHODE	ANODE

$$\boxed{\begin{array}{l} Cu^{2+}(aq) \\ H^+(aq) \end{array}}$$

$$\boxed{\begin{array}{l} SO_4{}^{2-}(aq) \\ OH^-(aq) \end{array}}$$

both migrate to the cathode. Cu^{2+}, being lower in the electrochemical series, discharges in preference to H^+.

$$Cu^{2+}(aq) + 2e^- \rightarrow Cu(s)$$

The copper deposits as a brown layer.

Platinum or carbon anode

See the exactly similar case of dilute sulphuric acid (page 155). See Figure 78 on page 160.

Summary
Oxygen given off.
Solution becomes acidic with sulphuric acid.

Copper anode

With this anode, there are three possibilities:
1. Discharge of $SO_4{}^{2-}$ ⎫
2. Discharge of OH^- ⎬ by loss of
3. Conversion of Cu ⎭ electrons
   atom to Cu^{2+}

The last of these occurs most readily. $SO_4{}^{2-}$ and OH^- are not discharged. Copper passes into solution from the anode as Cu^{2+} ions.

$$Cu(s) - 2e^- \rightarrow Cu^{2+}(aq)$$

Summary
Copper deposited.

Summary
Copper passes into solution as ions. The total concentration of the solution in $SO_4{}^{2-}$ (not discharged) and Cu^{2+} (copper is depositing on the cathode) is constant. The electrolysis merely transfers copper from anode to cathode.

It will be seen that, with platinum or carbon electrodes, the colour of the solution will fade as the copper is deposited. With copper electrodes the colour will not change.

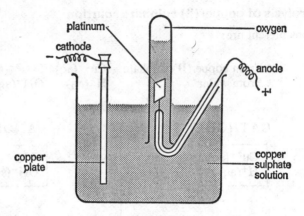

Figure 78

Summary of the effects of electrolysis

Solution electrolysed	Cathode of	Anode of	At cathode	At anode
Hydrochloric acid (conc.)	Carbon or Platinum	Carbon	Hydrogen one volume	Chlorine one volume
Sulphuric acid (dil.)	Platinum	Platinum	Hydrogen 2 volumes; decrease of acidity	Oxygen 1 volume; increase of acidity
Caustic soda (sodium hydroxide)	Platinum	Platinum	Hydrogen 2 volumes; increase of alkalinity	Oxygen 1 volume; decrease of alkalinity
Common salt (sodium chloride)	Platinum or Carbon	Carbon	Hydrogen 1 volume; sodium hydroxide solution	Chlorine 1 volume
Copper(II) sulphate	Copper	Platinum or Carbon	Copper deposited	Oxygen and sulphuric acid
Copper(II) sulphate	Copper	Copper	Copper deposited	Copper dissolved

Cells

We have seen how it is possible for an electric current to produce chemical change when it is passed through an electrolyte solution. The next experiment is designed to show that chemical reactions can produce an electric current.

Experiment 37
The production of an electric current from a chemical reaction

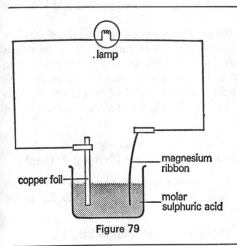

Figure 79

Set up the circuit shown in Figure 79 using a 1.25 V, 0.25 A lamp bulb, and a beaker two-thirds full of a molar solution of sulphuric acid. Into the beaker place strips of magnesium ribbon and copper foil, thoroughly cleaned by abrasive paper, and connect them up to the circuit using the crocodile clips on the ends of the connecting wires. A reaction will take place quite rapidly and the bulb will light, indicating the flow of electric current in the circuit.

The arrangement of two different metals in an electrolyte as in the above experiment is known as a *galvanic couple*, and the whole system when supplying an electric current constitutes a *chemical cell*. To operate effectively the two metals must be widely separated in the electrochemical series, *e.g.*, *copper* and *magnesium*. If rods of these metals are placed in the electrolyte, *dilute sulphuric acid*, and the metals are pure, nothing appreciable will occur other than the dissolving of the magnesium so long as the metals are not in contact. If, however, they are joined by an electrical conductor such as a metallic wire (not in contact with the electrolyte), it will be found that current will flow in the wire from copper ($+$) to magnesium ($-$), the magnesium will dissolve in the electrolyte and hydrogen will appear as bubbles on the copper. Magnesium becomes a cathode and copper an anode in this voltaic cell.

To produce this result, magnesium, the more electropositive of the two metals, ionizes by electron loss and the electrons pass from magnesium to copper through the wire. This is equivalent to the flow of conventional current in the opposite direction. At the copper surface, the electrons reduce hydrogen ions from the electrolyte.

$$Mg(s) \rightarrow Mg^{2+}(aq) + 2e^-$$
$$2H^+(aq) + 2e^- \rightarrow H_2(g)$$

Theoretically, current can continue to flow as long as materials last; in practice, bubbles of hydrogen adhere to the copper, cut off much of its contact with the electrolyte, and so 'polarize' the cell.

A cell of this kind is a device for converting chemical energy into electrical energy. As the above equations show, the whole process in the cell corresponds to the change:

$$Mg(s) + 2H^+(aq) \rightarrow Mg^{2+}(aq) + H_2(g)$$

This occurs when magnesium dissolves in dilute sulphuric acid and, normally, chemical energy is made available from this change as *heat*; in this cell, much of the chemical energy is converted to *electrical* energy as electrons flow from magnesium to copper through the connecting wire.

Leclanché cell

This cell is a practicable form of 'primary' cell for producing electric current, *i.e.*, it operates by using up the chemicals of which it is composed and it cannot be recharged.

The *cathode* is a *zinc* rod or sheet; the *anode* is a *carbon* rod and the electrolyte in which they are immersed is *ammonium chloride* solution. When the cell is operating, zinc ionizes by electron loss and zinc ions dissolve in the electrolyte; the electrons pass round the external circuit, performing useful work (*e.g.*, lighting a small bulb or ringing a bell), and are absorbed at the anode. This causes discharge of ammonium ions from the electrolyte, with formation of *ammonia* (in solution) and *hydrogen*.

$$\text{At cathode} \qquad\qquad \text{At anode}$$
$$\text{Zn(s)} \longrightarrow \text{Zn}^{2+}\text{(aq)} + 2e^- \qquad 2\text{NH}_4^+\text{(aq)} + 2e^- \longrightarrow 2\text{NH}_3\text{(g)} + \text{H}_2\text{(g)}$$

The anode is immersed in a porous pot containing manganese(IV) oxide to oxidize the hydrogen. Otherwise, bubbles of hydrogen adhering to the anode would polarize the cell.

$$2\text{Mn}^{4+}\text{(aq)} + \text{H}_2\text{(g)} + 2\text{OH}^-\text{(aq)} \longrightarrow 2\text{Mn}^{3+}\text{(aq)} + 2\text{H}_2\text{O(l)}$$

In the so-called 'dry' Leclanché cell, the electrolyte is gelatinized to prevent spilling and the cathode is usually sheet zinc, also acting as the cell case. This cell will yield a small current continuously without serious polarization. Much larger currents can be obtained intermittently, intervals being required to allow the cell to recover from polarization.

Lead accumulator

The lead accumulator is a secondary cell, *i.e.*, it has to be 'charged' by passage of direct current (usually rectified mains current) through it, after which it will 'discharge' yielding direct current for use where required, *e.g.*, in the electrical system of an automobile. These processes of charge and discharge can be repeated many times.

The cathode and anode of the accumulator are both grids of lead–antimony alloy. At discharge, both grids carry a *lead sulphate* filling. The electrolyte is sulphuric acid suitably diluted with water.

During charge, the following changes occur.

$$\text{At cathode} \qquad\qquad\qquad \text{At anode}$$
$$\begin{cases} \text{Pb}^{2+}\text{(aq)}+2e^- \longrightarrow \text{Pb(s)} \\ \text{SO}_4^{2-} \text{ into solution} \end{cases} \begin{cases} \text{Pb}^{2+}\text{(aq)}+2\text{H}_2\text{O(l)}-2e^- \longrightarrow \text{PbO}_2\text{(s)}+4\text{H}^+\text{(aq)} \\ \text{SO}_4^{2-} \text{ into solution} \end{cases}$$

That is, the *cathode* grid acquires a filling of *spongy lead* and the *anode* grid one of *lead(IV) oxide*. Passage of ions into solution (from equations above) in the proportion of 2SO_4^{2-} to 4H^+ increases the concentration and density of the acid. At full charge, the E.M.F. is a little above 2 volts and the acid density is 1.25 g cm^{-3}.

During the discharge the cell yields electrical energy by the following changes.

At cathode	*At anode*
$Pb(s) \rightarrow Pb^{2+}(aq) + 2e^-$	$PbO_2(s) + 4H^+(aq) + 2e^- \rightarrow Pb^{2+}(aq) + 2H_2O(l)$
From solution, SO_4^{2-}	From solution, SO_4^{2-}
$PbSO_4$ deposits	$PbSO_4$ deposits

Electrons available from lead at the cathode pass round the *external* circuit performing the electrical work required and are absorbed at the anode. Absorption of ions ($4H^+ : 2SO_4^{2-}$) from the electrolyte decreases the concentration and density of the acid. The E.M.F. falls to 2 volts soon after discharge begins and stays constant until it is almost complete, then falling to 1.8 volts. At this point, recharging is required.

Laws of electrolysis

The laws expressing the quantitative results of electrolysis were first stated by Faraday. Expressed in modern terms they assert that the amount (expressed in moles) of an element liberated during electrolysis depends upon:

(1) the time of passing the steady current
(2) the magnitude of the steady current passed
(3) the charge on the ion of the element.

The dependence of the amount of substance liberated during electrolysis upon these factors may be investigated through the experiments which follow. Factors (1) and (2) may be incorporated into the same experiment since the product of the current (measured in amperes) and the time (measured in seconds) gives a measure of electricity known as the *quantity* of electricity, which is measured in units called *coulombs*. Thus:

$$\text{quantity of electricity} = \text{current} \times \text{time}$$
$$\text{(coulombs)} \qquad \text{(amps)} \qquad \text{(seconds)}$$

Experiment 38
Investigation of the amount of substance liberated in electrolysis by different quantities of electricity

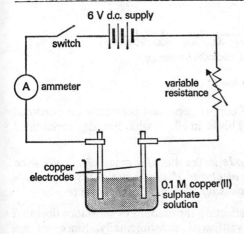

Figure 80

Set up the circuit shown in Figure 80. Fill the beaker two-thirds full of 0.1M copper(II) sulphate solution. Take a piece of clean, dry copper foil, approximately 5 cm × 3 cm, and weigh it. This is to be the cathode in the circuit, and should therefore be attached to the negative pole of the d.c. supply. A second, similar, copper electrode should be placed in the copper(II) sulphate solution and both electrodes connected into the circuit. Simultaneously start a stopclock and complete the circuit using the switch, and then *quickly* adjust the variable resistance to obtain a steady current of approximately 0.2 A. It may be necessary to alter the variable resistance from time to time to maintain the current at this steady value.

After about fifteen minutes switch off the current and note the time on the stopclock.

Remove the copper cathode, wash it carefully in distilled water, then in a little ethanol, and finally in propanone (acetone). When it is completely dry weigh the cathode, and determine the increase in mass of the electrode. This is the mass of copper deposited in the time for which the current flowed.

Replace the cathode in the beaker, start the stopclock and switch on the current. After a further fifteen minutes switch off the current, and determine the increase in mass of the copper cathode. Repeat this procedure for a further half hour (*i.e.*, two more measurements).

Specimen results

These are summarized in the table below.

Current/ amps	Time/seconds	Quantity of electricity/coulombs	Mass of copper deposited/grammes
0.21	15 × 60 = 900	900 × 0.21 = 189	0.063
0.21	30 × 60 = 1800	1800 × 0.21 = 378	0.129
0.21	45 × 60 = 2700	2700 × 0.21 = 576	0.187
0.21	60 × 60 = 3600	3600 × 0.21 = 756	0.250

The relationship between the amount of copper deposited and the quantity of electricity passed may be seen by inspection of the last two columns in the Table, and the relationship may be demonstrated graphically, as illustrated in Figure 81. From the shape of the graph, which is a straight line passing through

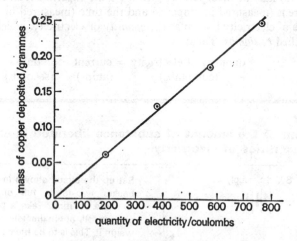

Figure 81

the origin, it is clear that the mass of copper deposited is directly proportional to the quantity of electricity passing. This is, in effect, what Faraday formulated in his First Law.

Faraday's First Law of Electrolysis states that *the mass of a substance liberated at (or dissolved from) an electrode during electrolysis is proportional to the quantity of electricity passing through the electrolyte.*

The third factor listed on page 16 affecting the amount of substance liberated during electrolysis may also be investigated experimentally. Since we are interested in the effect of the charge on the ions present in solution, we need to keep the quantity of electricity fixed whilst varying the types of ion in solution. This may be achieved by passing the same quantity of electricity through two cells, with ions of different charges in each, and is described in the following experiment.

Experiment 39
Comparison of the amounts of different substances liberated by the same quantity of electricity

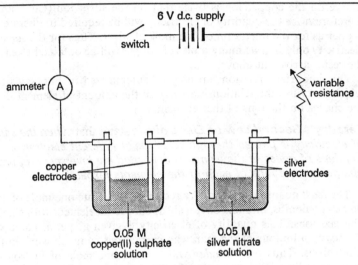

Figure 82

Set up the circuit of Figure 82, containing a copper voltameter and a silver voltameter (a voltameter is a vessel containing two electrodes immersed in a solution of ions through which a current is to be passed). Identify the copper and silver cathodes, clean and dry them as in Experiment 38, and after weighing them replace them in their respective voltameters. Pass a current of about 0.5 A for twenty to thirty minutes, after which the cathodes should be removed, cleaned and dried, and reweighed. Compare the masses of copper and silver deposited. Note that care must be taken in removing the silver cathode from the solution as the metal does not always adhere well to the cathode.

Specimen results

Current flowing $= 0.45$ A
Duration of current flow $= 25$ minutes
Mass of copper deposited $= 0.221$ grammes
Mass of silver deposited $= 0.755$ grammes

It can be seen that the masses of copper and silver deposited in the experiment are different, but a meaningful comparison of the amounts of each of the elements deposited can be made only by calculating the number of moles of atoms of each of the elements formed.

Thus:

$$\text{Amount of copper deposited} = \frac{0.221}{63.5} \text{ mole} = 0.0035 \text{ mole}$$

$$\text{Amount of silver deposited} = \frac{0.755}{107.8} \text{ mole} = 0.0070 \text{ mole}$$

and it can therefore be seen that twice as many atoms of silver are deposited as atoms of copper. This arises because of the third factor listed on page 163, where the charge on the ion of the element concerned must be considered.

Since the process at the cathode is that of the addition of electrons to the cations:

$$X^{n+}(aq) + ne^- \longrightarrow X(s)$$

it can be seen that the quantity of electricity required to liberate one mole of X will depend upon the charge, $n+$, on the ion. In the case of copper and silver the charge on the copper ion is twice that on the silver ion (Cu^{2+} and Ag^+) and therefore *twice* the quantity of electricity will be required to liberate one mole of copper as for the liberation of one mole of silver. Thus for the *same* quantity of electricity only half as many moles of copper will be obtained as silver, which is the result arrived at above.

This important relationship may be summarized in Faraday's Second Law which describes the relationship between the amount of element deposited and the charge on the ions of that element.

Faraday's Second Law of Electrolysis states that *when the same quantity of electricity is passed through solutions of different electrolytes the relative numbers of moles of the elements deposited are inversely proportional to the charges on the ions of each of the elements respectively.*

The *least* quantity of electricity required to liberate one mole of an element is 96 500 coulombs, and this obviously applies to an element with singly positively charged ions. This quantity of electricity is given a special name which is *the Faraday*, in honour of the man who carried out so much work in the study of electrolysis. Thus one *Faraday* will liberate one mole of hydrogen ions, one mole of sodium ions, one mole of potassium ions, and in general since

$$M^+(aq) + e^- \longrightarrow M(s)$$

represents the cathode reaction for a univalent metal M, it follows that for one mole of unipositive ions to be discharged the charge carried by **one mole of electrons** will be required.

Hence,

one Faraday is equal to one mole of electrons.

Estimation of relative atomic masses

Returning to the results of Experiment 39, we may estimate the relative atomic masses of copper and silver as follows.

Quantity of electricity passing when a current of 0.45 amps flows for 25 minutes (25 × 60 seconds) is given by:

$$\text{Quantity of electricity} = 0.45 \times 25 \times 60 \text{ coulombs}$$
$$= 675 \text{ coulombs}$$

For silver

$$675 \text{ coulombs liberate } 0.755 \text{ grammes silver}$$

$$\therefore 96\,500 \text{ coulombs liberate } \frac{0.755 \times 96\,500}{675} \text{ grammes silver}$$

$$= 108.0 \text{ grammes silver}$$

Since silver forms unipositive ions, Ag^+, one mole of electrons (one Faraday) will liberate the relative atomic mass (mole) of atoms. Hence, relative atomic mass of silver $= 108$.

For copper

$$675 \text{ coulombs liberate } 0.221 \text{ grammes copper}$$

$$\therefore 96\,500 \text{ coulombs liberate } \frac{0.221 \times 96\,500}{675} \text{ grammes copper}$$

$$= 31.6 \text{ grammes copper}$$

Since copper forms dipositive ions, Cu^{2+}, one mole of electrons will liberate half the relative atomic mass of atoms. Hence, relative atomic mass of copper $= 2 \times 31.6 = 63.2$.

Electroplating

Electroplating is the electrical precipitation of one metal on another to secure improved appearance or greater resistance to corrosion.

In *silver plating*, articles such as table-ware or cake dishes, made of base alloy, *e.g.*, cupronickel, are made the *cathode* in a plating bath of potassium (or sodium) argentocyanide $KAg(CN)_2$ solution. This contains some silver ions, Ag^+.

$$KAg(CN)_2(aq) \rightleftharpoons K^+(aq) + Ag^+(aq) + 2CN^-(aq)$$

The *anode* is pure *silver*. When direct current passes, the following occurs.

At *cathode*:

$$Ag^+(aq) + e^- \longrightarrow Ag(s)$$
Silver deposits

At *anode*:

$$Ag(s) - e^- \longrightarrow Ag^+(aq)$$
Silver dissolves

In correct conditions, the silver layer deposited on the cathode article is coherent and tough and can be highly polished.

Chromium plating is much used to improve the appearance of steel parts and protect them from rusting. The steel is usually plated first with nickel or copper, because chromium does not adhere well on to a steel surface. The object is made the *cathode* in a plating bath which contains chromium compounds (*e.g.*, sulphate and oxide) in sulphuric acid and water. A lead anode is usual. When direct current passes, chromium deposits on the article at the cathode as a bright coherent layer.

$$Cr^{3+}(aq) + 3e^- \longrightarrow Cr(s)$$

This layer resists rusting and gives a bright 'silvery' appearance.

Questions

Relative atomic masses will be found on page 499.

1. What do you understand by the term 'electrolyte'? Describe experiments to demonstrate the products formed in the electrolysis of solutions of (*a*) sulphuric acid; (*b*) sodium sulphate; (*c*) copper(II) sulphate. (O. and C.)

2. State Faraday's Laws of Electrolysis. Describe carefully what happens when copper(II) sulphate solution is electrolysed between (*a*) platinum and (*b*) copper electrodes, and when sodium chloride *solution* is electrolysed between (*a*) platinum, and (*b*) carbon electrodes. (L.)

3. Give a general but concise account of the phenomena which occur when a salt is dissolved in water and the solution is electrolysed.

Describe briefly two instances of the practical application of electrolysis. (L.)

4. When an aqueous solution of sodium nitrate is electrolysed with inert electrodes, the products are hydrogen (2 vol.) as the cathode and oxygen (1 vol.) at the anode. Also, the cathodic liquid becomes alkaline and the anodic liquid acidic. Explain these results and write ionic equations in illustration.

5. Explain why solid sodium chloride is a very poor electrical conductor while, if melted, it conducts electric current readily. State the products of electrolysis of molten sodium chloride and give ionic equations to account for them. State and explain the different products obtained by electrolysis of a concentrated solution of sodium chloride in water.

6. Draw a fully labelled diagram of the apparatus you would use to electrolyse a concentrated solution of sodium chloride and measure the volumes of the gases produced at the electrodes. Name these gases and state their relative volumes. Say how they can be made to react with one another: name the product and give its relative volume. What will be the actual volume at s.t.p. of this product, if 96 500 coulombs were used in the electrolysis, assuming that complete combination of the reacting gases occurred? (C.)

7. What mass of silver, and what volume of oxygen (measured at s.t.p.) would be liberated in electrolysis by 9650 coulombs of electricity? (S.)

8. How many Faradays of electricity are required to produce by electrolysis: (a) 27 grammes of aluminium, (b) 8 grammes of oxygen? (S.)

9. Two plates, one of zinc and one of copper, held apart by a separator and connected to a small light bulb, are dipped into dilute sulphuric acid. The bulb lights up but the light soon becomes dim.

(a) What would be observed at the copper plate? Write an ionic equation for the reaction occurring.

(b) What would happen at the zinc plate? Write an ionic equation for the reaction.

(c) Explain why the light fades after a short time.

(d) If the zinc plate were replaced by an iron plate, would the lamp glow more or less brightly?

(e) If the zinc plate were retained but the copper plate were replaced by a silver plate, would the lamp glow more or less brightly? (J.M.B.)

10. Draw a fully labelled diagram of a voltameter suitable for the electrolysis of water acidified with dilute sulphuric acid, showing how the gaseous products are collected. Give the names and relative proportions of the gases evolved, and the names, materials and polarities of the electrodes. Represent the reactions taking place at the electrodes by ionic equations.

Explain why the same result would be obtained if the water was made alkaline by adding sodium hydroxide solution instead of acidifying it with dilute sulphuric acid.

State and explain the effect of electrolysing a dilute solution of sodium sulphate in the same voltameter. (J.M.B.)

11. (a) Name the product at each electrode and write equations for the reactions which occur during the electrolysis of copper(II) sulphate (cupric sulphate) solution, using platinum electrodes. (b) Calculate the mass of each product of electrolysis if the current were stopped after the passage of 0.01 Faraday.

(c) Describe what would be seen at each electrode if the direction of the current were then reversed and the current allowed to flow until a further 0.02 Faraday had passed. Write an equation for any reaction which occurs in (c) but which does not occur in (a).

Outline a method of obtaining copper from copper(II) sulphate in which electrolysis is not used. (J.M.B.)

12. What do you understand by the term electrolyte? Illustrate your answer by describing the passage of an electric current between platinum electrodes immersed in copper(II) sulphate solution. Give ionic equations to show what happens at each electrode, and state what would be observed.

What different observations would be made if the electrodes were made of copper?

Explain what connection there is between the term electrolyte and the term 'electrovalency'. (S.)

13. A solution of copper(II) sulphate, acidified with sulphuric acid, is electrolysed using copper electrodes. (a) Give the formulae of the ions present in the solution before electrolysis. (b) What changes, if any, are observed at the cathode, at the anode, and in the solution? (c) Explain the reactions which take place at the electrodes. (d) What use is made of this process in industry?

You are asked to investigate the connection between the quantity of electricity passed through a solution of copper(II) sulphate and the mass of copper deposited. (i) Sketch the electrical circuit you would use. (ii) State the measurements you would make.

In such an experiment 1930 coulombs liberated 0.64 g of copper. When the same quantity of electricity was passed through a solution containing silver ions, Ag^+, 2.16 g of silver were liberated. How do you explain these results? (A.E.B.)

14. 0.02 Faradays of electricity were passed through a solution of sodium hydroxide using platinum electrodes.

(a) Give the names of the gases evolved, and the names or signs of the electrodes at which they were produced.

(b) Draw a labelled diagram of a suitable apparatus for this electrolysis and for the collection of the products.

(c) Represent the reactions taking place at the electrodes by ionic equations.

(d) Calculate the number of moles of each gas produced and also the volume which each gas would occupy at s.t.p.

(e) Calculate the time required to complete the passage of 0.02 Faradays if a current of 2 amps were passed through the solution.

(f) Write an equation to represent the reaction which would take place if the volumes of gases mentioned in (d) were mixed and ignited. State the number of moles of the product which would be formed.

(The gramme-molecular volume of a gas at s.t.p. is 22.4 dm^3. 1 Faraday = 96 500 coulombs.) (J.M.B.)

15. (a) A steady current of 1 amp is passed between copper electrodes through a solution of copper sulphate for 1 hour. The cathode gains in mass by 1.19 g and the anode loses mass by the same amount.

In another experiment, a steady current of 2 amps is passed for 1 hour between copper electrodes in a warm alkaline solution of sodium chloride. The anode is found to lose in mass by 4.76 g. Calculate the number of Faradays of electricity required to cause 1 mole of copper (63.6 g) to dissolve from the anode in each case, and comment on the answer.

(1 Faraday = 96 500 coulombs = 26.8 amp hours; 1 coulomb = 1 amp for 1 sec.)

(b) Why is it inadvisable to examine the acid levels in a car battery by the light of a burning candle?

(c) The elements magnesium, nickel, hydrogen, copper, and silver stand in that order in the electrochemical series (magnesium most active).

(i) What would you expect to happen at the cathode if an aqueous solution of nickel nitrate and silver nitrate is electrolysed with platinum electrodes?

(ii) Magnesium is placed in an aqueous solution of copper(II) sulphate in one beaker. Copper is placed in an aqueous solution of magnesium sulphate in another beaker. What would you expect to happen in each beaker? (L.)

16. An element X has a relative atomic mass of 88. When a current of 0.5 amp was passed through the fused chloride of X for 32 minutes 10 seconds, 0.44 g of X was deposited at the cathode. (1 Faraday = 96 500 coulombs.) (a) Calculate the number of Faradays needed to liberate 1 mole of X. (b) Write the formula for the X ion. (c) Write the formula for the hydroxide of X. (J.M.B.)

17. 0.2 Faradays of electricity were passed through solutions of (a) copper(II) sulphate and (b) dilute sulphuric acid.

Calculate the mass of copper liberated in (a) and the volume of hydrogen evolved at s.t.p. in (b). (S.)

18. What mass of (a) copper, (b) aluminium, and what volume at s.t.p. of (c) chlorine and (d) oxygen will be liberated during electrolysis by a charge of one Faraday? (S.)

19. Make a fully labelled diagram of the apparatus you would use to electrolyse acidified water. Name the acid used and the material of the electrodes. Assuming that the electrolyte is freshly made immediately before switching on the current, explain why the relative volumes of the gaseous products are not at first in a simple ratio to one another. How would you find out whether any of the acid you have used in making up the electrolyte is consumed during the action?

Give *one* example of the electrolysis of a fused salt, naming the products at the electrodes. (C.)

20. By means of a labelled diagram, show how you would carry out the electrolysis of molten lead(II) bromide, indicating the names and polarities of the electrodes and giving ionic equations for the actions taking place at the electrodes. State the approximate voltage applied to the cell. Describe in detail *one* chemical test to identify the product at the cathode. (J.M.B.)

21. What do you understand by the term *electrolysis*?

(a) Give an explanation of the fact that hydrochloric acid is a better conductor than acetic acid of equivalent molar concentration, and give the cathode reaction which is common to both solutions.

(b) (i) Give *one* explanation for the fact that metals other than copper are not deposited on the cathode during the electrolytic purification of copper.

(ii) Compare the quantity of electricity required to deposit one mole of aluminium with that required to deposit one mole of copper. (S.)

22. A copper rod stands in a dilute solution of copper(II) sulphate in a beaker. A zinc

rod stands in a second beaker containing a dilute solution of zinc sulphate.

(*a*) Draw a labelled diagram of this apparatus, adding to it any items necessary to enable it to act as a voltaic cell.

(*b*) Show, on your diagram, the direction of flow of *electrons* when this cell is connected to an external circuit.

(*c*) Describe carefully all changes that will take place in each beaker while current flows. Explain these changes.

(*d*) Give *two* reasons why the flow of electrons will, after a time, diminish. (S.)

23. Two pieces of lead foil are dipped, side by side but not touching, in a beaker of fairly concentrated sulphuric acid. After some days' immersion the plates are coated with a white deposit. The two pieces of lead are now connected to a 6 V d.c. supply, and a current passes. After some hours, one piece of lead has lost its white deposit and bubbles of hydrogen are streaming off it. The other piece of lead now appears to be coated with a dark brown deposit, replacing the white deposit.

(*a*) Draw a labelled diagram, showing the apparatus when current is being passed through it.

(*b*) Suggest a name for the white deposit first formed on each plate.

(*c*) Suggest a name for the dark brown deposit. If some of this deposit could be scraped off, describe how you would attempt to identify it.

(*d*) Suggest any change that you would expect to occur in the liquid during the passage of the current. (O. and C.)

24. What do you understand by the term *electrolyte*?

A is a liquid and *B* is a solid and each becomes an electrolyte when dissolved in water: *C* is a solid and becomes an electrolyte when melted. Name three different substances *A*, *B*, and *C* which become electrolytes under the conditions mentioned, and give a brief explanation in each case.

Figure 83 represents a network for passing direct current simultaneously through four cells wired in parallel: the contents of each cell is indicated and each cell is wired in series with a lamp. If the current is adjusted so that the lamp *L2* glows dimly, state what you think will be the relative brightness of all four lamps and give reasons for differences. (Individual electrode reactions need not be quoted.)

Name (i) a pure liquid, and (ii) a solution which could be used instead of ethanol without changing the current flow in the network. (O.)

25. A steady current is passed through two voltameters connected in series containing copper(II) sulphate solution and dilute sulphuric acid respectively. All electrodes are of platinum. What is the total volume of gases liberated in the latter (measured at s.t.p.) when 6.35 grammes of copper are deposited in the former? (S.)

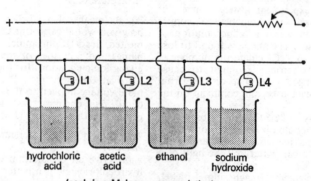

hydrochloric   acetic   ethanol   sodium
acid   acid   hydroxide

(each is a Molar aqueous solution)

Figure 83

17 Oxidation and Reduction

The concepts of oxidation and reduction are very good examples of the way in which ideas about the nature of chemical reactions develop as science makes advances in a whole variety of seemingly unrelated fields. Thus oxidation was originally thought of in extremely simple terms, which have since been modified in the light of the increase in our knowledge of the basis of chemical reactions. In order to understand oxidation and reduction more fully, we shall briefly examine the various stages through which the definitions of these terms have passed, and ultimately arrive at a concise definition in modern terms.

Oxidation as the addition of oxygen or removal of hydrogen

Originally, *oxidation* was simply a reaction in which oxygen combined with another substance, either wholly or in part. *Reduction* was, therefore, a reaction in which a compound was deprived of all, or part, of the oxygen it contained.

Thus in the reaction

$$CuO(s) + H_2(g) \rightarrow Cu(s) + H_2O(l)$$

the hydrogen has been oxidized to water (since oxygen has been added to it) and the copper(II) oxide has been reduced to copper (since oxygen has been removed from it). In such a reaction the copper(II) oxide is described as the *oxidizing agent*, and the hydrogen as the *reducing agent*.

An important aspect of such reactions which was immediately obvious once the terms oxidation and reduction had been defined as above, is that oxidation cannot take place in a reaction without reduction simultaneously occurring. Thus the idea of oxygen *transfer* (from the oxidizing agent to the reducing agent) is fundamental to this initial definition.

The readiness with which hydrogen combines with oxygen to form water caused hydrogen to be regarded as a kind of 'chemical opposite' of oxygen; so the term oxidation was extended to include reactions in which a compound gave up some or all of its hydrogen as well as those in which it combined with oxygen. (This idea of oxidation is analogous to the idea of enriching a man by relieving him of his debts.) Thus when chlorine reacts with hydrogen sulphide to produce sulphur:

$$H_2S(g) + Cl_2(g) \rightarrow 2HCl(g) + S(s)$$

although oxygen does not appear in any of the substances involved the reaction is capable of interpretation in oxidation and reduction terms. Because the chlorine molecule effectively *removes* hydrogen from the hydrogen sulphide the chlorine is, by our latest definition, an oxidizing agent. In a similar way since the hydrogen sulphide has in effect added hydrogen to the chlorine then it may be regarded as a reducing agent.

Another example may illustrate this point further. Consider the reaction

$$3Fe(s) + 4H_2O(g) \rightarrow Fe_3O_4(s) + 4H_2(g)$$

in which steam is passed over red-hot iron to produce iron oxide. Since oxygen has been added to the iron, the iron has been oxidized by the water (which is therefore the oxidizing agent in this reaction), and the iron is the reducing agent since it has removed oxygen from the water.

A very important point is that it is most dangerous to regard any particular substance as 'an oxidizing agent' or as 'a reducing agent' without at the same time stating the reaction under discussion. Examine the following reactions.

(1) $$H_2O(g) + C(s) \rightarrow H_2(g) + CO(g)$$

(2) $$2H_2O(l) + 2F_2(g) \rightarrow 4HF(aq) + O_2(g)$$

According to our definition so far, in reaction (1) the water is behaving as an oxidizing agent, since it transfers oxygen to the carbon. In reaction (2), however, the water transfers *hydrogen* to the fluorine, and it is therefore behaving as a reducing agent. Hence we must not regard the property of 'oxidizing agent' or 'reducing agent' as being fixed for any given substance, but we must examine each reaction to decide in every single case what the behaviour of any particular substance is.

Summary

Oxidation is the addition of oxygen to a substance,
or the removal of hydrogen from a substance.

Reduction is the addition of hydrogen to a substance,
or the removal of oxygen from a substance.

An oxidizing agent is a substance which transfers oxygen *to* another substance,
or removes hydrogen *from* that substance.

A reducing agent is a substance which transfers hydrogen *to* another substance,
or removes oxygen *from* that substance.

Oxidation as the addition of electronegative elements

We have already seen (Chapter 9) that some elements have a tendency to attract electrons to themselves (to become, ultimately, negative ions) and others have a tendency to donate electrons to other substances (to become positive ions). Examples of electronegative elements which tend to attract electrons are the non-metallic elements, for example fluorine, chlorine, oxygen, sulphur, and bromine. The so-called electropositive elements, those which tend to donate electrons, are often the metallic elements, for example sodium, potassium, magnesium, calcium, zinc, and aluminium. Now because oxygen is an electronegative element, and because there are many reactions in which neither oxygen nor hydrogen is present, the concept of oxidation was extended to include not only the addition of oxygen, but also the addition of 'oxygen-like' elements – the electronegative elements. In a similar way the concept of reduction was extended to include 'hydrogen-like' elements – the electropositive elements. Therefore in the reaction:

$$C_2H_4(g) + Cl_2(g) \rightarrow C_2H_4Cl_2(g)$$

the compound ethene (C_2H_4) has been oxidized because chlorine, an 'oxygen-like' element, has been added to it. Consider also the reaction:

$$Mg(s) + 2HCl(aq) \rightarrow MgCl_2(aq) + H_2(g)$$

where the element magnesium, having removed chlorine from HCl, is behaving as a reducing agent.

In order to operate such a definition of oxidation and reduction it is important to know the relative electronegativities of the elements concerned. This is indicated below, both for metallic and for non-metallic elements.

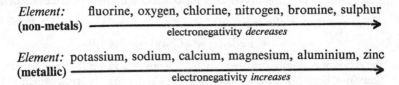

Element: fluorine, oxygen, chlorine, nitrogen, bromine, sulphur
(non-metals) ————————————————————————————————→
electronegativity *decreases*

Element: potassium, sodium, calcium, magnesium, aluminium, zinc
(metallic) ————————————————————————————————→
electronegativity *increases*

Using such information it would therefore be possible to decide, in some fairly difficult cases, which substance is acting as the oxidizing agent in a reaction. For example sulphur reacts with chlorine under certain conditions as follows,

$$S(l) + Cl_2(g) \rightarrow SCl_2(l)$$

and since both sulphur and chlorine are electronegative elements it would not be easy to decide, on the basis of the original definitions of oxidation and reduction, which substance is behaving as the oxidizing agent. However, since chlorine is much more electronegative than sulphur, chlorine is regarded as the oxidizing agent in this particular case.

Summary

Oxidation is the addition of an electronegative element to a substance
       or the removal of an electropositive element from a substance

Reduction is the addition of an electropositive element to a substance
       or the removal of an electronegative element from a substance.

Oxidation and reduction in terms of a change in oxidation number

The state of oxidation of an element in a compound is sometimes indicated by an **oxidation number**. The oxidation number of an element is a form of chemical book-keeping which can be applied to nearly all chemical substances, provided a set of simple rules concerning the idea is followed.

(1) All elements in the free state (that is, uncombined with any other elements) have an oxidation number of zero.
(2) In the case of a simple ion, the element has an oxidation number with the same size and sign of the charge on the ion. For example, the ion Cu^{2+} has an oxidation number of $+2$ and the ion S^{2-} has an oxidation number of -2.
(3) The sum of all the oxidation numbers of the elements in a compound is zero. For example in the compound $FeSO_4$

(oxidation number of Fe) + (oxidation number of S)
           + 4 × (oxidation number of oxygen) = 0

This rule enables us to work out the oxidation number of an element of which we are uncertain, *e.g.*, the sulphur in $FeSO_4$. In nearly all of its compounds the oxidation number of oxygen is -2. Therefore, using the above relationship

$$(+2) + (\text{oxidation number S}) + 4(-2) = 0$$

i.e., $+2 + \text{oxidation number of sulphur} - 8 = 0$

∴ oxidation number of sulphur $= +6$

With these simple rules in mind the concepts of oxidation and reduction may now be redefined.

Oxidation: when oxidation occurs the oxidation number of the element **increases.**
Reduction: when reduction occurs the oxidation number of the element **decreases.**

For example in the reaction

$$FeCl_2(s) + \tfrac{1}{2}Cl_2(g) \rightarrow FeCl_3(s)$$

the oxidation number of iron changes from $+2$ in $FeCl_2$ to $+3$ in $FeCl_3$, and therefore $FeCl_2$ has been oxidized to $FeCl_3$ by the chlorine, which is therefore the oxidizing agent. In a similar way the oxidation number of the chlorine changes from 0 in the free element Cl_2 to -1 in $FeCl_3$, and the chlorine gas has therefore been reduced by the $FeCl_2$.

The charge on a complex ion is the algebraic sum of the oxidation numbers of its constituent atoms. In the ion MnO_4^- manganese is in the oxidation state of $+7$ (corresponding to oxide, Mn_2O_7) and oxygen of -2. The expression, $7 + 4(-2)$, gives the residual ionic charge of -1. In SO_4^{2-}, the sulphur has to be taken as positive relative to oxygen and existing in the $+6$ oxidation state (corresponding to SO_3). Then the ionic charge is given by the expression:

$$S^{+6} + 4O^{-2} \rightarrow SO_4^{2-}$$

(which should **not** be interpreted as a normal equation). Notice that the sign for oxidation numbers *always comes before the figure*. There is no such thing as an S^{6+} ion, but by adding up the oxidation numbers of the sulphur and the oxygen, the correct charge on the sulphate ion is given.

Oxidation of an ion (or compound) raises the oxidation number of one at least of its constituent elements, while reduction brings a corresponding fall.

Summary

Oxidation is an *increase* in oxidation number.

Reduction is a *decrease* in oxidation number.

Oxidation in terms of electron transfer

The most recent definition of oxidation and reduction has brought together the ideas expressed in terms of electronegativity and of oxidation number, and it expresses the changes which take place during the process of oxidation through a consideration of the transfer of electrons from element to element in the course of the chemical reaction. From the electronic point of view the following are the definitions of oxidation and reduction.

Oxidation is the process of electron loss.

Reduction is the process of electron gain.

An oxidizing agent is an acceptor of electrons.

A reducing agent is a donor of electrons.

Oxidation and reduction always occur together; they are complementary processes of electron loss and electron gain respectively and must occur simultaneously. The electrons lost by the reducing agent must be accepted by the oxidizing agent present. The following examples illustrate these ideas.

EXAMPLE 1. When magnesium is oxidized by combination with oxygen, the metal is *oxidized* by losing two electrons per atom. These electrons are accepted by oxygen atoms, which are *reduced* as a result. Magnesium (giving out electrons) is the reducing agent; oxygen (accepting electrons) is the oxidizing agent.

$$Mg - 2e^- \rightarrow Mg^{2+}; \quad \tfrac{1}{2}O_2 + 2e^- \rightarrow O^{2-}$$

Magnesium oxide is a collection of Mg^{2+} and O^{2-} ions in equal numbers.

Notice that the combination of magnesium with chlorine or sulphur is a similar process, as:

$$Mg(s) + Cl_2(g) \rightarrow Mg^{2+}.2Cl^-(s)$$
$$Mg(s) + S(s) \rightarrow Mg^{2+}.S^{2-}(s)$$

Chlorine and sulphur must, like oxygen, rank as oxidizing agents; as before, magnesium is a reducing agent in these reactions. From this point of view, any conversion of a metal to its ions is oxidation, *i.e.*, electron loss; correspondingly, any conversion of a non-metal to its ions is reduction, *i.e.*, electron gain.

EXAMPLE 2. If a metallic ion, *e.g.*, the iron(II) ion, Fe^{2+}, is so treated that it loses a further electron, it is oxidized and is a reducing agent. The process is:

$$Fe^{2+} - e^- \rightarrow Fe^{3+}$$

An iron(III) ion is formed. An agent must be present, *e.g.*, chlorine, to accept the electrons made available by the ferrous ions. It acts as the oxidizing agent (electron acceptor) and is reduced.

$$\tfrac{1}{2}Cl_2 + e^- \rightarrow Cl^-$$

The complete reaction can be represented:

$$Fe^{2+}(aq) + \tfrac{1}{2}Cl_2(g) \rightarrow Fe^{3+}(aq) + Cl^-(aq)$$

It will be observed that the 'valency' of the metal increases from 2 to 3 during the oxidation.

EXAMPLE 3. The 'removal of hydrogen' aspect of oxidation is interpreted in the following way. Consider the oxidation of hydrogen sulphide by chlorine. Hydrogen sulphide is slightly ionized as:

$$H_2S \rightleftharpoons 2H^+ + S^{2-}$$

The sulphide ion parts with its two electrons and is, therefore, oxidized, acting as a reducing agent.

$$S^{2-} - 2e^- \rightarrow S$$

The electrons are accepted by chlorine atoms, so that chlorine acts as an oxidizing agent and is reduced.

$$Cl_2 + 2e^- \rightarrow 2Cl^-$$

Adding the two equations,

$$S^{2-}(g) + Cl_2(g) \rightarrow S(s) + 2Cl^-(g)$$

The hydrogen ion of the hydrogen sulphide is unchanged.

EXAMPLE 4. The reduction of hot copper(II) oxide by hydrogen is given as:

$$Cu^{2+}.O^{2-}(s) + H_2(g) \rightarrow Cu(s) + H_2O(g)$$

It is clear that the copper(II) ion is reduced by electron gain, as:

$$Cu^{2+} + 2e^- \rightarrow Cu$$

The two electrons are made available by the reaction between the oxide ion O^{2-} and hydrogen:

$$O^{2-} + H_2 \rightarrow H_2O + 2e^-$$

By combining with oxygen in this way and supplying electrons to the metallic ion, hydrogen exercises reducing properties. The oxide ion is oxidized by electron loss and the oxygen atom remains in combination with hydrogen as water.

EXAMPLE 5. Nitric acid can operate as an oxidizing agent, by accepting electrons, in several different ways. The two of greatest importance are:

$$4HNO_3(aq) + 2e^- \rightarrow 2NO_3^-(aq) + 2H_2O(l) + 2NO_2(g) \qquad (i)$$

$$8HNO_3(aq) + 6e^- \rightarrow 6NO_3^-(aq) + 4H_2O(l) + 2NO(g) \qquad (ii)$$

The electrons are supplied by a reducing agent. A metal often acts in this way, e.g.,

$$Cu \rightarrow Cu^{2+} + 2e^-$$

To supply the two electrons needed in equation (i), one copper atom is required. This yields the reaction:

$$Cu(s) + 4HNO_3(aq) \rightarrow Cu^{2+}.(NO_3^-)_2(aq) + 2H_2O(l) + 2NO_2(g)$$

This is the chief reaction occurring when copper reacts with concentrated nitric acid. The products are copper(II) nitrate, water, and nitrogen dioxide.

To supply the six electrons needed in reaction (ii), three copper atoms are required. This yields the reaction:

$$3Cu(s) + 8HNO_3(aq) \rightarrow 3(Cu^{2+}.2NO_3^-)(aq) + 4H_2O(l) + 2NO(g)$$

This is the principal reaction when copper reacts with a mixture of concentrated nitric acid and water in equal volumes, and is the recognized laboratory preparation of nitrogen oxide, NO. This gas is also the product when iron(II) sulphate solution is warmed with nitric acid of suitable concentration:

$$6Fe^{2+}(aq) + 8HNO_3(aq) \rightarrow 6Fe^{3+}(aq) + 6NO_3^-(aq) + 4H_2O(l) + 2NO(g)$$

The iron(II) ions are oxidized (by electron loss) to the iron(III) state.

The table below gives a list of some common oxidizing and reducing agents, and their usual mode of operation in terms of electron exchange.

Oxidizing agents	*Reducing agents*
Oxygen (page 285) $\frac{1}{2}O_2 + 2e^- \rightarrow O^{2-}$	Hydrogen sulphide (page 400) $H_2S \rightleftharpoons 2H^+ + S^{2-}$ $S^{2-} \rightarrow S + 2e^-$
Chlorine (page 369) $\frac{1}{2}Cl_2 + e^- \rightarrow Cl^-$	Sulphur dioxide (aqueous) (page 405) $SO_2 + H_2O \rightleftharpoons H_2SO_3 \rightleftharpoons 2H^+ + SO_3^{2-}$ $SO_3^{2-} + H_2O \rightarrow SO_4^{2-} + 2H^+ + 2e^-$
Ozone (page 294) $O_3 + 2e^- \rightarrow O_2 + O^{2-}$	Hydrogen (with heated metallic oxides) (page 52) $O^{2-} + H_2 \rightarrow H_2O + 2e^-$
Hydrogen peroxide (page 296) $H_2O_2 + 2H^+ + 2e^- \rightarrow 2H_2O$ The H^+ is supplied by water or dilute acid present in the liquid	Carbon monoxide (with heated metallic oxides) (page 316) $O^{2-} + CO \rightarrow CO_2 + 2e^-$
Nitric acid (page 438) $4HNO_3 + 2e^- \rightarrow$ $\quad 2NO_3^- + 2H_2O + 2NO_2$ $8HNO_3 + 6e^- \rightarrow$ $\quad 6NO_3^- + 4H_2O + 2NO$	Carbon (with heated metallic oxides) (page 310) $\quad O^{2-} + C \rightarrow CO + 2e^-$ or $2O^{2-} + C \rightarrow CO_2 + 4e^-$

Notice that all the oxidizing agents are electron acceptors; all the reducing agents are electron donors. Examples of their oxidizing or reducing action will be found on the pages quoted.

Evidence for the transfer of electrons during an oxidation–reduction reaction (commonly known as a 'redox' reaction) is given in the following experiment, in which iron(III) sulphate solution and potassium iodide solution are made to react together in two quite distinct ways.

Experiment 40
Electron transfer in the reaction between iron(III) sulphate solution and potassium iodide solution

---

(a) To a few cm³ of iron(III) sulphate solution in a test-tube add a few drops of potassium iodide solution. The brown colour of iodine will be seen to form throughout the solution.

$$Fe_2(SO_4)_3(aq) + 2KI(aq) \rightarrow$$
$$2FeSO_4(aq) + K_2SO_4(aq) + I_2(aq)$$

Expressed in ionic form to illustrate the electron transfer this becomes:

$$Fe^{3+}(aq) + e^- \rightarrow Fe^{2+}(aq)$$
$$I^-(aq) \rightarrow \tfrac{1}{2}I_2(aq) + e^-$$

in which an electron is transferred from the iodide ion to the iron(III) ion. The iodide has reduced the iron(III) compound.

(b) The same reaction may be carried out without mixing the two solutions, as follows.

Set up the arrangement shown in Figure 84, in which the iron(III) sulphate solution and the potassium iodide solution are contained in separate beakers, each of which has a platinum (or similar) electrode dipping into it, the two electrodes being joined to a sensitive voltmeter. On connecting the two beakers with a strip of

filter paper soaked in common salt solution (this is known as a 'salt bridge', and completes the circuit) observe carefully the changes taking place in the two beakers.

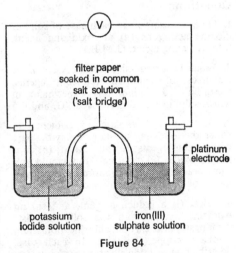

Figure 84

The appearance of the brown colour of iodine in the potassium iodide beaker indicates that the iodide ions are being oxidized to iodine. Since there is no source of d.c. supply in the circuit there is no electrolysis taking place, nor do the two solutions come into contact. The salt solution in the filter paper does not react with either solution (try separate test-tube experiments to convince yourself of this).

Because the salt bridge completes a full electrical circuit the electrons *lost* by the iodide ions are able to transfer to the iron(III) ions in the other beaker, and thereby carry out the reduction of Fe^{3+} ions. The appearance of iodine around the *electrode* in the potassium iodide solution is a further indication of the part played by the electrons in this process, as is the reading on the voltmeter in the external circuit.

Oxidation and reduction in electrolysis

It should be noted here (see also page 152) that a cathode, *as a source of electrons*, is equivalent to a reducing agent, and that discharge of positively charged ions at a cathode is, chemically, a reduction. The hydrogen ion, for example, is reduced (by electron gain) to the hydrogen atom: $H^+ + e^- \rightarrow H$. The electron is supplied by the cathode. Similarly, discharge of metallic ions at a cathode is, in all cases, a reduction to the ordinary atomic state of the metal, as:

$$Na^+ + e^- \rightarrow Na \quad \text{or} \quad Al^{3+} + 3e^- \rightarrow Al$$

Correspondingly, an anode, *as an electron acceptor*, is equivalent to an oxidizing agent. Discharge of negatively charged ions at an anode is an oxidation. For example, chloride ion, Cl^-, is oxidized to the atom by electron loss, the electron being accepted by the anode, as:

$$Cl^- - e^- \rightarrow Cl$$

Because the potential difference between the electrodes in electrolysis can be made large, the *cathode* can become a **very strong reducing agent** in effect. This is what happens in many cases of the extraction of metals from their ores, where conventional reducing agents (such as hydrogen, carbon, or carbon monoxide) are not powerful enough (see Chapter 36).

Questions

1. Give *one* example in each case of sulphur dioxide acting as (*a*) an oxidizing agent, (*b*) a reducing agent. (J.M.B.)

2. What is a reducing agent? Give *three* examples of common reducing agents. Describe and explain any experiment in which sulphuric acid is reduced. (O. and C.)

3. Describe and explain experiments in which the following substances play the part of oxidizing agents: (*a*) nitric acid; (*b*) copper(II) oxide; (*c*) chlorine. How would you show practically that oxidation has occurred in *two* of these cases you select? (O. and C.)

4. Select (i) a reducing agent, and (ii) an oxidizing agent, from the following list: sodium sulphite, carbon, chlorine, potassium sulphate, copper(II) oxide. In each case, describe one reaction illustrating the oxidizing or reducing property. (O. and C.)

5. How would you make a solution of iron(II) chloride starting with iron wire? Name *three* reagents which would convert the solution containing iron(II) ions into one containing iron(III) ions, and write an equation for the reaction using *one* of these reagents and stating what operations you would carry out. Describe tests you would perform on the solutions to show the presence of the appropriate iron ions. (O.)

6. What is meant by the terms *oxidation* and *reduction*?

A white powder is known to be either an oxidizing agent or a reducing agent. Describe *two* tests that you could carry out to determine which it is.

State which reactant is oxidized in each of the following reactions and, in each case, give a reason:

(*a*) $2H_2S + SO_2 \rightarrow 2H_2O + 3S$
(*b*) $2FeCl_2 + Cl_2 \rightarrow 2FeCl_3$
(*c*) $SO_2 + H_2O + NaClO \rightarrow$
$$NaCl + H_2SO_4$$
(*d*) $2K + 2H_2O \rightarrow 2KOH + H_2$ (S.)

7. Define oxidation in terms of (*a*) oxygen, (*b*) hydrogen, (*c*) electrons.

Oxygen, manganese(IV) oxide (manganese dioxide), potassium permanganate, chlorine, hydrogen peroxide, and sodium hypochlorite.

From the list given above select (i) a gaseous oxidizing agent, (ii) an oxidizing agent which is commonly used in solution, (iii) an oxidizing agent which can be used in its solid form.

For each of the examples selected, describe a reaction which illustrates its use as an oxidizing agent, and state why you regard the effect as oxidation. (J.M.B.)

8. Describe the reactions which occur between the following pairs of substances. In each case state the necessary conditions, name the products, and indicate which reactant is oxidized. (*a*) Chlorine and phosphorus. (*b*) Hydrogen peroxide and potassium iodide solution. (*c*) Nitric acid and iron(II) sulphate. (*d*) Aluminium and iron(III) oxide. (J.M.B.)

9. Describe and explain, in terms of reduction/oxidation, the following reactions: (*a*) hydrogen sulphide reacting with chlorine; (*b*) a *named* metal reacting with dilute sulphuric acid; (*c*) a *named* metallic oxide converted into a metal; (*d*) chlorine reacting with potassium iodide solution; (*e*) a *named* metal reacting with copper(II) sulphate solution to give a precipitate of copper.

State clearly in each reaction which substance is oxidized and which is reduced.

Give an ionic explanation where appropriate. (O. and C.)

10. State what you would see, and explain the reactions which occur, when an excess of hydrogen peroxide reacts with (i) a solution of potassium iodide in dilute hydrochloric acid, (ii) a solution of iron(II) sulphate in dilute sulphuric acid, (iii) lead(II) sulphide.

Describe what you would see, and state whether hydrogen peroxide is acting as an acid, oxidizing agent, or reducing agent in each of the following:

(iv) the reaction with silver(I) oxide,
$$Ag_2O + H_2O_2 \rightarrow$$
$$2Ag \downarrow + H_2O + O_2$$

(v) the reaction with barium hydroxide solution,
$$Ba(OH)_2 + H_2O_2 \rightarrow$$
$$BaO_2 \downarrow + 2H_2O. \text{ (J.M.B.)}$$

11. What do you understand by the term *reduction*?

Each of the following substances can behave as a reducing agent: carbon monoxide, hydrogen sulphide, carbon, iron(II) ion. Give *one* example of such behaviour for each substance, giving the equation and stating clearly what has been reduced.

Quote and explain *one* example of a reduction taking place during electrolysis. (O.)

12. Oxidation is often defined as increase in the positive valency of an element due to loss of electrons, and reduction as the reverse of oxidation. Explain how this statement applies in the case of the following reactions:

(a) the conversion of hydrogen ions to hydrogen molecules;

(b) the conversion of iron(II) ions into iron(III) ions;

(c) the liberation of sulphur from hydrogen sulphide;

(d) the conversion of sulphite ion into sulphate ion.

In each case describe an experiment in which these reactions occur, stating what reagents you use, and what you observe. (S.)

13. What do you understand by the term *reduction*?

Outline how you could convert (one method in each case):

(a) iron(II) ions (Fe^{2+}) into iron(III) ions (Fe^{3+});

(b) ammonium ions (NH_4^+) into ammonia;

(c) copper(II) ions (Cu^{2+}) into copper;

(d) zinc into zinc ions (Zn^{2+}).

Hydrogen peroxide solution (H_2O_2) reacts with acidified potassium iodide according to the following equation:

$$H_2O_2 + H_2SO_4 + 2KI \rightarrow$$
$$K_2SO_4 + 2H_2O + I_2$$

Calculate the mass of iodine liberated when a solution containing 13.6 grammes of hydrogen peroxide is added to an excess of potassium iodide solution acidified with dilute sulphuric acid. (S.)

14. 'By defining oxidation and reduction in terms of electron transfer we are able to bring under one heading such apparently diverse kinds of reaction as: (i) the combination of magnesium with oxygen; (ii) the displacement of bromine from aqueous bromide by chlorine; (iii) precipitating copper from aqueous solutions of its salts by adding iron; and (iv) the electrolysis of lead(II) bromide.'

(a) By explaining these reactions in terms of electron transfer, illustrate the truth of the above statement.

(b) Discuss briefly and with the aid of equations, *two* reactions which you are sure are oxidation–reduction reactions, but which are not readily seen as involving electron transfer. (N.I.)

18 Chemical Reactions and Equilibrium

Chemical reactions may be classified according to certain types of phenomena which accompany them (see Chapter 1). They can be further subdivided into classes of reactions, each of which has its own characteristics, and a few of these will be considered briefly.

Combination

This takes place when two or more substances combine to form a single substance.

EXAMPLE 1. Iron and sulphur *combine* when heated to form iron(II) sulphide.
$$Fe(s) + S(s) \longrightarrow FeS(s)$$

EXAMPLE 2. If warm lead(IV) oxide is lowered into a gas-jar of sulphur dioxide, the two compounds *combine* and lead(II) sulphate is formed.
$$PbO_2(s) + SO_2(g) \longrightarrow PbSO_4(s)$$

Decomposition

This occurs when a compound splits up into simpler substances. This change usually takes place without the necessity for the presence of a second substance, and very often the action of heat is sufficient to cause the reaction to take place.

EXAMPLE 1. If calcium carbonate (for example, marble) is heated in an *open* crucible to bright red heat, the calcium carbonate *decomposes* into calcium oxide (lime) and carbon dioxide.
$$CaCO_3(s) \longrightarrow CaO(s) + CO_2(g)$$

EXAMPLE 2. If potassium chlorate is heated strongly it *decomposes* into potassium chloride and oxygen.
$$2KClO_3(s) \longrightarrow 2KCl(s) + 3O_2(g)$$

Displacement

This occurs when one element (or group) takes the place of another element (or group) in a compound.

EXAMPLE 1. If zinc is placed in copper(II) sulphate solution, copper is *displaced* by the zinc and zinc sulphate is left in solution.
$$Zn(s) + CuSO_4(aq) \longrightarrow ZnSO_4(aq) + Cu(s)$$

EXAMPLE 2. If chlorine is bubbled into potassium bromide solution, the chlorine *displaces* bromine and a red bubble of bromine is formed. A solution of potassium chloride is left.

$$2KBr(aq) + Cl_2(g) \rightarrow 2KCl(aq) + Br_2(l)$$

Double decomposition

This name has been given to reactions in which two compounds take part, both are decomposed and two new substances formed *by an exchange of radicals*. Double decomposition reactions are always of the type:

$$A.B + C.D \rightarrow A.D + C.B$$

For example:

$$Cu.SO_4 + H_2.S \rightarrow Cu.S + H_2.SO_4$$

Commonly, both the original compounds used in the reaction are soluble in water, while, of the products formed, one (sulphuric acid) is soluble and one (copper(II) sulphide) is not. Usually the precipitated compound is the one which is wanted, for it can easily be separated and purified by filtration and washing. Less frequently, the important product of a double decomposition reaction is more volatile than the other compounds concerned and is driven off either as a gas or, by heating, as the vapour of a volatile liquid. For example:

$$NaCl + H_2SO_4 \rightarrow NaHSO_4 + HCl\uparrow$$
$$\text{gas}$$

$$KNO_3 + H_2SO_4 \rightarrow KHSO_4 + HNO_3$$
$$\text{volatile}$$
$$\text{liquid}$$

It must be observed, however, that, from the modern point of view, many reactions of 'double decomposition', especially those occurring in solution, are regarded as taking place between compounds which are already fully ionized, so that no decomposition takes place. The situation is merely that, if the ions present in the mixture can form an *insoluble* combination, they will do so, and the corresponding compound will precipitate. For example, if a solution containing sodium chloride (that is, the ions Na^+ and Cl^-) is mixed with one containing silver nitrate (that is, the ions Ag^+ and NO_3^-), silver chloride precipitates. The essential change is represented in the form:

$$Ag^+(aq) + Cl^-(aq) \rightarrow AgCl(s)$$

Ions Na^+ and NO_3^- remain in solution. Such reactions of *ion aggregation* are virtually irreversible because the very low solubility of the precipitate suppresses any possible reverse with the dissolved ions.

Reversible reactions

So far, most of the reactions we have considered proceed quite definitely in a certain direction, and it is possible to identify substances in the reaction which are called *reactants* and other substances, the result of the reaction, which are called *products*. Now there exists a group of reactions in which the direction of chemical change can be easily reversed by changing the conditions under which the reaction is taking place. For example, when hydrated copper(II) sulphate is heated the blue colour of the crystals changes to the white appearance of the anhydrous salt. The change may be represented by the following equation:

$$CuSO_4.5H_2O(s) \rightarrow CuSO_4(s) + 5H_2O(g)$$
$$\text{hydrated salt} \quad \text{anhydrous salt}$$
$$\text{(blue)} \quad \text{(colourless)}$$

However, anhydrous copper(II) sulphate may be changed to the blue hydrated form simply by taking a sample of the anhydrous salt and adding water to it (this is the familiar test made for the presence of water).

$$CuSO_4(s) + 5H_2O(l) \rightarrow CuSO_4.5H_2O(s)$$

anhydrous salt   hydrated salt
(colourless)   (blue)

It is clear that we have managed to carry out the reaction which is the *reverse* of the one stated above. Because the reaction *can* be easily reversed it is known as a *reversible reaction* and this is designated in a special way when the equation for the reaction is written,

$$CuSO_4.5H_2O(s) \rightleftharpoons CuSO_4(s) + 5H_2O(g)$$

the sign $\rightleftharpoons$ indicating that the reaction may proceed in one direction or the other according to the conditions under which it is carried out.

Another example of a reversible reaction is given in the following experiment.

Experiment 41
A reaction which goes both ways – the colour of bromine water in acid and alkali solutions

Into a 100 cm³ beaker pour approximately 20 cm³ bromine water, and stand the beaker on a sheet of white card or paper. Using a teat pipette add 2M sodium hydroxide solution slowly to the bromine water with occasional stirring and observe the change in colour which takes place. Immediately after the colour change, add drops of 2M sulphuric acid from a second teat pipette until the original colour is restored. Repeat the alternate addition of drops of alkali and acid several times.

This reaction may be made to go either in one direction or the other by changing the pH of the solution, that is, by making the solution either predominantly acidic or alkaline (see page 264). It is therefore an example of a reversible reaction.

Chemical equilibrium

In the reversible reactions we have considered so far we have arranged the conditions such that one set of substances is converted *completely* into another set of substances, which in turn may be *completely* converted back by a change in conditions. Clearly the idea of 'reactants' and 'products' is a confusing one in such circumstances, where the 'products' of the reaction in one direction become the 'reactants' of the same reaction in the other direction! Since it is possible for the reaction to proceed in either direction is it also possible that it may do so at the same time? That is, can the change taking place in one direction occur at the same time as the reverse reaction is also taking place? If this did happen, then it is possible that the reaction might come to some kind of 'balance' in which 'reactants' and 'products' are all present simultaneously.

Suppose that n molecules of a substance A can react with m molecules of a substance B to produce x and y molecules respectively of products C and D, the system being homogeneous (*i.e.*, entirely liquid or entirely gaseous).

$$nA + mB \rightleftharpoons xC + yD$$

As soon as a little of C and D is formed, a reverse reaction will begin. At first the forward reaction will predominate but, as C and D accumulate, the reverse reaction will build up until an *equilibrium* position is reached with forward and reverse reactions proceeding at the same rate. The composition of the mixture will then appear constant, though it is the net result of the two opposing reactions, not a static situation.

Although the state of affairs in such a reaction is described as an *equilibrium* situation, it is not the common idea of equilibrium normally encountered in everyday life. Thus, when a book is placed upon a table, it may be said to be in a state of equilibrium; when two children balance one another on a see-saw, they may be said to be in equilibrium with one another; and a ball balanced on the top of a long stick held in the hand of a juggler is also in an equilibrium state. But chemical reactions in equilibrium are quite different from all of these examples. Since chemical equilibrium involves the balancing of two reactions which are proceeding at the same time in opposite directions it is said to be a *dynamic equilibrium* – that is, it is an equilibrium involving the constant interchange of particles in *motion*.

Whereas in the case of the book, the children, and the ball it is possible to *observe* the fact that they are all in equilibrium, it is difficult to do so in the case of chemical reactions for we cannot see the individual particles involved. However, it is possible to 'label' some of the particles taking part in a chemical reaction by making them radioactive, and by following the reaction with a Geiger–Müller counter the dynamic nature of the equilibrium can be demonstrated. This is achieved by the following experiment, in which the reaction under investigation is the dissolving of lead(II) chloride in water.

$$PbCl_2(s) \rightleftharpoons Pb^{2+}(aq) + 2Cl^-(aq)$$

Experiment 42
Dynamic equilibrium of lead(II) chloride in water

The arrangement for this experiment is illustrated in Figure 85. The experiment is carried out in three parts.

(*a*) Prepare a saturated solution of lead(II) chloride by adding 2M hydrochloric acid drop by drop to a 10 cm³ 0.2M lead(II) nitrate solution. When precipitation is complete, filter (or centrifuge) off the solid lead(II) chloride, and

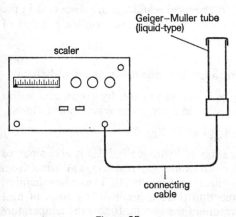

Geiger–Muller tube (liquid-type)

scaler

connecting cable

Figure 85

transfer the *filtrate* to the Geiger–Müller tube (specially designed to accept liquids), and take the reading on the Scaler over a period of five minutes. Work out the number of counts per minute; this is known as the 'background' count.

(*b*) Prepare a sample of solid lead(II) chloride containing radioactive lead by the following method. To 10 cm³ of 0.1M thorium nitrate solution in a beaker, add 5 cm³ 0.2M lead nitrate solution, and then add 2M hydrochloric acid drop by drop until precipitation is complete. The solid, which contains radioactive lead ions, should be filtered.

(*c*) Take two spatula measures of the radioactive lead(II) chloride from (*b*) and add it to the saturated solution of lead(II) chloride in (*a*), which has been removed from the Geiger–Müller tube and placed in a beaker. Stir the contents of the beaker for approximately fifteen minutes with a glass rod. After this period of time, filter off the solid, and transfer the filtrate (which should be perfectly clear) to the Geiger–Müller tube. Take the reading on the Scaler over a period of five minutes and work out the counts per minute.

The count rate in the second measurement will be found to be several times greater than the 'background' count. Can we explain this fact? The reaction we are investigating is the equilibrium

$$PbCl_2(s) \rightleftharpoons Pb^{2+}(aq) + 2Cl^-(aq)$$

in which solid lead chloride is in equilibrium with its ions in solution. The 'background' count, measured in Experiment 42(a), gave a measure of the radioactivity due to the Pb^{2+} and Cl^- ions in solution, and to any 'natural' radio-activity in the laboratory. This was very low since neither of the two ions was radioactive. However, on the addition of *solid* radioactive lead(II) chloride, the solution was eventually found to be quite highly radioactive (Experiment 42(b)), indicating that by some means or other radioactive lead ions had entered the solution. However, we know that before the solid radioactive lead(II) chloride was added the solution was already *saturated* with respect to lead ions; that is, it could not possibly dissolve any more Pb^{2+}. If a chemical equilibrium is a static one then it would be impossible for the radioactive Pb^{2+} ions to enter into the solution, but if the equilibrium is a dynamic one in which lead ions are entering solution from the solid at the same rate as lead ions in solution are becoming attached to the solid, then the radioactive lead ions *can* pass into solution. Since the solution eventually becomes radioactive, then the dynamic nature of chemical equilibrium is demonstrated.

Conditions affecting the balance of chemical equilibrium

The conditions affecting chemical equilibrium will now be considered in relation to a few reversible reactions of industrial or experimental importance. In the consideration of such reactions, a very valuable contribution can be made by a rule known as **Le Chatelier's Principle.** It is very widely applied to physical and chemical situations and, for the latter, takes the following form.

If a chemical system is in equilibrium and one of the factors involved in the equilibrium is altered, the equilibrium will shift so as to tend to annul the effect of the change.

The statement above simply summarizes the idea that whenever a system which is in equilibrium is affected by the increase or decrease in one of the factors tending to keep the system in equilibrium then the other factors will change their values in order to restore the balance or *position* of equilibrium. Some of the factors which commonly contribute to chemical equilibria are discussed in the following sections, and the effect of changing these factors on the position of equilibrium is discussed.

The effect of temperature change on the position of equilibrium

In a reaction which proceeds entirely in the gas phase, hydrogen and iodine combine together to form hydrogen iodide in a reversible reaction as follows:

$$H_2(g) + I_2(g) \rightleftharpoons 2HI(g)$$

in which the forward reaction (formation of hydrogen iodide) is accompanied by the evolution of heat (it is therefore an *exothermic* reaction, as distinct from an *endothermic* reaction which absorbs heat; see Chapter 20). Thus when chemical equilibrium is established for this reaction there is a constant output of heat taking place, maintaining the temperature of the system. If now the temperature of the vessel in which the reaction is taking place is suddenly increased the balance of the reaction is momentarily disturbed, and Le Chatelier's Principle requires the reaction to respond to oppose this change, that is to *lower* the temperature. This can be achieved if the back reaction (the formation of hydrogen and iodine), which is endothermic, is allowed to predominate over the forward reaction, which is exothermic; in such a case the position of balance of the reaction is disturbed, and we say that the position of equilibrium has been shifted from right to left. This is equivalent to saying that the *new* equilibrium mixture will have less hydrogen iodide and more hydrogen and iodine than the 'original' equilibrium mixture. Hence we may summarize the effect of temperature on a chemical equilibrium as follows.

Forward reaction (*left to right*)	Change in temperature	Effect on position of equilibrium
EXOTHERMIC	increase	new equilibrium has more of substances on left (reactants in forward reaction)
	decrease	new equilibrium has more of substances on right (products in forward reaction)
ENDOTHERMIC	increase	new equilibrium has more of substances on right (products in forward reaction)
	decrease	new equilibrium has more of substances on left (reactants in forward reaction)

A common example of the effect of temperature on the position of equilibrium is the equilibrium between water and steam.

$$H_2O(l) \rightleftharpoons H_2O(g)$$

Here the forward reaction is accompanied by an absorption of heat, *i.e.*, it is an endothermic reaction. Thus if the temperature of the system is raised this will tend to produce more steam (equilibrium position moves to the right), whereas if the temperature is lowered more water will be formed (equilibrium position moves to the left) and the steam condenses. Notice that the condensation of steam is an exothermic process: which is why a steam scald is usually worse than one caused by water.

The effect of pressure change on the position of equilibrium

The effects of pressure changes are much more noticeable in reactions occurring in the gaseous state than those occurring between solids and liquids, since the volumes of solids and liquids are very little affected by even quite large changes in pressure.

Consider the gas phase reaction involving the decomposition of dinitrogen tetroxide into nitrogen dioxide.

$$N_2O_4(g) \rightleftharpoons 2NO_2(g)$$

Under any given set of conditions there will be present a mixture of the two compounds in a definite proportion. If all the factors involved in such an equilibrium are maintained, with the exception that the pressure upon the mixture of gases is altered, then the position of equilibrium will correspondingly change. This happens because, when the total pressure is increased, Le Chatelier's Principle demands that the equilibrium position of the reaction should change in order to restore the balance, and this can take place in the above reaction by a decrease in volume (since the total capacity of the reaction vessel is fixed a decrease in the volume of the gases is equivalent to a decrease in pressure). In the case of the reaction under consideration, increase in total pressure would result in the appearance of a greater proportion of N_2O_4 in the equilibrium mixture. In practice this is found to be so, for at relatively high pressures a sample of N_2O_4 is a very pale colour, indicating a high proportion of N_2O_4 present (which is colourless), whereas at relatively low pressures the colour of the sample is a dark brown, indicating that there is a high proportion of NO_2 present.

The effect of a change in pressure on an equilibrium reaction may be summarized as follows.

Type of reaction	Effect of **increase** in total pressure	Effect of **decrease** in total pressure
Increase in number of molecules left to right $e.g.,\ 2O_3(g) \rightleftharpoons 3O_2(g)$	Position of equilibrium moves to the left $e.g.,$ more O_3 in equilibrium mixture	Position of equilibrium moves to the right $e.g.,$ more O_2 in equilibrium mixture
Decrease in number of molecules left to right $e.g.,\ N_2(g) + 3H_2(g) \rightleftharpoons 2NH_3(g)$	Position of equilibrium moves to the right $e.g.,$ more NH_3 in equilibrium mixture	Position of equilibrium moves to the left $e.g.,$ more N_2 and H_2 in equilibrium mixture
No change in number of molecules, left to right $e.g.,\ H_2(g) + I_2(g) \rightleftharpoons 2HI(g)$	No effect Position of equilibrium maintained	No effect Position of equilibrium maintained

It will be seen from the table that change in total pressure has no effect on those reactions in which there is no change in the number of molecules as a result of the reaction.

The effect of concentration change on the position of equilibrium

If the concentration of one of the substances present in an equilibrium reaction is changed without change in any of the other conditions then, by Le Chatelier's Principle, the position of equilibrium will move to decrease the concentration of the added substance. Thus in the reaction

$$N_2(g) + O_2(g) \rightleftharpoons 2NO(g)$$

contained in a reaction vessel at a given temperature, if extra oxygen was pumped into the vessel the position of equilibrium would move from left to right, *i.e.*, there would be a greater proportion of NO in the new equilibrium mixture. In a similar way if a substance taking part in a chemical equilibrium is removed by some means, then the equilibrium position will change to produce more of that substance; this is again in accordance with Le Chatelier's Principle.

The following experiment also illustrates this effect.

Experiment 43
Reaction of water with bismuth(III) chloride

Arrange a test-tube rack with five empty test-tubes placed in it. Into the first tube place a spatula measure of bismuth(III) chloride and dissolve it in a small volume (about 2 cm³) of concentrated hydrochloric acid.

Fill each of the remaining test-tubes approximately two-thirds full of distilled water, and then add carefully, using a teat pipette, five drops of concentrated hydrochloric acid to the second test-tube, ten to the third, fifteen to the fourth, and twenty to the fifth.

Using a different teat pipette add five drops of the bismuth(III) chloride solution (in the first test-tube) to each of the other tubes. Compare the different appearances of the white precipitate in each of the tubes.

The reaction under investigation in this experiment is the equilibrium:

$$BiCl_3(aq) + H_2O(l) \rightleftharpoons BiOCl(s) + 2HCl(aq)$$

The turbidity of the solution, due to the white precipitate, occurs because of the formation of BiOCl, bismuth oxychloride. It is therefore to be expected from Le Chatelier's Principle that the greatest turbidity will occur in that test-tube with the lowest concentration of HCl, since addition of hydrochloric acid will drive the position of equilibrium from the right to the left. Compare this conclusion with that obtained from Experiment 43.

A further example of the effect of change in concentration of substances on position of equilibrium can be seen in the reaction between iron and steam.

This reaction is usually performed by boiling water in a flask and passing steam over iron filings in an iron tube at bright red heat. It produces hydrogen and tri-iron tetroxide and the reaction is reversible.

$$3Fe(s) + 4H_2O(g) \rightleftharpoons Fe_3O_4(s) + 4H_2(g)$$

In the conditions stated above, the reaction goes almost to completion from left to right in the sense that the iron can be converted almost entirely to the oxide. This is so because incoming steam maintains its concentration at a high level (forcing the reaction from left to right) while hydrogen is swept out of the reaction tube and its concentration continually tends towards zero (keeping the reverse action slight).

If a current of hydrogen is passed over red-hot triferric tetroxide, the conditions are reversed and the oxide can be almost completely reduced to iron.

If the iron tube is sealed, an equilibrium will ultimately be set up with all four materials present and the forward and reverse reactions maintaining the equilibrium. The composition of the mixture at equilibrium is determined

Change in concentration of substance	Effect on equilibrium position of reaction $A + B \rightleftharpoons C + D$
Increase in concentration of A or B	proportion of C and D increased *i.e.*, equilibrium shifts to right
Decrease in concentration of A or B	proportion of C and D decreased *i.e.*, equilibrium shifts to left
Increase in concentration of C or D	proportion of A and B increased *i.e.*, equilibrium shifts to left
Decrease in concentration of C or D	proportion of A and B decreased *i.e.*, equilibrium shifts to right

mainly by the temperature employed. Pressure is without influence on the equilibrium position because there is no change in the number of *gaseous* molecules present as a result of the reaction.

The table on page 187 summarizes the effect of changes in concentration of substances on the position of equilibrium in a chemical reaction, represented by

$$A + B \rightleftharpoons C + D$$

Some important industrial applications of chemical equilibrium

The ability to change the position of equilibrium in a chemical reaction by altering the conditions under which the reaction takes place is a most important feature of the considerations which a manufacturing chemist must take into account when deciding how the plant is to be set up. Since the manufacturer is anxious to obtain the maximum quantity of the product for the minimum cost, he has to arrive at the optimum conditions under which the reaction will produce the greatest proportion of desired product in the shortest possible time. The way in which this is achieved in practice is described briefly for two important industrial reactions, the Haber Process for the production of ammonia, and the Contact Process for the manufacture of sulphuric acid.

(1) Synthesis of ammonia by the Haber Process

Haber's process uses the reaction:

$$N_2(g) + 3H_2(g) \rightleftharpoons 2NH_3(g)$$

1 vol.　　　3 vol.　　　　2 vol.　at constant t. and p.

The reaction (from left to right) is exothermic, *i.e.*, liberates heat.

Effect of pressure. Ammonia is produced from its elements with *reduction of* volume. Therefore, if the system is in equilibrium and the pressure is then raised, the equilibrium must shift so as to tend to lower the pressure (Le Chatelier's Principle). To do this, the volume must be reduced by production of more ammonia. That is, *high pressure favours production of ammonia.*

Effect of temperature. The formation of ammonia from its elements is an exothermic change. If the system is in equilibrium and the temperature is then lowered, the equilibrium must shift so as to tend to raise the temperature again (Le Chatelier's Principle). That is, heat must be liberated by the production of more ammonia. That is, *low temperature favours the production of ammonia.* But lowering of temperature reduces the rate of reaction, so it is necessary to introduce a *catalyst* which will give a sufficient reaction rate in spite of a relatively low temperature.

Effect of concentration. If the system is in equilibrium and more nitrogen is then added to increase its concentration, Le Chatelier's Principle requires the equilibrium to shift so as to tend to reduce the nitrogen concentration. That is more ammonia will be produced to use up nitrogen. This increases the yield of ammonia relative to hydrogen, and vice versa if the hydrogen concentration is increased. However, in practice, there is no particular advantage in using excess of either material; the gases are used in the theoretical proportion of nitrogen to hydrogen, 1 : 3 by volume.

The conditions chosen in accordance with the above discussion are:

very high pressure (200–500 atm)
temperature about 450 °C
catalyst: finely divided reduced iron, usually
'promoted' by alumina

Hydrogen is manufactured from partial combustion of hydrocarbons, and nitrogen from the air. They are mixed in the required 3 : 1 proportion by volume and dried (*e.g.*, by silica gel). They are pre-heated by gases leaving the catalyst chamber and are then passed over the catalyst at 450 °C. The ammonia produced is absorbed in water or liquefied by refrigeration and the unused gases are recirculated.

(2) Sulphuric acid by the Contact Process

The first step in the production of sulphuric acid is the conversion of sulphur dioxide to sulphur trioxide according to the reaction

$$2SO_2(g) + O_2(g) \rightleftharpoons 2SO_3(g)$$

which is an exothermic one (from left to right).

Effect of pressure. Pressure changes are significant only in reactions concerning *gases*, because liquids and solids are almost incompressible. Take the Contact Process reaction; suppose that, in a system in equilibrium in the gaseous state, pressure is increased (with no other change). Le Chatelier's Principle requires the system to react to oppose the change, *i.e.*, to reduce the pressure towards its former value. Since 2 molecules of sulphur dioxide and 1 molecule of oxygen produce 2 molecules of sulphur trioxide, a total 3 volumes of sulphur dioxide and oxygen are converted to 2 volumes of sulphur trioxide (by Avogadro's Law). That is, the system must convert more sulphur dioxide and oxygen to sulphur trioxide to reduce pressure by reducing volume. In general, if a given (entirely gaseous) reaction proceeds with a *reduction* in the number of molecules present, it is favoured by *high* pressure. (Conversely, gaseous reactions involving *increase* in the number of molecules present are favoured by *low* pressure.) In practice, the Contact Process can reach a satisfactory 98% yield of sulphur trioxide (calculated on the SO_2 used) without recourse to pressure above atmospheric.

Effect of temperature change. Suppose the temperature of the reacting system to be lowered (with no other change). Le Chatelier's Principle requires the system to react so as to oppose this change, *i.e.*, to raise the temperature again. To do this, heat must be liberated by causing more sulphur trioxide to be produced, *i.e.*, *lower temperature is favourable to production of sulphur trioxide*. At once the difficulty arises that lowering of temperature reduces the *rate* of reaction so increasing the time required for the production of sulphur trioxide. For this reason a *catalyst* must be introduced, *e.g.*, vanadium pentoxide V_2O_5, which increases the rate of reaction and makes it viable at a relatively low temperature (450 °C) which is favourable to the production of sulphur trioxide. In general, *low* temperature gives an equilibrium favourable to an *exothermic* reaction but catalysis is needed to give a favourable reaction *rate*.

Concentration of reactants. Suppose that, in the Contact Process reaction, equilibrium has been reached in certain conditions and then the concentration of oxygen is raised (relative to sulphur dioxide). Le Chatelier's Principle requires the system to react to oppose this change, *i.e.*, to reduce the oxygen concentration towards its former level. This can only be done by combining it with sulphur dioxide to form sulphur trioxide, *i.e.*, increased concentration of oxygen favours conversion of more sulphur dioxide to trioxide. (Correspondingly, increased concentration of sulphur dioxide favours conversion of more oxygen to sulphur trioxide.) In practice, the sulphur dioxide and air mixture used contains about three times as much oxygen as is theoretically required for the sulphur dioxide content. (Use of more air renders the product too dilute in sulphur trioxide.)

The sulphur trioxide is removed from the equilibrium mixture by dissolving it in fairly concentrated sulphuric acid, forming 'oleum' which is then diluted to produce sulphuric acid of the required concentration (see page 411).

Thermal dissociation

A **thermal dissociation** is a reversible reaction brought about by the application of heat. Examples are the thermal dissociations of *ammonium chloride* and *dinitrogen tetroxide*.

$$NH_4Cl(s) \rightleftharpoons NH_3(g) + HCl(g)$$
$$N_2O_4(g) \rightleftharpoons 2NO_2(g)$$

In both cases, complete dissociation *doubles* the number of molecules and the volume (temperature and pressure constant). Consequently, since the mass of material is unchanged and the volume is doubled, the vapour density of the product is ultimately halved. For example, at about 350 °C, the vapour density of ammonium chloride is 14.5 against a required value of $\frac{1}{2}NH_4Cl$ or 26.75 for NH_4Cl molecules. This represents about 85% dissociation. Similarly, the vapour density of dinitrogen tetroxide which is close to 46 just above its boiling point (22 °C), corresponding to molecules of N_2O_4 (relative molecular mass 92), gradually falls with temperature rise until, at about 150° C, it is about 23, corresponding to NO_2 molecules (relative molecular mass 46). At intermediate temperatures, mixtures of N_2O_4 and NO_2 molecules are in equilibrium and give vapour densities between 46 and 23. The gas also changes colour from pale yellow through reddish-brown to almost black as temperature rises, and reverses the colours with cooling.

The above data are for one atmosphere pressure. If an equilibrium position exists at a certain temperature and pressure and then the pressure is *increased*, Le Chatelier's Principle requires both systems to oppose this change, *i.e.*, to reduce pressure towards its former value. This they can do by reducing the degree of dissociation. This reduces the number of molecules present and so reduces volume and pressure. Thus, in both cases, dissociation is reduced by increased pressure at constant temperature; conversely, reduced pressure increases dissociation.

Thermal decomposition is the name given to the break-up of a compound by heat without any recombination on cooling.

Questions

1. Define *reversible reaction, thermal decomposition*, and *thermal dissociation*. Describe any experiment you have seen to demonstrate thermal dissociation.

What happens when (a) lead(II) nitrate; (b) mercury(II) oxide; (c) nitrogen dioxide; (d) ammonium nitrate are heated? In each case state to which of the above classes the reaction belongs, giving your reasons. (J.M.B.)

2. Explain what is meant by (i) a reversible reaction, (ii) chemical equilibrium. Indicate *three* reversible reactions by appropriate equations. Choose *one* of these and describe in detail experiments by which the reaction can be shown to be reversible. (Mere statement of formation of a different substance is insufficient; some description or identification of the substance is necessary.) (A.E.B.)

3. Chemical reactions can be classified into several types. Study the following reactions and then indicate to which type they belong.
   (a) The action of heat on lead(II) nitrate.
   (b) The addition of dilute hydrochloric acid to silver nitrate solution.
   (c) The interaction of iron and steam in a closed vessel.
   (d) The action of chlorine on iron(II) chloride solution. (S.)

4. Write short notes on each of the following, stating the reagents used, the conditions required for the reaction to take place, and the products formed: (a) the Haber process; (b) the contact process. (J.M.B.)

5. What is meant by a 'reversible reaction'? Give *three* examples. In *one* case, describe how you would show experimentally that the reaction is reversible. (L.)

6. Sulphuric acid is manufactured by the Contact Process which makes use of the equilibrium reaction

$$2SO_2 + O_2 \rightleftharpoons 2SO_3$$

Heat is given out in the formation of sulphur trioxide.

State what effect there would be on the equilibrium concentration of sulphur trioxide if (i) the pressure were increased, (ii) the temperature were raised.

The actual conditions used in the process vary somewhat, but a rough average appears to be (a) an excess of air, (b) a temperature of about 450 °C, (c) a pressure slightly in excess of one atmosphere, and (d) a catalyst. Explain why each of the conditions (a) to (d) is necessary.

Sulphur trioxide reacts with water to give sulphuric acid.

$$SO_3 + H_2O \rightarrow H_2SO_4$$

What is the disadvantage of terminating the process in this way and how are the final stages accomplished in industry? (J.M.B.)

7. Give a different equation in each case to illustrate the following types of reaction: (a) thermal decomposition, (b) catalytic decomposition. Name a catalyst which could be used in (b). (J.M.B.)

8. The equation for the reaction by which ammonia is manufactured is

$$N_2 + 3H_2 \rightleftharpoons 2NH_3$$

(a) What would be the effect on the equilibrium concentration of ammonia of (i) increasing the pressure, (ii) increasing the nitrogen concentration? (b) The equilibrium concentration of ammonia increases as the temperature is lowered. Is heat evolved or absorbed when ammonia is formed? (c) Why is a catalyst used in this reaction? (J.M.B.)

9. A good draught is essential for a satisfactory yield of calcium oxide from calcium carbonate in a lime-kiln. Explain why this is so. (O. (part).)

10. (a) The equation for the dissociation of calcium carbonate is

$$CaCO_3 \rightleftharpoons CaO + CO_2$$

$\Delta H = +179 \text{ kJ mol}^{-1}$ (heat taken in)

What will be the effect on the proportion of calcium carbonate in the equilibrium mixture of (i) increasing the temperature, (ii) increasing the pressure? What conditions would be suitable for manufacturing calcium oxide from calcium carbonate on a large scale?

(b) At 400 °C, the three gases, hydrogen, iodine, and hydrogen iodide exist together in equilibrium. The equation for the reactions is

$$H_2 + I_2 \rightleftharpoons 2HI$$

What effect will an increase in pressure have on (i) the rate of reaction of hydrogen with iodine and (ii) the position of equilibrium? Name the product formed when hydrogen iodide is dissolved in water and describe what would happen if chlorine were passed into the solution. (L.)

11. The reaction represented by the following equation is said to be *reversible*:

$$Ag^+(aq) + Fe^{2+}(aq) \rightleftharpoons Ag(s) + Fe^{3+}(aq)$$

This question is concerned with the experiments which might be carried out to decide whether this reaction really is reversible.

  (i) What chemicals would you mix to make the reaction take place from left to right, and what would you expect to observe if they do react?

 (ii) Outline briefly any further tests and observations which could be made on the reaction mixture to decide whether reaction had taken place, at least partly, from left to right.

(iii) What chemicals would you mix to make the reaction take place from right to left, and what would you expect to observe if they do react?

(iv) Outline briefly any further tests and observations which could be made on this second reaction mixture to decide whether reaction had taken place, at least partly, from right to left. (L.)

12. The reaction

$$3Fe + 4H_2O \rightleftharpoons Fe_3O_4 + 4H_2$$

is reversible in conditions in which iron and its oxide are solid and the other reagents gaseous. Describe and explain what occurs (a) if hydrogen is passed over red-hot tri-iron tetroxide in an open tube, (b) if steam is passed over red-hot iron in an open tube. If an equilibrium system exists among all these four reagents in a closed vessel at red heat, what will be the (qualitative) effect, if any, on the equilibrium of (c) adding more of the oxide of iron, (d) doubling the total pressure, (e) doubling the partial pressure of hydrogen, temperature being constant throughout? Briefly explain your reasoning in each case.

13. The reaction $N_2 + O_2 \rightleftharpoons 2NO$ is reversible and (from left to right) endothermic, all reagents being gaseous. Explain the term *endothermic reaction*. If the above system is in equilibrium (at a temperature which allows quite rapid reaction), what, if any, will be the effect on the equilibrium of (a) doubling the total pressure, (b) doubling the pressure of nitrogen, (c) lowering the temperature slightly? Explain briefly why this reaction (when used industrially) was not catalysed. What alternative was available?

19 Rate of Reaction

Rate of change in chemical reactions

We are already familiar with the fact that some chemical reactions take a much longer time to complete than others. Indeed, in Chapter 18 we saw that some chemical reactions never go to completion; that is, they never convert *all* the reactants into products, but stop at an intermediate, equilibrium, position. Some reactions take a very short time to come to completion (the precipitation of insoluble salts from their constituent ions in aqueous solution, for example), some take a little longer (the rusting of a nail left in the open air would be an example), while some occur extremely slowly (like the formation of stalactites and stalagmites in caves).

This chapter is concerned with the measurement of the rates of some chemical reactions and with a discussion of the factors which influence those rates.

By the *rate* of a chemical reaction is simply meant the amount of reaction which occurs in unit time, and therefore if we wish to measure the rate of a chemical reaction we must choose some property of the reaction which will indicate how far the reaction has changed, and observe the way in which the magnitude of that property varies with time. It is advantageous to choose a property of the reaction which can be observed without disturbing the reaction itself, and in practice this is done whenever possible. Thus, in a reaction where a colourless substance is changing to a coloured one, the rate at which the intensity of the colour increases would give the rate of reaction. In a similar way, for a reaction in which a gas is liberated from solution, it may be possible to collect the gas and measure the way in which the volume of gas increases with time, in order to obtain the rate of reaction. The experiments contained in this chapter illustrate a variety of methods for measuring the rates of chemical reactions, and also illustrate the effects which various factors have on those rates. Among the factors to be considered are the surface area of the reactants, the concentration of the reactants, the temperature of the reaction mixture, the pressure of the reactants (for gases), the effect of light, and the addition of a catalyst.

The field of study concerned with rates of reaction is known as 'chemical kinetics', and it is on the basis of the kinetic theory that the explanations for many of the effects which are described in this chapter are founded. You should therefore make certain that you are familiar with the fundamental ideas of the kinetic theory, as outlined in Chapter 3, before proceeding with this work.

The effect of concentration of reactants on the rate of reaction

On the basis of elementary kinetic theory a chemical reaction will occur only if the particles of reacting substances are allowed to come into contact. The rate of the reaction would therefore be expected to depend upon the frequency with which the particles collide, which in turn will depend (among other factors) on their density, *i.e.*, on their concentration. The more 'crowded' the particles are, the more often we should expect them to bump into one another. Thus concentration changes might reasonably be expected to lead to changes in the rate of reaction. The following experiment is designed to investigate this.

Experiment 44
To investigate the effect of concentration on the rate of reaction

The reaction to be studied is that between dilute hydrochloric acid and sodium thiosulphate solution to produce a precipitate of sulphur. Since this reaction produces a precipitate from two colourless solutions, the intensity of the precipitate at any given moment in time represents the extent of the reaction. The way

or note the time on a watch with a second hand. Swirl the beaker carefully a couple of times and place it on a white card with a cross drawn on it. Observe the cross by looking down through the solution from above the beaker and stop the clock (or take the time) the moment the cross is no longer visible (Figure 86).

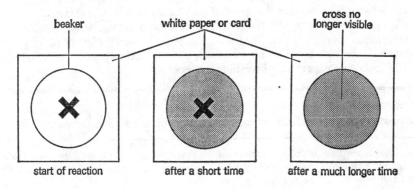

Figure 86

in which the intensity of the precipitate is measured is by carrying out the entire reaction in a beaker, and by placing the beaker and its contents on a white piece of paper with a cross marked on it. The time for the disappearance of the cross when the contents of the beaker are viewed from above will give a measure of the time taken for a certain fraction of the reaction to occur (see Figure 86).

Place 50 cm³ of sodium thiosulphate solution (containing 40 g dm⁻³ of the compound) in a 100 cm³ beaker. Add 5 cm³ 2M hydrochloric acid and at the same time start a stopclock,

The concentration of the thiosulphate solution may be varied by taking 40, 30, 20, and 10 cm³ of original thiosulphate solution and making the total volume up to 50 cm³ each time. (This is to ensure the same depth of solution in the beaker for each reaction, so that the time of disappearance of the cross is made at the same stage in each of the reactions.)

Draw up a table showing the way in which the time taken for the cross to disappear varies with the concentration of the thiosulphate solution.

You will notice from your table of results that as the concentration of the thiosulphate solution increases the time taken for the disappearance of the cross decreases. Now the rate of the reaction is inversely proportional to the time taken (a *fast* reaction takes a *short* time, but a *slow* reaction takes a *long* time), and therefore we may conclude that as the concentration of the sodium thiosulphate solution is increased, so the rate of reaction increases.

This conclusion seems to bear out our earlier supposition, based on the kinetic theory. If the reagents involved in a chemical reaction have their concentrations increased in a homogeneous mixture, more frequent molecular collisions will occur and the reaction rate will be thereby increased. A simple case of this occurs when addition of some concentrated acid increases the rate of liberation of hydrogen by the action of zinc on dilute hydrochloric acid. In the case of mixed *gases*, increase of *pressure* implies increase of concentration and tends to increase the rate of reaction between the gases because of more frequent molecular collisions. Reactions in the liquid or solid state are very little influenced by pressure because liquids and solids are almost incompressible.

The effect of temperature on the rate of reaction

We have supposed that chemical reactions occur because, as a first requirement, the atoms, molecules, or ions concerned in the reaction collide together. We further decided that any factor which increases the rate at which the particles collide would be expected to increase the rate of the reaction. When the temperature of a reaction is increased, heat is supplied to the particles involved in the reaction, and since heat is a form of energy we should expect that energy to go at least partly to increase the speed at which the particles travel (*i.e.*, the particles acquire kinetic energy). If the particles travel at a greater speed when the temperature is increased, then they will collide with one another at more frequent intervals, and we could therefore expect the reaction to proceed at a faster rate. The following experiment investigates this effect.

Experiment 45
To investigate the effect of temperature on the rate of reaction

We shall investigate the effect of temperature on the reaction between sodium thiosulphate solution and hydrochloric acid, since we have established a convenient method of measuring the rate of this reaction in Experiment 44.

Measure out 10 cm³ of sodium thiosulphate solution (concentration 40 g dm⁻³) into a 100 cm³ beaker. Add 40 cm³ water and warm the solution gently by placing the beaker on top of a gauze (supported on a tripod stand) and heat it with a small Bunsen flame. When the temperature is a little above 20 °C add 5 cm³ 2M hydrochloric acid solution to the beaker, simultaneously starting the stopclock and noting the temperature of the mixture. Stop the clock when the cross is no longer visible (see Figure 86).

Repeat the experiment several times, such that measurements are taken at temperatures over the range 20 °C to 60 °C.

Construct a table illustrating the way in which the time for disappearance of the cross varies with the temperature of the solution. If the temperature of the solution falls during the course of the measurement, the *average* temperature over the time concerned should be used in the table of results.

Your results should clearly indicate that as the temperature of the reaction is *increased* the time of disappearance of the cross decreases, which in turn means that the rate of the reaction *increases*. In general, rise of temperature tends to increase the rate of a chemical reaction by two factors: molecules of the reactants move more rapidly at higher temperature, producing both more frequent and more energetic collisions. For example, a mixture of hydrogen and oxygen (2 : 1 by volume) remains unchanged indefinitely at ordinary temperature but the introduction of a local source of high temperature, *e.g.*, an electric spark, starts an explosive combination of the two gases.

The effect of light on the rate of reaction

Light is a source of energy and can influence the rate of some chemical reactions considerably by energizing some of the molecules involved. For example, the reaction between chlorine and hydrogen at ordinary pressure is negligible in darkness, slow in daylight, but explosive in sunlight (at room temperature). This is thought to be due to the production of chlorine atoms from the chlorine molecule, the energy for which comes from the incident light:

$$Cl_2(g) \rightarrow Cl\cdot(g) + Cl\cdot(g)$$

after which the chlorine atoms (or 'free radicals' as they are often called), being very reactive, combine rapidly with the hydrogen molecule in a so-called chain reaction. The combination of methane and chlorine is similarly affected by light, and occurs by a similar mechanism.

Light is also vital to the photosynthetic production of starch by plants from carbon dioxide and water (page 236).

One of the commonest examples of the effect of light on the rate of a chemical reaction is in photography. Many solids are sensitive to light but the silver halides are among the most sensitive. This may be demonstrated by the following simple experiment.

Experiment 46
To investigate the effect of light on the decomposition of silver bromide

To approximately 20 cm³ potassium bromide solution in a large test-tube add a few cm³ of silver nitrate solution. An immediate precipitate of silver bromide will be formed. Quickly divide the precipitate into three parts, and put them into three separate test-tubes. Put one of the test-tubes immediately into a dark cupboard, the second may be left out on the bench, and the third may be placed near a source of strong light (*e.g.* in direct sunshine, or near to a lamp). Examine the colour of the precipitates at fairly regular intervals.

The change in colour of the precipitate from pale yellow to grey is due to the formation of metallic silver by interaction with the light. Thus the appearance of the test-tube placed in the dark cupboard remains virtually unchanged, whereas the test-tubes placed in the light change to a degree dependent upon the intensity of the light falling upon them.

The effect of surface area of reactants on the rate of reaction

The reactions studied so far in this chapter have been so-called *homogeneous* reactions, which means that all the reactants in any given reaction are present in the same phase, either all in solution, all in the gas phase, or, in the case of silver bromide, in the solid state. There are, however, many reactions which are classified as *heterogeneous*, because all the reactants are not in the same phase. An example of a heterogeneous reaction is that between solid zinc and dilute hydrochloric acid solution,

$$Zn(s) + 2HCl(aq) \rightarrow ZnCl_2(aq) + H_2(g)$$

and another would be that between magnesium and oxygen when the metal burns in air.

$$2Mg(s) + O_2(g) \rightarrow 2MgO(s)$$

When the reaction is a heterogeneous one, then particle size may influence the rate of the reaction. For example, aluminium foil reacts moderately with bench sodium hydroxide solution when warmed but powdered aluminium reacts rapidly from cold and will usually froth out of the test-tube spontaneously.

$$2Al(s) + 2NaOH(aq) + 6H_2O(l) \rightarrow 2NaAl(OH)_4(aq) + 3H_2(g)$$

This occurs because, for a given mass of metal, powder offers a much greater area to the reacting liquid than does foil. At the limit, subdivision to actual atoms or molecules (such as occurs in solution) offers maximum opportunity for reaction.

Experiment 47
To investigate the effect of surface area on the rate of reaction

The heterogeneous reaction chosen is that between dilute hydrochloric acid and marble, to liberate carbon dioxide.

$CaCO_3(s) + 2HCl(aq)$
$\rightarrow CaCl_2(aq) + CO_2(g) + H_2O(l)$

Since carbon dioxide is evolved during the course of this reaction the reaction flask containing the marble chips and hydrochloric acid solution will decrease in mass with time. It is the rate of decrease in mass of the flask and its contents which may be taken as the rate of the reaction. For this experiment it would therefore be desirable to have a direct-reading top-pan balance available, but this is not essential provided that weighings can be taken quickly.

Place 40 cm³ 2M hydrochloric acid in a 100 cm³ conical flask. To the conical flask add 20 g marble chips and quickly insert a loose plug of cotton wool in the neck of the flask. Weigh the flask and its contents and continue to take readings every half-minute for approximately quarter of an hour. The purpose of the cotton wool plug is to prevent the escaping carbon dioxide gas from carrying acid spray out of the neck of the flask, so damaging the balance. Construct a table giving the mass of the flask and contents at each half-minute interval.

The experiment should be repeated with 20 g of marble chips which have been crushed into much smaller pieces, so increasing the surface area for reaction.

The results of the two experiments may most conveniently be interpreted by plotting a graph of *mass of flask and contents* against *time*, giving the results of both experiments on the same graph. The appearance of the graph should be similar to that in Figure 87. It can be clearly seen that the smaller the pieces of marble the greater the rate of reaction, which can be attributed to the greater surface area of marble for attack by the acid in the case of the smaller pieces. Why do each of the curves A and B level off at the same point?

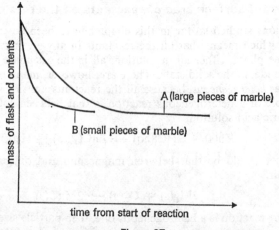

Figure 87

The effect of a catalyst on the rate of reaction

Catalysis is said to occur when the rate of a chemical reaction is altered by an agent (the *catalyst*) which is left unchanged in amount and in chemical nature at the end of the reaction.

A catalyst usually *increases* the rate of a reaction and this is called *positive* catalysis. It is found that, in a reversible reaction, a given catalyst influences the rate of the forward and reverse reactions equally. It has no influence on the actual equilibrium position which is reached; it only enables the equilibrium to be attained more rapidly.

***Definition.** A catalyst is a substance which, although often present in small proportions, alters the rate of a chemical reaction, but remains chemically unchanged at the end of the reaction.*

A catalyst may change its physical nature during a reaction, *e.g.*, coarsely powdered manganese dioxide may become fine powder (in the first example below) but it must be left *chemically unchanged* at the end of the reaction.

In general, a catalyst will function even though it is present in only minute proportion, *e.g.*, the rate of decomposition of hydrogen peroxide is measurably increased by the introduction of one ten-millionth of its mass of finely divided platinum.

Examples of catalysts for some common reactions are given in the following table.

Reaction	Catalyst
1. Heating of potassium chlorate $2KClO_3(s) \longrightarrow 2KCl(s) + 3O_2(g)$	Manganese(IV) oxide MnO_2
2. Synthesis of sulphur trioxide $2SO_2(g) + O_2(g) \rightleftharpoons 2SO_3(g)$	Vanadium(V) oxide V_2O_5 or platinum (powder)
3. Synthesis of ammonia $N_2(g) + 3H_2(g) \rightleftharpoons 2NH_3(g)$	Reduced iron (powder)
4. Decomposition of hydrogen peroxide $2H_2O_2(aq) \longrightarrow 2H_2O(l) + O_2(g)$	Manganese(IV) oxide MnO_2 or platinum powder

The following experiment illustrates the first of the catalysed reactions in the table above, the thermal decomposition of potassium chlorate which is catalysed by manganese(IV) oxide.

Experiment 48
The catalytic effect of manganese(IV) oxide on the thermal decomposition of potassium chlorate

The reaction under investigation is represented by the equation

$$2KClO_3(s) \longrightarrow 2KCl(s) + 3O_2(g)$$

and evidence that the reaction is proceeding can be obtained by inserting a glowing splint into the reaction vessel to test for the oxygen.

Mix a little manganese(IV) oxide with about four times its bulk of potassium chlorate and place in an ignition tube. Into each of two other tubes put an approximately equal bulk of manganese(IV) oxide and potassium chlorate respectively. Surround each with sand on a sand tray so that they are close together and vertical. (See Figure 88.) Start to heat and test at intervals for oxygen by lowering a glowing splint into each test-tube. After about one minute oxygen is freely evolved from the mixture, with no signs of gas from either of the other tubes.

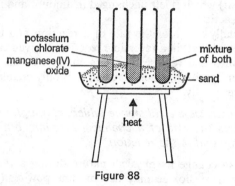

potassium chlorate
manganese(IV) oxide
mixture of both
sand
heat

Figure 88

In order to be quite certain that the oxygen was coming from the chlorate and not merely from the manganese(IV) oxide, it would be necessary to show that the mass of manganese(IV) oxide was the same after the experiment as before it. This could be done by dissolving the chlorate and chloride of potassium in water and recovering the manganese(IV) oxide by filtering, washing with water, and drying in an oven.

An inhibitor *decreases* the rate of a chemical reaction, and this is occasionally used to suppress an unwanted reaction (e.g. 2% ethanol in trichloromethane acts as an inhibitor for the oxidation of the trichloromethane to poisonous products by the air). Inhibitors are used less frequently than catalysts.

One of the outstanding features of catalysts is their specificity, and this makes it important that whenever an example of a catalyst is required the reaction catalysed is also quoted. It is true that some substances will catalyse more than one reaction, but it is not the case that catalysts can be interchanged between reactions at will.

The specific nature of catalysts is even more evident for catalysts in biological processes; they are known as *enzymes*. An enzyme will, as a rule, catalyse only one specific reaction and even then only within a very small temperature range. Most enzymes function most efficiently between 35 °C and 45 °C for biological processes in animals, and around 25 °C for processes in plants. The enzyme is usually only effective over a small range of pH, and most function in either neutral or slightly acidic media.

Summary of factors affecting the rate of reaction

The rate of a chemical reaction may, in general, be increased by any of the following factors:

(*a*) an increase in the temperature of the system;
(*b*) an increase in the concentration of the reactants;
(*c*) the addition of a catalyst, carefully selected;
(*d*) an increase in the surface area of the reactants (for heterogeneous reactions);
(*e*) an increase in the total pressure of the reactants (for reactions in the gas phase).

Light also affects the rate of some chemical reactions.

Questions

1. Describe fully one reaction to illustrate each of the following. Four *different* reactions are required. (*a*) A reaction whose rate is increased by raising the temperature. (*b*) A reaction whose rate is increased by raising the pressure. (*c*) A reaction whose rate is affected by the presence or absence of light. (*d*) A reaction in which varying the concentration of one of the reactants causes a different product to be formed. (A.E.B.)

2. Write an equation for a reaction which can be catalysed by manganese(IV) oxide (manganese dioxide). What would be the effect on the rate of this reaction of (i) adding more manganese(IV) oxide, (ii) using the same mass of catalyst but increasing the particle size?

The rate of reaction between hydrogen and chlorine depends on the intensity of light falling on the reagents. For another reaction whose rate also depends on the intensity of the light, (i) name the reagents and (ii) *either* name the product *or* state the importance or use of the reaction. (J.M.B.)

3. The following statements are made in a textbook.

The rates of most chemical reactions are approximately doubled by raising the temperature at which the reactions are carried out by 10 °C.

The rate at which a chemical substance reacts is directly proportional to its concentration.

(i) Describe the experiments you would carry out to test the truth of these two statements when applied to *either* the reaction between a metal and a dilute acid *or* the decomposition of hydrogen peroxide catalysed by manganese dioxide (manganese(IV) oxide).

(ii) Explain simply, in terms of the ions or molecules present, why the rate of a reaction is increased both by raising the temperature and also by increasing the concentration of the reagents. (C.)

4. The rate of a chemical reaction increases with increasing temperature. Select one chemical reaction and describe how you would demonstrate that its rate is greater at a higher temperature.

Ammonium chloride has a vapour density of 13.5 at 150 °C, whereas from its formula (NH_4Cl, formula mass 53.5) the vapour density would be expected to be 26.7. What is the explanation for this discrepancy? (L.)

5. Describe an experiment that you have either seen or performed to show how temperature affects the rate of *either* the reaction of a metal with a dilute acid *or* the decomposition of hydrogen peroxide solution. Include in your description the apparatus used, the experimental procedure, and the measurements made.

An excess of dilute hydrochloric acid is added to 10 g of marble (calcium carbonate) and left at constant temperature until the reaction has stopped. (i) Calculate the maximum volume of carbon dioxide at s.t.p. formed when the reaction has stopped. (ii) Draw a graph to show approximately how you would expect the volume of carbon dioxide formed to vary with *time* during the experiment. (C.)

6. Manganese(IV) oxide is a *catalyst* in the decomposition of aqueous hydrogen peroxide to water and oxygen. Using this as an example, explain the term *catalyst*. With the aid of a diagram, describe an apparatus which you could use to measure the volume of oxygen evolved in a given time. How would you show that the manganese(IV) oxide had acted as a catalyst? Describe *one* reaction, apart from the decomposition of hydrogen peroxide, in which a metal is used as a catalyst. (O. and C.)

7. Hydrogen peroxide, a colourless liquid, decomposes slowly in aqueous solution. The equation for the decomposition is given below.

$$2H_2O_2(aq) \rightarrow 2H_2O(l) + O_2(g)$$

This decomposition is said to be catalysed by certain metal oxides, including copper(II) oxide and chromium(III) oxide. Outline the experiments you would carry out to investigate whether copper(II) oxide and chromium(III) oxide do catalyse this decomposition. How would you decide which of these two oxides is the better catalyst? (L.)

8. An aqueous solution of hydrogen peroxide decomposes according to the equation

$$2H_2O_2 \rightarrow 2H_2O + O_2$$

This reaction occurs rapidly at room temperature (20 °C) in the presence of a suitable catalyst.

You are asked to measure the rate at which oxygen is evolved at room temperature when 0.5 g of a catalyst is mixed with 50 cm³ of 0.1M hydrogen peroxide solution. (*a*) Name a suitable catalyst for this reaction. (*b*) Sketch the apparatus you would use. (*c*) Briefly describe how you would carry out the experiment. (*d*) Sketch a graph showing how the volume of oxygen (plotted along the vertical axis) might vary with time. Label this graph A. (It is *not* necessary to use graph paper.) (*e*) Explain the form of this graph. (*f*) Using the same axes, sketch the graphs that might be obtained if the experiment was repeated with the hydrogen peroxide solution maintained at (i) 10 °C, (ii) 30 °C. Label these graphs B and C respectively. (*g*) In (*f*) the factor varied was the temperature of the hydrogen peroxide solution. Suggest *four* other factors that might affect the rate of the reaction. (*h*) How would you show that the catalyst was unchanged chemically at the end of the original reaction? (A.E.B.)

9. A small flask was connected to a gas syringe by means of a stopper and delivery tube. 30 cm³ of water and 0.5 g of manganese(IV) oxide (manganese dioxide) were placed in the flask. 5 cm³ of hydrogen peroxide were added, the flask was quickly stoppered and readings of the volume of gas in the syringe were recorded every 10 seconds.

Time (sec)	0	10	20	30	40	50	60	70	80
Volume (cm³)	0	18	30	40	48	53	57	58	58

(*a*) Plot a graph of the volume shown on the syringe against the time in seconds. Label this curve A. When the reaction was complete, the stopper was removed and the syringe was emptied of gas. Without emptying the flask, another 10 cm³ of water and 5 cm³ of hydrogen peroxide were added, and the experiment repeated exactly as before. Using the same axes as in (*a*), sketch a second graph to show how the volume of gas collected would vary with time in this second experiment. Label this second curve B. Explain the reasons for choosing the line you have drawn.

(*b*) In this reaction, the manganese(IV) oxide acts as a catalyst. Explain fully how you would attempt to prove this.

(*c*) Write an equation and state the conditions for another reaction of manganese(IV) oxide in which it does not act as a catalyst. What part is played by the oxide in this reaction? (J.M.B.)

10. In the laboratory, oxygen is obtained from hydrogen peroxide, using manganese(IV) oxide as a *catalyst*. (i) Define the term *catalyst*. Describe how you would confirm that this substance does behave as a catalyst in this reaction. (ii) With the aid of a diagram, describe how you would measure the volume of oxygen evolved in a given time. (iii) Write down the equation for the reaction. (iv) Calculate the volume of oxygen at s.t.p. which theoretically could be obtained from 50 cm³ of a solution of hydrogen peroxide which contains 68 g dm⁻³. (O. and C.)

11. The rate of decomposition of a solution of hydrogen peroxide is increased by the presence of a catalyst. Write down an equation for the decomposition and name a suitable catalyst. Describe, with the aid of a diagram, an experiment to measure the rate of decomposition of a solution of hydrogen peroxide.

What simple modification of your apparatus would enable you to investigate the effect of temperature change on the rate of decomposition of the hydrogen peroxide? In your answer, state clearly what measurements you would make and the units in which you would express the rate of decomposition. (C.)

12. You are informed by a friend, whose chemical knowledge is sometimes suspect, that black powders are frequently good catalysts. Describe carefully how you would investigate this theory by studying the effect of copper(II) oxide on the thermal decomposition of potassium nitrate or potassium chlorate. (N.I.)

13. Study the following list of catalysts and then decide which one is suitable for each of the reactions illustrated by an equation. Each catalyst can only be used once.

Manganese(IV) oxide, platinum, nickel, iron, vanadium(V) oxide.

(i) $2H_2 + O_2 \rightarrow 2H_2O$
(ii) $N_2 + 3H_2 \rightarrow 2NH_3$
(iii) $C_2H_4 + H_2 \rightarrow C_2H_6$
(iv) $2KClO_3 \rightarrow 2KCl + 3O_2$ (S.)

14. Define the term *catalyst*. Name one substance which catalyses the decomposition of hydrogen peroxide, and calculate the volume of oxygen (measured at s.t.p.) released in this reaction from a solution containing 0.1 mole hydrogen peroxide. Compare this volume with that of the oxygen released (measured at s.t.p.) by the reaction of 0.1 mole hydrogen peroxide with acidified potassium permanganate according to the equation

$$2KMnO_4 + 5H_2O_2 + 3H_2SO_4 \rightarrow$$
$$K_2SO_4 + 2MnSO_4 + 5O_2 + 8H_2O$$

What would you expect to see during this permanganate reaction?

What are the products of the electrolysis of water acidified with dilute sulphuric acid using platinum electrodes? Give equations for the electrode reactions. (O.)

20 Energy Changes in Chemical Reactions

All matter possesses energy in one form or another. Some substances have energy as a result of the fact that the particles of those substances are moving; such energy is known as *kinetic energy*. Other substances have energy stored within themselves, and this energy is a result of the positions of the particles contained in the substance relative to one another; this stored energy is known as *potential energy*. But determination of the magnitude of the total energy which any given substance possesses is an extremely difficult task, and is made especially so by the fact that this energy can be present in so many different forms. It is an easier task, however, to determine the *energy change* which takes place when that substance interacts with another. The energy *change* can also be recognized in many different forms; sometimes it is in the form of light (as in the burning of magnesium in oxygen), sometimes in the form of sound (as in the explosion of hydrogen and oxygen), sometimes in the form of electricity (as in a cell reaction), and nearly always in the form of heat (as in the burning of coal in a fire), and very often it is a mixture of several of these.

The most common form of energy change in chemical reactions is the *heat change*, and it is with this that the majority of this chapter will be concerned.

Exothermic and endothermic reactions

The great majority of chemical reactions are accompanied by a marked heat change. Two types of heat change are distinguished. Those reactions which proceed with an *evolution* of heat to the surroundings are said to be *exothermic* reactions, and those in which an *absorption* of heat from the surroundings takes place in passing from reactants to products are said to be *endothermic* reactions.

An exothermic reaction is one during which heat is liberated to the surroundings.

An endothermic reaction is one during which heat is absorbed from the surroundings.

Units of heat change

For many years the unit of heat change for chemical reactions has been the *kcal*.

A *kcal* is a *kilocalorie* and is the heat required to raise the temperature of one kilogramme of water by 1 °C. This is one thousand times as large as the ordinary calorie which raises the temperature of one gramme of water by 1 °C.

J is the symbol for the *joule*, which is now replacing the calorie as the international unit of energy. The joule is expressed in electrical terms; thus, if one coulomb of electricity passes at an electrical pressure of one volt, the energy involved is one joule. In general,

$$\text{volts} \times \text{coulombs} = \text{joules}$$

The relation between the calorie and the joule is:

$$1 \text{ calorie} = 4.18 \text{ joule}$$

The symbol, kJ, represents 1000 joule, and is now the internationally acceptable unit for heat change in chemical reactions.

The ΔH notation

Just as it is possible to express in a shorthand way, through the chemical equation, the reaction which is taking place, so a convention has been established by which energy changes in chemical reactions may be expressed.

The symbol for the heat change is ΔH (Δ = change in), and it is always written in connection with a chemical equation representing the reaction to which the energy change refers. Furthermore, unless otherwise stated, the magnitude of ΔH corresponds to the *mole* reacting quantities as expressed in the equation.

The usual convention now used is to refer the heat change to the reacting system, *i.e.*, if heat is lost by the reacting system (an *exothermic* reaction), ΔH is *negative*; if heat is taken in by the reacting system (an *endothermic* reaction), ΔH is *positive*.

Thus for the exothermic reactions involving the burning of hydrogen or carbon:

$$H_2(g) + \tfrac{1}{2}O_2(g) \rightarrow H_2O(l); \quad \Delta H = -286 \text{ kJ}$$
$$C(s) + O_2(g) \rightarrow CO_2(g); \quad \Delta H = -406 \text{ kJ}$$

and these equations may be interpreted as follows.

'When one mole of hydrogen gas is completely burned in oxygen it forms one mole of liquid water with the evolution of 286 kJ of heat.'

'When one mole of solid carbon is completely burned in oxygen to form one mole of gaseous carbon dioxide, 406 kJ of heat are liberated.'

In a similar way endothermic reactions are represented as follows.

$$\tfrac{1}{2}N_2(g) + \tfrac{1}{2}O_2(g) \rightarrow NO(g); \quad \Delta H = +90.3 \text{ kJ}$$
$$C(s) + 2S(s) \rightarrow CS_2(l); \quad \Delta H = +117 \text{ kJ}$$

These equations should be read as follows.

'When one mole of gaseous nitrogen monoxide is formed from the gaseous elements, 90.3 kJ of heat are absorbed from the surroundings.'

'When one mole of solid carbon reacts with two moles of solid sulphur to produce one mole of liquid carbon disulphide, 117 kJ of heat are absorbed from the surroundings.'

The importance of giving the state symbols in the equation should now be clear, since it will be important whether or not the substances are in the solid, liquid, or gaseous states so far as the energy change is required.

If in the reaction

$$H_2(g) + \tfrac{1}{2}O_2(g) \rightarrow H_2O(l); \quad \Delta H = -286 \text{ kJ}$$

the water had finally been produced in the form of steam, then the heat evolved would be considerably less, since energy is required to convert liquid water into gaseous steam. The change would then be

$$H_2(g) + \tfrac{1}{2}O_2(g) \rightarrow H_2O(g); \quad \Delta H = -242 \text{ kJ}$$

and you should work out for yourself from these two equations how much energy is required to convert one mole of liquid water into one mole of gaseous steam.

Standard conditions for energy changes

The value of an energy change, ΔH, for a reaction is not a fixed value independent of the conditions under which the measurement is made. We have already seen that it is important to indicate the physical state of the substances involved by inserting the state symbols into the equation for the reaction. It is also true that the value of ΔH depends upon other factors such as the temperature at which the measurement is made and (for reactions involving gases) the pressure upon the system.

For these reasons it is necessary when making accurate comparisons of energy changes for different reactions to choose the conditions under which the measurements are made very precisely. These conditions are known as the *Standard* conditions for the reaction, and have been chosen as follows.

Standard temperature $= 298$ K (25 °C)
Standard pressure $= 1$ atmosphere

These standard conditions should not be confused with s.t.p., which applies to the measurement of gas volumes. Whenever a heat change is measured under the standard conditions stated above it is given a special symbol, $\Delta H^{\oplus}$, the superscript $^{\oplus}$ indicating that the energy change is one measured under standard conditions.

Sources of heat change

As already stated, particles of a chemical substance will possess energy by virtue of the fact that they are *moving* (kinetic energy) and also because there is energy *stored* within them (potential energy).

The kinetic energy of the particles will be made up of different kinds of motion. The particles will be travelling through space and will therefore possess *translational* energy. They will also be rotating in several different ways, according to the shape of the molecule, and they will therefore possess *rotational* energy. For those particles with more than one atom there will also be vibrations within the molecule, and the substance will therefore possess *vibrational* energy. Now it is possible to calculate the total kinetic energy of a substance (translational + rotational + vibrational) from theoretical considerations, and it seems that the total kinetic energy of a mole of particles at room temperature is approximately 12 to 15 kJ.

Now if the source of the heat change in a chemical reaction was the kinetic energy of the particles, then those heat changes would be very small indeed, for the maximum possible value for ΔH would correspond to a transfer of *all* the kinetic energy from the substance, which would only give $\Delta H = -15$ kJ! However, we know that in practice ΔH values are of a much greater magnitude than this – often of several hundred kilojoules per mole. It seems likely therefore that the source of the heat change must be the *potential* energy which the substance possesses.

We have already seen (Chapter 9) that elements combine together to form compounds in a variety of ways, but whenever a compound is formed from its elements, bonds are made between the constituent atoms or ions. These bonds represent a storing of energy in the substance and are the source of potential energy, and hence of the heat change during chemical reaction. It is always necessary to *provide* energy in order to break a bond in a chemical compound, and in the same way energy is always *evolved* when a new bond is created from the constituent atoms. Thus:

bond breaking requires the **absorption** of heat **from** the surroundings (**endothermic** process);

bond making involves the **liberation** of heat **to** the surroundings (**exothermic** process).

Now, since during a chemical reaction bonds are *broken* in the reactants and bonds are *formed* in the products, it is quite clear that the balance between the endothermic process of bond breaking and the exothermic process of bond making will determine the overall sign of ΔH for the reaction. This may be summarized as follows.

Energy involved in bond breaking $>$ energy involved in bond making
$\Rightarrow$ ENDOTHERMIC REACTION

Energy involved in bond breaking $<$ energy involved in bond making
$\Rightarrow$ EXOTHERMIC REACTION

Types of heat change

Just as it has been possible to classify chemical reactions into certain types, so it is possible to identify energy changes in chemical reactions according to the type of reaction to which they refer. Some of these reactions will now be discussed, with an indication of how the measurement for ΔH can be made.

Heat of combustion

The process of combustion involves the burning of a substance in oxygen, and since many industrial and domestic sources of energy concern the burning of materials to provide heat, the heat of combustion is clearly an important quantity. In fact, the heat of combustion of a substance is *always* negative, that is, heat is always evolved when substances are burned in oxygen.

The heat of combustion of a substance is defined as *the heat change which takes place when one mole of the substance is completely burned in oxygen.*

The values of the heats of combustion for a few substances are given in the following table.

Substance	Heat of combustion/kJ mol^{-1}
$H_2(g)$	-242
$C(s)$	-394
$CO(g)$	-283
$H_2(l)$	-286
$CH_4(g)$ methane	-1560
$C_8H_{18}(l)$ octane	-6125
$C_2H_5OH(l)$ alcohol (ethanol)	-1367
$C_{12}H_{22}O_{11}$ sugar	-5865

Some of the substances in the above table are used as fuels in one form or another, but their relative effectiveness as fuels is not usually expressed in terms of their heats of combustion. It has been traditional to refer to the *calorific value* of the fuels, in which the units of measurement are Btu lb^{-1} (British thermal units per pound). Since a British thermal unit is the quantity of heat required to raise the temperature of one pound of water by 1 °F, the unit Btu lb^{-1} is a much larger unit than the kJ mol^{-1} in which laboratory measurements are made. This merely reflects the greater quantities of heat which are involved when fuels are used on an industrial scale. Since food is also a kind of biological fuel, it has been conventional to refer to the relative effectiveness of different foods in providing energy in terms of the number of 'calories' which each unit mass of food liberates. People on a diet are often said to be 'counting their calories',

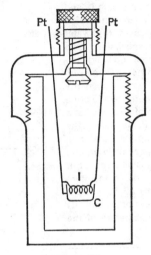

Figure 89

which simply means that they are regulating the quantity of food taken in according to the amount of energy which will be liberated by the biological combustion taking place in the body.

The accurate determination of the heat of combustion of a substance is carried out using a bomb calorimeter (Figure 89).

It is usually made of steel (for strength), nickel-plated on the outside and pro-tected from oxidation by a coating of enamel or (better) gold leaf on the inside. A known mass of a compound is placed in the platinum cup, *C*. Air is displaced by oxygen which is then allowed to reach 20–25 atm pressure and the bomb is closed by a screw valve.

The bomb is immersed in a known quantity of water in a well-lagged calori-meter fitted with stirrer and accurate thermometer. Electric current (through insulated platinum leads) heats the iron wire, I, which fires the experimental compound. This causes a sudden rise of temperature which is read on the ther-mometer (or, better, expressed on a temperature–time curve with readings at 15 second intervals). If the thermal capacity of the whole system (bomb calor-imeter, water, thermometer, calorimeter, and stirrer) is known and the tempera-ture rise is measured, the total heat evolved may be calculated (a correction must be made for the heat evolved by the iron wire, I). The heat of combustion then corresponds to the heat evolved when one mole of the compound is completely burned.

A much simpler, though not so accurate, method of determining the heat of combustion of a substance is described in Experiment 49.

Experiment 49
Estimation of the heat of combustion of ethanol

The liquid alcohol may be burned in an improvised lamp, made simply out of a specimen bottle with a plastic cap, through which may be inserted a wick which dips into the liquid contained in the bottle (see Figure 90). The wick passes through the cap of the bottle in a short piece of glass tube.

Choose a thin-walled tin can, weigh it, and place approximately 100 g water in it and weigh again. Record the exact mass of the water placed in the can. Fill the specimen bottle approximately half full of ethanol and weigh the bottle complete with top and neck. Clamp the tin can above the specimen bottle, insert the thermometer, and record the initial temperature of the water. Light the wick of the lamp and arrange the tin can so that the flame touches the bottom of the can and heats up the water in it. It may be necessary to arrange a shield around the lamp so that the flame remains steady. Stir the water frequently, and when the temperature has increased by approximately 25 °C, put out the flame and reweigh the bottle and its contents. Record the final temperature of the water.

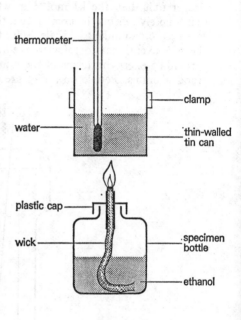

Figure 90

Specimen results

Initial temperature of water	= 17.7 °C
Final temperature of water	= 41.2 °C
Mass of water in tin can	= 101 g
Mass of lamp + ethanol before burning	= 29.974 g
Mass of lamp + ethanol after burning	= 29.592 g

Calculation

We must first find how much energy has been transferred to the water from the flame.

Energy evolved = mass water × heat capacity water × temperature rise

Taking the heat capacity of water as 4.2 J g^{-1},

$$\text{energy evolved} = 101 \times 4.2 \times (41.2 - 17.7) \text{ J}$$
$$= 101 \times 4.2 \times 23.5 \text{ J}$$
$$= 9970 \text{ J}$$

But mass of ethanol burned is (29.974 − 29.592) g
$$= 0.382 \text{ g}$$

Since the relative molecular mass of $C_2H_5OH = 46$, the number of moles ethanol burned $= \dfrac{0.382}{46} = 0.0083$ mole

Hence when 0.0083 mole ethanol are completely burned, 9970 J of energy are evolved.

Therefore when one mole of ethanol is completely burned,

$$\text{energy evolved} = \frac{9970}{0.0083}\,J = 1.185 \times 10^6\,J$$

$$= 1185\,kJ$$

Hence the heat of combustion of ethanol is $-1185\,kJ\,mol^{-1}$

Comparison of the value obtained in this experiment with that in the table on page 204 shows that the experiment has not been very accurate. Can you make a list of the possible sources of error in Experiment 49? Compare the method used in Experiment 49 with that described for the bomb calorimeter.

Although the method is an approximate one only, it may be conveniently used to *compare* the heats of combustion of different liquids. Thus the heats of combustion of the liquids methanol, ethanol, and propan-1-ol may be compared by this method, for example.

Heat of neutralization

Whenever acids and bases react together in a neutralization reaction (see Chapters 14 and 24), heat energy is always evolved. We already know that each hydrogen ion provided by the acid eventually appears as a molecule of water at the end of the reaction, in accordance with the generalization:

$$\text{acid} + \text{base} \longrightarrow \text{salt} + \text{water}$$

Now since different acids provide different numbers of hydrogen ions for neutralization, it is impossible to express the heat change for the reaction in terms of the number of moles of acid or base. Instead the energy change is referred to the number of moles of water formed in the neutralization process and the quantity is defined as follows.

The heat of neutralization of an acid or a base is the heat evolved when that amount of acid or base needed to form one mole of water is neutralized.

Thus the heat of neutralization is essentially the energy change for the reaction

$$\underset{\substack{\text{from} \\ \text{acid}}}{H^+(aq)} + \underset{\substack{\text{from} \\ \text{base}}}{OH^-(aq)} \longrightarrow H_2O(l)$$

Heats of neutralization may be determined quite simply in the laboratory as follows.

Experiment 50
Heat of neutralization of hydrochloric acid by sodium hydroxide solution

The 'reaction vessel' for this experiment can be any thermally insulated container. A plastic coffee cup, or plastic bottle, will serve the purpose well.

Pipette 20.0 cm³ 2M hydrochloric acid into the plastic bottle. Take the temperature of the acid solution, and also that of a 2M solution of sodium hydroxide solution in a beaker (wash and dry the thermometer between the two temperature measurements). *As rapidly as you possibly can,* transfer 20.0 cm³ of the 2M sodium hydroxide solution to the plastic bottle using a pipette, and using the thermometer to stir the mixture, record the maximum temperature attained.

Specimen results

Volume of 2M hydrochloric acid solution	$= 20.0 \text{ cm}^3$
Volume of 2M sodium hydroxide solution	$= 20.0 \text{ cm}^3$
Initial temperature of hydrochloric acid	$= 15.0 \,°C$
Initial temperature of sodium hydroxide solution	$= 15.4 \,°C$
Final (maximum) temperature of mixture	$= 28.2 \,°C$

Calculation

We first of all need to calculate the total heat evolved in the reaction.

The final volume of solution is $(20 + 20) = 40 \text{ cm}^3$, and within the accuracy of the experiment as we are performing it, we may assume the total mass of the solution is 40 g. Since the starting temperatures of the two solutions were not the same, and since the two solutions are present in equal volumes, we may average the initial temperatures to find the initial temperature of the mixture, which would therefore be $\dfrac{15.0 + 15.4}{2} \,°C = 15.2 \,°C$. Hence the temperature rise for the solution will be $(28.2 - 15.2) \,°C = 13.0 \,°C$. Taking the heat capacity of water as 4.2 J g^{-1} we have:

$$\begin{aligned} \text{heat evolved} &= \text{mass} \times \text{heat capacity} \times \text{temperature rise} \\ &= 40 \times 4.2 \times 13.0 \text{ J} \\ &= 2184 \text{ J} \end{aligned}$$

Now in 20 cm³ 2M hydrochloric acid there is $\dfrac{20 \times 2}{1000}$ mole HCl $= 0.04$ mole HCl (in a similar way, there is 0.04 mole NaOH).

The equation for the neutralization is

$$\text{HCl(aq)} + \text{NaOH(aq)} \to \text{NaCl(aq)} + \text{H}_2\text{O(l)}$$

and since one mole HCl produces one mole water

$$0.04 \text{ mole HCl will produce } 0.04 \text{ mole water}$$

Hence,

0.04 mole water is produced with the evolution of 2184 J heat energy

$\therefore$ 1 mole water will be produced with the evolution of $\dfrac{2184}{0.04}$ J heat energy

$$\begin{aligned} &= 54\,600 \text{ J} \\ &= 54.6 \text{ kJ.} \end{aligned}$$

Therefore the heat of neutralization is $-54.6 \text{ kJ mol}^{-1}$ water formed

The value obtained is again rather lower than the accepted value for this reaction (which is $-57.3 \text{ kJ mol}^{-1}$). A more accurate method of performing the experiment is illustrated in Figure 91.

The apparatus is allowed to stand to attain a steady temperature (which is read on the accurate thermometer). The test-tube is then broken, the acid–alkali solutions mix and are stirred, and there is a temperature rise which is read on the thermometer. (Better, the temperature can be read at 15 sec intervals and the results rendered as a graph.)

Whenever these more accurate determinations are carried out it is found that the heats of neutralization of all strong acids and all strong bases are constant; *i.e.*, $\Delta H = -57.3 \text{ kJ}$. This result is discussed in Chapter 24, on page 269.

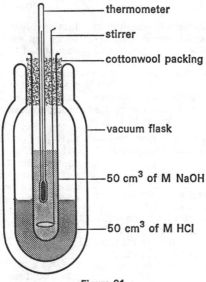

Figure 91

Conservation of energy

All energy changes which occur during chemical and physical changes of the kind considered above must conform to the Law of Conservation of Energy; that is, energy can only be changed from one form into its equivalent of another form with no total loss or gain.

To take the combustion of sucrose (cane sugar) to carbon dioxide and water as products (at ordinary temperature and pressure), we have the equation:

$$C_{12}H_{22}O_{11}(s) + 12O_2(g) \rightarrow 12CO_2(g) + 11H_2O(l); \quad \Delta H = -5685 \text{ kJ}$$

During this reaction, covalent bonds between carbon, hydrogen, and oxygen atoms in the sugar molecules are replaced by covalent bonds between carbon and oxygen atoms in carbon dioxide and between hydrogen and oxygen atoms in water. These changes involve alterations in electron orbits and, therefore, energy changes. Also, the forces holding sugar molecules together in crystals are overcome and replaced by forces holding water molecules together as liquid. At the same time, molecules of carbon dioxide are left as gas and molecules of water as liquid, both types of molecule possessing energy of motion. If these varying energy changes (and any others we may have forgotten) are allotted their correct signs and quantities, and account is similarly taken of the heat energy liberated during the combustion, the *total* energy change (obtained by adding these quantities together) must be *zero*.

The dissolution of a solid in water may be a purely physical change, *i.e.*, the separation of ions or molecules from crystals and their dispersal into the water. This requires a supply of energy to overcome inter-ionic forces (as in a Na^+Cl^- lattice) or van der Waals forces (as in many organic solids). This energy is taken from the water (as heat) and the temperature falls, *i.e.*, dissolution of a solid in water is often *endothermic*. In some cases, however, chemical reaction occurs, *e.g.*, ions may be hydrated, as with anhydrous copper(II) sulphate:

$$Cu^{2+}(s) + 4H_2O(l) \rightarrow (Cu.4H_2O)^{2+}(aq)$$

In such a case, this heat of hydration (added to the true heat of solution) may make the *total* change *exothermic*.

Questions

1. When hydrogen chloride gas dissolves in water there is an appreciable rise in temperature. What type of reaction is suggested and how do you account for the temperature change? How could you confirm your suggestion?

Hydrogen chloride dissolves in trichloromethane without any apparent change in temperature. What does this suggest? (S.)

2. The formation of methanol from hydrogen and carbon monoxide can be represented by

$$CO + 2H_2 \rightleftharpoons CH_3.OH \quad \Delta H = +91 \text{ kJ}$$

What mass of hydrogen would react to cause a heat change of 91 kJ?

What would be the effect on the equilibrium concentration of methanol in this endothermic reaction if (i) the temperature was increased, (ii) the pressure was increased, (iii) the hydrogen concentration was increased? (J.M.B.)

3. Producer gas is made by blowing air through red hot coke. Water gas is made by blowing steam through white hot coke. (a) Write equations for the reactions involved in these two processes. Do not describe the reactions. (b) In industry the two gases are made alternately, first producer gas and then water gas. Explain why this is so. (c) Which do you consider to be the better fuel, producer gas or water gas? Give the reason for your answer. (d) Design an apparatus and describe how you would use it in an attempt to prove that your answer to (c) was correct. You may assume that you are provided with each gas under pressure in a steel cylinder. (J.M.B.)

4. Explain why the heat of neutralization of sodium hydroxide by hydrochloric acid is the same as the heat of neutralization of potassium hydroxide by nitric acid.

$$NaOH + HCl \longrightarrow NaCl + H_2O;$$
$$\Delta H = -57.3 \text{ kJ} \quad \text{(Heat given out)}$$
$$KOH + HNO_3 \longrightarrow KNO_3 + H_2O;$$
$$\Delta H = -57.3 \text{ kJ} \quad \text{(Heat given out)}$$

(J.M.B. (part).)

5. Describe laboratory experiments (*one* in each case) which illustrate the following statements. (a) 'The decomposition of some compounds is speeded up by using a catalyst.' (b) 'The synthesis of compounds from their elements is often noticeably exothermic.'

Outline *one* industrial process in which a catalyst is employed. (You must not use any reaction given in your answers to the previous parts of this question.) (A.E.B.)

6. What is the composition of (a) water gas, (b) producer gas? Describe how these gases are made. Which of the methods of prepara-

tion involves an exothermic reaction and which an endothermic one? Explain the meaning of these two terms.

What do you understand by the term 'photosynthesis'? Explain briefly how this reaction takes place naturally. Mention *one* other chemical reaction which can be brought about by means of light energy. (S.)

7. An important gaseous fuel is butane C_4H_{10}. For the combustion of butane $\Delta H = -2880 \text{ kJ mol}^{-1}$. (i) Write down the equation for butane burning in excess air, including the heat change. (ii) Calculate the quantity of heat which would be evolved when 16 dm³ of butane, measured at room temperature and pressure, are burnt. (The gramme–molecular volume of a gas may be taken as 24 dm³ at room temperature and pressure.) (L.)

8. Give a careful description of the method you would use to prepare crystals of common salt from sodium hydroxide and hydrochloric acid.

The interaction of sodium hydroxide and hydrochloric acid is an 'exothermic' reaction. What do you infer from this statement, and how, in the course of your preparation, could you demonstrate its truth? Give one example in each instance of exothermic reactions occurring between (a) a solid and a gas, (b) two gases. (S.)

9. (a) Describe how you would attempt to find by experiment the quantity of heat liberated when one gramme of the liquid methanol, CH_3OH, is burnt in air.

(b) The table shows the heat liberated when 1.0 g of each of three alcohols is burnt in air.

	Heat evolved
Methanol CH_3OH	22.6 kJ
Ethanol C_2H_5OH	29.7 kJ
Propanol C_3H_7OH	33.4 kJ

For each of these alcohols, calculate the heat of combustion in kJ per mol.

(c) From your results in (b), estimate the heat of combustion of butanol, $C_4H_{10}O$.

(d) The substance dimethyl ether has the same molecular formula, C_2H_6O, as ethanol, but its heat of combustion is different. Suggest a reason for this difference. (L.)

10. The number of joules required to increase the temperature of 100 g of certain elements by 1 °C is given in the table opposite.

(a) How many moles are present in 100 g of each of these elements?

(b) Plot a graph of number of moles

Element	Relative atomic mass	Number of joules to raise 100 g by 1 °C
Magnesium	24	103
Aluminium	27	90
Copper	64	39
Molybdenum	96	27
Platinum	195	13

present in 100 g against the number of joules required to raise 100 g of the element by 1 °C. Use a scale of 2 inches (or 4 cm) for 1 mole along the x axis, and 1 inch (or 2 cm) for 20 joules along the y axis (the vertical axis).

(c) 54 joules are required to raise the temperature of 100 g of the element scandium by 1 °C. If a pattern has emerged from your graph, use this to estimate the mass of 1 mole of scandium.

(d) 100 g of scandium are found to combine with 236 g of chlorine (relative atomic mass 35.5). How many moles are present in these masses of scandium and chlorine? What information does this give about the formula of scandium chloride? (L.)

11. Two widely used fuels are natural gas and water gas. Natural gas is mainly methane, CH_4, with smaller amounts of the next two members of the alkane series (plus a little hydrogen). Water gas is an equimolar mixture of carbon monoxide and hydrogen. The heats of combustion of carbon monoxide and hydrogen are given.

$$CO(g) + \tfrac{1}{2}O_2(g) = CO_2(g);$$
$$\Delta H = -286.0 \text{ kJ}$$
$$H_2(g) + \tfrac{1}{2}O_2(g) = H_2O(g);$$
$$\Delta H = -244.0 \text{ kJ}$$

(a) From these figures calculate the value of ΔH in the complete combustion of 22.4 dm³ of water gas (measured at s.t.p.). (b) Write an equation for the complete combustion of methane in oxygen. (c) The complete combustion of one molar volume of methane (measured at s.t.p.), in oxygen liberates 617.0 kJ of heat more than the complete combustion of the same volume of water gas under the same conditions. Calculate the heat of combustion of methane in kJ. (Scottish.)

12. The heat change in dilute aqueous solution for the exothermic reaction

is
$$NaOH + HCl \longrightarrow NaCl + H_2O$$
$$\Delta H = -57.3 \text{ kJ}.$$

State the heat changes for the following reactions in dilute aqueous solutions and give reasons for your answers:

(a) $KOH + HNO_3 \longrightarrow KNO_3 + H_2O$
(b) $2NaOH + H_2SO_4 \longrightarrow$
$$Na_2SO_4 + 2H_2O. \text{ (J.M.B.)}$$

21 Radioactivity

Radioactivity was first noticed and investigated in 1896 by Becquerel, who found that uranium salts emitted rays with the following properties. They affected a photographic plate in the same way as rays of light, discharged a gold-leaf electroscope, and caused phosphorescence in certain materials, *e.g.*, zinc sulphide. Becquerel did not follow this investigation very far, but his discoveries aroused interest and activity in other workers. In particular, Marie and Pierre Curie, working in Paris about 1900, detected more intense radioactivity in certain mineral specimens, such as Bohemian pitchblende, and were able to isolate two radioactive elements, to which they gave the names *polonium* (in honour of Marie Curie's native Poland) and *radium*. Many other radioactive elements are now known.

Kinds of radiation

Becquerel found that the radiation emitted by uranium was not uniform but could be separated into three different types under the influence of an electrostatic field. The three types were named *alpha-*, *beta-*, and *gamma-rays*.

Alpha-rays were deflected towards the *negative* plate in the electrostatic field and so, since unlike charges attract each other, must carry a *positive* charge themselves. The deflection was not very marked, from which it was concluded that the moving particles constituting the rays were relatively massive. It has since been shown that alpha-rays are, in fact, helium ions in rapid motion. Helium ions are formed from helium atoms by loss of two electrons per atom. They then carry a double positive charge resulting from their possession of two nuclear protons and have a mass of about 4 units on the scale of $^{12}C = 12$.

$$He \rightarrow He^{2+} + 2e^-$$

Beta-rays showed a very marked deflection towards the *positive* plate of the electrostatic field. From this, it was concluded that the particles of which beta-rays are composed carry a *negative* charge and are relatively small in mass. Beta-rays have since been shown to consist of electrons in rapid motion. Beta-radiation is simply a moving stream of electrons.

Gamma-rays were not deflected at all by an electrostatic field; that is, gamma-rays carry no electrical charge. They are not material particles but consist of electromagnetic waves of very short wavelength (about 5×10^{-9} m). That is, they are of the same nature as light but of higher frequency. (See Figure 92.)

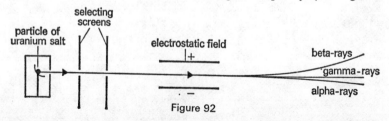

Figure 92

Nature of radioactivity

It was soon noticed that radioactivity is very different from ordinary chemical change. It is, for example, unaffected by temperature changes of the kind occurring in common scientific operations, being as rapid at the temperature of liquid air (about $-180\ °C$) as at red heat (about $850\ °C$), while chemical change is usually greatly accelerated by marked rise of temperature.

After a period of general uncertainty, Rutherford and Soddy put forward the idea, now universally accepted, that radioactivity arises from the spontaneous disintegration of unstable atomic nuclei. In fact, all atomic nuclei more complex than the nucleus of the bismuth atom (that is, containing more than 83 protons) are unstable. By disintegration, all such complex nuclei give rise to radioactivity. For example, the polonium atomic nucleus has 84 protons, the radium nucleus has 88 protons and the uranium nucleus 92 protons. To take a particular example, the radioactivity of radium is caused by the emission of an alpha-particle (helium ion, He^{2+}) from the nucleus of a radium atom. The helium ion carries away with it two protons (as well as two neutrons). This leaves behind an atomic nucleus containing 86 protons and this is the nucleus of an atom of the noble gas, *radon*. This change is accompanied by a very large evolution of energy (about 4×10^8 kJ per mole of radium) so that a solution of a radium salt maintains itself at a temperature above that of its surroundings. This radioactive change can be expressed by the equation:

$$^{226}_{88}Ra \longrightarrow {}^{222}_{86}Rn + {}^{4}_{2}He$$

Equations of the above kind must balance by having equal totals of atomic numbers and of mass numbers on each side. For the mass numbers, 226 on the left is balanced by $(222 + 4)$ on the right; for the atomic numbers, 88 on the left is balanced by $(86 + 2)$ on the right. In a similar way, the following equation expresses the fact that a sulphur atom can absorb a neutron and then emit a proton, so producing an atom of phosphorus.

$$^{32}_{16}S + {}^{1}_{0}n \longrightarrow {}^{32}_{15}P + {}^{1}_{1}p$$

Both the proton and the neutron have a mass of approximately 1 on the scale of $^{12}_{6}C = 12$. This accounts for the upper figures. The neutron has no charge, therefore has atomic number 0; the proton, with its single positive charge, has atomic number 1. This accounts for the lower figures and the equation balances as required.

Results of radioactive changes

It has been found that there are two important kinds of radioactive change.

Emission of alpha-particle

In this case, an atomic nucleus emits an alpha-particle, that is, a helium ion containing two protons and two neutrons. In this way the nucleus loses 4 units of mass. The two lost protons reduce the atomic number of the element by two units, consequently the product of the change is an atom with a mass number reduced by 4 units and an atomic number reduced by two units in comparison with the original atom. An example of this is:

$$^{226}_{88}Ra \longrightarrow {}^{222}_{86}Rn + {}^{4}_{2}He$$
(alpha-
particle)

Emission of beta-particle (or electron)

This kind of radioactive change is equivalent to the splitting of a neutron in the atomic nucleus of the radioactive element into an electron and a proton, the proton being then retained by the nucleus while the electron is emitted as beta-radiation. The electron is a very light particle and its loss makes no significant change in the atomic mass of the atom concerned. The retention of the proton raises the atomic number of the element by one unit. That is, the product has the same mass number as the original atom and an atomic number one unit greater. For example, a radioactive isotope of lead can lose a beta-particle to produce an atom of bismuth.

$$^{210}_{82}\text{Pb} \longrightarrow ^{210}_{83}\text{Bi} + _{-1}^{0}e$$

In terms of atomic number, a beta-particle (electron) being negatively charged counts as (-1), just as a proton positively charged counts as $(+1)$.

These types of change continue till a stable atomic nucleus is reached. It is now recognized that there are several series of radioactive change, named after the elements from which they originate in nature, *e.g.*, the *uranium* series, the *thorium series*, and the *actinium* series. The changes can be very complex. In the uranium series, the original uranium atom passes through radium as an intermediate and finishes as an atom of non-radioactive lead.

$$^{238}_{92}\text{U} \longrightarrow ^{226}_{88}\text{Ra} \longrightarrow ^{206}_{82}\text{Pb}$$

The uranium atom loses 10 protons in all, reducing its atomic number by ten units. It also loses 32 units of atomic mass.

These changes occur in pitchblende, a uranium ore from Bohemia, and the source of the radium isolated by the Curies. The ore also contains helium gas occluded in it. This gas is derived from the alpha-particles (helium ions) emitted during the radioactive changes.

Stability of radioactive elements

The mathematics of radioactive change is rather complex, but it leads to the conclusion that a convenient way of expressing the degree of stability of a radioactive element is to state its *half-life*, that is the time required for its radioactivity to fall to half its observed value at any given instant. A half-life may vary from a fraction of a second for very unstable elements to millions of years. Radium has a half-life of 1620 years; that is, any mass of radium existing in 1900 will have diminished to half that mass by about the year 3520.

Artificial nuclear transmutation

From what has been written above, it is obvious that radioactivity transforms the radioactive element into a different element. This change is called *transmutation* of elements. The early radioactive changes observed were all spontaneous and uncontrollable, but efforts were soon made to bring about artificial transmutations by experimental means. The first success was that of Rutherford (1919) who transmuted nitrogen into oxygen. The nitrogen was subjected to the action of swift alpha-particles, derived from a radium salt. The nucleus of a nitrogen atom captured one of these alpha-particles and then emitted a proton, leaving an oxygen atom as the product.

$$^{14}_{7}\text{N} + ^{4}_{2}\text{He} \longrightarrow ^{17}_{8}\text{O} + ^{1}_{1}p$$

Many other transmutations have since been achieved. Some of them are mentioned later when radioactive isotopes are considered.

Nuclear energy

It has already been noticed that radioactive changes can be accompanied by very great evolutions of energy in the form of heat, *e.g.*,

$$^{226}_{88}Ra \longrightarrow \, ^{222}_{86}Rn + \, ^{4}_{2}He; \quad \Delta H = -(4 \times 10^8) \, kJ$$

It was soon realized that if arrangements could be made to isolate radioactive materials and then control their activity, a new and very important source of energy would be made available for the human race. Unfortunately the first supplies of energy from this source were used for destructive purposes in two atomic bombs at the close of the World War of 1939–45. This development depended on the following circumstances.

In 1939, the German nuclear scientist, Hahn, discovered that the atomic nucleus of the isotope of uranium, $^{235}_{92}U$, could absorb a neutron and then break into two roughly equal parts. This is called *atomic fission*. The total products of the fission have a mass slightly less than that of the total initial material. This difference of mass is radiated as energy, the amount of energy being stated by the equation of Einstein,

$$E = mc^2$$

where E is the liberated energy in joules

m is the mass in kilogrammes

c is the velocity of light in m s^{-1} ($c = 3 \times 10^8$ m s^{-1})

The factor, c, is so large, especially when squared, that the loss of a small quantity of mass brings the liberation of an enormous quantity of energy.

This process of nuclear fission of $^{235}_{92}U$ can be made into a chain reaction in suitable conditions (Figure 93). This is because, while fission occurs as a result

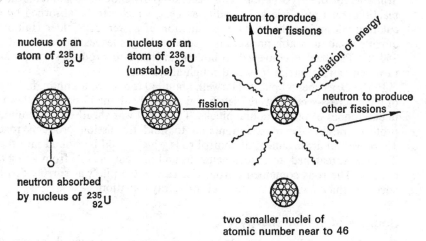

Figure 93

of neutron absorption by the uranium nucleus, the process of fission also emits neutrons. These are available for absorption to produce new fissions, and so on. If the mass of uranium is small, the neutrons may escape from it at such a rate that the chain reaction cannot be maintained, but, above a certain *critical mass*, the uranium absorbs the neutrons so as to produce very rapid nuclear fission of at least a substantial proportion of it. The energy liberated then produces a terrific explosion, with a temperature of several million degrees centigrade immediately after its occurrence.

The essential ideas relating to the production of the first atomic bomb were the following.

(1) A sufficient amount of $^{235}_{92}U$ had to be isolated from a natural source of uranium, which contains about 0.7% of $^{235}_{92}U$ and 99.3% of another isotope, $^{238}_{92}U$. (Both isotopes have atomic number 92.) The separation of the isotopes was brought about by producing the fluorides, UF_6, which are gaseous. $^{238}_{92}U$ produces a fluoride of higher molecular mass than $^{235}_{92}U$. The $^{235}_{92}U$ fluoride consequently diffuses more rapidly than the $^{238}_{92}U$ fluoride (by Graham's Law). The two fluorides were separated in the U.S.A. by very elaborate apparatus for fractional diffusion, and then the $^{235}_{92}U$ was obtained pure by decomposing the fluoride into its elements by heat.

(2) A mechanism was constructed to keep the $^{235}_{92}U$ apart in quantities below the critical mass until the decisive moment and then shoot them rapidly together to form one unit greater than the critical mass. A source of neutrons was also required to initiate the fission.

(3) A tamping device in the form of suitably shaped, thick steel sheet was placed round the uranium to reflect escaping neutrons back into the fissionable material. The isotope of plutonium, $^{239}_{94}Pu$, will also undergo nuclear fission and produce a similar chain reaction. It was also used in an atomic bomb in 1945.

The problem of adapting fission processes for useful purposes is that of releasing the same energy as in the fission bombs, but steadily over a relatively long period and at a temperature convenient for making use of the energy. The British Calder Hall reactor (1956) was based upon experience with an experimental atomic pile at Harwell. To avoid the very expensive separation of $^{235}_{92}U$, natural uranium was employed. By the use of pure graphite, fast neutrons were slowed to a velocity such that they are not absorbed by the $^{238}_{92}U$ present, but are available for the $^{235}_{92}U$ fission. This occurs rapidly enough to maintain the chain reaction. Energy is released steadily as heat, which can be absorbed by a circulating gas and applied for the production of power. At Calder Hall, carbon dioxide at about 7 atm pressure is used and the gas leaves the reactor at about 350 °C. The heat is conveyed to boilers and used to produce steam, which in turn generates electrical power through alternators.

In the experimental pile at Harwell, about 40 tonnes of uranium, in long bars of 6 cm² cross-section, were distributed regularly about 17 cm apart both ways in 800 tonnes of pure graphite blocks. The metal was sheathed in aluminium to protect uranium from oxidation and to hold the fission products together. There was an arrangement of control rods which could be wound into the pile, if danger threatened, to absorb neutrons and prevent the pile from going out of control. The rods contained boron. For safety, the pile was surrounded by an adequate thickness of concrete to absorb stray radiation.

Sources of stellar energy

Development of knowledge about atomic energy has revived interest in the sources of energy in the stars, especially the sun. The output of energy in the sun is maintained by the conversion of hydrogen into helium. By this process, four hydrogen nuclei (protons) are converted to one helium nucleus (alpha-particle) with the liberation of very large amounts of energy. This energy appears because the helium atom (4.0026) has a mass slightly less than that of four hydrogen atoms (4.0319) on the scale of $^{12}_{6}C = 12$. The lost mass is radiated as energy in accordance with Einstein's equation mentioned earlier, $E = mc^2$. The conversion of hydrogen to helium does not take place directly. A scheme has been suggested by which carbon intervenes in the process as 'catalyst', in a series of nuclear changes which also involve $^{13}_{7}N$, $^{15}_{8}O$, and other isotopes, to bring about the net change of hydrogen to helium. Temperatures of the order of millions of degrees centigrade are necessary and, even then, the conversion of hydrogen to helium is slow. The sun is estimated to contain enough hydrogen to continue radiating at its present rate for, at any rate, several thousand million years. A process of this kind in which several atoms produce a more complex atom is called *atomic fusion*.

The *hydrogen bomb* uses an adaptation of the synthesis of helium from hydrogen which occurs in the sun. In one version of this bomb, the two heavier isotopes of hydrogen, *deuterium* $^{2}_{1}D$, and *tritium* $^{3}_{1}T$, are fused to produce helium and neutrons.

$$^{2}_{1}D + ^{3}_{1}T \longrightarrow ^{4}_{2}He + ^{1}_{0}n$$

During this process, about 0.4% of the original mass is converted to energy. So great is this release of energy that an approximate tonne of material used in the hydrogen bomb is as destructive as several million tonnes of ordinary military explosive such as TNT. Deuterium can be obtained from water. Tritium is a product of the action of slow neutrons on lithium.

$$^{6}_{3}Li + ^{1}_{0}n \longrightarrow ^{3}_{1}T + ^{4}_{2}He$$

The temperature required for their fusion is several million degrees centigrade and is supplied by the action of a uranium fission bomb round which the fusion materials are placed.

Controlled nuclear fusion is now the subject of intense world-wide research. If it should be harnessed, the deuterium in sea water would provide all the earth's energy needs for the foreseeable future. The problem is largely one of containing ionized gas (plasma) at many millions of kelvin in a magnetic field so that it does not touch the walls of the apparatus and cool down. An alternative approach using high-powered laser beams is now under study.

Artificial elements

The most complex atomic nucleus known on earth up to about 1940 was that of uranium (atomic number 92). More recently, more complex nuclei have been assembled. The first two of these can be produced by exposing $^{238}_{92}U$ to neutron bombardment. $^{238}_{92}U$ captures a neutron to give the isotope, $^{239}_{92}U$. This passes, by a stage of beta-decay, into *neptunium* and then, by a similar stage, into *plutonium*. These elements have atomic numbers of 93 and 94 respectively and were previously unknown. Both are radioactive and plutonium is fissionable.

$$^{238}_{92}U + ^{1}_{0}n \longrightarrow ^{239}_{92}U$$
$$^{239}_{92}U \longrightarrow ^{239}_{93}Np + _{-1}^{0}e$$
$$^{239}_{93}Np \longrightarrow ^{239}_{94}Pu + _{-1}^{0}e$$

Other, more complex, nuclei have also been produced in recent years to atomic number 103 (Lawrencium Lr), though some of them have occurred in very minute amounts, such as a few hundred atoms, and some are very unstable.

It will be seen from the above account that the alchemists' dream of converting one element into another has now become a commonplace of physics. Their idea of converting base metal into gold is, however, somewhat beside the point. To do this would either liberate or absorb great quantities of energy, depending on the nature of the base metal used. In the first case, the value of the energy would dwarf that of the gold; in the second case, the cost of the gold in energy would be prohibitive.

Radioactive isotopes of common elements

Radioactivity was discovered in the most complex atoms, such as those of uranium, polonium, and radium, and, in nature, is restricted mainly to atoms of this kind. It is possible, however, by methods outlined below, to produce radioactive isotopes of many common elements and to employ them usefully in various ways. Such radioactive isotopes usually have a short half-life (a few days or weeks) and so decay to vanishing point quite rapidly. This is why they are not found in nature to any noticeable extent.

Two examples of the production of radioactive isotopes of common elements are the following.

(1) If a sodium salt, such as the chloride or carbonate, is irradiated by an intense concentration of neutrons in a uranium pile for a few days, some of the sodium atoms (mass number 23) will absorb a neutron each to give radio-sodium atoms (mass number 24), which are radioactive.

$$_{11}^{23}\text{Na} + _0^1 n \rightarrow _{11}^{24}\text{Na}$$

(2) In conditions similar to the above, sulphur atoms will absorb one neutron each, then radiate a proton and finish as a radioactive form of phosphorus.

$$_{16}^{32}\text{S} + _0^1 n \rightarrow _{15}^{32}\text{P} + _1^1 p$$

Ordinary phosphorus is $_{15}^{31}\text{P}$.

Use of radioactive isotopes

These uses are principally curative, general industrial uses, and uses in scientific research.

Curative uses

Cancerous growths may sometimes be eradicated by exposure to gamma-rays (*i.e.*, electromagnetic waves of very high frequency), which are derived from radioactive materials. Both healthy and diseased tissue are destroyed by uncontrolled exposure, but by regulating the dosage and directing it suitably, healthy tissue can be protected.

Radium was formerly used for this purpose but radio-cobalt has taken its place. This material is made by irradiating ordinary cobalt, $_{27}^{59}\text{Co}$, with neutrons.

$$_{27}^{59}\text{Co} + _0^1 n \rightarrow _{27}^{60}\text{Co}$$

The product, cobalt-60, is radioactive, emitting very penetrative gamma-rays. It is placed at the centre of a sphere of lead which suppresses most of the radiation, which would be dangerous to operators. A narrow radial boring in the sphere allows a pencil of the radiation to escape and it is directed at the diseased tissue. Radioactive phosphorus, $_{15}^{32}\text{P}$, has been tried in compounds for treatment of leukaemia, and radio-iodine for treating thyroid disease.

Radio-strontium, or strontium-90, $_{38}^{90}\text{Sr}$, also emits destructive radiation. (Ordinary strontium is mainly $_{38}^{88}\text{Sr}$ and is not radioactive.) Strontium-90 is contained in the fall-out from certain kinds of atomic bomb tests. If it is absorbed into bone tissue in man above a certain concentration, its radiation may set up a severe, possibly fatal, disease of the blood by destroying bone-marrow and red blood corpuscles. This is one of the objections to the testing of such bombs in quantity.

Uses in scientific research

These uses are mainly examples of tracer technique. Living organisms do not distinguish between ordinary atoms and their radioactive isotopes. A radioactive form of an element can, therefore, be used, highly diluted, as a marker to trace, by its radioactivity, what happens to this element during the growth of the organism. A notable example of this technique has been the use (by Calvin in the U.S.A.) of radio-carbon, $^{14}_{6}C$, as carbon dioxide, to trace the course of photosynthesis in plants. Many other similar cases have been reported and the method is increasingly used (Figure 94).

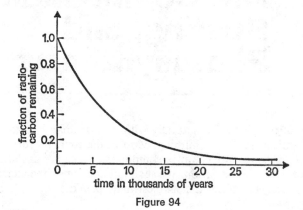

Figure 94

Radio-carbon also plays a part in the dating of archaeological specimens. This form of carbon occurs in the carbon dioxide of the atmosphere in a constant proportion. It is taken into the tissue of growing trees in this proportion. When the timber is cut, the radioactivity of the radio-carbon decays at a known rate (half-life of radio-carbon is 5760 years). Consequently, if the present, reduced radioactivity of an ancient sample of wood is determined, the age of the wood can be estimated. Such estimates agree closely with datings from other (historical) evidence where it is available, so can be used for dating fresh specimens of ancient tools, domestic articles, and weapons.

Industrial uses

These uses of radioactivity are many and increasing. Radio-sodium, employed as carbonate, is used for rapid detection of leaks in water-mains. The point of leakage can be found by the effect of radio-sodium on a Geiger counter. Similarly, when the grade of oil is changed in a pipe-line, radioactive material can be added at the change-point and is easily detected at the receiving end. Other uses of radioactive materials are in the inspection of welded steam-pipes and boiler-plates for the detection of faults; and for the dissipation of electrostatic charges built up by friction on moving bands. Such charges can cause sparking, which is dangerous in certain environments.

Questions

1. Alchemists spent many years trying to make gold from 'base' metals such as lead, without success. Say how you would make gold using nuclear reactions.

2. Write an essay about the advantages and disadvantages of nuclear reactors as a source of power.

3. What are the different forms of radiation that can be emitted from a radioactive substance? What do the different forms of radiation consist of, and how could you tell one form from another?

4. Radioactive phosphorus and radioactive sodium are made by the reactions

$$^{23}_{11}Na + {}^{1}_{0}n \longrightarrow {}^{24}_{11}Na$$

and $\qquad ^{32}_{16}S + {}^{1}_{0}n \longrightarrow {}^{32}_{15}P + {}^{1}_{1}p$

Using tables of accurate isotope weights and Einstein's relation $E = mc^2$, find out how much energy is released or absorbed during these reactions.

22 Air, Combustion, Rusting, and Photosynthesis

A study of the air begins naturally with an examination of that most familiar of all chemical reactions – burning. Most of the common combustible materials (coal, wood, petrol) are complex compounds which are unsuitable as the starting-point of our investigation. We shall fall back upon the metals, which are all elements, and, therefore, the simplest materials known.

Experiment 51
Effect of heating metals in air

(a) Copper

Take up a piece of copper foil in a pair of tongs and hold in a Bunsen flame. The metal becomes red hot and, on cooling, is covered with a black layer. This is **black copper oxide**. If the metal is scraped, the surface layer is obtained as a powder and the fresh copper exposed can be similarly treated. In time, a quantity of the black powder can be obtained.

(b) Lead

Heat a little lead foil on a crucible lid. It melts to shining beads of molten lead. Stir the beads. The metal gradually changes to a yellow powder called **massicot**.

(c) Magnesium

Hold one end of a length of magnesium ribbon in tongs and place the other end in a Bunsen flame. The ribbon burns with a dazzling flame (it is dangerous to look at it for any length of time) and leaves a white ash – **magnesium oxide**.

These experiments leave no doubt that the metals concerned have undergone a drastic change. The products left after heating them are quite different from the original metals.

The nature of the change

It is hardly necessary to describe experiments to prove that air is concerned in this change. This can, however, quite readily be shown by taking the most combustible of these metals, magnesium, placing it in a crucible, filling the crucible with sand, well pressed down to exclude air, and heating the crucible to redness. On cooling, the magnesium is unchanged.

Since air is concerned in the change, two possibilities have to be considered. The metals may have combined with something from the air or they may have lost something, which has been taken up by the air. (A third possibility is that the change in the metals may be due to some rearrangement of their material without loss or gain, but this is unlikely since such a change could presumably occur without air.) Clearly, in the first case, the material which had combined would make the mass of the product greater than that of the metal, while, in the second case, the material lost would have the reverse effect. We have only to weigh the metal before burning and the product after burning to decide this point.

Experiment 52
To find whether there is any change in mass when magnesium burns in air

Weigh a crucible (with lid) containing about 0.5 g of magnesium. Set up the apparatus as in Figure 95. Remove the lid and heat the crucible. When the magnesium begins to burn, put the lid on the crucible. Raise it occasionally to allow air to enter to burn the magnesium but, as far as possible, avoid losing any 'smoke' (fine particles of magnesium oxide) which would tend to make the final mass too low. When all the magnesium has burned allow the crucible and lid to cool. Then weigh them again. There will be a *gain* in mass.

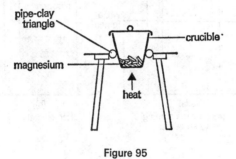

Figure 95

The products of combustion of a substance always weigh more than the original substance

This is even true for coal. If we could collect all the ash, soot smoke, and the gases (carbon dioxide and steam) which escape up the chimney, they would weigh more than the original coal. (See page 23.) This gain in mass, which occurs no matter what the substance is that burns, at once establishes the point that, during burning, the burning material combines with something. We now have to show whether the 'something' comes from the air.

To do this, it will be necessary to find whether, when a substance burns in air, the air decreases in amount. We must devise an experiment to test this point. It would be absurd to carry out the experiment in the open laboratory into which air could leak from outside to replace loss; we must secure a sample of air in a confined space, that is, in a closed vessel, and, if possible, any change in the amount of air should be automatically shown to us by the apparatus. A very simple arrangement satisfies this requirement. We shall confine the sample of air in a bell-jar over water (see Figure 96). The water forms a flexible base to the bell-jar and will move up or down inside the jar to show us what is happening to the amount of air inside. As a matter of mere convenience we shall choose yellow phosphorus for our burning substance this time. (Yellow phosphorus takes fire very readily and must be treated with great care. Never touch it with your fingers. The heat of them may start it burning and the burns it will cause are very severe and difficult to heal. Yellow phosphorus is always kept under water because of the ease with which it catches fire.)

Experiment 53
To find whether there is a diminution in the volume of the air when phosphorus burns in it

Float a small porcelain dish on water in a pneumatic trough and put in it a piece of yellow phosphorus about as big as a pea. Place over it a bell-jar and adjust the water to level A (Figure 96 (a)), the stopper of the bell-jar will then be found to *rise* and, when the bell-jar is cold, the water-level will stand at mark B on the jar (Figure 96 (b)). Clearly, the rise of water inside the bell-jar means that, during the burning of the phosphorus, some of the air was used up

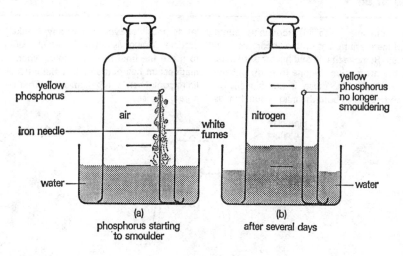

(a)
phosphorus starting
to smoulder

(b)
after several days

Figure 96

being removed. The bell-jar above A is graduated into five equal portions. Heat a long iron needle in a Bunsen flame, touch the phosphorus with it, withdraw the needle quickly, and insert the stopper of the bell-jar. The phosphorus burns with a bright yellow flame, giving off dense white fumes of phosphorus pentoxide, which fill the jar. After a time the phosphorus no longer burns. The water-level inside the bell-jar

to combine with the phosphorus.

(While the phosphorus is burning, the level of water inside the bell-jar will fall. This is an expansion effect of the heated air. It is also necessary to pour water into the trough until the levels of water inside and outside the bell-jar are equal. If this is omitted we are not measuring the volumes under the same conditions.)

After a time, the white fumes dissolve in the water, leaving the bell-jar clear. It will then be seen that some unburnt phosphorus remains in the porcelain dish. This is very significant. **The flame was not extinguished for lack of phosphorus.** We can see that some gas still remains in the bell-jar from the mark B upwards. This gas must be different from ordinary air because it will not allow phosphorus to burn in it; it must also be different from the part of the air which has combined with the phosphorus because it will not do this. **If a lighted splint is plunged into the residual gas the splint is extinguished.** We are forced to conclude, therefore, that the air is not a single substance. It must contain at least two gases – one which supports the combustion of phosphorus and one which does not. Further, we may conclude that the gas which is active in supporting the combustion of the phosphorus constitutes about one-fifth of the air by volume (this represents the rise of the water from A to B) and the other gas about four-fifths. These two gases have names. The one which supports the combustion of phosphorus is called **oxygen,** the other **nitrogen.**

Let us sum up in a few sentences what we have learnt so far.

The principal gases in air are **oxygen** and **nitrogen.** Oxygen constitutes about one-fifth of the air by volume and nitrogen about four-fifths. During the combustion of a substance, it combines chemically with the oxygen of the air and the chemical combination is accompanied by the evolution of light and heat. The combination with oxygen causes a gain in mass. Nitrogen will not support combustion.

By an experiment similar to the above, it can be shown that the material of a burning candle combines with oxygen and causes the water-level inside the bell-jar to rise. The candle will not, however, remove all the oxygen.

It is important to realize, however, that Experiment 53 is *not* an accurate method of determining the percentage of oxygen in air.

We may now give the equations for the chemical reactions considered in this chapter.

$$2Cu(s) + O_2(g) \rightarrow 2CuO(s)$$
copper(II) oxide

$$2Pb(s) + O_2(g) \rightarrow 2PbO(s)$$
lead(II) oxide

$$2Mg(s) + O_2(g) \rightarrow 2MgO(s)$$
magnesium
oxide

$$P_4(s) + 5O_2(g) \rightarrow P_4O_{10}(s)$$
phosphorus
pentoxide

The smouldering of phosphorus

Phosphorus smoulders in air. The chemical effects of the smouldering are very similar to those of the active burning of phosphorus except for the time factor.

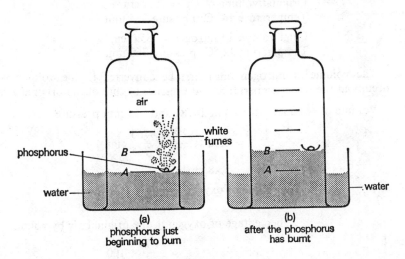

(a)
phosphorus just
beginning to burn

(b)
after the phosphorus
has burnt

Figure 97

This can be shown by an experiment for which Figure 97 is sufficient explanation.

Experiment 54
More accurate determination of the proportion of oxygen in air by volume

In this experiment, we make use of the smouldering of phosphorus to absorb the oxygen from a measured volume of air. (See also above.)

Take a graduated glass tube, closed at one end, fill it to a depth of about 5 cm with water, close the open end with a thumb and invert the tube in a deep jar of water. (If possible, allow the tube to stand like this for several hours so that the air is saturated with water-vapour.) Adjust the level of water in the graduated tube to be the same as the level in the jar. The air inside the tube is then at atmospheric pressure. Read off the volume of air. Now push up inside the tube a flexible wire carrying a piece of yellow phosphorus (Figure 98 (a)). Read the temperature of the laboratory and the barometer and set the apparatus aside until the phosphorus no longer smoulders. The water-level inside the tube will have risen to *C*, the remaining gas being nitrogen (Figure 98(b)). Remove the phosphorus and then lower the graduated tube until *C* is at the level of water in the jar, giving atmospheric pressure again inside the tube. Read off the volume of nitrogen. Take the temperature of the laboratory and read the barometer.

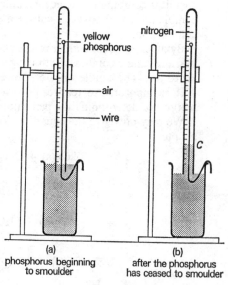

(a)
phosphorus beginning
to smoulder

(b)
after the phosphorus
has ceased to smoulder

Figure 98

Specimen calculation

Original volume of air $= 70.5$ cm^3
Temperature $14°$ C; pressure 755 mm

Final volume of nitrogen $= 55.0$ cm^3
Temperature 12 °C; pressure 760 mm

The volume of nitrogen must first be converted to the volume it would occupy at the same temperature and pressure as that of the original air.

Volume of residual nitrogen at 14 °C and 755 mm pressure

$$= 55.0 \times \frac{287}{285} \times \frac{760}{755} \text{ cm}^3$$

$$= 55.8 \text{ cm}^3$$

$$\therefore \text{ volume of oxygen} = (70.5 - 55.8) \text{ cm}^3$$
$$= 14.7 \text{ cm}^3$$

$\therefore$ percentage of oxygen in the original air by volume

$$= \frac{14.7}{70.5} \times 100$$

$$= 20.8$$

In dry air, the correct percentage of oxygen is 20.9 by volume.

Other gases present in air

Carbon dioxide

Carbon dioxide is present in air to the extent of 0.03% by volume. It is formed during the combustion of all the common fuels – coal, coke, coal-gas, water-gas, petrol, paraffin oil – all of which contain carbon.

$$C(s) + O_2(g) \rightarrow CO_2(g)$$

It is also breathed out as a waste product by all animals.

In spite of the enormous quantities poured into the atmosphere in this way, the proportion of it remains constant, partly because carbon dioxide is taken up by the leaves of plants and converted to complex starchy compounds (for a more detailed discussion of this subject see page 237) and partly because it dissolves in the water of the oceans.

The presence of carbon dioxide in the air can be shown by aspirating air

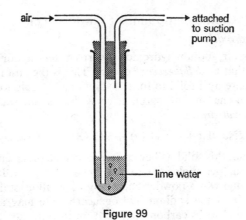

Figure 99

through a boiling-tube containing a little lime water. After a time the lime water will go turbid, showing the presence of carbon dioxide (Figure 99).

$$Ca(OH)_2(aq) + CO_2(g) \rightarrow CaCO_3(s) + H_2O(l)$$

Water-vapour

This substance is always present in the air in varying quantities. It is given off by evaporation from the oceans, rivers, and lakes.

Its presence may be demonstrated by exposing some deliquescent substance, say calcium chloride, to the air on a clock-glass. A solution of the compound will be obtained after a day or two. If this is distilled (see page 11), the colourless liquid obtained may be proved to be water by the tests given on page 24.

The noble gases

About 1% of the air by volume is made up of the noble gases. The most abundant of them is argon, the others being neon, xenon, krypton, and helium. The proportion of each of the last four is very minute. Argon and neon have found a use in 'gas-filled' electric light bulbs and coloured 'neon' electrical signs. They are obtained from liquid air.

Impurities

The air always contains small traces of many gases – hydrogen sulphide, sulphur dioxide, and others – especially in industrial areas. They are given off during the combustion of coal and fuels derived from it. The tarnishing of silver is chiefly due to the formation of a layer of black silver sulphide on it by the action of traces of atmospheric hydrogen sulphide.

Figure 100 summarizes the composition of the air.

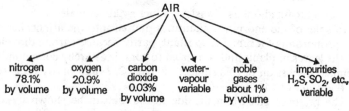

Figure 100

Exposure of some compounds to ordinary air

Sodium hydroxide

If exposed to air, sodium hydroxide absorbs water-vapour from the air and forms a solution; that is, *deliquescence* occurs. This is the absorbing of moisture from the atmosphere by a solid to form a solution. The solution then absorbs carbon dioxide from the air and forms a crystalline solid, washing soda, or *sodium carbonate decahydrate*.

$$2NaOH(s) + CO_2(g) + 9H_2O(l) \rightarrow Na_2CO_3.10H_2O(s)$$

If left to stand, this solid will eventually lose nine of its ten molecules of water of crystallization (per molecule) to the air. That is, the decahydrate *effloresces*, falling to a fine white powder consisting of small crystals of the monohydrate $Na_2CO_3.H_2O$. Some sodium hydrogencarbonate may also be formed as the monohydrate absorbs carbon dioxide from the air.

$$Na_2CO_3.H_2O(s) + CO_2(g) \rightarrow 2NaHCO_3(s)$$

Calcium chloride

Calcium chloride deliquesces in air and forms a solution. This tendency to absorb water-vapour explains the use of calcium chloride as a drying agent for gases (not ammonia, with which it combines). In this context, however, the relative amount of moisture is small and the anhydrous calcium chloride is only hydrated to the solid hydrate, $CaCl_2.6H_2O$. This hydrate is also deliquescent.

Phosphorus(v) oxide (*phosphorus pentoxide*)

Phosphorus pentoxide deliquesces in air, forming a colourless solution containing *metaphosphoric acid*,

$$P_4O_{10}(s) + 2H_2O(l) \rightarrow 4HPO_3(aq)$$

The pentoxide is also used as a drying agent for gases, but not for alkaline gases like ammonia.

Concentrated sulphuric acid

Concentrated sulphuric acid absorbs water from the air, diluting itself, usually up to about three times the original volume. The acid is said to be *hygroscopic* (not deliquescent, this term being reserved for *solids* which absorb water from the air to form solutions).

The efflorescence of washing soda was noticed above. Another efflorescent compound is Glauber's salt, *sodium sulphate decahydrate* $Na_2SO_4.10H_2O$. On exposure to air it loses the whole of its water of crystallization.

Iron(II) sulphate heptahydrate

Iron(II) sulphate heptahydrate $FeSO_4.7H_2O$ is also efflorescent, *i.e.*, loses water of crystallization on exposure to air. In addition, it undergoes oxidation by oxygen of the air, acquiring brown patches of basic *iron(III)* sulphate.

$$12FeSO_4(s) + 6H_2O(l) + 3O_2(g) \rightarrow 4\{Fe_2(SO_4)_3.Fe(OH)_3\}(s)$$

Composition of air by weight

This estimation was carried out by Dumas, 1841. Air was drawn through the following apparatus in the order shown (Figure 101).

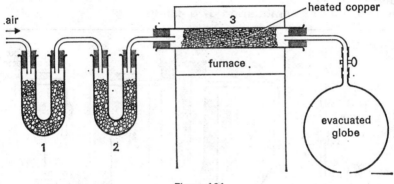

Figure 101

1. U-tubes containing potassium hydroxide to remove carbon dioxide (only one shown in the figure).
2. U-tubes containing concentrated sulphuric acid (on pumice) to remove water-vapour (only one shown in the figure).
3. A heated, weighed tube containing finely divided copper to absorb oxygen;

and, finally, the remaining 'atmospheric nitrogen' (still containing the noble gases) entered a weighed evacuated globe.

The increase in mass of the copper gave the mass of oxygen, and the increase in mass of the globe, the mass of nitrogen (and noble gases).

Neglecting carbon dioxide, the percentage of oxygen by mass in dry, pure air is 23.2%, the remainder being nitrogen and noble gases.

Air – mixture or compound?

The evidence on which a decision on this question can be reached is given below:

(1) The composition of air is very nearly, but not quite, constant. Small, but definite, differences of composition have been detected when samples of air from different parts of the earth have been analysed.

(2) (*a*) If air is dissolved in water and boiled out again (see page 243), the percentage by volume of oxygen in the air boiled out is increased from 21% to about 30%. No chemical reaction is involved here; the composition of air has been altered by a physical method, which depends merely on the fact that oxygen is twice as soluble in water as is nitrogen.

(*b*) If liquid air is allowed to evaporate, nitrogen evaporates more quickly, leaving almost pure oxygen. Here again, the gases of air are separated by a purely physical process.

This evidence alone is sufficient to decide that **air is a mixture.** Confirming it are the following facts:

(3) If nitrogen, oxygen, carbon dioxide, water-vapour, and the noble gases are mixed in appropriate proportions, there is no explosion, evolution of heat, volume change, or other evidence of chemical combination, but the product resembles ordinary air in every way.

(4) The composition of air corresponds to no simple chemical formula such as it would be expected to possess if it were a compound.

Experiment 55
The products of combustion of a candle

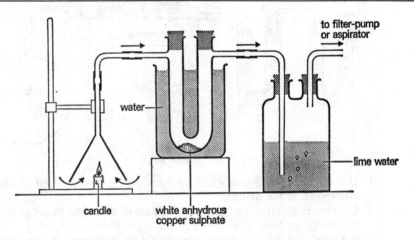

Figure 102

The products of combustion of a candle can be shown by the apparatus of Figure 102.

The products from the burning candle are drawn up the funnel and through the apparatus.

Drops of liquid will condense in the U-tube and the anhydrous copper sulphate will change to blue hydrated crystals. This proves that one of the products when a candle is burned is *water.* The lime water will rapidly turn milky (the milkiness is caused by a fine precipitate of chalk). This proves that *carbon dioxide* is also given off from the candle.

Candle-wax contains carbon and hydrogen. During the burning, these elements combine with oxygen of the air, forming carbon dioxide and water.

$$C(s) + O_2(g) \rightarrow CO_2(g)$$
$$2H_2(g) + O_2(g) \rightarrow 2H_2O(g)$$

Then
$$\underset{\text{white}}{CuSO_4(s)} + 5H_2O(l) \rightarrow \underset{\substack{\text{hydrated crystals;} \\ \text{blue}}}{CuSO_4.5H_2O(s)}$$

$$Ca(OH)_2(aq) + CO_2(g) \rightarrow \underset{\text{chalk}}{CaCO_3(s)} + H_2O(l)$$

Most of the common fuels – coal, coke, coal-gas, petrol, paraffin oil, water-gas – contain one or both of the same elements as candle-wax and their products of combustion consist mainly of carbon dioxide and water.

Though they pass off as gases, the products of combustion of a candle should weigh more than the candle-wax, which has burnt.

Experiment 56
To show that the products of combustion of a candle weigh more than the candle-wax burnt

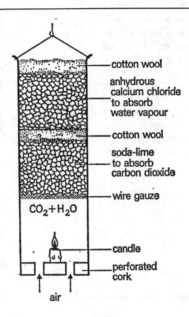

Figure 103

The apparatus is described by Figure 103.

Suspend the apparatus from the balance hook of a large, rough balance, and add weights to counterpoise it. Light the candle. The apparatus will quickly gain in mass and will depress the pan of the balance to which it is attached.

The gain in mass is the mass of oxygen from the air with which the carbon and hydrogen of the candle-wax have combined during burning.

Other kinds of combustion

Combustions in oxygen or air are so common that it becomes almost habitual to use the word 'combustion' as if it referred to this kind of reaction alone. Actually, it may be applied to any chemical combination accompanied by light and heat in which one or more of the reactants are gaseous. You will find, for example, in this book, accounts of the combustion of hydrogen, phosphorus, and copper in chlorine gas.

An interesting reversal of the usual state of affairs, in which air is acting as the combustible material, is in the burning of gas in a Bunsen flame. When considering such a flame in air we usually refer to the gas coming from the burner as the 'combustible material' and the air as the 'supporter of combustion'. This is mainly because we live in an atmosphere of air (which contains oxygen, a very reactive gas) which surrounds any burning material. If, by accident, this atmosphere of ours were to consist of such gas, it would be possible to produce the flame of a Bunsen burner by pumping air down the pipe usually connected to the gas supply. Experiment 57 illustrates this.

Experiment 57
To burn air in an atmosphere of coal-gas

Set up the apparatus shown in Figure 104, which consists of a lamp chimney with a square of asbestos (containing a round hole) resting on the top of the chimney. The cork at the base is fitted with a glass tube for a gas inlet and a mica or metal tube at *B*. The gas is turned on, and the aperture at *A* is closed momentarily whilst the gas issuing from *B* is ignited. The aperture at *A* is now opened and the gas is lighted. On inspection, it will be seen that there is now a flame at the top of the mica tube which consists of air burning in an atmosphere of coal-gas.

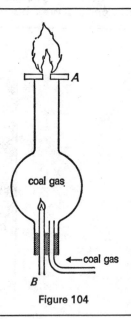

Figure 104

Structure of flame

The structure of a flame varies according to the chemical composition of the gas which is burning. The general conical shape is brought about by several factors, the more important of which are the effects of convection and the necessity for further supplies of gas to search for air with which to burn. The gas burning immediately on issue from the burner, uses up the air in that region and the gas following has to seek its supply of air from more distant sources. At the same time the convection currents set up confine this search to an upward direction. The blue zone seen at the base of a Bunsen or candle flame (see Figure 106) is caused by the upward stream of air impinging on the base of the issuing cone of gas. The inner zone near the burner contains unburnt gas, since the outside of the flame has obviously the best opportunity to come in contact with the air, and the inside of the flame, the least.

We will consider the structure of some well-known flames:

The flame of hydrogen burning in air

This flame is, as we should expect, a simple one. Two zones only are produced, a zone of unburnt gas surrounded by a zone in which combination of oxygen and hydrogen takes place. The flame is almost invisible in dust-free air (see Figure 105).

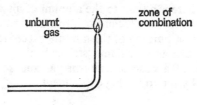

Figure 105

The candle flame

Candle-wax consists of hydrocarbons, both the carbon and the hydrogen being combustible materials. It is assumed that both elements do not burn completely to form water and carbon dioxide except at the very outside of the flame where there is plenty of air. In the bright yellow zone there is incomplete combustion and particles of carbon raised to a white heat are present in it to give the flame most of its luminosity, whereas the outside zone where combustion becomes complete is only faintly visible. The zone round the wick consists of unburnt gas, and the blue zone at the base is a zone of rapid burning caused by the upward rush of air (due to convection) first meeting the combustible gases. (See Figure 106.)

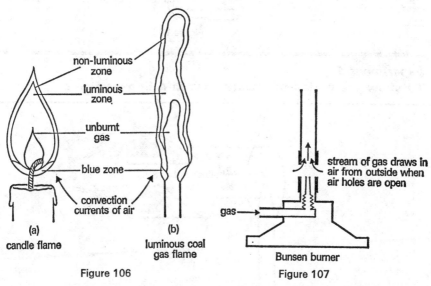

Figure 106
Figure 107

The luminous Bunsen flame (air holes closed) gives a parallel with the candle flame, the gases in the town-gas being methane (a hydrocarbon), hydrogen, and carbon monoxide.

The Bunsen flame

The flame of a burning hydrocarbon or the luminous Bunsen flame is not a very hot flame and deposits soot on a cold article held in it. Bunsen devised a simple burner (Figure 107) to ensure more complete combustion by introducing a supply of air entering with the gas, so that this supply of air together with the external supply is sufficient to produce complete combustion. Hence the flame with the air holes open is more compact, much hotter, and non-luminous. The structure of this flame is shown in Figure 108, and it will be seen that the luminous zone has been replaced by a zone in which the gas burns with the internal supply of air. There is a limit to the amount of air which can be supplied by the holes at the base, for if sufficient is introduced to cause almost complete combustion, the rate of burning of the mixture exceeds the speed at which the gas is moving up the tube and the flame 'strikes back'.

The existence of a zone of unburnt gas and zones of varying temperatures may be shown by the following experiment.

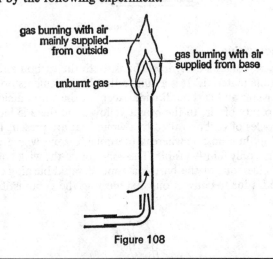

Figure 108

Experiment 58
To show the region of unburnt gas in a Bunsen flame

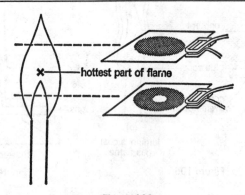

Figure 109

Hold a piece of asbestos paper (fairly thin) horizontally in the Bunsen flame with a pair of tongs at various levels as shown in Figure 109.

The paper will glow at the hottest points of the flame.

The causes of luminosity of flame

The luminosity of a flame is affected by alteration of temperature and pressure of the burning gases and also by the presence or absence of solid particles.

The presence of solid particles in a flame increases luminosity

Sprinkle a few iron filings into the non-luminous Bunsen flame. Sparks are formed as each particle is raised to a white heat. Similarly, platinum foil or a porcelain rod become white hot and emit light when heated to a high temperature in the non-luminous Bunsen flame.

Hold a cold evaporating dish by means of tongs for a minute in the luminous Bunsen flame. On removal, the dish will be found to be blackened by carbon. No such effect is observed if the experiment is repeated using the non-luminous flame. The presence of the particles of carbon is thought to be responsible for the luminosity of this type of flame.

Effect of increase of pressure

Increase of pressure of the gases taking part in the combustion increases the luminosity of the flame.

The effect of temperature on luminosity

Luminosity increases with increase in temperature.

Experiment 59
Effect of temperature on luminosity

Set up the apparatus shown in Figure 110, with a silica tube extension fitting over a Bunsen burner tube. Open the air holes and obtain the non-luminous flame at the end of the silica tube. Now heat the silica tube strongly with a second Bunsen burner. As the silica tube is heated, the flame at the end of the silica tube gradually becomes luminous and luminosity diminishes as the tube is allowed to cool.

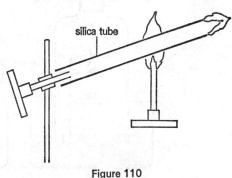

silica tube

Figure 110

Explosions

Explosions result from very rapid, exothermic chemical reactions. Consider a mixture of hydrogen, two volumes, and oxygen, one volume. If the mixture is sparked, the gases are heated to very high temperature near the spark. They combine and liberate heat.

$$H_2(g) + \tfrac{1}{2}O_2(g) \rightarrow H_2O(l); \quad \Delta H = -286 \text{ kJ}$$

This heat raises the temperature of neighbouring gas, which combines liberating more heat, and so on with great rapidity. The mass of gas is raised to incandescence in a fraction of a second and the consequent great expansion produces a pressure wave in the air. The whole effect is called an explosion.

The rusting of iron

The important facts connected with the rusting of iron may be ascertained by the following experiments:

L

Experiment 60
To find if there is any change in mass when iron rusts

Weigh a clock-glass containing some iron borings. Damp the borings and set them aside to rust. When rusted, place them in an oven to dry thoroughly, then weigh the clock-glass again. There will be a *gain in mass.*

In this respect rusting is analogous to burning. We may now try an experiment to see whether the air is similarly concerned in both.

Experiment 61
To find whether iron combines with anything from the air while rusting

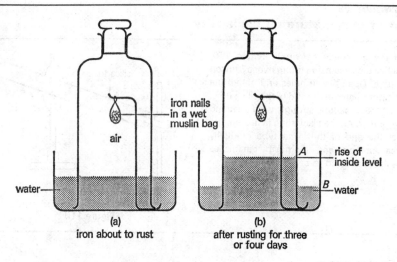

Figure 111

This experiment is described by Figure 111.

To show the character of the gas left in the bell-jar in Figure 111(b), fill up the trough till level *B* rises to level *A*. Then remove the stopper of the bell-jar and insert a lighted taper. It will be extinguished; the remaining gas is *nitrogen.* During rusting the iron has combined with the *oxygen* of the air. This accounts for the rise in the water-level (Figure 111(b)) and the gain in mass (see above).

Rusting and burning are similar

It is important to understand clearly that, from the chemical standpoint, rusting and burning are the same process. During burning, magnesium combines with oxygen of the air, forming magnesium oxide.

$$2Mg(s) + O_2(g) \rightarrow 2MgO(s)$$

During rusting, iron combines with oxygen of the air in the presence of water to form brown hydrated iron(III) oxide, 'rust'.

$$4Fe(s) + 3O_2(g) \rightarrow 2Fe_2O_3(s)$$

The only difference is in the time required for the two processes. Heat is generated during rusting just as it is during burning, but it is dissipated to the surroundings without attracting notice because of its much slower rate of production.

It is very unfortunate for mankind that iron, which possesses so many useful properties, should be so readily attacked by the oxygen of the air. To protect iron from rusting, it is painted or galvanized (see page 461). Very large sums of money are spent throughout the world on the various protective processes.

We will now investigate further the conditions under which iron rusts. We have already seen that during rusting, iron combines with oxygen of the air, and it is common knowledge that iron rusts more readily when plenty of water is present. It will be of interest to separate these agents and find their individual effects. To do this, we must expose iron to the action of air in the complete absence of water (to water-free air) and to water in the complete absence of air (to air-free water).

Experiment 62
To find the effect of exposing iron to air and water separately

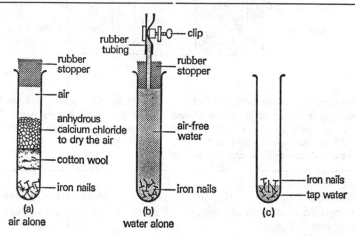

Figure 112

(a) *Exposure of iron to air separately*

To do this, set up apparatus as in Figure 112(a). The figure sufficiently explains the experiment.

(b) *Exposure of iron to water separately*

Boil about 350 cm³ of water rapidly in a beaker for at least half an hour. This will boil all the air out of it. Put a few iron nails into a test-tube and fill it to the brim with the boiled water. Press into the mouth of the test-tube a rubber stopper carrying a glass tube and rubber tube as shown in Figure 112(b). (Note that the glass tube must be flush with the bottom of the stopper.) The

water will rise into the rubber tube. Place a clip in position to exclude all air.

Leave the test-tubes for several days. In neither case will rusting occur. This means that iron will not rust in the presence of air alone or of water alone; both are needed together to rust the iron.

To make sure that the nails you have used are actually capable of rusting, put them into a test-tube with a little water, leave the test-tube open to the air, and notice the result after a day or two (Figure 112(c)).

What we have done in the above experiment is to take two substances, air and water, which normally act on iron *together* and test the effect of each *singly*. This is the favourite device of scientists. It is only by finding out the effects produced by one agent at a time that reliable information can be obtained.

Oxygen and carbon dioxide in life-processes. Photosynthesis

The following experiment is designed to show that carbon dioxide is present in the air expelled from the lungs of a human being.

***

Experiment 63
To detect carbon dioxide in expelled air from the lungs

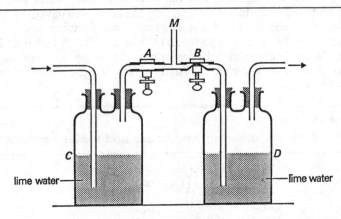

Figure 113

Set up the apparatus shown in Figure 113.

With the clip *A* open and the clip *B* closed, breathe in air through the mouthpiece *M*. The air bubbles through the lime water in *C*. Then close clip *A*, open clip *B*, and breathe out the air so that it passes through the lime water in *D*. Repeat this several times. The lime water in *C* remains unaffected while that in *D* is rapidly turned milky.

***

This must mean that during its occupation of the lungs, the air has increased its proportion of carbon dioxide, which reacts with the lime water in *D* to form a precipitate of chalk.

$$Ca(OH)_2(aq) + CO_2(g) \rightarrow CaCO_3(s) + H_2O(l)$$

At the same time, oxygen is absorbed from the air into the blood forming a loose compound with the haemoglobin (the red colouring matter) of the red blood corpuscles.

The use of the oxygen and the presence of carbon dioxide may be explained briefly as follows. The process of digestion converts our food materials into compounds which are either soluble in water or easily emulsified with it. The soluble compounds are absorbed into the blood as the food passes through the small intestine. They are carried round by the blood-stream and used either to replace wastage and maintain growth in the body-tissues or to supply energy for movement of the body and to maintain its temperature at about 37 °C, a temperature usually considerably higher than that of surrounding objects. This energy is supplied by oxidation of the soluble products of digestion. Consider, for example, a sugar. It is well known that if a sugar is thrown on to a fire it burns vigorously, giving off heat, and forming, as products of combustion, carbon dioxide and water. The same oxidation process, giving the same products, occurs in the body, the oxygen being taken from its loose combination with haemoglobin to oxidize the dissolved sugar. For a given quantity of sugar oxidized, the same amount of heat is given out whether it is burnt rapidly on the fire or oxidized more slowly in the body.

$$C_6H_{12}O_6(s) + 6O_2(g) \rightarrow 6CO_2(g) + 6H_2O(l); \quad \Delta H = -5865 \text{ kJ mol}^{-1}$$

This is the source of the body's heat. The waste product, carbon dioxide, is carried round in the blood (chiefly as hydrogencarbonate) and is liberated as the blood passes through the lungs, from which it is breathed into the air. The breathing process of animals is similar.

It has already been pointed out that all common fuels give off carbon dioxide when burnt. Living animals also discharge carbon dioxide as we have just seen. There must obviously be some agency at work using up these vast quantities of carbon dioxide and restoring the oxygen absorbed from the air during combustion, for otherwise the composition of the air would change appreciably.

This agency is the plant life of the world which is at work restoring the balance. With the help of energy from the sun and with chlorophyll (the green colouring matter of leaves) as catalyst, plants are continually building up carbohydrates (*e.g.*, starch and sugar) from water and carbon dioxide of the air.

Carbon dioxide + water + energy → starch + oxygen

Oxygen is liberated and passes into the air. Notice that this process (*photosynthesis*) is endothermic. The energy absorbed is released again when carbohydrates are used as food and oxidized in the blood-stream to carbon dioxide and water (see above). It is obvious that the waste product, carbon dioxide, breathed out by animals and given off by carbonaceous fuels, is the raw material from which plants build up carbohydrates. The waste product, oxygen, from the plant is the gas essential for animal breathing and combustion of fuels. These opposing processes, at work together, keep the composition of the air constant.

In addition to carrying on the above feeding process, plants also breathe in the same way as animals, but, relatively, their breathing process is of small account. On balance, a plant uses up carbon dioxide and liberates oxygen. It may be mentioned that, biologically, the characteristic difference between animals and plants is not connected with size or movement or similar factors, but lies in the fact that plants build up (synthesize) complex starchy food materials from the simple compounds, carbon dioxide and water, while animals must have these complex materials available as food and break them down by digestion into simpler substances.

Experiments illustrating the above brief outline will be found described in any elementary textbook of biology.

Questions

1. State whether a sample of each of the following would increase or decrease in mass if left exposed to the air. Give a reason for your answer in each case: (*a*) ammonia solution, (*b*) concentrated sulphuric acid, (*c*) concentrated hydrochloric acid, (*d*) calcium oxide. (J.M.B.)

2. What simple experiments would you perform to show that (*a*) calcium sulphate is slightly soluble in water, (*b*) ammonium chloride dissociates when it is heated, (*c*) a candle is composed of compounds containing carbon and hydrogen, (*d*) iron requires both air and moisture for rusting? (J.M.B.)

3. Describe with the aid of sketches simple experiments you would set up to show that

the presence of both air and water is necessary for iron to rust. (O.)

4. Give *three* reasons in each case why (*a*) air is considered to be a *mixture* of nitrogen and oxygen, (*b*) water is considered to be a *compound* of hydrogen and oxygen.

Draw a diagram of the apparatus you would use to obtain a sample of the air dissolved in tap-water. How would you determine the proportion of oxygen in the air so obtained? How and why would your result differ from the proportion of oxygen in ordinary air? (O. and C.)

5. Name the *two* main gases present in each of the following mixtures, and give the approximate proportion by volume of each

gas in the mixture: (*a*) air, (*b*) water gas, (*c*) the gaseous mixture obtained by warming concentrated sulphuric acid with oxalic acid. (J.M.B.)

6. Give *two* conditions necessary for iron to rust. Give also *two* ways of preventing the formation of rust. (S.)

7. Give the conditions necessary for iron to rust. Explain why iron is often coated with zinc to prevent rusting. What is the name of this process? (S.)

8. Give *one* reason in each instance why (*a*) air is said to be a mixture and not a compound, (*b*) ethene is said to be unsaturated, (*c*) washing soda crystals are said to be efflorescent, (*d*) the addition of concentrated sulphuric acid to water is said to be exothermic. (S.)

9. (*a*) Design and describe a quantitative experiment to find out whether a new alloy is oxidized by the atmosphere. What results would you expect to observe if the alloy were (i) easily oxidized, (ii) rust resistant?

(*b*) If the alloy were attacked, what further experiments would you set up to discover which parts of the air cause the reaction?

(*c*) How is air prevented from attacking iron on (i) the blades of a lawn mower, (ii) a metal dustbin, (iii) cutlery? (J.M.B.)

10. (*a*) Explain what you understand by the term *photosynthesis*. (*b*) Name *two* non-metallic oxides which cause pollution of the atmosphere. Indicate the sources of this pollution and mention steps being taken to minimize its effects. (*c*) Compare the two processes 'combustion' as applied to fuels, and 'respiration' as applied to animals. (O. and C.)

11. What happens when the following substances are exposed to the air: (*a*) quicklime, (*b*) solid sodium hydroxide, (*c*) washing soda, and (*d*) anhydrous calcium chloride? How would you confirm experimentally your answer in *one* case? (J.M.B.)

12. What do you understand by the term combustion? Give two examples of combustion in which oxygen plays no part. Describe experiments which illustrate the resemblance between the combustion of a candle and the respiration process in an animal, giving sketches of any apparatus you would use.

13. An ordinary paraffin candle consists of carbon and hydrogen in combination. State what becomes of the candle when it burns in air, and describe an experiment in support of your statement. Make a sketch of the apparatus you would use to show that the mass of the products of combustion is greater than the mass of the candle consumed. (J.M.B.)

14. Air is a mixture of various gases. An experiment was carried out, using the apparatus shown in Figure 114, to investigate the composition of air and to obtain a sample of nitrogen from it. The metal in *E* was strongly heated and then water was slowly run into *A* to displace air. The first bubbles of gas escaping at *F* were not collected.

(*a*) Why were the first bubbles of gas escaping at *F* not collected?

(*b*) The lime-water in *B* changes in appearance during the experiment. Describe this change and explain why it occurs.

(*c*) What is the reason for using aqueous sodium hydroxide in bottle *C*?

(*d*) Concentrated sulphuric acid is frequently used to dry a gas. Explain why it would not be used in the U-tube *D*.

(*e*) Name a suitable drying agent which *could* be used in the U-tube *D*.

(*f*) Copper is a suitable metal to use in *E*. What conclusion would you come to if the copper turned black during the experiment?

(*g*) Explain why platinum and mercury are unsuitable metals to use in the silica tube in *E*.

(*h*) Nitrogen is a diatomic gas. What does diatomic mean?

(*i*) What is the density of pure nitrogen at room temperature and pressure in g dm^{-3}?

(*j*) The sample of nitrogen collected in the experiment has a greater density than expected. What conclusion can be made about the gas?

(*k*) In a further experiment the nitrogen produced was then passed over heated magnesium, which combines with nitrogen to form a non-volatile solid. A very small volume of gas was then collected. What do you think this gas could be? (L.)

15. When copper is heated in air, a black substance is formed. How would you prove that the copper had combined with another element to form this new substance, and how could you determine what this element is? (J.M.B.)

16. If a green plant is placed in a closed volume of air in daylight, what changes would you expect to result in the composition of the air? How would you demonstrate that these changes occur? To what do you attribute them? (O. and C.)

17. (*a*) Give briefly *three* reasons for regarding air as a mixture and not a compound.

(*b*) How would you determine the mass of oxygen in a given volume of air? Sketch the apparatus you would use and show, using an imaginary case, how you would work out the calculation.

(*c*) Describe briefly how you would determine the *volume* of oxygen in a sample of air confined over water in an eudiometer tube. (J.M.B.)

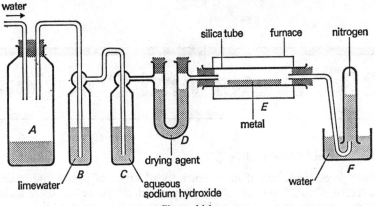

Figure 114

18. Describe in detail how you would proceed in order to remove completely the oxygen, carbon dioxide, and water-vapour from ordinary air. How would you prove that the residual gas was really free from these substances? (L.)

19. (a) A pupil investigating the effect of dissolved substances on the rate of corrosion of iron carried out the following experiment. He polished three similar pieces of sheet iron with steel wool and placed one in each of three test-tubes set up as shown in Figure 115.
He then added 1 cm³ of a solution of potassium ferricyanide (potassium hexacyanoferrate(III)) to each of the tubes. On inspecting the tubes a few hours later he found that corrosion was most marked in tube A, but

that there was no appreciable difference in the extent to which the iron in tubes B and C had corroded.
(i) Why was it necessary to polish the pieces of iron before the experiment? (ii) How did the addition of potassium ferricyanide help him to compare the rates of corrosion? (iii) Suggest a reason why the presence of sodium chloride in the water should speed up corrosion, whereas that of glucose does not.
(b) Plumbers installing hot water systems should not use copper or brass nuts to connect pipes to galvanized iron tanks, nor should they allow any copper trimmings from the pipes to be left in such tanks. (i) What is meant by galvanized iron? (ii) Explain why the above precautions should be taken. (Scottish.)

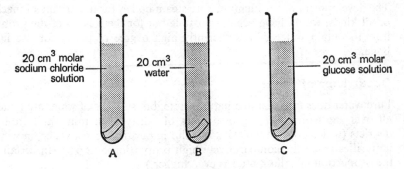

Figure 115

23 Water and Solution

Water is essential to life

Water is of fundamental importance to all kinds of plants and animals and therefore to man. It is of equal importance with the air we breathe in maintaining the vital processes necessary to life and growth, but since it is not everywhere available its provision has, from the earliest times, limited the setting up of villages and towns to the places where a water supply existed. Not only is water used all over the world in vast quantities for drinking purposes, but it is used in even greater quantities for washing, bleaching, dyeing, cooling, raising steam to drive engines or turbines to generate electricity, and as a solvent in industrial processes far too numerous to mention. It is the concern of the chemist to ensure that a supply of water is maintained which is suitable for all these purposes.

If the water is too soft (see page 248) it will attack the lead of the pipes in which it may be carried. If acids are present from decaying organic matter, sufficient lead may be dissolved by the water to cause harm to those who drink it. If the water supply for a town is too hard, because of the high percentage of dissolved solid matter which it contains, an individual firm may not consider establishing a new water-dependent industry there because of the enormous extra expense which it will incur in softening the water.

Purification plants used in swimming baths, by the use of chlorine, keep the water comparatively free from the bacteria which carry many infectious diseases. The development of the high-pressure steam boiler, for the driving of machinery of all kinds, would have been impossible but for the solving of the problem of how to obtain water of a sufficiently high degree of purity for use in these boilers.

Occurrence of water

Pure water does not exist in a natural state, but supplies of water are obtainable all over the world, varying in degrees of purity from rain water from clean districts (which contain 0.0005% of solid impurities) to sea water, in which the impurities reach the comparatively high proportion of 3.6%. (In certain lakes the proportion of solid matter is even higher.)

Purification

A sample of fairly pure water can be made from rain water, tap water, or river water by the process of distillation. The apparatus used is that illustrated in Figure 4 (page 11), the impure water being placed in the distillation flask and heated to boiling. The steam comes off (together with gaseous impurities) whilst the solid impurities are left behind. The steam is condensed to water in the condenser. A *pure* liquid distils at a constant temperature which is its boiling point at the prevailing pressure (for water 100 °C at 760 mm). This behaviour is a recognized test for purity of a liquid. For *fractional distillation*, see page 259.

Pure water

The preparation of perfectly pure water (or as near to perfectly pure water as its unusual solvent powers will allow) is a matter of much difficulty. It is prepared for conductivity experiments by as many as twenty successive distillations from a pure tin or platinum retort into a receiver which has been cleaned by having the purest water then obtainable kept in it for ten years! Potassium permanganate is added to the impure water in the earlier stages to oxidize organic impurities.

Properties

Water is a clear colourless liquid with an insipid taste. It is usually recognized in the laboratory by its capacity to turn anhydrous copper(II) sulphate (white) to a blue colour.

$$CuSO_4(s) \quad + \quad 5H_2O(l) \rightarrow CuSO_4.5H_2O(s)$$

copper(II)   water   hydrated copper(II)
sulphate   sulphate (blue)

This test, of course, merely denotes the **presence** of water and not the **absence** of everything else except water, *e.g.*, a dilute solution of sulphuric acid would turn anhydrous copper sulphate from white to blue. **Pure** water has the following properties:

(*a*) It freezes at 0 °C.
(*b*) It boils at 100 °C, when the barometer stands at 760 mm, and pure water will boil away completely with no change in temperature.
(*c*) Its maximum density is 1 g cm^{-3} at 4 °C.
(*d*) It is neutral to litmus.

Water as the universal solvent

There are few substances which do not dissolve in water to some extent. Even when you drink a glass of water, you are drinking a little of the glass as well. It is true you need not get alarmed, for the amount is very small indeed, but for certain experiments ordinary glass vessels cannot be used as containers for water because of this solvent effect.

Tap water can easily be shown to contain a considerable quantity of both dissolved solids and gases by the following experiments:

Experiment 64
To show tap water contains dissolved solids

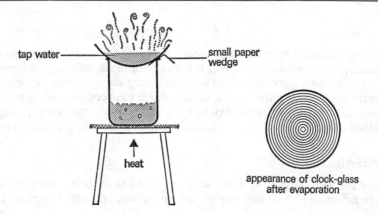

tap water

small paper wedge

heat

appearance of clock-glass after evaporation

Figure 116

Fill a large clean clock-glass with tap water and evaporate it down to dryness on a steam bath as shown in Figure 116. On holding the glass up to the light or against a sheet of white paper, you will observe a large number of concentric rings of solid matter left as the water gradually evaporated.

Experiment 65
To show that tap water contains dissolved gases

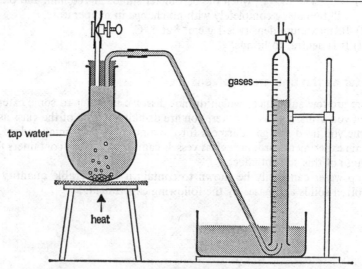

gases

tap water

heat

Figure 117

Use in this experiment a burette into which will fit the absorption cup shown in Figure 118.

Fill a flask with water and put in a few pieces of porous pot. Insert a two-holed rubber stopper to which are fitted a delivery tube and a short piece of glass tubing which can be closed by a rubber tube and clip and attached to the tap in the initial stages (the end of the delivery tube should not project beyond the surface of the stopper). By attaching the small piece of glass tubing to the tap by means of a rubber tube the whole apparatus (including the delivery tube) is filled with water (Figure 117). As the water is heated, bubbles of gas are seen to rise and these will collect and be carried over into the burette. Boil the water until no more gas is given off. The gas can be shown to differ from air in that its oxygen content is much higher, and the gas boiled out will rekindle a glowing splint.

Experiment 66
To determine the volume of oxygen in the air boiled out of water

Fill the cup with crystals of pyrogallol and add water to fill up the air spaces. (See Figure 118.) Have ready a piece of solid caustic soda (make certain that it will be able to enter the burette), insert it into the burette and follow it quickly with the absorption cup. Invert the burette several times and release the cup under water. Transfer the burette to a deep gas-jar, lower it until the levels are the same inside and out, and note the volume of gas absorbed. The percentage of oxygen in this air will be more than 30%.

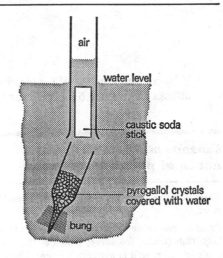

Figure 118

Value of these dissolved gases to fish life

The oxygen of the air dissolves in water to the extent of only 4 volumes of oxygen in 100 volumes of water (*i.e.*, 1 dm^3 of water contains only a maximum of 40 cm^3 [or 0.06 g] of oxygen). Although this amount is only very small it is of utmost importance to fish life. The fish (with a few exceptions) rely on this oxygen for breathing, in just the same way as we rely on the air around us.

Chemical value of the solvent properties of water

Use is made in the laboratory of this exceptional property to bring into very close contact the particles of reacting substances. When the particles dissolve in water they have an opportunity for movement which they do not have in the solid state. Under these circumstances many reactions take place which do not take place if the reactants are solids (see ionization), *e.g.*,

$$AgNO_3(aq) + NaCl(aq) \rightarrow AgCl(s) + NaNO_3(aq)$$

The above reaction will not take place if common salt and silver nitrate are ground together in a mortar, because the average distance of the particles from one another is too great.

In the above cases the water acts as a medium in which the action takes place. It does not as a rule react with the substances. The chemical actions of water, however, as an oxide and as a hydroxide producer are extensive.

Action of water on metals

$$\left.\begin{matrix}\text{K}\\\text{Na}\\\text{Ca}\end{matrix}\right\}\text{Attack}\atop\text{water}$$

$$\left.\begin{matrix}\text{Mg}\\\text{Al}\\\text{Zn}\\\text{Fe}\end{matrix}\right\}\text{Attack}\atop\text{steam}$$

$$\left.\begin{matrix}\text{Pb}\\\text{Cu}\\\text{Hg}\\\text{Ag}\\\text{Au}\end{matrix}\right\}\text{Do not attack}\atop\text{water or steam}$$

By an examination of the electrochemical series it is easily shown that water attacks the metals to a degree varying with their position in the series.

Experiment 67
Action of potassium on water

Note: Goggles must be worn for this experiment.

Place a small piece of potassium on water in a large dish (notice the silvery gleam of the un-oxidised metal as it is cut with a knife). *Under*

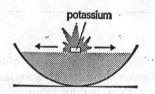

potassium

Figure 119

no circumstances should the piece of metal be larger than a 3 mm cube as the action is very vigorous. Stand well back from the dish as the

action proceeds and protect yourself in case of violent reaction. The potassium melts to a silvery ball, darts about the water, and a gas is given off (hydrogen) which burns spontaneously with a lilac flame (the colour is due to the burning of small quantities of potassium vapour) – see Figure 119.

$$2K(s) \;+\; 2H_2O(l)$$
potassium   water

$$\longrightarrow 2KOH(aq) + H_2(g)$$
potassium   hydrogen
hydroxide

If a few drops of phenolphthalein indicator are placed in the solution it will turn red because of the presence of potassium hydroxide, which is an alkali.

Experiment 68
Action of sodium on water

Note: Goggles must be worn for this experiment.

Perform Experiment 67 with sodium. The sodium melts to a silvery ball, but does not burn unless it is restricted in movement. Effervescence occurs, a gas is liberated, and if a light is applied it burns with a yellow flame (the yellow colour is from the sodium). If the sodium is packed tightly into a sodium spoon (see Figure 120), the gas can be collected and shown to be hydrogen.

$$2Na(s) + 2H_2O(l) \longrightarrow 2NaOH(aq) + H_2(g)$$

sodium   water   sodium   hydrogen
hydroxide

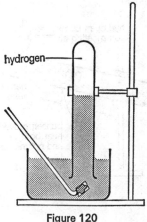

Figure 120

Experiment 69
Action of calcium on water

Drop a piece of calcium (a grey metal) into a dish of water and invert over it a boiling-tube full of water. The calcium sinks, unlike the potassium and sodium; there is effervescence, and a gas (hydrogen) is given off which explodes if mixed with air and a flame applied (Figure 121).

The calcium gradually disappears and a white milky suspension is produced. This is because the calcium hydroxide formed is only slightly soluble. If the suspension is carefully filtered to give a clear solution, carbon dioxide can then be blown through to give the usual suspension of calcium carbonate.

$$Ca(s) + 2H_2O(l)$$
calcium   water

$$\longrightarrow Ca(OH)_2(aq) + H_2(g)$$
calcium   hydrogen
hydroxide

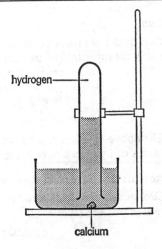

calcium

Figure 121

Experiment 70
Action of magnesium on steam

Note: Goggles should be worn for this experiment.

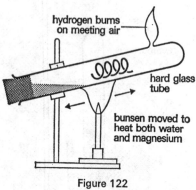

hydrogen burns on meeting air

hard glass tube

bunsen moved to heat both water and magnesium

Figure 122

Blow a small hole in a *hard glass* test-tube by applying a small blowpipe flame until the tube is soft and blowing at the open end whilst the glass is being heated. Put a coil of 15 cm magnesium ribbon in the tube, add 2 or 3 cm³ of water, insert the cork and clamp the apparatus by the cork at the angle shown (Figure 122). Heat, gently at first, with a Bunsen burner, keeping the latter moving to maintain an atmosphere of steam. Continue to pass the steam until all the air has been expelled from the tube. Finally, heat strongly when the magnesium will burn and simultaneously the liberated hydrogen will burn as it meets the outside air.

$$Mg(s) + H_2O(g) \longrightarrow MgO(s) + H_2(g)$$

Zinc

Zinc does not attack hot or cold water. If zinc is heated to redness in a current of steam, hydrogen is formed.

$$Zn(s) + H_2O(g) \rightarrow \underset{\substack{\text{zinc} \\ \text{oxide}}}{ZnO(s)} + H_2(g)$$

Iron

Iron does not attack water (rusting takes place only when air is present as well) but is readily attacked by **excess** of steam at red heat.

$$\underset{\substack{\text{iron} \\ \text{(red hot)}}}{3Fe(s)} + \underset{\text{steam}}{4H_2O(g)} \rightarrow \underset{\substack{\text{triferric} \\ \text{tetroxide} \\ \text{(black oxide} \\ \text{of iron)}}}{Fe_3O_4(s)} + \underset{\text{hydrogen}}{4H_2(g)}$$

The above reaction can be made to proceed in the reverse direction by passing **excess** of hydrogen over heated black oxide of iron (see page 187).

Action of water on non-metals

Carbon attacks steam at a white heat, forming carbon monoxide and hydrogen (water gas, see page 311).

$$C(s) + H_2O(g) \rightarrow CO(g) + H_2(g)$$

Chlorine acts on water to form a mixture of two acids.

$$\underset{\text{chlorine}}{Cl_2(g)} + \underset{\text{water}}{H_2O(l)} \rightleftharpoons \underset{\substack{\text{hypochlorous} \\ \text{acid}}}{HOCl(aq)} + \underset{\substack{\text{hydrochloric} \\ \text{acid}}}{HCl(aq)}$$

Action of water on oxides

Potassium oxide is attacked by water with the formation of potassium hydroxide.

$$\underset{\substack{\text{potassium} \\ \text{oxide}}}{K_2O(s)} + \underset{\text{water}}{H_2O(l)} \rightarrow \underset{\substack{\text{potassium} \\ \text{hydroxide}}}{2KOH(aq)}$$

Sodium oxide is similarly attacked by water with the formation of sodium hydroxide.

$$Na_2O(s) + H_2O(l) \rightarrow 2NaOH(aq)$$

sodium   water   sodium
oxide   hydroxide

Calcium oxide (*quicklime*). Place a piece of quicklime in a dish and add water a few drops at a time. For a little while nothing is observed and then water-vapour is seen to come off, whilst a hissing sound as the water drops on indicates that the mass is becoming hot. It commences to expand and crack, and finally crumbles to a powder, known as *slaked lime*.

$$CaO(s) + H_2O(l) \rightarrow Ca(OH)_2(s)$$

calcium   water   calcium
oxide   hydroxide
(quicklime)   (slaked lime)

The above three hydroxides are soluble in water (slaked lime only slightly) and together with ammonia solution form the common alkalis.

Sulphur dioxide reacts readily with water to form sulphurous acid.

$$SO_2(g) + H_2O(l) \rightarrow H_2SO_3(aq)$$

sulphurous acid

Similarly, other acidic oxides form acids with water, *e.g.*,

$$SO_3(g) + H_2O(l) \rightarrow H_2SO_4(aq)$$

sulphur trioxide   sulphuric acid

$$2NO_2(g) + H_2O(l) \rightarrow HNO_2(aq) + HNO_3(aq)$$

nitrogen dioxide   nitrous acid   nitric acid

$$CO_2(g) + H_2O(l) \rightarrow H_2CO_3(aq)$$

carbon dioxide   carbonic acid

Action of water on chlorides

See Chlorides, page 385.

Action of water on certain metallic carbides

Calcium and aluminium carbides react with water forming the hydroxide of the metal and hydrocarbons.

$$CaC_2(s) + 2H_2O(l) \rightarrow Ca(OH)_2(aq) + C_2H_2(g)$$

ethyne

$$Al_4C_3(s) + 12H_2O(l) \rightarrow 4Al(OH)_3(aq) + 3CH_4(g)$$

methane

Composition of water

The results of experiments already performed indicate clearly that water contains hydrogen and oxygen. If hydrogen is burnt in oxygen or exploded with it, water is produced (page 304) and nothing else. It remains to find the number of atoms of each element which is present in the molecule.

Volume composition of steam. The experiment described on page 126 shows two volumes of steam to be formed from one volume of oxygen and two volumes of hydrogen. It follows that the formula for steam is H_2O.

Electrolysis of water. The above is the synthesis of water from its elements whilst electrolysis is the analysis. On electrolysis of water (see page 155), it is found that the volume of hydrogen liberated is twice that of the oxygen liberated, thus confirming the above experiment.

Hardness of water

There are many types of natural water found on the earth's surface.

Rain water is the purest natural water, and if collected in a country district contains oxygen, nitrogen, carbon dioxide (dissolved as the rain drops pass through the atmosphere), and only a small amount of dissolved solids (0.0005%).

River water, from which many domestic supplies are obtained, will obviously contain the same gaseous impurities and also any solids which the water has dissolved as it passed over the soil. The amount and kind of impurity will depend, therefore, on the type of soil over which the water runs. If the water runs over impervious material such as granite, the river water may be nearly as pure as rain water. In actual practice many springs and rivulets feed the large river from which a town's supply is obtained, and hence the impurities are often the same as those of spring water.

Spring water is water which has made its way downwards through the soil and contains solid impurities.

Sea water is the reservoir into which all the impurities eventually go, and hence the solid content of sea water is usually high (3.6%).

The solids which are found in the natural waters are mainly the sulphates and bicarbonates (hydrogencarbonates) of calcium and magnesium together with smaller amounts of sodium chloride, silicates, nitrates, ammonium salts, as well as the gaseous impurities already mentioned as being present in rain water, *i.e.,* oxygen, nitrogen, and carbon dioxide.

Of the solid impurities the most important are calcium sulphate and calcium bicarbonate (hydrogencarbonate).

Calcium sulphate is present because many rocks and soils contain gypsum $(CaSO_4.2H_2O)$ which is slightly soluble in water (1 : 500) and hence some of it dissolves in any water with which it comes into contact.

Calcium hydrogencarbonate is present because water which contains carbon dioxide is capable of very slowly dissolving limestone or chalk (which is found in large quantities in some soils and in small quantities in practically all soils). This reaction is capable of affecting the geography of large areas, particularly those consisting of limestone or chalk hills. It is responsible for the formation of caves (see page 251) and often spectacular gorge scenery.

$$CaCO_3(s) + H_2O(l) + CO_2(g) \rightarrow Ca(HCO_3)_2(aq)$$

| limestone (insoluble) | water | carbon dioxide | calcium hydrogencarbonate |

It will be obvious that all the more soluble substances present on the face of the earth were washed away in the course of past ages, except where rainfall has been negligible – *e.g.,* the nitre deposits in the Atacama desert of north Chile (see page 439).

Definition. *A hard water is one which will not readily form a lather with soap.*

The nature and method of manufacture of soap

Soap is the sodium salt of an organic* acid. One of the commonest soaps is the sodium salt of stearic acid (octadecanoic acid) $C_{17}H_{35}CO_2H$, and has the formula $C_{17}H_{35}CO_2Na$, sodium stearate. It is important to notice that, for all its complex formula, soap is of exactly the same chemical nature as common salt NaCl. Both are sodium salts of acids. The complex group $C_{17}H_{35}CO_2$— of sodium stearate corresponds to the Cl of sodium chloride. For convenience, the formula of sodium stearate is often written NaSt, the St being used as a substitute for the stearate group $C_{17}H_{35}CO_2$—.

Soap is manufactured by heating vegetable oils (such as palm oil or olive oil) or animal fats with caustic soda solution. The oils or fats are compounds formed from glycerol and certain complex organic acids, such as stearic acid, mentioned above. The caustic soda liberates glycerol from the fat and forms the sodium salt of the acid, which is the soap (caustic soda is *sodium hydroxide*).

$$\text{Fat or oil} + \text{caustic soda} \rightarrow \text{soap} + \text{glycerol}$$

Soap is manufactured by steam-heating the fat and caustic soda solution in large pans. Common salt is added later and assists in the separation of the soap which, when cool, sets as a hard cake on the surface of the liquid. It is removed and purified, dyes and perfumes being added to produce toilet soaps.

The manufacturing process can be illustrated in the laboratory, using mutton fat or lard. Put the lard into an evaporating dish and add to it a solution of caustic soda to which methylated spirit has been added to quicken the action. Heat the dish on a steam bath. When all the liquid has evaporated off, a yellowish solid will be left. It is impure soap.

The soap cleanses by dissolving in the water, loosening the particles of dirt, and the whole (soap, water, and dirt) can then be washed away.

Cause of hardness

Now if there happens to be a calcium compound dissolved in the water, the soap is precipitated in the form of calcium stearate (which appears as a 'curd'), the latter being insoluble, *e.g.*,

$$2\text{NaSt(aq)} + \text{CaSO}_4\text{(aq)} \rightarrow \text{Na}_2\text{SO}_4\text{(aq)} + \text{CaSt}_2\text{(s)}$$

sodium stearate (soluble)	calcium sulphate (soluble)	sodium sulphate (soluble)	calcium stearate (insoluble)

Until the whole of the calcium compound has been acted upon by the soap, none of the latter can form a lather. Thus, with a hard water, a large amount of soap is used to precipitate and remove the calcium, and only a small extra amount to cause a lather.

In this way the valuable stearate group, which loosens the dirt, is lost completely, since it would be just as useful to try to wash with insoluble calcium stearate as any other substance insoluble in water, for example, marble, or iron.

This is not the only reason why hardness must be removed. Where water is used for boilers, certain of these solid substances (calcium sulphate and magnesium silicate being the worst offenders) are left behind as 'scale' on the inside of the pipes of the boiler as the water is evaporated off. As this scale increases in

---

* An organic compound is a compound of carbon (see page 328). The chemistry of carbon is so complex that it is convenient to treat it separately from that of other elements as 'organic chemistry'.

thickness, the bore of the pipe becomes smaller, and the walls of which the pipes are made become heated to a higher temperature than is normal, causing the pipes to weaken and finally burst.

Synthetic detergents have been introduced to replace soap in domestic and laundry work. A typical detergent is made from a complex *alkene*, i.e., a hydrocarbon of the type, C_nH_{2n}, where *n* is between 12 and 20. The alkene is first sulphonated by concentrated sulphuric acid; the sulphonate is then converted to its sodium salt by sodium hydroxide solution. The sodium salt is the detergent. An example is the following.

$$C_{15}H_{30} + H_2SO_4 \rightarrow C_{15}H_{31}.SO_4H$$
$$C_{15}H_{31}.SO_4H + NaOH \rightarrow C_{15}H_{31}.SO_4Na + H_2O$$

This detergent resembles sodium stearate (soap) in possessing a long fat-soluble carbon chain, $C_{15}H_{31}$—, and a water-soluble end-group, —SO_4Na.

Detergents of this kind have two advantages over soap; they are more soluble in water and are not affected by hardness in water. This is because the calcium salt of the detergent is soluble in water, whereas calcium stearate is insoluble. (See following sections.)

Removal of hardness

Distillation will remove all solid matter. This method is, as a rule, far too costly to be employed. Many high-pressure boilers supply steam to drive turbines, and the steam which comes from the turbines is condensed to water, which is actually distilled water. In some works, the water is fed back again into the boiler and small amounts of distilled water artificially prepared are added to make up the losses which are inevitable.

Removal of calcium as calcium carbonate (*chalk*). Many of the methods of rendering a hard water soft have as their object the conversion of a soluble calcium salt into the insoluble carbonate. In this way the calcium is removed, since the calcium carbonate, being insoluble, takes no further part in the reaction. An insoluble calcium salt cannot cause hardness.

Temporary hardness

Hardness which is due to the presence of calcium hydrogencarbonate can be removed by heating the water to boiling for a few minutes. Heat decomposes the calcium hydrogencarbonate into calcium carbonate (chalk) and carbon dioxide is expelled.

$$\underset{\substack{\text{calcium} \\ \text{hydrogencarbonate}}}{Ca(HCO_3)_2(aq)} \rightarrow \underset{\substack{\text{calcium} \\ \text{carbonate} \\ \text{(insoluble)}}}{CaCO_3(s)} + \underset{\text{water}}{H_2O(l)} + \underset{\substack{\text{carbon} \\ \text{dioxide}}}{CO_2(g)}$$

Because it can be removed merely by boiling, the name 'temporary' is given to this type of hardness. This method of removal would be expensive on the large scale.

Furring of kettles

In a district where the water contains calcium hydrogencarbonate the insides of kettles become coated with a layer of calcium carbonate caused by the decomposition of the hydrogencarbonate according to the equation shown above.

Stalagmites and stalactites

These pillars of almost pure calcium carbonate are made by water containing dissolved calcium hydrogencarbonate dripping from the roof on to the floor of a cavern. Some of the calcium hydrogencarbonate decomposes, giving off carbon dioxide into the atmosphere of the cave, and depositing calcium carbonate a little at a time on the roof and floor. This deposition causes a stalactite to grow downwards from the top of the cave and a stalagmite to grow upwards from the floor of the cave, until after a time the two meet. The growth varies very much from small fractions of a cm to 100 cm or more per year.

If a sample of stalactite is available, its chemical nature can be demonstrated by the tests given below.

(1) To a piece of the stalactite on a watch-glass, add a few drops of pure concentrated hydrochloric acid. Take up a little of the mixture on a platinum wire (sealed into the end of a glass tube to act as handle). Put the wire into the lower half of a Bunsen flame. A brick-red coloration shows *calcium* present. (At first, the yellow flame of sodium may interfere, because sodium compounds are very common impurities.)

(2) Put two or three small pieces of stalactite into a narrow test-tube and add dilute hydrochloric acid. Effervescence will occur. If the gas is passed into limewater and the liquid is shaken, it will show a white turbidity. That is, carbon dioxide is evolved, proving the test material to be a *carbonate*. (A hydrogencarbonate also gives off carbon dioxide with this acid, but only two common hydrogencarbonates exist as solid in ordinary conditions—$NaHCO_3$ and $KHCO_3$.)

$$CO_3^{2-}(aq) + 2H^+(aq) \rightarrow H_2O(l) + CO_2(g)$$

That is, the stalactite contains *calcium carbonate*, probably slightly impure.

In cross-section, stalactites usually show a ring structure similar to that of a tree trunk. This occurs because the summer rate of deposition of chalk is more rapid than the winter rate, temperature being lower in winter.

Removal of temporary hardness by addition of slaked lime

Temporary hardness can be removed by the addition of the calculated quantity of slaked lime (excess of lime would cause hardness on its own account). The amount of lime is calculated from a knowledge of the hardness of the water and of the capacity of the reservoir (Clark's method).

$$Ca(OH)_2(s) + Ca(HCO_3)_2(aq) \rightarrow 2CaCO_3(s) + 2H_2O(l)$$

| slaked lime (slightly soluble) | calcium hydrogen-carbonate | calcium carbonate (insoluble) | water |

A third method of removal of temporary hardness is to add sodium carbonate. (Permanent hardness is also removed at the same time.)

$$Na_2CO_3(aq) + Ca(HCO_3)_2(aq) \rightarrow 2NaHCO_3(aq) + CaCO_3(s)$$

Permanent hardness

This is due mainly to the presence of dissolved calcium sulphate, which cannot be decomposed by boiling, and hence the name 'permanent' is given to this type of hardness.

It is most easily removed by the addition of washing soda crystals to the water, when a precipitate of insoluble calcium carbonate is thrown down and thus prevents the calcium from interfering. (You should make yourself quite certain of the reason why soluble calcium sulphate can cause hardness, whereas insoluble calcium carbonate cannot. See page 248.)

$$CaSO_4(aq) + Na_2CO_3(aq) \rightarrow CaCO_3(s) + Na_2SO_4(aq)$$

calcium   sodium   calcium
sulphate   carbonate   carbonate
(soluble)   (insoluble)

Experiment 71
To make temporarily hard water

Bubble carbon dioxide into lime-water in a flask for about 20 minutes. The precipitate at first formed will have dissolved to form a temporarily hard water.

$$Ca(OH)_2(aq) + CO_2(g)$$

lime-water   carbon
              dioxide

$$\rightarrow CaCO_3(s) + H_2O(l)$$

       calcium   water
       carbonate

$$CaCO_3(s) + H_2O(l) + CO_2(g)$$

calcium   water   carbon
carbonate   dioxide

$$\rightarrow Ca(HCO_3)_2(aq)$$

         calcium
         hydrogencarbonate

To make permanently hard water. Add a little gypsum or anhydrous calcium sulphate to water in a flask. Shake and allow to stand, decant off the clear liquid; you will find it will be permanently hard water.

To make a soap solution. Scrape shavings off a tablet of toilet soap. Weigh out about 6 g of shavings and add to about 100 cm³ of methylated spirit in a beaker on a water bath. Warm and stir. When dissolved, transfer to a one dm³ flask and add water to make up about one dm³. Shake and allow to stand.

N.B. Since the composition of soap is variable the above solution need not be made up with great accuracy, nor are the readings to be taken as a mathematical comparison of hardness. **In these experiments the hardness of various waters is being compared.**

Fill a burette with the soap solution and place 25 cm³ of distilled water by means of a pipette into a conical flask. Run in 1 cm³ of the soap solution at a time, and, between additions, cork up the flask and shake. **When a lather is obtained, which persists unbroken for two minutes, the titration is completed.** You will easily see the difference between a 'curd' (which is formed when soap solution is run into a hard water) and a lather, as the former breaks very quickly and the latter consists of many tiny bubbles which reflect the light from the windows. When you have run 30 cm³ of soap solution into a hard water, you need not proceed further. The water is hard. Having obtained the result for distilled water perform the following experiments.

To show that boiling softens temporarily hard water. Take 25 cm³ of the temporarily hard water made as indicated above and titrate against soap solution. Notice the curd which forms.

$$Ca(HCO_3)_2(aq) + 2NaSt(aq)$$

$$\rightarrow CaSt_2(s) + 2NaHCO_3(aq)$$

         calcium
         stearate
         (curd)

Take *another* 25 cm³ and heat to boiling on a tripod and gauze. Notice that a milkiness appears (chalk). Cool, titrate against soap solution again, and note your result when a lather is obtained which remains unbroken for two minutes.

$$Ca(HCO_3)_2(aq) \rightarrow$$

$$CaCO_3(s) + H_2O(l) + CO_2(g)$$

Experiment 72
To show that washing soda crystals soften permanently hard water

Titrate 25 cm³ of the permanently hard water and record your result. Take *another* 25 cm³ of the hard water, add a few crystals of washing soda, and shake until they dissolve. Again you will notice a turbidity due to a precipitate of chalk.

$$Na_2CO_3(aq) + CaSO_4(aq) \rightarrow$$

$$Na_2SO_4(aq) + CaCO_3(s)$$

Titrate again and notice your results. They should speak for themselves. A typical set of results is given as follows.

	Volume of soap solution necessary to produce a lather to last unbroken for two minutes

25 cm³ distilled water

2nd reading	3.0 cm³	}	2.0 cm³
1st reading	1.0 cm³		

25 cm³ temporarily hard water

2nd reading	33.0 cm³	}	30.0 cm³*
1st reading	3.0 cm³		

25 cm³ temporarily hard water after boiling

2nd reading	40.0 cm³	}	7.0 cm³
1st reading	33.0 cm³		

25 cm³ permanently hard water

2nd reading	31.0 cm³	}	30.0 cm³*
1st reading	1.0 cm³		

25 cm³ permanently hard water to which washing soda was added

2nd reading	36.0 cm³	}	5.0 cm³
1st reading	31.0 cm³		

* Indicates no lather produced. No more soap solution was added — the water is extremely hard.

Permutit method of softening water

The above methods of softening water (*i.e.*, boiling and adding 'soda') are used mainly in the home for softening small amounts of water. In the treatment of larger supplies of water (but not so large as to be treated by the lime method) the Permutit process is used. Many 'water softeners' sold for domestic use work on this principle.

Permutit is a complex substance (hydrated sodium aluminium silicates) but we can regard it as Na_2Y $(Y = Al_2Si_2O_8.xH_2O)$. When a dissolved calcium salt runs over it, ion-exchange occurs.

$$(Na^+)_2Y^{2-}(s) + Ca^{2+}SO_4^{2-}(aq) \rightarrow Ca^{2+}Y^{2-}(s) + (Na^+)_2SO_4^{2-}(aq)$$
$$\text{insoluble}$$

The sodium permutit will finally become a calcium permutit, and it can be made fresh again by running concentrated common salt solution over it and washing away the soluble calcium chloride formed.

$$CaY(s) + 2NaCl(aq) \rightarrow Na_2Y(s) + CaCl_2(aq)$$

Organic resins (beads of polymers such as polystyrene) are now increasingly used for water softening by ion-exchange.

Water of crystallization

When a few crystals of copper(II) sulphate in a test-tube are heated gently, a copious evolution of water-vapour takes place (which condenses and runs back and cracks the tube if the mouth of the tube is not held lower than the bottom). The colour and shape of the crystals disappears and in place of blue crystals of hydrated copper sulphate, a white powdery mass of anhydrous copper sulphate forms.

$$CuSO_4.5H_2O(s) \rightarrow CuSO_4(s) + 5H_2O(g)$$

If water is slowly added to the white, anhydrous powder, hissing is heard and steam is evolved, showing that heat is generated. A blue solid is left. This liberation of heat energy occurs because the anhydrous copper sulphate is hydrated by the water and this is a chemical action, *i.e.*, combination of H_2O molecules with the ions of the salt. The heat generated is called *heat of hydration*. The reaction shown in the above equation is reversed.

Water of crystallization is necessary to the crystalline shape of some crystals, and is that definite amount of water with which the substance is associated on crystallizing out from an aqueous solution. The crystals cannot form in these cases without the presence of water with which to form a loose compound. It is sometimes termed 'water of hydration'.

Experiment 73
Determination of water of crystallization

Barium chloride crystals $BaCl_2.xH_2O$ provide a suitable case. A clean dry crucible is weighed with lid. Two or three grammes of barium chloride crystals are added and the whole is weighed. The crucible, with lid, is then heated on a pipe-clay triangle on a tripod, gently at first and later strongly, to drive off water of crystallization. The whole is allowed to cool in a desiccator (to exclude moisture) and weighed. It is then heated again, cooled as before, and reweighed. This is repeated until a *constant* mass is reached, which shows that all water has been expelled.

Calculation

Mass of crucible and lid	$= a$ g
Mass of crucible and lid and barium chloride crystals	$= b$ g
Mass of crucible and lid and barium chloride anhydrous	$= c$ g
Mass of water expelled	$= (b - c)$ g
Mass of barium chloride anhydrous	$= (c - a)$ g

Then

$$\frac{xH_2O}{BaCl_2} = \frac{(b - c)}{(c - a)}$$

and, inserting molar masses,

$$\frac{18x}{208} = \frac{(b - c)}{(c - a)} \quad \text{and} \quad x = \frac{(b - c)}{(c - a)} \times \frac{208}{18}$$

Substances which contain water of crystallization	Substances which do not contain water of crystallization
Sodium carbonate crystals ($Na_2CO_3.10H_2O$)	Sodium chloride NaCl
Sodium sulphate crystals ($Na_2SO_4.10H_2O$)	Potassium permanganate $KMnO_4$
Copper(II) sulphate crystals ($CuSO_4.5H_2O$)	Potassium nitrate KNO_3
Iron(II) sulphate crystals ($FeSO_4.7H_2O$)	Ammonium sulphate $(NH_4)_2SO_4$

Solution

Many solids possess the property of dissolving in water. For instance, if we shake up a few small crystals of copper(II) sulphate (crush them in a mortar first) with water in a test-tube, the water will turn blue and the crystals will finally disappear.

The particles of copper(II) sulphate in the solution must be very small, for we cannot see them with the naked eye, nor even with the most powerful microscope made. Furthermore, they will pass through the pores of a filter paper with ease. The solution (as this mixture is called) of copper(II) sulphate in water is uniform in blue colour and in composition (after stirring) and, in a sealed tube, remains so indefinitely. In general, to produce such a solution, the ordered crystal lattice of the solid (the *solute*) built up from positive and negative ions (see NaCl, page 80) or from molecules (see I_2, page 90), must first be broken down so that its particles can diffuse between the molecules of the liquid (the *solvent*) and move among them. Since the particles in the solid lattice exercise electrical attraction on each other, energy must be supplied to separate them. This energy is taken from the heat energy of the liquid; consequently, there is usually a fall of temperature as a solid dissolves in water. If, however, some form of chemical action occurs at the same time, *e.g.*, hydration of ions, this action may liberate enough heat to cause an overall rise in temperature, *e.g.*, when anhydrous copper(II) sulphate is dissolved in water.

Liquids other than water may act as solvents, *e.g.*, trichloromethane, tetrachloromethane, and carbon disulphide. These liquids, and many others, are *covalent* compounds and dissolve a great number of *covalent* solids, *e.g.*, trichloromethane dissolves iodine, carbon disulphide dissolves sulphur. Ionic compounds, *e.g.*, Na^+Cl^-, do not usually dissolve in covalent solvents but a great many are soluble in water, in which their ions can dissociate.

Chemical solution

This is a term often employed to indicate the apparent solution of a solute in a solvent, together with chemical action. For example, zinc appears to dissolve in dilute sulphuric acid and neither zinc nor dilute sulphuric acid can be recovered by evaporation or distillation, since the solid residue on evaporation would be zinc sulphate. Actually the processes of chemical action and solvent action follow one another. The zinc attacks the acid to form zinc sulphate, which then dissolves in the water present.

Suspension

The above properties of a solution help us to differentiate between a solution and a suspension. A solid is said to be in suspension in a liquid when small particles of it are contained in the liquid but are not dissolved in it.

If the mixture is left undisturbed the solid particles will slowly settle to the bottom of the containing vessel, leaving the pure liquid above them.

Muddy water is a typical suspension. The mud would settle after a time if left undisturbed, leaving a brown residue on the bottom of the containing vessel and clear water above. The particles of mud would be retained by a filter paper whilst the water (and any solids in solution) would pass through.

Saturated solution

If we add half a gramme of sodium chloride to 100 g of water in a beaker the salt will dissolve. We could go on for a time adding salt half a gramme at a time, and, by stirring vigorously after each addition, bring about solution of the sodium chloride, but with increasing difficulty. Finally, there would come a time when no more sodium chloride would dissolve **at that particular temperature,** and, no matter how long we left it or how vigorously we stirred, no more sodium chloride would dissolve. The solution is then said to be saturated with sodium chloride at the particular temperature.

Definition. A saturated solution of a solute at a particular temperature is one which contains as much solute as it can dissolve at that temperature, in the presence of the crystals of the solute.

The concentration of a saturated solution varies with the solute, the solvent, and also with the temperature. Thus sulphur is almost insoluble in water yet readily dissolves in carbon disulphide, and a rise in temperature will cause more to dissolve.

This is generally true for solids; for example, potassium nitrate is at least seven times more soluble in water at 80 °C than it is in water at 10 °C.

Determination of solubility

To give a quantitative meaning to solubility, it is necessary to fix the amount of the solvent and to state the temperature under consideration. The amount of solvent is usually fixed at 100 g.

Definition. The solubility of a solute in a solvent at a particular temperature is the number of grammes of the solute necessary to saturate 100 g of the solvent at that temperature.

It denotes a limit, that is, the *maximum* amount which can normally be held in solution. Solubility is also sometimes expressed in grammes of solute per dm^3 of solution at a given temperature.

Experiment 74
To determine the solubility of potassium nitrate in water at the temperature of the laboratory

This determination must be carried out in two stages. It is first necessary to prepare a saturated potassium nitrate solution at laboratory temperature and then to find the proportions of potassium nitrate and water in it.

To make the saturated solution. The rate of solution of a solid in cold water is generally so slow that it is almost impossible to obtain a saturated solution of it in a reasonable time by merely shaking the solid with the water. The quicker and more certain way is to crystallize from a warm solution by cooling.

Half fill a boiling-tube with water and dissolve in it some potassium nitrate. Warm and shake well. Pour off a small sample into a test-tube and cool it under the tap. If no crystals appear, return the sample to the boiling-tube and add more solute. Test another sample and continue in this way till a sample gives crystals. Then cool the whole solution. When the crystals have separated and the solution is quite cold, take the temperature of it. Then filter it through a *dry* filter paper and funnel into a *dry* receiver to avoid diluting it. The filtrate is a saturated solution of potassium nitrate at the observed temperature.

To obtain the solubility of potassium nitrate, using this solution. Weigh a clean dry dish and add some of the saturated solution to it. Weigh again. **Once having weighed be careful not to lose any portion of the solution.** Place the dish on a steam bath (page 9), and evaporate until the potassium nitrate is left quite dry. (The dish on a gauze may be warmed very gently over the Bunsen flame for a few minutes to complete the removal of water.) Allow the dish to cool and weigh it. Calculate the mass in grammes of the potassium nitrate which would have dissolved in 100 g of water as in the following calculation.

Alternative method of evaporation to dryness. The following method is quicker than the one suggested above with little loss of accuracy if carefully performed. Weigh a dish (75 mm diameter is suitable) with a clock-glass to fit over it. Weigh again, having added some satur- ated solution of potassium nitrate and replaced the clock-glass. Evaporate to dryness over a medium flame increasing the size of the flame towards the end. Do not heat the solid suffi- ciently strongly to decompose it. Allow to cool and weigh.

Specimen results

$$
\begin{aligned}
\text{Mass of dish} &= 14.32 \text{ g } (a) \\
\text{Mass of dish and solution} &= 35.70 \text{ g } (b) \\
\text{Mass of dish and potassium nitrate} &= 18.60 \text{ g } (c) \\
\text{Temperature of saturated solution} &= 15\,°C
\end{aligned}
$$

Hence:

17.10 g $(b - c)$ of water dissolve 4.28 g $(c - a)$ of potassium nitrate

so, 100 g of water dissolve $\dfrac{4.28}{17.10} \times 100$ g of potassium nitrate

$$= 25.0 \text{ g}$$

That is, solubility of potassium nitrate is 25.0 g per 100 g of water at 15 °C.

Generally, increase in temperature increases the solubility of a solute in water. This is because most solutes dissolve in water with absorption of heat. So, if a solution and solid crystals of the solute are in equilibrium at a certain temperature and the mixture is then heated, the system will alter so as to tend to lower the temperature again (Le Chatelier's Principle, page 184). That is, more crystals will dissolve because their dissolution absorbs heat.

By finding the solubilities of a solute at varying temperatures a graph can be plotted to show how the solubility alters with increase of temperature, with many interesting results. This is called a *solubility curve* of the solute.

Experiment 75
To determine the solubility of potassium nitrate at 50 °C

This is the method employed to determine the solubility at any temperature above laboratory temperature. At these higher temperatures, the rate of solution of the potassium nitrate is greatly increased, and a saturated solution may be made directly.

Crush some crystals of potassium nitrate in a mortar, place some of them in a boiling-tube and add a little water (to make the tube about half-full). Put the boiling-tube in a beaker of water and warm the latter up to a temperature of about 55 °C (Figure 123). Whilst warming the solution keep adding potassium nitrate crystals to the boiling-tube, and stir all the time. Add potassium nitrate until some remains un-dissolved at the bottom of the tube. Remove the flame when the temperature, as read by the thermometer, reaches 55° C, and allow the apparatus to cool, stirring all the while and always maintaining some undissolved potassium

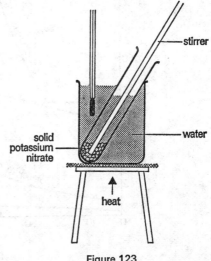

Figure 123

nitrate at the bottom of the tube. Just before the temperature falls to 50 °C, remove the stirrer, allow the solid potassium nitrate to settle, and put the dry thermometer into the potassium nitrate solution. When the temperature is exactly 50 °C, rapidly decant a little of the saturated solution into a weighed dish, leaving all solid potassium nitrate behind. Weigh the dish again and evaporate to dryness as in the previous experiment. Calculate the mass of potassium nitrate dissolved in 100 g of water at 50 °C in a similar way.

By repeating the experiment at varying temperatures several values can be obtained from which a curve can be plotted. The following figures were actually obtained by a middle school form. (The figures in brackets are the accepted accurate values.) Plot the graph on squared paper.

Temperature	11 °C	15 °C	30 °C	40.5 °C	50 °C	57 °C
Grammes of KNO_3 per 100 g water	23.6 g (10 °C; 20 g)	25.1 g (25 g)	43.3 g (45 g)	63 g (40 °C; 63 g)	84 g (85 g)	102 g (106 g)

Experiment 76
Alternative method (using any substance which does not exhibit super-saturation)

Weigh 4.5 g of potassium chlorate into a boiling-tube and run in 10 cm³ (g) of water from a burette. Warm until dissolved, remove from the flame, insert a thermometer and allow to cool, stirring with the thermometer. Note the temperature at which crystals appear. This will be the temperature at which the solubility is 45 g per 100 g water. Add a further 10 cm³ of water and repeat the experiment. Continue the addition, determining the temperature at which the solution is just saturated until 60 cm³ of water have been added. Construct the graph.

Graph of solubilities

Look at the accompanying graph of the solubilities of a few familiar salts (Figure 124). Answer the following questions.

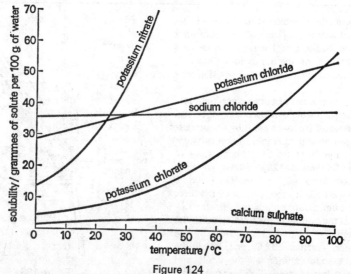

Figure 124

(a) For which salt does the solubility increase most rapidly with rise in temperature?

(b) Given a mixture of equal masses of potassium chlorate and potassium chloride, how would you obtain some pure chlorate?

(c) For which salt is there a decrease in solubility with increase in temperature?

Super-saturation

Definition. *A solution is said to be super-saturated when it contains in solution more of the solute than it could hold at that temperature if crystals of the solute were present.*

It is important to notice that 'super-saturated' solutions are in an unstable condition. They can only be obtained to any marked extent:

(a) From a few compounds, for example: Glauber's salt $Na_2SO_4.10H_2O$, 'hypo' $Na_2S_2O_3.5H_2O$.

(b) By excluding all dust. The dust particles might act as centres of crystallization.

(c) By cooling the solution slowly. 'Hypo' is exceptional in giving a super-saturated solution when cooled quickly.

(d) By avoiding all shaking or disturbance of the solution.

A solution cannot be super-saturated if in contact with crystals of the solute.

Experiment 77
To prepare a super-saturated solution of sodium thiosulphate

Fill a 150 × 25 mm boiling-tube to a depth of about 25 mm with water, then fill it up with crystals of 'hypo', sodium thiosulphate $Na_2S_2O_3.5H_2O$. Heat the tube gently and shake until the crystals are all dissolved. Holding the boiling-tube quite still, cool the solution under the tap. Even when the solution is quite cold, no crystals separate (Figure 125(a)). Now select a single pin-head-size crystal of sodium thiosulphate and drop it into the solution. At once, white crystals separate, growing slowly downwards through the solution and starting from the added crystal as centre (Figure 125(b)). The contents of the boiling-tube become almost a solid mass of crystals and only a very little solution can be poured off. Note the rise of temperature as the crystals separate.

After the crystals have separated, the solution left must be still saturated. Before they separated, it must have been 'more than saturated', or 'super-saturated', as it is called.

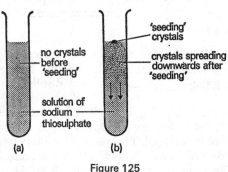

'seeding' crystals

crystals spreading downwards after 'seeding'

no crystals before 'seeding'

solution of sodium thiosulphate

(a)　　　　(b)

Figure 125

Fractional distillation

If two liquids have reasonably different boiling points, *e.g.*, water 100 °C, ethanol (alcohol) 78 °C at standard pressure, they can be separated (though not always completely) by *fractional distillation*. Various forms of fractionating column can be used (Figure 126). Their general purpose is to provide surfaces, *e.g.*, flat discs, on which ascending vapour can condense. This produces a succession of liquid films in which, as the column is ascended, there is an *increasing* concentration of the *more* volatile liquid (of *lower* boiling point).

Ascending vapour comes into approximate equilibrium with these liquid films. Consequently, the vapour which leaves the column and passes to the condenser is far richer in the more volatile constituent than the original mixture.

For example, a mixture of equal volumes of 'methylated spirit' and water will not burn in air. If it is distilled as in Figure 126, and a few cm³ of the early distillate are put into an evaporating dish, the liquid will burn readily on the application of a light. That is, the proportion of alcohol is significantly increased.

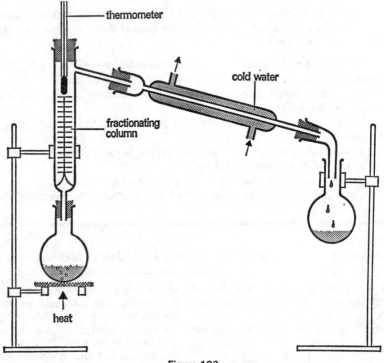

Figure 126

An efficient industrial still can deliver alcohol of about 96% concentration from a very dilute alcohol wash produced by fermentation, but this mixture has a lower boiling point than any other water–alcohol mixture and distillation cannot purify it further.

An industrial fractionating column (as used, *e.g.*, in distillation of crude petroleum) will be cylindrical, made of steel and up to, perhaps, 60–70 m high. Hot vapour ascends through the middle of the column. Trays at regular intervals round the walls collect layers of condensed liquid through which ascending vapour is forced to pass by baffles. The column delivers fractions of distillate from various points, with decreasing boiling points higher up the column. See also page 334.

Questions

1. Explain the difference between (*a*) temporary, and (*b*) permanent hardness in water. Give the names of *one* substance causing temporary and of *two* substances causing permanent hardness in water. Explain how permanent hardness may be removed, giving equations for the chemical changes involved. How would you test for the presence of a chloride in tap water? (J.M.B.)

2. What are solubility curves? Of what use are they? What experiments would you make in order to construct a solubility curve for potassium chlorate? (O. and C.)

3. What are the substances which cause (a) temporary; (b) permanent 'hardness' in water?

Explain the effect of adding the following substances to 'hard' water: (a) soap; (b) lime-water; (c) washing soda.

What is the 'fur' deposited in the kettle? Briefly explain how stalactites and stalagmites are formed in caves.

4. What are the conditions, and what are the products, for the reaction of the following substances with water or steam: (a) iron; (b) calcium; (c) charcoal; (d) chlorine; (e) calcium carbide?

5. Imagine that you wish to prove experimentally to someone that 18 g of water contain 2 g of hydrogen and 16 g of oxygen. Give a labelled sketch of the apparatus you would use; indicate the two chief precautions you would adopt to ensure an accurate result; show how you would use the data you obtain to prove the above statement. (J.M.B.)

6. Ordinary tap water always contains some air in solution. Describe in detail how you would collect a quantity of this air from tap water. How could you find the proportion by volume of oxygen in such air? Explain why the composition of this dissolved air will be different from that of ordinary air. (J.M.B.)

7. State and explain what happens when carbon dioxide gas is passed for a long time through lime-water. State and explain what happens when soap solution is shaken with the final product (a) before it has been boiled; (b) after it has been boiled. (J.M.B.)

8. What is meant by the terms 'saturated solution' and 'super-saturated solution'? Describe exactly how you would proceed to determine the solubility of potassium nitrate in water at 15 °C. (J.M.B.)

9. What experiments would you make to find out whether (a) a given sample of water is hard or soft; (b) a gas-jar contains di-nitrogen oxide or a mixture of nitrogen and oxygen; (c) a given solid is sodium sulphide or sodium sulphate? (O. and C.)

10. Define the term solubility. What are the effects of temperature and pressure on the solubility of (a) gases; (b) solids, in water? Give the requisite practical details for constructing a solubility curve of a salt, *e.g.*, potassium nitrate. State two of the uses of a solubility curve.

11. A colourless liquid (X) is either pure water, or water containing some dissolved solid. Describe carefully how to discover which it is by observations of the temperature at which it (a) boils; (b) freezes. Sketch the apparatus you would use in each case.

Assume that a supply of pure water is provided. (J.M.B.)

12. What do you observe when carbon dioxide is passed for a long time through lime-water? Give equations representing the reactions which take place.

What happens when the final solution obtained is (a) treated with soap solution; (b) boiled; (c) treated with a solution of sodium carbonate.

Describe the 'permutit' process for softening water and state how the permutit is restored (or revivified). (W.)

13. Describe in detail an experiment by which you could determine the mass of oxygen which combines with 1 g of hydrogen to form water. What is meant by the term *hardness of water*? Describe *two* methods by which all the calcium ions present in a sample of water can be removed. (S.)

14. How, and under what conditions, does (a) zinc, (b) iron, (c) copper, (d) magnesium, react with (i) water or steam, (ii) dilute sulphuric acid? If in a particular instance there is no reaction, this should be clearly stated.

Nickel is a metal which reacts only slowly with steam at high temperature, and which also reacts slowly with dilute sulphuric acid. Place the metals named in the first paragraph and nickel in the order in which they appear in the reactivity series, placing the most reactive first. (J.M.B.)

15. (a) 'Belfast water is soft, Dover water is hard.' Describe what is meant by this statement and explain in chemical terms the difference it implies. Mention *one* disadvantage of hard water and give *one* method of removing hardness from water.

(b) Figure 127 shows the solubility of potassium nitrate in water at various temperatures, solubility being expressed in grammes of solute per 100 grammes of solvent. 16 grammes of potassium nitrate are heated with 10 grammes of water in a boiling-tube, by means of a water bath, until a clear solution is obtained, stirring being continuous.

(i) At what temperature would a clear solution first be obtained? (ii) If the resultant solution is cooled to 50 °C, what mass of solute will have been deposited out of solution? (iii) The contents of the boiling-tube are now reheated to 80 °C and 6 grammes of water added at this temperature. The solution is again allowed to cool: at what temperature will crystals of solute first appear if supercooling does not occur? (N.I.)

16. You are asked to determine the solubility (in g/100 g water) of potassium chlorate at room temperature. The method to be used involves the evaporation of a measured

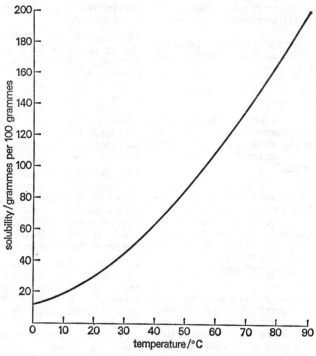

Figure 127

portion of a solution which is saturated at room temperature. Describe how you would prepare the solution and use it to carry out the solubility determination. Mention the precautions you would take to ensure accuracy, and indicate how you would calculate the result.

Draw a solubility curve for potassium chlorate from the following data. (Plot temperature along the horizontal axis, taking 1 cm to represent 5 °C, and solubility along the vertical axis, taking 1 cm to represent 5 g/100 g.)

Temperature, °C	10	30	50	70	90
Solubility, g/100 g H_2O	5	10	18	30	46

Use your graph to find (*a*) the solubility of potassium chlorate at 20 °C; (*b*) the mass of solid which would crystallize if 130 g of a solution saturated at 70 °C was cooled to 20 °C. (A.E.B.)

17. (*a*) Two separate 100 cm³ samples of tap water containing calcium ions were taken in conical flasks labelled *A* and *B*. *A* was left untouched; a few crystals of sodium carbonate were added to *B*. Each sample was now titrated against some soap solution which contained dissolved sodium stearate and the following observations were made:

Flask *A*: A precipitate was formed and 26 cm³ of soap solution were needed to give a lather.

Flask *B*: No precipitate was formed and only 2 cm³ of soap solution were needed to give a lather.

Explain these results as fully as you can.

(*b*) Draw a simple diagram to show the apparatus you would use to electrolyse a sample of water acidified with dilute sulphuric acid. The diagram should show how any gases produced are collected. Give the names and relative volumes of the gaseous products at the named electrodes, and write equations for the reactions occurring at these electrodes. How would the results differ if you replaced the acidified water with brine (concentrated aqueous sodium chloride)?

(*c*) 4.10 g of hydrated magnesium sulphate crystals $MgSO_4.xH_2O$ gave 2.00 g of anhydrous salt on careful heating. Calculate the formula of the hydrated salt. (L.)

18. Determine the solubility of substance *S* in water at room temperature from the following data.

mass of evaporating basin = 25 g
mass of evaporating basin
+ saturated solution of *S* = 55 g
mass of evaporating basin
+ solid *S* = 30 g

(S.)

19. (*a*) Explain, with examples, the meaning of the terms deliquescence, efflorescence, super-saturated solution. You are provided with a solution containing a mixture of two salts which can be separated by crystallization. Indicate: (i) the nature of the solubility curves of the two salts; (ii) how you would attempt to separate them.

(*b*) The order of reactivity of the following elements is zinc, lead, hydrogen, copper in descending order of reactivity. Describe the experiments you would carry out to confirm this order, indicating carefully what observations you would expect to make. (Oxides and salts of the above may be assumed to be readily available.) (S.)

20. What is meant by the term 'solubility of a salt'? Choose any well-known soluble salt, and describe experiments you would perform in order to investigate the effect of temperature on its solubility in water. Give a very simple sketch of the kind of graph you would expect to get if you plotted your results (*i.e.*, solubility against temperature). From your results do you deduce that it would be easy to crystallize this salt from its solution? Explain your answer. (S.)

21. (*a*) Calcium carbonate, present in rocks or soil, is one of the causes of hardness in water. Explain why this is so.

(*b*) Explain the use of (i) calcium hydroxide, and (ii) sodium carbonate, in the softening of hard water.

(*c*) Why does the presence of dissolved sodium carbonate not make water hard?

(*d*) A copper boiler used in the preparation of distilled water is encrusted with a layer of white scale caused by the hardness in the water used. Explain how this scale was formed from the hard water. If supplies of dilute sulphuric, hydrochloric, and nitric acids were available, which of these acids would you use to remove the scale from the boiler? Give the reasons for your choice. (C.)

22. Describe briefly how you would determine whether a crystalline solid which apparently did not dissolve in water was in fact slightly soluble. A solid acid can be represented by the formula H_2X, and it has a formula weight of 130. From a reference book it was learned:

at 25 °C, 10 g of the acid would dissolve in 100 g of water
at 85 °C, 30 g of acid would dissolve in 100 g of water

The following experiments were carried out to determine the solubilities at two other temperatures:

I 25 g of a saturated solution of the acid in water at 70 °C were evaporated to dryness. The residue of acid had a mass of 5 g.
II 25 g of a saturated solution of the acid in water at 40 °C needed 50 cm³ of molar (1M) sodium hydroxide for complete neutralization reaction is:

$$H_2X + 2NaOH \rightarrow Na_2X + 2H_2O$$

(*a*) Determine the mass of acid which would dissolve in 100 g of water at 70 °C and at 40 °C.

(*b*) Plot a graph of the number of grammes of acid dissolving in 100 g of water against the temperature. Use a scale of 1 inch (or 2 cm) for 10 °C along the *x* axis (the horizontal axis), and 1 inch (or 2 cm) for 5 g along the *y* axis.

(*c*) Use your graph to predict: (i) the maximum mass of acid which would dissolve in 100 g of water at 0 °C, (ii) the mass of acid which would separate out if 100 g of water saturated with acid at 55 °C were cooled to 10 °C. (J.M.B.)

23. 50 cm³ of tap water require the addition of 10 cm³ of soap solution before a persistent lather is formed on shaking. A fresh sample of 50 cm³ of water from the same tap is gently boiled for ten minutes and allowed to cool. This water now requires only 3 cm³ of the soap solution to form a lasting lather. A further sample of tap water is distilled and it is found that 50 cm³ of the distillate require only one drop of soap solution to form a lather.

Copy the following sentences, writing in the missing words:

(*a*) The tap water contains both ... and ... hardness.

(*b*) When the water is boiled, ... changes to ... which is precipitated.

(*c*) The boiled water contains dissolved ... and these form a scum with soap until the ... ions are removed from solution. (S.)

24. Give explanations of the following:
(*a*) When a sample of tap water is boiled for some seconds a white precipitate is observed.
(*b*) When carbon dioxide is bubbled through a suspension of a small quantity of powdered chalk in water, a clear solution is obtained.

A sample of water is found to contain, in every dm³, 0.111 g of calcium chloride $CaCl_2$, and 0.1205 g of magnesium sulphate $MgSO_4$. Explain how this water can be softened by the addition of sodium carbonate, and calculate the mass of anhydrous sodium carbonate that would be required for 1 dm³ of the water. (O. and C.)

24 Acids, Bases, and Salts

Acids

Acids have a variety of properties (as stated later), but all these properties are derived from a single type of behaviour. This behaviour is the production of the hydrogen ion, H^+, by acids when dissolved in water. Consequently, an acid can be defined in the following way:

Definition. An acid is a compound which, when dissolved in water, produces hydrogen ions, $H^+(aq)$, as the only positive ion.

Examples are:

Hydrochloric acid	$HCl \rightleftharpoons H^+ + Cl^-$
Sulphuric acid	$H_2SO_4 \rightleftharpoons 2H^+ + SO_4^{2-}$
Nitric acid	$HNO_3 \rightleftharpoons H^+ + NO_3^-$

It may be mentioned here that the situation is not quite as simple as the above implies. Pure water-free HCl shows no acidity because it contains no ions. It is a covalent compound. The aqueous solution contains ions. The ion responsible for acidity, however, is not the simple H^+ ion, but a hydrated form of it, the hydroxonium ion H_3O^+, produced by a reaction between HCl and water:

$$H_2O + HCl \rightleftharpoons H_3O^+ + Cl^-$$

Similarly, water-free H_2SO_4 and HNO_3 are probably not acidic (being un-ionized), but ionize in water and then become strongly acidic:

$$H_2SO_4 + 2H_2O \rightleftharpoons 2H_3O^+ + SO_4^{2-}$$
$$HNO_3 + H_2O \rightleftharpoons H_3O^+ + NO_3^-$$

For ordinary purposes the hydration of the hydrogen ion is often ignored and H^+ is used. (*For HCl in toluene (methylbenzene), see page 279.*)

If, as in the above three dilute acids, the ionization is almost *complete*, the acid is said to be a *strong* acid. If the ionization is only *slight*, the acid is said to be *weak*. For example, bench acetic (ethanoic) acid is about 0.4% ionized, that is, only four out of every thousand molecules of the acid pass into the ionized state.

$$CH_3COOH \rightleftharpoons H^+ + CH_3COO^-$$

Per thousand molecules	996	4	4

Some properties of acids

Taste. Most dilute acids have a sour taste. This is true of the three common *mineral* acids, sulphuric, hydrochloric, and nitric, and of many others. The sour taste of many unripe fruits, lemons, and sour milk is caused by the acids in them.

Action on litmus. Most acids turn blue litmus to red. Some of the weaker acids, however, for example carbonic acid, are so feebly acidic that they can only turn litmus to claret colour.

Corrosive action. The man in the street connects the term *acid* with the idea of a corrosive, 'burning' liquid. This is because two of the commonest acids – sulphuric acid (oil of vitriol) and nitric acid (aqua fortis) – are actually corrosive liquids. Acids are not, however, generally corrosive and most of them are solids.

Action with metals. Metals which are much more electropositive than hydrogen (page 474), react with dilute hydrochloric or dilute sulphuric acid (or both) to liberate hydrogen. The metals, *e.g.*, Zn, Mg, Al (with hot concentrated HCl only), and Fe, supply electrons which are taken up by H^+ ions of the acid, as:

$$Zn(s) \rightarrow Zn^{2+}(aq) + 2e^-$$
$$2H^+(aq) + 2e^- \rightarrow H_2(g)$$

A metal such as lead, which is only slightly more electropositive than hydrogen, reacts with neither of these dilute acids, but will react in a similar way with hot, concentrated hydrochloric acid.

Nitric acid is too strongly oxidizing to allow the liberation of hydrogen. Oxides of nitrogen are obtained. For a further account of this, see page 438.

Action with carbonates. Almost all acids (only the very weakest are exceptions) liberate carbon dioxide from a carbonate. The carbonate supplies the ion, $CO_3{}^{2-}$ which, with H^+ from the acid, gives the reaction:

$$2H^+(aq) + CO_3{}^{2-}(s) \rightarrow H_2O(l) + CO_2(g)$$

Action with bases and alkalis. This is very important but must be postponed (to page 268) till the nature of bases and alkalis has been considered.

Methods of preparation of acids

(1) By the reaction between an acid anhydride (the acidic oxide of a non-metal) and water.

Examples of this method are the preparation of sulphurous acid and meta-phosphoric acid by the action of sulphur dioxide and phosphorus(V) oxide with cold water.

$$SO_2(g) + H_2O(l) \rightleftharpoons H_2SO_3(aq); \quad P_4O_{10}(s) + 2H_2O(l) \rightarrow 4HPO_3(aq)$$

(2) By displacing a weaker or more volatile acid from its salt by a stronger or less volatile acid.

For example: (*a*) displacement of the more volatile hydrogen chloride from metallic chloride by the less volatile concentrated sulphuric acid.

$$NaCl(s) + H_2SO_4(aq) \rightarrow NaHSO_4(aq) + HCl(g)$$

(*b*) displacement of the weaker boric acid from borax by sulphuric acid.

$$Na_2B_4O_7(s) + H_2SO_4(aq) + 5H_2O(l) \rightarrow Na_2SO_4(aq) + 4H_3BO_3(aq)$$
borax · boric acid

(3) By precipitating an insoluble sulphide from a metallic salt by hydrogen sulphide.

$$Pb(C_2H_3O_2)_2(aq) + H_2S(g) \rightarrow PbS(s) + 2CH_3COOH(aq)$$
lead ethanoate · · · · · · · · · · · · · · · · · ethanoic acid

Basicity of an acid

We have seen that it is characteristic of an acid to yield H^+ in aqueous solution. The number of H^+ ions produced per molecule of the acid is called its *basicity*.

Definition. *The basicity of an acid is the number of hydrogen ions, H⁺(aq), which can be produced by one molecule of the acid.*

Acid	Basicity
$HCl \rightleftharpoons H^+ + Cl^-$	Monobasic
$H_2SO_4 \rightleftharpoons 2H^+ + SO_4{}^{2-}$	Dibasic
$H_3PO_4 \rightleftharpoons 3H^+ + PO_4{}^{3-}$	Tribasic

Notice that the basicity of an acid is not necessarily the number of hydrogen atoms contained in one molecule of it. For example, ethanoic acid $C_2H_4O_2$, though containing four hydrogen atoms per molecule, is only *mono*basic. Three of the four hydrogen atoms are so combined as to be incapable of ionizing.

$$C_2H_4O_2(aq) \rightleftharpoons H^+(aq) + C_2H_3O_2{}^-(aq)$$

Bases

A little earlier, an acid was characterized as a producer of H^+ in aqueous solution. To correspond with this, a base may be defined in the following way:

Definition. *A base is a substance which can combine with hydrogen ion, H⁺(aq).*

A number of bases are mentioned below and their combination with H^+ is shown. The reactions are reversible. If, as the equations are written below, the equilibrium lies strongly to the *left*, *i.e.*, few molecules are formed, the base is said to be *weak*; if the equilibrium lies strongly to the *right*, *i.e.*, few ions remain, the base is said to be *strong*.

Base	Base	Acid	Strength of base
Chloride ion	$Cl^- + H^+$	$\rightleftharpoons HCl$	very weak
Nitrate ion	$NO_3^- + H^+$	$\rightleftharpoons HNO_3$	very weak
Sulphate ion	$SO_4{}^{2-} + 2H^+$	$\rightleftharpoons H_2SO_4$	very weak
Ethanoate ion	$CH_3COO^- + H^+$	$\rightleftharpoons CH_3COOH$	strong
Hydroxyl ion	$OH^- + H^+$	$\rightleftharpoons H_2O$	very strong

It is obvious that a base exists corresponding to each acid. The corresponding pairs can be called *conjugates* and, if the acid is weak, the base is strong, and vice versa.

At the chemical level for which this book is intended, the most important of all the bases is the hydroxyl ion, OH^-. It is a very strong base indeed, and when it combines with hydrogen ion, the water formed is almost completely unionized, *i.e.*, from this point of view, water acts as a very weak acid. Metallic oxides and hydroxides are usually electrovalent compounds, containing the ion, O^{2-} (oxides), or OH^- (hydroxides), in addition to the metallic ion. The ion, O^{2-}, can enter the equilibrium: $O^{2-} + H_2O \rightleftharpoons 2OH^-$ with water. Consequently, metallic oxides and hydroxides of this kind can produce the base, OH^-, and are known as *basic* oxides or hydroxides. By reacting with an acid, *i.e.*, with H^+, they produce molecules of water.

$$H^+(aq) + OH^-(aq) \rightleftharpoons H_2O(l)$$

At the same time, the metallic ion from the oxide and the negative ion from the acid remain, and together constitute a *salt*. Examples of this are:

$$Na^+OH^-(aq) + H^+Cl^-(aq) \rightarrow Na^+Cl^-(aq) + H_2O(l)$$
$$Ca^{2+}(OH^-)_2(aq) + 2H^+NO_3^-(aq) \rightarrow Ca^{2+}(NO_3^-)_2(aq) + 2H_2O(l)$$

Those ions which take no part in the reaction are known as 'spectator ions'.

From these considerations, the following definitions can be derived:

Definition. *A basic oxide (or hydroxide) is a metallic oxide (or hydroxide) which contains ions, O^{2-} (or OH^-), and will react with an acid to form a salt and water only.*

It is vital to realize the importance of the word *only* in these definitions. If it was omitted, certain compounds, which are quite different from basic metallic oxides and hydroxides, would be included under the definition of base. Thus, lead(IV) oxide reacts with hydrochloric acid to produce lead(II) chloride (a salt) and water, but the word *only* excludes it from the class of bases because chlorine is also produced.

$$PbO_2(s) + 4HCl(aq) \rightarrow PbCl_2(aq) + 2H_2O(l) + Cl_2(g)$$

Compare this with the equations above. Lead(IV) oxide is clearly not a base.

The nature of the hydroxides of the metals varies according to the position of the metal in the reactivity series, as illustrated below.

K Na Ca	The hydroxides of these metals are *soluble* in water and are alkalis.	Hydroxides of sodium and potassium not decomposed by heat.
Mg Al Zn Fe Pb	These metals form hydroxides which are insoluble in water. They are also amphoteric excepting the two hydroxides of iron and that of magnesium.	Decomposition into oxide and water when hydroxides of these metals are heated.
Cu	Hydroxide is insoluble in water.	
Hg Ag Au	Hydroxides of these metals do not exist.	

Metallic hydroxides are electrovalent compounds, composed of metallic ions, which are positively charged, and hydroxyl ions, OH^-. The number of OH^- ions associated with one metallic ion is equal to the valency of the metal, *e.g.*,

$$Na^+OH^-; \quad \text{sodium is univalent}$$
$$Ca^{2+}(OH^-)_2; \quad \text{calcium is divalent}$$

The metallic hydroxides form a very important series of compounds. The soluble hydroxides (that is, the alkalis) are particularly important. They have many uses both in the laboratory and industry.

Alkalis

A definition of *basic hydroxide* has been given above. If such a hydroxide is soluble to a considerable extent in water, it is known as an *alkali*.

Definition. *An alkali is a basic hydroxide which is soluble in water.*

Only a few alkalis are known. The common ones are sodium hydroxide (caustic soda) NaOH, potassium hydroxide (caustic potash) KOH, calcium hydroxide (slaked lime) $Ca(OH)_2$, and ammonia solution which strictly speaking is not a hydroxide. (The ammonia molecules form dative bonds with protons from the equilibrium $H_2O(l) \rightleftharpoons H^+(aq) + OH^-(aq)$, causing more water to ionize and more hydroxyl ions to be present in the solution.) This property of solubility in water is the only difference between the select little group of alkalis and the basic hydroxides generally, but it is a very important difference. It puts the alkalis at our service in hundreds of reactions in solution and for a great many purposes for which the insoluble basic hydroxides are quite useless. It must be clearly realized that the alkalis have all the properties of basic metallic hydroxides in general, but possess also the property of dissolving to a substantial extent in water. Slaked lime is the least soluble of the common alkalis (about 0.15 g in 100 g of water at room temperature).

Properties of alkalis

In addition to the very important property of *neutralizing acids* (see below) alkalis also have a bitter taste and turn red litmus blue. Further, the two caustic alkalis, NaOH and KOH, have a powerful corrosive action on the skin and should be treated with care. Just as a non-volatile acid displaces a volatile acid from its salts, a non-volatile alkali displaces the volatile ammonia from ammonium salts. If an ammonium salt is warmed with an alkali (in the presence of water), ammonia gas is liberated. The essential reaction is:

$$NH_4^+(aq) + OH^-(aq) \rightarrow NH_3(g) + H_2O(l)$$

Expressed in molecular form, two examples of this are:

$$NaOH(aq) + NH_4Cl(aq) \rightarrow NaCl(aq) + H_2O(l) + NH_3(g)$$
$$\text{ammonia}$$
$$Ca(OH)_2(aq) + (NH_4)_2SO_4(aq) \rightarrow CaSO_4(s) + 2H_2O(l) + 2NH_3(g)$$

Neutralization

The reactions between basic oxides, or hydroxides, and acids are very important and are called *neutralizations*. Since the metallic ions and anions from the acid do not change (see the two equations above), the essential reaction of a neutralization is always the formation of unionized molecules of water from hydroxide and hydrogen ions (or hydroxonium ions).

$$H^+(aq) + OH^-(aq) \rightleftharpoons H_2O(l) \quad \text{or} \quad H_3O^+(aq) + OH^-(aq) \rightleftharpoons 2H_2O(l)$$

For practical purposes, these reactions are complete from left to right. This leads to the following definition.

Definition. Neutralization is the formation of molecules of water from hydroxide ions, OH^-, and hydroxonium ions, H_3O^+. A salt is formed at the same time.

The following are examples of neutralizations. Notice that the metallic ions and the negative ions from the acids remain to produce the salts.

Acid		Basic hydroxide		Salt		Water
$H_3O^+Cl^-$	+	K^+OH^-	$\rightarrow$	K^+Cl^-	+	$2H_2O$
$H_3O^+NO_3^-$	+	Na^+OH^-	$\rightarrow$	$Na^+NO_3^-$	+	$2H_2O$

Since both these reactions reduce to $H_3O^+ + OH^- \rightarrow 2H_2O$, the energy change accompanying both should be the same. This is, in fact, so. If aqueous acidic solutions are made up containing one mole of HCl, HNO_3, and $\frac{1}{2}H_2SO_4$ in $1\,dm^3$, and alkaline solutions containing one mole of $NaOH$ and KOH in $1\,dm^3$ (*i.e.*, *strong* acids and alkalis), any neutralization between these solutions liberates the same amount of heat energy, *viz.*, 57.3 kJ (13.7 kcal). This was surprising in an earlier 'molecular' context because all the compounds appeared to have separate identities. Ionic ideas, however, require all *strong* acids and alkalis (and the salts they produce) to be completely ionized in dilute solution. Consequently, the only significant change between them in neutralization is:

$$H_3O^+(aq) + OH^-(aq) \rightarrow 2H_2O(l); \quad \Delta H = -57.3\ kJ$$

Salts

We have encountered the term *salt* on page 268, applied to a product of the process of neutralization. A salt was seen to consist, from this point of view, of an aggregation of metallic ions (positively charged) and acidic ions (negatively charged). Consequently, a salt is derived from the acid to which it corresponds by replacing the H^+ of the acid by an equivalent number of metallic ions. Examples are:

Acid	Salt
H^+Cl^-	Na^+Cl^- sodium chloride
$(H^+)_2SO_4{}^{2-}$	$Cu^{2+}SO_4{}^{2-}$ copper(II) sulphate
$H^+NO_3{}^-$	$K^+NO_3{}^-$ potassium nitrate
$(H^+)_2CO_3{}^{2-}$	$Ca^{2+}CO_3{}^{2-}$ calcium carbonate

Notice that the salt is electrically neutral by a balancing of the oppositely charged ions.

Normal and acid salts

When an acid can produce more than one hydrogen ion per molecule, it is possible for the replacement of H^+ by metallic ion to occur in stages. Thus, one of the H^+ ions of sulphuric acid $(H^+)_2SO_4{}^{2-}$ may be replaced by the ion, Na^+, to give sodium hydrogensulphate $Na^+HSO_4{}^-$, after which a second, similar replacement may yield sodium sulphate $(Na^+)_2SO_4{}^{2-}$. Salts like $Na^+HSO_4{}^-$ behave like a salt because they contain a metallic ion and a negative ion derived from an acid; they behave like an acid because the negative ion is capable of further ionization to yield H^+, as: $HSO_4{}^- \rightleftharpoons H^+ + SO_4{}^{2-}$. Having this dual nature, they are called *acid salts*. Salts like sodium sulphate, in which all the H^+ of the acid has been replaced by metallic ion, are known as *normal salts*.

Definition. A salt is a compound consisting of positive metallic ions and negative ions derived from an acid; if the negative ions are capable of further ionization to yield H^+ the salt is an acid salt, but if not, the salt is a normal salt.

Examples:

Acid	Acid salt	Normal salt
$(H^+)_2SO_4{}^{2-}$	$Na^+HSO_4{}^-$ (sodium hydrogensulphate)	$(Na^+)_2SO_4{}^{2-}$ (sodium sulphate)

Acid	Acid salt	Normal salt
$(H^+)_2CO_3{}^{2-}$	$Na^+HCO_3{}^-$ (sodium hydrogencarbonate)	$(Na^+)_2CO_3{}^{2-}$ (sodium carbonate)
$(H^+)_2S^{2-}$	Na^+HS^- (sodium hydrogensulphide)	$(Na^+)_2S^{2-}$ (sodium sulphide)

A method of preparing the normal and acid sodium salt of sulphuric acid is given on page 412, of the normal and acid sulphides of sodium on page 402, and of sodium carbonate and hydrogencarbonate on pages 320 and 322.

Basic salts

Basic oxides and hydroxides (page 266) contain the ions O^{2-} and OH^-, respectively. Salts in which these ions are retained, together with metallic ions and the negative ions of acids, are known as *basic salts*.

Basic hydroxide	Basic salt	Normal salt
$Zn^{2+}(OH^-)_2$	$Zn^{2+}OH^-Cl^-$	$Zn^{2+}(Cl^-)_2$
$Mg^{2+}(OH^-)_2$	$Mg^{2+}OH^-Cl^-$	$Mg^{2+}(Cl^-)_2$

The basic salts quoted in the table are known as *basic zinc* (or *magnesium*) *chloride*. *White lead* is a well-known basic salt and is the basic carbonate of lead $3Pb^{2+}.2OH^-.2CO_3{}^{2-}$. Notice that the ionic charges balance so that the salt as a whole is electrically neutral.

Methods of preparing salts

Several general methods are available for preparing salts. The method chosen for preparing any particular salt depends largely on whether it is soluble in water or not. It is necessary, therefore, for you to become quite familiar with the simple rules of solubility indicated in the table below. Knowing the solubility of the salt, you will then be able to decide at once what type of method to use.

Soluble salts are usually prepared by methods which involve **crystallization.** **Insoluble** salts are usually prepared by methods which involve **precipitation.**

Soluble	Insoluble
1. All common *salts* of *sodium, potassium,* and *ammonium.*	
2. All common *nitrates* of metals.	
3. All common *chlorides* except:	silver chloride, mercury(I) chloride, and lead chloride.
4. All common *sulphates* except:	barium sulphate and lead(II) sulphate. (Calcium sulphate sparingly soluble.)
	5. All common *carbonates* except those of sodium, potassium, and ammonium.

Preparation of soluble salts by the action of an acid upon a metal

This is usually carried out using a *dilute* acid and a metal. The salt formed then passes into solution in the water of the dilute acid and can be obtained, with the necessary precautions to give purity, by crystallization. This method is suitable for preparing *soluble* salts.

Experiment 78
Preparation of crystalline zinc sulphate from metallic zinc

Half-fill a beaker with bench dilute sulphuric acid and add granulated zinc. Effervescence occurs; if it is slow, add a little copper(II) sulphate to quicken the action and warm gently. The gas is hydrogen (test: explosion on applying a lighted taper). If the action ceases because of the disappearance of zinc, add more zinc to make sure that the acid is not left in considerable excess. The reason for this is that later we shall evaporate the solution and any excess acid would then tend to become concentrated. When the effervescence slows down and there is still plenty of zinc left, filter to remove insoluble impurities such as excess zinc and particles of carbon which were an impurity in it.

The colourless solution will contain zinc sulphate together with a little sulphuric acid. There will probably be too much water present to allow the crystals to separate, so we must remove some of it. Place the solution in an evaporating dish and heat. At intervals, pour a little of the solution into a test-tube and cool it under the tap, shaking well. After evaporating for a time, you will see small crystals in one of the cooled samples. This shows that the solution will give crystals when cool, for the small sample is typical of the whole. Pour the solution into a beaker and allow it to cool and

crystallize. In this case the crystals will be colourless needles.

Filter off the crystals, wash them two or three times with a small quantity of cold distilled water, and place them on a porous plate or between filter papers to dry. The washing with distilled water is to remove the surface solution from the crystals and replace it with pure water which, as it dries off, will not deposit impurities as would the solution. The porous plate or filter papers are used to absorb water from the surfaces of the crystals. To purify still further, dissolve the crystals in a little very hot water and repeat the crystallization process. The impurities will then be carried off dissolved in the filtrate and a smaller quantity of purer crystals will be left.

$$Zn(s) + H_2SO_4(aq) \longrightarrow$$
$$ZnSO_4(aq) + H_2(g)$$

$$(or\ Zn(s) + 2H^+(aq) \longrightarrow$$
$$Zn^{2+}(aq) + H_2(g))$$

$$ZnSO_4(s) + 7H_2O(l) \longrightarrow$$
$$ZnSO_4.7H_2O(s)$$

Other salts may be similarly prepared, for example, iron(II) sulphate and magnesium sulphate, from the respective metals.

Occasionally this method is applied using a metal and a concentrated acid. The only case of this you will encounter, though it is an important one, is the following.

Experiment 79
Preparation of crystalline copper(II) sulphate from metallic copper

Note especially that this preparation cannot be carried out like the one above, because dilute sulphuric acid will not act upon copper.

Into a beaker put some concentrated sulphuric acid and copper turnings (**care!**). Put the

beaker on a tripod and gauze in the fume cupboard and warm gently. After a time *effervescence* will begin, the gas evolved being sulphur dioxide, which must not be allowed to escape into the laboratory as it is injurious. The action will probably become vigorous, and the flame should then be removed. After effervescence has ceased there will be left a dark brown mass which may contain the following substances:

1. Solid copper sulphate as a **precipitate**. (This is copper(II) sulphate.)
2. Solid dark brown copper(I) sulphide, formed in small amount, by a side reaction.
3. Excess copper.
4. Excess concentrated sulphuric acid.

Our problem is to prepare from this mixture a sample of pure copper(II) sulphate crystals. We cannot work in the presence of a large amount of concentrated sulphuric acid, so the first step must be to pour off as much liquid as possible. The liquid is simply a waste product. We are now left with a solid containing copper

sulphate with impurities 2 and 3, as given above. They are both insoluble in water, so, to remove them, add a considerable quantity of water and heat gently to boiling. Filter, leaving the two impurities, copper and copper(I) sulphide, on the filter paper, and obtaining, as filtrate, blue copper sulphate solution. From this, crystals can be obtained as previously described.

$$Cu(s) + 2H_2SO_4(aq) \longrightarrow$$
$$CuSO_4(s) + 2H_2O(l) + SO_2(g)$$

$$CuSO_4(s) + 5H_2O(l) \longrightarrow CuSO_4.5H_2O(s)$$

This process of heating copper with concentrated sulphuric acid is the best laboratory method of making sulphur dioxide, and copper sulphate can be prepared from the residue left in the flask after carrying out this preparation (page 404).

Nitrates of certain metals can be prepared by acting upon the metals with dilute or concentrated nitric acid and crystallizing as described (page 271). The nitrates of common heavy metals, except lead nitrate, are, however, very soluble in water and deliquescent. This makes it a matter of greater difficulty to prepare their crystals.

$$3Pb(s) + 8HNO_3(aq) \longrightarrow 3Pb(NO_3)_2(aq) + 4H_2O(l) + 2NO(g)$$
$$\underset{\text{copper}}{3Cu(s)} + 8HNO_3(aq) \longrightarrow \underset{\text{copper(II) nitrate}}{3Cu(NO_3)_2(aq)} + 4H_2O(l) + 2NO(g)$$

Chlorides of heavy metals are generally prepared in the anhydrous state by heating the metal in a current of dry chlorine or hydrogen chloride (see page 369).

The reason for this is that many of them crystallize with water of crystallization, and, if an attempt is made to drive off this water by heating, they hydrolyse to basic salts (see page 388).

Thus: $$ZnCl_2.H_2O(s) \xrightarrow{\text{heat}} Zn(OH)Cl(s) + HCl(g)$$

Preparation of salts by double decomposition

In a double decomposition reaction, we usually begin with two soluble compounds and use them to prepare one soluble and one insoluble compound. Of these, the one which is wanted is usually the insoluble compound for it can be easily separated by filtering. That is, insoluble salts are prepared by double decomposition. For a more complete discussion of double decomposition see page 181. In what follows, all *lead* salts are lead(II).

Experiment 80
Preparation of a sample of lead(II) sulphate

Here we must begin with a soluble lead salt to provide the lead ions, and a soluble sulphate to provide the sulphate ions of the lead sulphate. Suitable compounds will be lead nitrate and dilute sulphuric acid (a soluble sulphate).

One-third fill a beaker with lead nitrate solution. Heat it and add dilute sulphuric acid, stirring the mixture. There will be an immediate white precipitate of lead sulphate. Heat to boiling,* then filter. The lead sulphate is left on

* The chief reason for this is that a boiling solution filters more rapidly than a cold one.

the filter paper and the colourless filtrate contains dilute nitric acid, the other product of the double decomposition reaction. Wash the precipitate on the filter paper several times with hot distilled water to remove soluble impurities. To be sure that the process is complete, test the washings for sulphate (page 414) and continue till the test gives a negative result. Allow the precipitate to dry on the filter paper or on a porous plate. It will become a white powder.

$$Pb(NO_3)_2(aq) + H_2SO_4(aq) \longrightarrow$$
$$PbSO_4(s) + 2HNO_3(aq)$$

A solution of any soluble sulphate could have been used for this preparation; for example:

$$Pb(NO_3)_2(aq) + Na_2SO_4(aq)$$
lead nitrate   sodium sulphate
$$\longrightarrow PbSO_4(s) + 2NaNO_3(aq)$$
lead   sodium
sulphate   nitrate

ionically:

$$Pb^{2+}(aq) + SO_4^{2-}(aq) \longrightarrow PbSO_4(s)$$

The sodium nitrate, like the nitric acid above, would be removed in solution.

Other insoluble salts which can be prepared by double decomposition are:

Barium sulphate $BaCl_2(aq) + H_2SO_4(aq) \longrightarrow BaSO_4(s) + 2HCl(aq)$
barium
chloride

ionically: $Ba^{2+}(aq) + SO_4^{2-}(aq) \longrightarrow BaSO_4(s)$

*Lead chloride** $Pb(NO_3)_2(aq) + 2NaCl(aq) \longrightarrow PbCl_2(s) + 2NaNO_3(aq)$
ionically: $Pb^{2+}(aq) + 2Cl^-(aq) \longrightarrow PbCl_2(s)$

Calcium carbonate $CaCl_2(aq) + Na_2CO_3(aq) \longrightarrow CaCO_3(s) + 2NaCl(aq)$
calcium
chloride

ionically: $Ca^{2+}(aq) + CO_3^-(aq) \longrightarrow CaCO_3(s)$

The carbonates of other heavy metals can be prepared like calcium carbonate but, in some cases, the method gives a basic carbonate. This does not matter much, however, because the method is chiefly used as an intermediate stage in preparing the oxide of a metal from one of its soluble salts (page 288), and in this case it does not matter whether a true or a basic carbonate is precipitated.

In the cases of zinc carbonate, copper carbonate, and lead carbonate, a purer product is obtained if sodium hydrogencarbonate is used instead of sodium carbonate.

Preparation of salts by neutralization

Neutralization is an action between a *base* and an *acid* to produce a *salt* and *water* only (page 268). The actual method of application of this process depends on whether the base in question is soluble in water, that is, is an alkali (page 267) or not.

Preparation of a salt from an alkali (a soluble base)

Salts of *sodium, potassium,* and *ammonium* can be prepared by this method from *caustic soda* (sodium hydroxide), *caustic potash* (potassium hydroxide), and *ammonia* solution, respectively, using the appropriate acid.

---

*Lead chloride must be washed with *cold* distilled water, as it is appreciably soluble in hot water.

Experiment 81
Preparation of sodium sulphate from caustic soda

Into a conical flask, put 50 cm³ of bench caustic soda solution and add to it enough litmus solution to give a pale blue liquid. From a burette, add dilute sulphuric acid until the solution is purple in colour; that is, it contains no excess of either acid (which would make the litmus red), or alkali (which would make the litmus blue). The litmus is here the *indicator* (page 138). The solution now contains sodium sulphate solution together with litmus. To remove the litmus, add a little animal charcoal on the end of a spatula,

boil the mixture, and filter. The animal charcoal will be left on the filter paper together with litmus which it has absorbed, and a colourless solution of sodium sulphate will be left as filtrate. From this, crystals can be obtained as described for zinc sulphate (page 271).

$$2NaOH(aq) + H_2SO_4(aq) \longrightarrow$$
$$Na_2SO_4(aq) + 2H_2O(l)$$

$$Na_2SO_4(s) + 10H_2O(l) \longrightarrow$$
$$Na_2SO_4.10H_2O(s)$$

To obtain sodium chloride or nitrate, use caustic soda solution with the appropriate acid.

$$NaOH(aq) + HCl(aq) \longrightarrow NaCl(aq) + H_2O(l)$$
sodium   dilute   sodium
hydroxide   hydro-   chloride
           chloric acid

$$NaOH(aq) + HNO_3(aq) \longrightarrow NaNO_3(aq) + H_2O(l)$$
dilute   sodium
nitric acid   nitrate

To obtain the common potassium salts, use caustic potash solution with the appropriate acid.

$$KOH(aq) + HCl(aq) \longrightarrow KCl(aq) + H_2O(l)$$
potassium   potassium
hydroxide   chloride

$$2KOH(aq) + H_2SO_4(aq) \longrightarrow K_2SO_4(aq) + 2H_2O(l)$$
potassium
sulphate

$$KOH(aq) + HNO_3(aq) \longrightarrow KNO_3(aq) + H_2O(l)$$
potassium
nitrate

Ammonium salts can be similarly prepared from ammonia.

$$NH_4OH(aq) + HCl(aq) \longrightarrow NH_4Cl(aq) + H_2O(l)$$
ammonium
chloride

$$2NH_4OH(aq) + H_2SO_4(aq) \longrightarrow (NH_4)_2SO_4(aq) + 2H_2O(l)$$
ammonium
sulphate

$$NH_4OH(aq) + HNO_3(aq) \longrightarrow NH_4NO_3(aq) + H_2O(l)$$
ammonium
nitrate

The above equations indicate the substances to be used (all in solution) in the preparation of the salt by the method described for sodium sulphate.

Preparation of a soluble salt from an insoluble base

In this case the base, being insoluble in water, is not an alkali, so the method given above cannot be applied. In what follows, all *copper* and *lead* compounds are copper(II) and lead(II) compounds.

Experiment 82
Preparation of copper(II) sulphate crystals from the insoluble base copper(II) oxide

Heat some dilute sulphuric acid in a beaker and add to it, a little at a time, some black copper(II) oxide. Stir gently. A blue solution of copper(II) sulphate will be formed.

$$CuO(s) + H_2SO_4(aq) \rightarrow$$
$$CuSO_4(aq) + H_2O(l)$$

It is advisable not to leave an excess of acid (for reason see preparation of zinc sulphate from zinc), so continue the addition of the copper(II) oxide until a permanent black precipitate of this material is left, showing that no more acid is available to act with it. Filter off the precipitate, leaving a clear blue filtrate, copper(II) sulphate solution. From it, obtain crystals as described for zinc sulphate.

$$CuSO_4(s) + 5H_2O(l) \rightarrow CuSO_4.5H_2O(s)$$

Many salts of metals can be similarly prepared, using either the oxide or the hydroxide of the metal with the appropriate acid; for example,

Zinc sulphate

$$ZnO(s) + H_2SO_4(aq) \rightarrow ZnSO_4(aq) + H_2O(l)$$
zinc
oxide

$$Zn(OH)_2(s) + H_2SO_4(aq) \rightarrow ZnSO_4(aq) + 2H_2O(l)$$
zinc
hydroxide

Lead nitrate

$$PbO(s) + 2HNO_3(aq) \rightarrow Pb(NO_3)_2(aq) + H_2O(l)$$
lead
oxide

$$Pb(OH)_2(s) + 2HNO_3(aq) \rightarrow Pb(NO_3)_2(aq) + 2H_2O(l)$$
lead
hydroxide

Preparation of salts by the action of an acid on the carbonate of a metal

The carbonate of any metal will react with the mineral acids to give the corresponding salt of the metal, water, and carbon dioxide. For example:

$$ZnCO_3(s) + H_2SO_4(aq) \rightarrow ZnSO_4(aq) + H_2O(l) + CO_2(g)$$
zinc   zinc
carbonate   sulphate

$$CaCO_3(s) + 2HCl(aq) \rightarrow CaCl_2(aq) + H_2O(l) + CO_2(g)$$
calcium   calcium
carbonate   chloride

$$PbCO_3(s) + 2HNO_3(aq) \rightarrow Pb(NO_3)_2(aq) + H_2O(l) + CO_2(g)$$
lead(II)   lead(II) nitrate
carbonate

The only limitation to this rule is that the action is unsatisfactory and incomplete if the carbonate is insoluble in water and, by its action on the acid, produces a salt which is also insoluble. In this case, the salt which is formed precipitates on the unchanged carbonate and stops the action. If, for example, dilute sulphuric acid is added to marble, there is rapid effervescence for a few seconds, but the action quickly stops.

$$CaCO_3(s) + H_2SO_4(aq) \rightarrow CaSO_4(s) + H_2O(l) + CO_2(g)$$

The very slightly soluble calcium sulphate has precipitated on the marble, stopping the action. With this limitation, this method is of general application. *Lead* compounds below are lead(II) compounds.

Experiment 83
Preparation of lead(II) nitrate crystals from lead carbonate

Half-fill a beaker with dilute nitric acid, warm it gently on a tripod and gauze, and add lead carbonate at intervals. There will be effervescence with evolution of carbon dioxide. (Test: lime-water turned milky.) Continue the addition of the carbonate until a slight permanent precipitate shows that no acid is left, then filter to remove insoluble materials. The colourless solution contains lead(II) nitrate, and crystals can be obtained from it as described for zinc sulphate (page 271).

$$PbCO_3(s) + 2HNO_3(aq) \rightarrow Pb(NO_3)_2(aq) + H_2O(l) + CO_2(g)$$

Note. When an insoluble salt has to be prepared from a compound which is also insoluble in water, it is not advisable to try to convert one into the other by a single process.

For example: To convert *lead(II) oxide* PbO (insoluble in water) into *lead(II) sulphate* (also insoluble in water), it is not advisable to attempt the change by the action of dilute sulphuric acid on the oxide. The reason is that the insoluble lead(II) sulphate will precipitate, as fast as it is formed, round the particles of oxide, and it is very difficult to make sure that all the oxide has reacted.

In such cases, it is better to prepare first a soluble compound and then to precipitate the required insoluble salt by double decomposition. In the example above, it is better to prepare first a solution of lead(II) nitrate by dissolving the oxide in dilute nitric acid.

$$PbO(s) + 2HNO_3(aq) \rightarrow Pb(NO_3)_2(aq) + H_2O(l)$$

Then, after filtering off any undissolved material, the required lead(II) sulphate can be precipitated by adding dilute sulphuric acid.

$$Pb(NO_3)_2(aq) + H_2SO_4(aq) \rightarrow PbSO_4(s) + 2HNO_3(aq)$$

Similarly, to convert *chalk* to *calcium sulphate*, do not attempt the conversion by using dilute sulphuric acid directly with the chalk. First dissolve the chalk in dilute nitric acid.

$$CaCO_3(s) + 2HNO_3(aq) \rightarrow Ca(NO_3)_2(aq) + H_2O(l) + CO_2(g)$$

Then filter, and, to the hot filtrate, add a concentrated solution of sodium sulphate.

$$Ca(NO_3)_2(aq) + Na_2SO_4(aq) \rightarrow CaSO_4(s) + 2NaNO_3(aq)$$

Filter the precipitated calcium sulphate (or lead sulphate above), wash it with hot distilled water, and allow it to dry.

Preparation of salts by direct combination of two elements

Certain binary salts can be prepared by direct combination of their two elements; for example,

$$2Fe(s) + 3Cl_2(g) \rightarrow 2FeCl_3(s) \quad \text{(see page 368)}$$
iron(III) chloride

$$Fe(s) + S(s) \rightarrow FeS(s) \quad \text{(see page 6)}$$
iron(II) sulphide

Measurement of acidity and alkalinity. The pH scale

The pH scale is a convenient means of expressing acidity and alkalinity in aqueous liquids. The term, pH, denotes *hydrogen ion index*, the p being derived from European use of terms such as *punkt* or *point* for the English *index*. The pH scale can be used with the minimum information contained in the next paragraph but students are advised to read the later material which explains more adequately the source of the scale and the significance of the numbers in it.

The number 7 on the pH scale represents the condition of pure water in relation to acidity and alkalinity, *i.e.*, the condition of exact neutrality. Numbers less than 7, *i.e.*, pH 6, 5, 4, etc., indicate *acidity increasing* as the numbers *decrease*; numbers greater than 7, *i.e.*, pH 8, 9, 10, etc., indicate *alkalinity increasing* as the numbers *increase*.

pH   1 2 3 4 5 6 7 8 9 10 11 12 13

$$\longleftarrow \qquad\qquad \uparrow \qquad \longrightarrow$$

increasing   neutral   increasing
acidity   alkalinity

Notice particularly that, below 7, *falling* pH values indicate *increasing* acidity; above 7, *rising* pH values indicate *increasing* alkalinity. For the mathematical explanation of this, see later. In a colourless liquid, a reasonably accurate value of pH can be obtained by adding a *universal indicator* (in quantity advised by the supplier) or by spotting the liquid on to universal indicator paper. In both cases, a colour will appear from which the pH of the liquid can be decided, either by verbal description or from the chart provided by the supplier for colour matching.

Derivation of the pH scale

Pure water is very slightly ionized and is, therefore, a very weak electrolyte.

$$H_2O(l) \rightleftharpoons H^+(aq) + OH^-(aq) \quad \text{or} \quad 2H_2O(l) \rightleftharpoons H_3O^+(aq) + OH^-(aq)$$

(The first of these equations is still in common use for simplicity though it is known that the hydrogen ion (proton) is hydrated in aqueous solution to give the hydroxonium ion, H_3O^+.) From conductivity measurements, it is known that, in mol dm^{-3}, the concentration of the two ions in water is 10^{-7}, so that their product is given by:

$$[H^+][OH^-] = 10^{-7} \times 10^{-7} = 10^{-14} \text{ mol}^2 \text{ dm}^{-6} = K_w \quad \text{(at 25 °C)}$$

where K_w is called the *ionic product* of water, and the square brackets around the ion symbol signify 'concentration'. This ionic product is maintained *constant* in all circumstances in water at 25 °C. When the concentrations of H^+ and OH^- are equal, the liquid is exactly neutral and both ion concentrations are 10^{-7} mol dm^{-3}. If the liquid is made acidic, H^+ concentration is raised above 10^{-7} mol dm^{-3}. Then, to maintain the value of K_w, OH^- concentration must be reduced correspondingly below 10^{-7} mol dm^{-3}. For example, in 0.01M strong monobasic acid, $[H^+] = 10^{-2}$, so that

$$10^{-2} \times [OH^-] = 10^{-14} \quad \text{and} \quad [OH^-] = 10^{-12}$$

If the liquid is made alkaline as, say, 0.001M sodium hydroxide solution, $[OH^-] = 10^{-3}$, so, to maintain the value of K_w $[H^+]$ must decrease to 10^{-11}. This gives the following situation:

Aqueous liquid	H^+ concentration ($mol\ dm^{-3}$)	OH^- concentration ($mol\ dm^{-3}$)
Neutral	10^{-7}	10^{-7}
Acidic	above 10^{-7}	below 10^{-7}
Alkaline	below 10^{-7}	above 10^{-7}

As a convenient means of expressing these situations, *the hydrogen ion index,* pH, has been introduced. This index is derived in the following way.

If, in an aqueous liquid, the concentration of H^+ is 10^{-x} mol dm^{-3}, then, for that liquid, pH $= x$.

From this it follows (above table) that neutrality is indicated by pH $= 7$. Notice that pH is a logarithmic index from which the negative sign has been omitted. Consequently, the *greater* the numerical value of pH the *lower* is the concentration of H^+ and the *less* the acidity.

It would be possible to use a hydroxyl ion index, pOH, to indicate alkalinity but, in fact, such a usage can be avoided. This is because, in all cases, at 25 °C,

$$[H^+][OH^-] = 10^{-14} = K_w$$

so that $$pH + pOH = 14 \quad \text{and} \quad pOH = 14 - pH$$

This constant relation enables pH to record alkalinity (as well as acidity) as in the following case. 0.01M potassium hydroxide solution contains 10^{-2} mol dm^{-3} of OH^-. This would give pOH $= 2$ and require pH $= 14 - 2 = 12$. That is, pH indicates alkalinity by values higher than 7.

Solution	pH
Acidic	below 7
Neutral	7
Alkaline	above 7

Universal indicator

A universal indicator is a mixture of indicators, often sold ready-made as a solution, which can indicate pH values, usually over a range of about 3–11, by successive changes of colour. A recognizable colour change takes place for every change of one unit of pH and, over certain ranges, only half a unit.

A given solution is tested for pH by adding the universal indicator to it in the proportion stated by the supplier, usually two drops of indicator solution to 10 cm^3 of test solution, with shaking. The colour which develops shows the pH of the solution by comparison with a chart provided by the supplier. This method is suitable for colourless liquids only (for obvious reasons). Knowledge of pH values is important for many industries, *e.g.*, for control of soil condition in farming and horticulture, for control of water supplies and sewage disposal.

An ordinary single indicator usually requires a change of about 2 units of pH to bring about its complete colour change. For example, litmus is fully red at pH 6, purple at pH 7, and fully blue at pH 8. The three common indicators for acid–alkali titration are the following.

Indicator	Acidic side	Neutral	Alkaline side
Methyl orange	pink	orange	yellow
Litmus	red	purple	blue
Phenolphthalein	colourless	colourless	pink
		—2 pH units increase——→	

Hydrogen chloride in methylbenzene (toluene)

The solvent methylbenzene $C_6H_5.CH_3$ readily dissolves hydrogen chloride in room conditions, but (in contrast to *aqueous* hydrogen chloride solution) the liquid is a non-conductor of electricity and has negligible acidic behaviour, *e.g.*, no reaction with metals (Zn, Mg, Fe) and only very slight reaction with carbonates. These differences are present because methylbenzene is not a proton (H^+) acceptor and the dissolved hydrogen chloride remains overwhelmingly in the molecular state as HCl; the solution is, therefore, electrically non-conducting for lack of ions.

Water, however, is a proton acceptor and, in dilute aqueous solution, hydrogen chloride can be taken as fully ionized.

$$HCl(g) + H_2O(l) \rightarrow H_3O^+(aq) + Cl^-(aq)$$

These ions allow ready electrical conductance through the liquid, and the hydroxonium ion liberates hydrogen with the more electropositive metals and carbon dioxide with carbonates.

$$Zn(s) + 2H_3O^+(aq) \rightarrow Zn^{2+}(aq) + 2H_2O(l) + H_2(g)$$
$$2H_3O^+(aq) + CO_3^{2-}(aq) \rightarrow 3H_2O(l) + CO_2(g)$$

These reactions are not given in methylbenzene for lack of H_3O^+. If ammonia (gas) is passed into it, a solution of hydrogen chloride in methylbenzene precipitates *ammonium chloride* (white solid).

$$NH_3(g) + HCl \rightarrow NH_4^+Cl^-(s)$$

Questions

1. Give the names and formulae of (*a*) an acid salt, (*b*) a dibasic acid, (*c*) an amphoteric oxide, (*d*) a hydrated salt. (S.)

2. Write ionic equations for: (*a*) the neutralization of an acid by an alkali, (*b*) the precipitation of calcium carbonate, (*c*) the reaction between zinc and a dilute acid. (S.)

3. You are provided with the following substances: water, calcium carbonate, anhydrous sodium sulphate, dilute nitric acid. Using a source of heat and the usual laboratory apparatus, but *no* other chemical substance, describe fully how you could prepare (i) a sample of calcium sulphate (calcium sulphate is almost insoluble in water), (ii) a solution of calcium hydrogencarbonate (calcium bicarbonate), (iii) crystals of hydrated sodium sulphate ($Na_2SO_4.10H_2O$). (C.)

4. Describe in detail how you would prepare pure dry specimens of (*a*) zinc sulphate from zinc carbonate, (*b*) lead(II) sulphate from lead(II) nitrate. Draw a fully labelled diagram of the apparatus you would use to separate and collect the products evolved when lead(II) nitrate is heated. (J.M.B.)

5. Give the name and formula of (*a*) an acid salt, (*b*) a hydrated salt, (*c*) a salt which decomposes on heating, (*d*) a salt which dissociates on heating. A different salt must be given in each case. (S.)

6. Describe how you would carry out the following practical operations in the laboratory: (*a*) obtain pure, dry barium sulphate (or some other insoluble salt) from common laboratory reagents, (*b*) obtain crystals of a hydrated salt of your choice from the corresponding metallic oxide. (Scottish.)

7. Describe in detail how you would prepare dry lead(II) nitrate crystals from lead(II) carbonate. Describe also how you would carry out the reverse process, that is, prepare a dry sample of lead(II) carbonate from lead (II) nitrate.

Describe what you would see and explain what happens when each of the following is heated: (a) lead(II) carbonate, (b) lead(II) nitrate. (J.M.B.)

8. Name and give the formula of an 'acid salt' and give one test by which you could distinguish it from the normal salt of the same acid and base. (S.)

9. (a) What are the products of the reaction of an acid with a base?

(b) Give *three* other characteristics of an acid and *three* other characteristics of a base.

(c) What kind of heat change occurs when an acid reacts with a base?

(d) Describe in detail how you could prepare pure solid sodium chloride by the action of an acid on a base.

(e) If acid containing 3.65 g of HCl was used, what mass of sodium chloride would be obtained? (A.E.B.)

10. What is meant by an *acid salt*? Give the name and formula of *one* example. Determine the relative molecular mass of the dibasic acid X from the following data: 1.35 grammes of X were dissolved in 50 cm^3 of a solution of sodium hydroxide which contained 40 grammes of the alkali per dm^3. The resulting solution required 10 cm^3 of hydrochloric acid solution which contained 73 grammes of the acid per dm^3 for neutralization.

5 grammes of a mixture of sodium carbonate and sodium chloride were found, after dissolving in water, to neutralize 20 cm^3 of a solution of hydrochloric acid containing 36.5 grammes per dm^3. Determine the percentage of sodium carbonate in the mixture. (S.)

11. What is meant by (i) an acid, (ii) a molar solution of an acid?

Ethanoic acid, CH$_3$COOH, is a monobasic acid. (i) What ions are present in an aqueous solution of ethanoic acid? (ii) How many grammes of ethanoic acid are present in 500 cm^3 of a molar solution? (iii) A molar solution of ethanoic acid conducts electricity but not so easily as a molar solution of hydrochloric acid. What explanation can you offer for this fact?

Calculate the minimum volume of molar hydrochloric acid required to react with 1 gramme of calcium carbonate and the volume of gas which would be evolved at standard temperature and pressure. (A.E.B.)

12. Describe briefly but with essential experimental detail how (i) starting with zinc carbonate you would prepare crystals of zinc sulphate, and (ii) starting with potassium hydroxide solution you would prepare crystals of potassium nitrate.

416 g anhydrous barium chloride were obtained when 488 g of the hydrated salt were heated. Calculate *n* in the formula BaCl$_2$.nH$_2$O. (O.)

13. The following is an outline of one method for the preparation of crystals of zinc sulphate ZnSO$_4$.7H$_2$O: 'An excess of zinc carbonate is added to 100 cm^3 of 2.0M sulphuric acid in a beaker and the mixture is warmed until no further reaction takes place. The mixture is filtered and the filtrate evaporated until the volume is about 25 cm^3. This liquid is left to cool and the crystals are separated, dried, and weighed.'

(a) Write the equation for the reaction.

(b) Why is an excess of zinc carbonate used?

(c) How can you tell when reaction has stopped?

(d) Why is it necessary to filter after the reaction has stopped?

(e) Why is the filtrate not evaporated to dryness?

(f) What is the molar mass of

$$ZnSO_4.7H_2O?$$

(g) What maximum mass of zinc sulphate crystals could be formed from 100 cm^3 of 2.0M sulphuric acid?

(h) It was found, in an actual preparation, that the mass of crystals obtained was much less than the maximum mass calculated in (g). Explain this result.

(i) Would a similar method be suitable for the preparation of lead(II) sulphate? Give reasons for your answer. (C.)

14. Salts can be prepared by the following methods, among others: (a) neutralization of an acid by a base, (b) action of a metal on an acid, (c) double decomposition.

For each of these three methods, select a salt from the list below for which the method would be suitable, and describe in detail how you would carry out the preparation. In each case you should say how you would obtain a pure dry specimen of the salt.

Barium sulphate, calcium carbonate, copper(II) chloride, magnesium sulphate, sodium hydrogencarbonate. (L.)

15. Write down the formula for the precipitate which is produced when pairs of aqueous solutions are mixed as follows: (i) Ba(NO$_3$)$_2$ and Na$_2$SO$_4$; (ii) H$_2$S and Pb(NO$_3$)$_2$; (iii) NH$_4$Br and AgNO$_3$; (iv) Al$_2$(SO$_4$)$_3$ and NH$_3$; (v) CaCl$_2$ and K$_2$CO$_3$. (N.I.)

16. Describe how you would carry out the reactions represented by the following ionic equations, giving full details of the reagents and conditions required, and naming any compound which could be separated from the final solution in each case.

(a) ZnO + 2H$^+$ → Zn^{2+} + H$_2$O.

(b) H$^+$ + OH$^-$ → H$_2$O.

(c) 2Na + 2H$^+$ → 2Na$^+$ + H$_2$.

(d) SO$_3$$^{2-}$ + 2H$^+$ → SO$_2$ + H$_2$O.

(J.M.B.)

17. When hydrogen chloride dissolves in water a reaction occurs which can be represented by

$$HCl + H_2O \rightarrow H_3O^+ + Cl^-$$

Which compound in this reaction is the acid? Which compound is the base? When the solution of hydrogen chloride is mixed with sodium hydroxide solution, another reaction occurs. What name is given to this second reaction? Write an ionic equation to represent it. (J.M.B.)

18. Given copper(II) oxide, aqueous solutions of sodium hydroxide and sulphuric acid, and a suitable indicator, describe how you would prepare reasonably pure samples of (*a*) sodium sulphate, (*b*) copper(II) sulphate.

Copper(II) sulphate crystals have the formula $CuSO_4.5H_2O$. What weight of crystals could theoretically be obtained from 8.0 g of copper(II) oxide? (O. and C.)

19. Starting from crystals of lead(II) nitrate, which are soluble in water, describe how you would prepare reasonably pure samples of the following substances: (*a*) lead(II) chloride, which is almost insoluble in cold water, (*b*) lead(II) oxide, (*c*) nitrogen dioxide, (*d*) lead(II) carbonate, which is insoluble in water. (C.)

20. A solution of hydrogen chloride in methylbenzene, an organic solvent, does not conduct electricity and has no apparent reaction with a solid metal oxide or a solid metal carbonate. A solution of hydrogen chloride in water conducts electricity and reacts with both a metal oxide and a metal carbonate.

What does this information indicate about the nature of the bonding in hydrogen chloride when it is in solution in (*a*) methylbenzene, (*b*) water?

Write equations, either molecular or ionic, for the reactions of an aqueous solution of hydrogen chloride with (i) magnesium(II) oxide, (ii) zinc(II) carbonate. (J.M.B.)

21. You are supplied with solutions A, B, C, D, and E, said to have pH values of (A) 2, (B) 7, (C) 7.5, (D) 13, (E) 6.5. Classify these solutions as *neutral, slightly* or *strongly acidic, slightly* or *strongly alkaline*. Describe a practical method by which you would verify *one* of the pH figures quoted. Which solution would be most likely to liberate hydrogen with magnesium powder? Which *other* solution would be most likely to liberate hydrogen with aluminium powder? Write equations for both reactions.

N

25 Oxygen and its Compounds

Discovery

Oxygen was first discovered by Scheele in 1772, but Priestley discovered it independently two years later by heating oxide of mercury. He called it 'dephlogisticated air', but Lavoisier called it oxygen (acid-producer) because he obtained acids by heating several non-metals in oxygen and dissolving the oxides in water.

Occurrence

Uncombined oxygen exists in the air, forming 21% by volume (or 23% by weight). Nearly half the mass of the Earth's crust consists of oxygen in a combined state in the form of water, silicates, many metallic and non-metallic oxides, and in the form of salts.

Experiment 84
Laboratory preparation of oxygen from potassium chlorate

Crush some potassium chlorate crystals (20 g) in a mortar and grind with them about one-quarter of their mass (5 g) of manganese(IV) oxide. Place the mixture in a hard glass tube and fit up the apparatus as shown (Figure 128). Heat the mixture and a gas will readily be given off which can be collected over water.

Since oxygen has about the same density as air, it cannot be collected by displacement of air. If required dry it can be dried by anhydrous calcium chloride and collected in a syringe.

Potassium chloride is left in the tube.

$$2KClO_3(s) \longrightarrow 2KCl(s) + 3O_2(g)$$
$$2ClO_3^-(s) \longrightarrow 2Cl^-(s) + 3O_2(g)$$

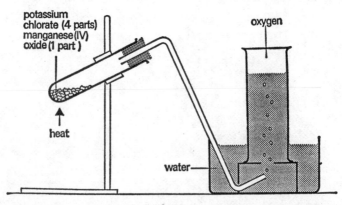

potassium chlorate (4 parts) manganese(IV) oxide (1 part)

oxygen

heat

water

Figure 128

Test for oxygen

It rekindles a glowing splint of wood. This distinguishes it from all gases except dinitrogen oxide, N_2O. It is distinguished from this gas:

(*a*) by having no smell (dinitrogen oxide has a sweet, sickly smell);
(*b*) with nitrogen oxide, oxygen produces brown fumes of nitrogen dioxide.

$$2NO(g) + O_2(g) \rightarrow 2NO_2(g)$$
nitrogen
dioxide

Dinitrogen oxide has no effect on nitrogen oxide.

In the above experiment the potassium chlorate was heated with manganese(IV) oxide in order to produce oxygen. If potassium chlorate is heated alone, it gives off oxygen, but only at a fairly high temperature (400 °C). If mixed with manganese(IV) oxide, the potassium chlorate gives off oxygen at a much lower temperature and much more steadily. On analysis of the residual mixture, it is found that the amount of manganese(IV) oxide is exactly the same at the end of the experiment as it was at the beginning.

A substance which can alter the rate of a chemical reaction in this way is called a *catalyst* (see Chapter 19).

Experiment 85
Laboratory preparation of oxygen from hydrogen peroxide

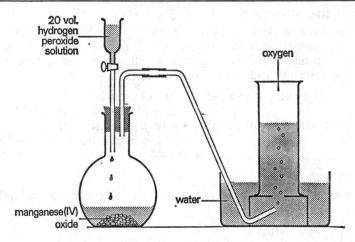

Figure 129

This is a convenient preparation because it requires no heat. Hydrogen peroxide solution (20 vol.) is added, drop by drop, to manganese(IV) oxide, which catalyses decomposition of the peroxide. Oxygen is collected over water (Figure 129).

$$2H_2O_2(aq) \rightarrow 2H_2O(l) + O_2(g)$$

An alternative preparation (in the same apparatus) is the drop by drop addition of hydrogen peroxide solution (20 vol.) to potassium permanganate in the presence of excess of dilute sulphuric acid. Oxygen is liberated until all permanganate is decomposed, by which time the mixture is colourless.

$$5H_2O_2(aq) + 2KMnO_4(aq) + 3H_2SO_4(aq) \rightarrow$$
$$K_2SO_4(aq) + 2MnSO_4(aq) + 8H_2O(l) + 5O_2(g)$$

Industrial preparation

Since oxygen exists to such a large extent in air, it is natural for attempts to be made to obtain it from this source. It is not easy to do this, since nitrogen is an unreactive element and cannot readily combine with anything and thus leave the oxygen pure. By far the best process for obtaining oxygen industrially is from liquid air.

Liquid air. Air is first compressed to about 200 atmospheres pressure, cooled, and allowed to escape from a small jet. Expansion cools the air further because heat energy is used in separating the molecules. The cooled air is allowed to flow away by passing round tubes containing the incoming compressed air. This cools the incoming air and these successive coolings are finally sufficient to liquefy the air. On evaporation of the liquid, nitrogen (b.p. 77 K) is first evolved, leaving a liquid very rich in oxygen (b.p. 90 K at 760 mm). This is a fractional distillation of the liquid air. Oxygen is sold for commercial use not as liquid but as gas compressed in strong steel cylinders at about 100 atmospheres pressure.

Uses

(1) As an aid to breathing where the natural supply of oxygen is insufficient, for example, in high-altitude flying or climbing, and also when anaesthetics are administered to a patient.

(2) In the oxyacetylene flame, which can be used for welding and for cutting even very thick steel plate. The temperature of the flame reaches about 2200 °C.

(3) In the new L-D process for making steel (page 464). A great and increasing oxygen tonnage is now used in this way.

Properties of oxygen

Oxygen is a colourless, odourless, neutral gas, is only slightly soluble in water, and has approximately the same density as that of air. It is an exceptionally active element, combining vigorously with many metals and non-metals, forming basic and acidic oxides respectively:

METALS + OXYGEN → METALLIC OXIDES
most of which are BASIC in character

NON-METALS + OXYGEN → NON-METALLIC OXIDES
most of which are ACIDIC in character

Action with metals

The manner in which oxygen reacts with metals is summarized in the list below.

K
Na
Ca
Mg
Al
Zn   } When heated in air these metals oxidize with a readiness indicated by
Fe   the order shown; that is, potassium most easily, copper least readily.
Pb
Cu

Hg
Ag   } These metals do not oxidize easily; their oxides *yield* oxygen on heating.
Au

The following experiment illustrates the difference in reactivity between a metal high in the list (magnesium) and a metal lower in the series (iron).

Experiment 86
Action of oxygen with metals

Magnesium. Lower a piece of burning magnesium ribbon by means of tongs into a gas-jar of oxygen. It burns with a more dazzling flame and forms a white ash, magnesium oxide.

$$2Mg(s) + O_2(g) \longrightarrow 2MgO(s)$$

or $\quad 2Mg(s) + O_2(g) \longrightarrow 2(Mg^{2+}O^{2-})(s)$

Here, and in similar cases, oxygen is acting as an oxidizing agent by accepting electrons (from the metal).

Iron. Attach a piece of iron wire to the end of a deflagrating spoon and dip the end of the wire in sulphur (to start the action). Warm the wire in the Bunsen flame until the sulphur begins to burn and then plunge it quickly into a gas-jar of oxygen which contains a little water. The iron wire burns, giving off a shower of sparks, and finally a molten bead of oxide drops into the water.

$$3Fe(s) + 2O_2(g) \longrightarrow Fe_3O_4(s)$$
$$\text{tri-iron}$$
$$\text{tetroxide}$$

Action with non-metals

Non-metals form oxides most of which are acidic, and this is illustrated in the following experiments in the case of the elements phosphorus, sulphur, and carbon.

Experiment 87
Action of phosphorus with oxygen

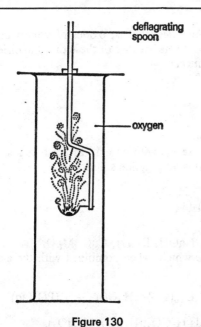

deflagrating spoon

oxygen

Figure 130

Place a small piece of yellow phosphorus in a deflagrating spoon, warm it until it begins to burn, and then plunge it into a gas-jar of oxygen (Figure 130) into which you have previously poured a little blue litmus solution. The phosphorus burns with a dazzling flame, emitting white clouds of oxides of phosphorus which dissolve in the water to form acids of phosphorus, which turn the litmus red.

$$P_4(s) + 5O_2(g) \longrightarrow P_4O_{10}(s)$$
$$\text{phosphorus(V)}$$
$$\text{oxide}$$

$$P_4(g) + 3O_2(g) \longrightarrow P_4O_6(s)$$
$$\text{phosphorus(III)}$$
$$\text{oxide}$$

On solution in water:

$$6H_2O(l) + P_4O_{10}(s) \longrightarrow 4H_3PO_4(aq)$$
$$\text{phosphoric(V)}$$
$$\text{acid}$$

$$6H_2O(l) + P_4O_6(s) \longrightarrow 4H_3PO_3(aq)$$
$$\text{phosphoric(III)}$$
$$\text{acid}$$

Experiment 88
Action of sulphur with oxygen

Sulphur. In a similar manner, lower a piece of burning sulphur into a gas-jar of oxygen containing blue litmus solution. Misty fumes of sulphur dioxide are given off as the sulphur burns more brightly with its characteristic blue flame, and this gas dissolves in the water to form sulphurous acid, which turns the litmus red.

$$S(s) + O_2(g) \longrightarrow SO_2(g)$$
sulphur oxygen sulphur
dioxide

$$SO_2(g) + H_2O(l) \longrightarrow H_2SO_3(aq)$$
sulphurous
acid

Experiment 89
Action of carbon with oxygen

Perform the same experiment with wood charcoal (carbon). The charcoal burns and emits a shower of sparks, combining vigorously with the oxygen to form a colourless gas, carbon dioxide, which dissolves in the water to form carbonic acid. This is only a very weak acid and the litmus is turned claret-coloured but not definitely red.

$$C(s) + O_2(g) \longrightarrow CO_2(g)$$
carbon dioxide

$$CO_2(g) + H_2O(l) \longrightarrow H_2CO_3(aq)$$
carbonic acid

If the above experiment is performed with lime water in the place of litmus solution, the lime water will become turbid because of the formation of a precipitate of chalk.

$$Ca(OH)_2(aq) + CO_2(g) \longrightarrow$$
lime water
(calcium
hydroxide)

$$CaCO_3(s) + H_2O(l)$$
chalk
(calcium
carbonate)

The above oxides are examples of *anhydrides* (that is, oxides of non-metals which react with water to form acids), and it was because of these that Lavoisier gave the gas the name oxygen (acid-producer).

Classification of oxides

Four important types of oxide are the following.

A *basic oxide* is a metallic oxide which reacts with an acid to produce a salt and water only; if soluble in water, it forms an alkaline solution, *e.g.*,

$$CaO(s) + 2HCl(aq) \rightarrow CaCl_2(aq) + H_2O(l);$$
$$CaO(s) + H_2O(l) \rightarrow Ca(OH)_2(aq)$$
alkali

Other examples are: Na_2O, K_2O (alkalis NaOH, KOH); CuO, MgO.

An *acidic oxide* is a non-metallic oxide which, when combined with the elements of water, produces an acid, *e.g.*,

$$SO_3(g) + H_2O(l) \rightarrow H_2SO_4(aq); \quad P_4O_{10}(s) + 2H_2O(l) \rightarrow 4HPO_3(aq)$$

Other examples are CO_2, SO_2, SiO_2 (acids H_2CO_3, H_2SO_3, H_2SiO_3).

An *amphoteric oxide* is a metallic oxide which can show both basic and acidic properties, *i.e.*, can react with both acid and alkali to produce a salt and water only.

Examples are ZnO and Al_2O_3.

basic $\qquad\qquad$ $ZnO(s) + H_2SO_4\,(aq) \rightarrow ZnSO_4(aq) + H_2O(l)$
$\qquad\qquad\qquad$ $Al_2O_3(s) + 6HCl(aq) \rightarrow 2AlCl_3(aq) + 3H_2O(l)$
acidic $\qquad$ $ZnO(s) + 2NaOH(aq) + H_2O(l) \rightarrow Na_2Zn(OH)_4(aq)$
$\qquad\qquad$ $Al_2O_3(s) + 2NaOH(aq) + 3H_2O(l) \rightarrow 2NaAl(OH)_4(aq)$

The salts from the acidic reactions are *sodium zincate* and *sodium aluminate*.

A *neutral oxide* is an oxide which shows neither basic nor acidic character (as defined above), *e.g.*, dinitrogen oxide, carbon monoxide.

The basic oxides

These oxides are the ones which correspond to the commonly occurring salts.

K Na Ca	} Oxides of these metals are soluble in water forming alkalis	} Oxides of these metals are not reduced to metal by hydrogen
Mg Al Zn Fe Pb Cu	} Oxides of these metals can be made from the metal by the action of nitric acid and then heat. (Al excepted)	
Hg Ag Au	} Oxides of these metals decompose when heated.	

Aluminium oxide Al_2O_3

This is a white solid. It is most conveniently prepared by first adding dilute ammonia solution to a solution of an aluminium salt. This precipitates aluminium hydroxide.

$$Al_2(SO_4)_3(aq) + 6NH_4OH(aq) \rightarrow 2Al(OH)_3(s) + 3(NH_4)_2SO_4(aq)$$

The precipitate is then filtered, washed, dried, and heated.

$$2Al(OH)_3(s) \rightarrow Al_2O_3(s) + 3H_2O(l)$$

If prepared at the lowest temperature possible, it shows both basic and acidic properties:

basic $\qquad\qquad$ $Al_2O_3(s) + 6HCl(aq) \rightarrow 2AlCl_3(aq) + 3H_2O(l)$
acidic $\quad$ $Al_2O_3(s) + 2NaOH(aq) + 3H_2O(l) \rightarrow 2NaAl(OH)_4(aq)$
$\qquad\qquad\qquad\qquad\qquad\qquad\qquad\qquad$ sodium
$\qquad\qquad\qquad\qquad\qquad\qquad\qquad\qquad$ aluminate

If strongly heated, it passes into a form which is insoluble in both acid and alkali.

Uses. The most important form of this oxide is *bauxite* $Al_2O_3.2H_2O$, from which the metal is extracted (page 459). It also occurs in an impure form as *emery* and is used as an abrasive.

Coloured by the presence of impurities, this oxide occurs as the gems, *ruby* (iron and titanium), *sapphire* (chromium), and *amethyst* (manganese).

Summary of preparation of the normal oxides of some common heavy metals from the metals or their soluble salts

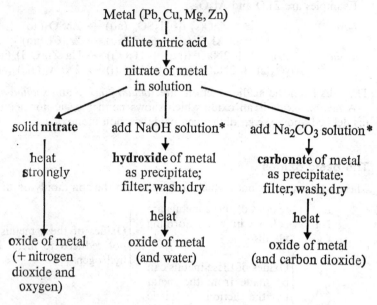

Metal (Pb, Cu, Mg, Zn)

↓ dilute nitric acid

nitrate of metal in solution

solid **nitrate** | add NaOH solution* | add Na₂CO₃ solution*

heat strongly

hydroxide of metal as precipitate; filter; wash; dry

carbonate of metal as precipitate; filter; wash; dry

heat | heat

oxide of metal (+ nitrogen dioxide and oxygen)

oxide of metal (and water)

oxide of metal (and carbon dioxide)

Oxides of sodium and potassium Na₂O and K₂O

These oxides are not much used in the laboratory or prepared in the usual course of experiment. They react vigorously with water to form sodium and potassium hydroxides.

$$Na_2O(s) + H_2O(l) \rightarrow 2NaOH(aq) \quad \text{or} \quad O^{2-}(s) + H_2O(l) \rightarrow 2OH^-(aq)$$

Calcium oxide (lime, quicklime) CaO

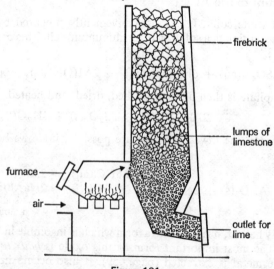

firebrick

lumps of limestone

furnace

air →

outlet for lime

Figure 131

* The hydroxide and carbonate may also be obtained by adding sodium hydroxide or sodium carbonate solution to a solution of any soluble salt of the metal.

Calcium oxide is made in industry by the action of strong heat upon limestone, calcium carbonate, the latter being placed in a kiln.

$$CaCO_3(s) \rightarrow CaO(s) + CO_2(g)$$

Very large quantities of calcium oxide are made in this way (Figure 131). In the laboratory, calcium oxide can be made by placing a piece of marble in a crucible and heating it strongly in a gas-heated muffle furnace. A high temperature is required and an hour or so is necessary to complete the action.

Experiment 90
Preparation of quicklime

Make a loop in a stout iron wire large enough to hold a piece of marble about the size of a pea. Place the piece of marble in the wire and so arrange it on a tripod that when a Bunsen burner is placed underneath it, the marble is just above the inner cone of the roaring Bunsen flame (see Figure 132). Leave in this position

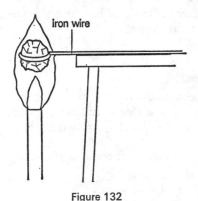

iron wire

Figure 132

for 5–10 minutes and then allow to cool until the solid can be comfortably held in the fingers.

The original substance (calcium carbonate) and the final product (calcium oxide) are very similar in appearance, both being white solids. The difference between them can be readily shown.

(a) *By the action of water*

(i) *On calcium carbonate* – no action.

(ii) *On calcium oxide.* Add water a drop at a time to the piece of calcium oxide in a dish. Great heat is developed (there may be hissing as the water drops on the calcium oxide),

steam is formed and the oxide cracks and puffs up and finally crumbles to a powder about three times as bulky. This is slaked lime (calcium hydroxide).

$$CaO(s) + H_2O(l) \rightarrow Ca(OH)_2(s)$$
$$\text{slaked lime}$$

or $$O^{2-}(s) + H_2O(l) \rightarrow 2OH^-(s)$$

Allow more water to fall on to the slaked lime until there is no further action. If desired, at this stage, the mixture of slaked lime and solution can be filtered and the filtrate shown to be lime water by expelling air from the lungs (containing carbon dioxide) through a glass tube into the solution.

(b) *By the action of dilute hydrochloric acid*

(i) *On calcium carbonate.* Effervescence is seen and the marble finally disappears. Carbon dioxide is evolved, which, if passed into lime water, turns the latter turbid.

$$CaCO_3(s) + 2HCl(aq) \rightarrow$$
$$CaCl_2(aq) + H_2O(l) + CO_2(g)$$

or $$CO_3^{2-}(s) + 2H^+(aq) \rightarrow$$
$$H_2O(l) + CO_2(g)$$

(ii) *On calcium oxide.* No evolution of carbon dioxide. The calcium oxide will first give a similar action to (a)(ii) [slaking] but will give finally a colourless solution of calcium chloride.

$$CaO(s) + H_2O(l) \rightarrow Ca(OH)_2(aq)$$

$$Ca(OH)_2(aq) + 2HCl(aq) \rightarrow$$
basic   acid
hydroxide $$CaCl_2(aq) + 2H_2O(l)$$
         salt   water

Properties of calcium oxide

Calcium oxide is a white solid. It is very refractory; that is, it will not melt even when heated to a very high temperature. It merely becomes incandescent and gives out a powerful light. It was at one time used for this purpose (lime-light).

It reacts vigorously with water (see above) to form slaked lime, which is an alkali. Its solution in water is called 'lime-water'. Since quicklime is basic in character and hygroscopic (that is, it absorbs water) it is used to dry ammonia gas. It is used, after slaking, in the building trade to make mortar, and for a very great number of operations needing a cheap alkali.

Zinc oxide ZnO

This compound is a white powder (yellow when hot) made in industry by distilling zinc and burning the vapour at a jet.

$$2Zn(s) + O_2(g) \rightarrow 2ZnO(s)$$
or
$$2Zn(s) + O_2(g) \rightarrow 2(Zn^{2+}O^{2-})(s)$$

It is made in the laboratory from zinc by dissolving the metal in dilute nitric acid, evaporating the zinc nitrate solution so formed to dryness, and heating the residue strongly.

$$3Zn(s) + 8HNO_3(aq) \rightarrow 3Zn(NO_3)_2(aq) + 4H_2O(l) + 2NO(g)$$
$$2Zn(NO_3)_2(s) \rightarrow 2ZnO(s) + 4NO_2(g) + O_2(g)$$

Zinc oxide is amphoteric (see page 286).

Zinc oxide (and the oxides already described) cannot be converted into the metal by heating the oxides in a stream of hydrogen.

Zinc oxide is used as a white pigment.

Iron(III) oxide Fe₂O₃

(Iron(II) oxide is not important.) This compound is a red powder known as 'jewellers' rouge'. It is used for polishing precious stones, and as a pigment.

It is found in the impure state as haematite. Iron(III) oxide is made in the laboratory:

(*a*) By heating iron(II) sulphate. (Note this action – it is a most unusual type.)

$$2FeSO_4(s) \rightarrow Fe_2O_3(s) + \underset{\substack{\text{sulphur} \\ \text{dioxide}}}{SO_2(g)} + \underset{\substack{\text{sulphur} \\ \text{trioxide}}}{SO_3(g)}$$

(See page 418.)

(*b*) By heating iron(III) hydroxide strongly.

$$2Fe(OH)_3(s) \rightarrow Fe_2O_3(s) + 3H_2O(g)$$

Iron(III) oxide is also the product formed if iron(II) hydroxide is heated strongly in the air. All iron(II) compounds tend to become oxidized to iron(III) compounds by the oxygen of the atmosphere.

It has the usual properties of an oxide. It can be reduced to metallic iron by being heated in a stream of hydrogen or carbon monoxide.

$$Fe_2O_3(s) + 3H_2(g) \rightarrow 2Fe(s) + 3H_2O(g)$$
$$Fe_2O_3(s) + 3CO(g) \rightarrow 2Fe(s) + 3CO_2(g)$$

Tri-iron tetroxide (magnetic oxide of iron) Fe₃O₄

This compound may be prepared by passing steam over red-hot iron or by burning iron in oxygen.

It occurs naturally as magnetite and as such is a natural magnet or 'lodestone'. On heating it in a stream of hydrogen it is reduced to iron (see page 303).

$$Fe_3O_4(s) + 4H_2(g) \rightleftharpoons 3Fe(s) + 4H_2O(g)$$

Lead(II) oxide PbO

This is a yellow powder. It can be made in the laboratory by heating lead(IV) oxide, red lead oxide, lead(II) nitrate, lead(II) carbonate, or lead(II) hydroxide. (For details see compounds concerned.) It is best made from the metal by the action of nitric acid, with subsequent evaporation and heating of the lead(II) nitrate. When prepared in the laboratory it usually ruins the test-tube in which it is prepared by fusing with the glass. It is, in fact, used to make lead glass – a glass of very high refractive index.

Although lead(II) oxide can be considered a typical base, the only common acid in which it will readily dissolve is nitric acid. The reason why it does not react quantitatively with the others is a purely mechanical one.

Action of lead(II) oxide on dilute sulphuric or hydrochloric acids. Lead(II) chloride and lead(II) sulphate are not formed quantitatively. These two substances are almost insoluble in water and, as the action can proceed only on the outside of a particle of oxide, the lead(II) chloride or sulphate forms as a layer on the outside. This layer of chloride or sulphate is not permeable to the acids and hence action stops before any appreciable amount of the salt has been formed. (For preparation, see under 'Chlorides' and 'Sulphates'.)

Lead(II) oxide is easily reduced to grey metallic lead by heating it in a stream of hydrogen, town-gas, or carbon monoxide.

$$PbO(s) + H_2(g) \rightarrow Pb(s) + H_2O(g)$$
$$PbO(s) + CO(g) \rightarrow Pb(s) + CO_2(g)$$

Lead(II) oxide is also an amphoteric oxide dissolving in caustic alkalis to form plumbites.

$$NaOH(aq) + PbO(s) + H_2O(l) \rightarrow NaPb(OH)_3(aq) \quad \text{or} \quad Na^+Pb(OH)_3{}^-$$
sodium
plumbite

Copper(II) oxide (black copper oxide) CuO

This is made in the laboratory by several methods which are given with full experimental details on page 24. Copper(II) oxide is hygroscopic, absorbing moisture from the air. It is a *basic* oxide and dissolves readily in warm dilute mineral acid, forming copper(II) salts, *e.g.,*

$$CuO(s) + H_2SO_4(aq) \rightarrow CuSO_4(aq) + H_2O(l)$$

If a solution of copper(II) oxide in concentrated hydrochloric acid is gently boiled with clean copper turnings in a fume cupboard, copper(II) chloride is *reduced* to copper(I) chloride (cuprous chloride). This precipitates (white) when the mixture is poured into cold, boiled-out water (which prevents re-oxidation).

$$CuO(s) + 2HCl(aq) \rightarrow CuCl_2(aq) + H_2O(l)$$
$$CuCl_2(aq) + Cu(s) \rightarrow 2CuCl(s)$$

By a similar reduction, a copper(II) sulphate solution, mixed with sodium hydroxide solution and sodium potassium tartrate (to prevent precipitation of copper(II) hydroxide), precipitates red copper(I) oxide, or cuprous oxide, when warmed with glucose (a mild reducing agent). This test, used frequently by biologists to identify reducing sugars, is known as *Fehling's test.*

$$2Cu^{2+}(aq) + 4OH^-(aq) + C_6H_{12}O_6(aq) \rightarrow Cu_2O(s) + 2H_2O(l) + C_6H_{12}O_7(aq)$$

Heated copper (II) oxide is reduced to copper by hydrogen or carbon monoxide.

$$CuO(s) + H_2(g) \rightarrow Cu(s) + H_2O(g) \quad \text{or} \quad CuO(s) + CO(g) \rightarrow Cu(s) + CO_2(g)$$

Like manganese(IV) oxide, copper(II) oxide catalyses the decomposition of potassium chlorate by heat.

Mercury(II) oxide (mercuric oxide) HgO

This red oxide yields, when heated, a mirror of mercury on the cooler sides of the test-tube, with oxygen evolved.

$$2HgO(s) \rightarrow 2Hg(s) + O_2(g)$$

The higher oxides

The higher oxides are oxides which contain more oxygen per molecule than the corresponding basic oxide.

Sodium peroxide Na$_2$O$_2$

This is made by heating sodium in excess of oxygen.

$$2Na(s) + O_2(g) \rightarrow Na_2O_2(s)$$

It is a yellow powder and is a vigorous oxidizing agent; it should never be allowed to come into contact with damp organic matter. With water, it liberates oxygen. It is used in confined spaces, *e.g.*, a submarine, where men are working, because it absorbs carbon dioxide and liberates oxygen at the same time.

$$2Na_2O_2(s) + 2H_2O(l) \rightarrow 4NaOH(aq) + O_2(g)$$
$$2Na_2O_2(s) + 2CO_2(g) \rightarrow 2Na_2CO_3(s) + O_2(g)$$

Sodium peroxide is a true peroxide, containing the O_2^{2-} ion, and yields hydrogen peroxide with dilute acids.

Lead(IV) oxide (lead dioxide) PbO$_2$

This oxide is a dark-brown powder and can be made in the laboratory as described in the following experiment.

Experiment 91
Preparation of lead(IV) oxide

Into a beaker put some dilute nitric acid and warm it. By means of a spatula, add dilead(II) lead(IV) oxide (red lead oxide) a little at a time. Care must be taken not to add too much red lead oxide or it will contaminate the product. As the red lead oxide reacts with the nitric acid, a brown powder is precipitated and lead(II) nitrate is formed in solution. The mixture is filtered, the residue in the filter paper is washed two or three times with hot distilled water and is allowed to dry on the filter paper, from which it may then be shaken.

$$Pb_3O_4(s) + 4HNO_3(aq) \rightarrow$$
red lead oxide

$$PbO_2(s) + 2H_2O(l) + 2Pb(NO_3)_2(aq)$$
lead(IV) oxide   lead(II) nitrate

The properties of lead(IV) oxide are summed up by the following experiments.

Experiment 92
Properties of lead(IV) oxide

(i) *Action of heat.* Heat a little lead(IV) oxide in a test-tube and hold a glowing splint in the mouth of the test-tube. The splint is rekindled, showing the presence of oxygen. Lead(II) oxide remains as a yellow solid, often fused into the glass.

$$2PbO_2(s) \longrightarrow 2PbO(s) + O_2(g)$$

(ii) *Action of concentrated hydrochloric acid.* Warm a little lead(IV) oxide with concentrated hydrochloric acid. A greenish-yellow gas which bleaches litmus is evolved (chlorine) and a white (often discoloured) solid, lead chloride, may be seen in the test-tube.

$$PbO_2(s) + 4HCl(aq) \longrightarrow$$
$$PbCl_2(s) + 2H_2O(l) + Cl_2(g)$$

(Compare manganese(IV) oxide.)

It will be seen from the above that lead(IV) oxide is an oxidizing agent. If warm, it is converted by sulphur dioxide into lead(II) sulphate (a white solid) and the mass glows as combination takes place.

$$PbO_2(s) + SO_2(g) \longrightarrow PbSO_4(s)$$

(iii) *Action of hot concentrated sulphuric acid.* Add lead(IV) oxide to concentrated sulphuric acid in a test-tube and warm gently. Effervescence occurs, oxygen is evolved (test as in (i) above) and a white precipitate of lead(II) sulphate is left.

$$2PbO_2(s) + 2H_2SO_4(aq) \longrightarrow$$
$$2PbSO_4(s) + 2H_2O(l) + O_2(g)$$

Red lead oxide (dilead(II) lead(IV) oxide) Pb_3O_4

Red lead oxide is prepared by heating lead(II) oxide for some time in the presence of air at a temperature of 450 °C.

$$6PbO(s) + O_2(g) \rightarrow 2Pb_3O_4(s)$$

This compound, in many chemical properties, acts as though it consists of lead(II) oxide and lead(IV) oxide. For example, in the experiment in preparing lead(IV) oxide (page 292):

$$Pb_3O_4(s) + 4HNO_3(aq) \rightarrow 2Pb(NO_3)_2(s) + 2H_2O(l) + PbO_2(s)$$

Compare $\underbrace{2PbO}_{\text{base}} + PbO_2 + \underbrace{4HNO_3}_{\text{acid}} \rightarrow \underbrace{2Pb(NO_3)_2}_{\text{salt}} + \underbrace{2H_2O}_{\text{water}} + PbO_2$

$$\underbrace{\qquad\qquad\qquad\qquad}_{\text{unchanged}}$$

Or, in the action of heat:

$$Pb_3O_4(s) \rightarrow 3PbO(s) + O(g)$$
$$\overline{2PbO + PbO_2 \rightarrow 2PbO + PbO + O}$$

or more correctly:

$$2Pb_3O_4(s) \rightleftharpoons 6PbO(s) + O_2(g)$$

Hence the action of strong heat on any oxide of lead is to leave lead(II) oxide.
Red lead oxide reacts with concentrated hydrochloric acid or concentrated sulphuric acid when warmed to produce the same observed effects as lead(IV) oxide (see above).

$$Pb_3O_4(s) + 8HCl(aq) \rightarrow 3PbCl_2(s) + 4H_2O(l) + Cl_2(g)$$
$$2Pb_3O_4(s) + 6H_2SO_4(aq) \rightarrow 6PbSO_4(s) + 6H_2O(l) + O_2(g)$$

Red lead oxide has been used for a long time as a pigment (the old name for red lead oxide, 'minium', gave the name 'miniature' to that type of picture) and is used today, with oil, as a jointing material for gas and water pipes and in the manufacture of glass.

Manganese(IV) oxide (manganese dioxide) MnO₂

See Oxygen (page 282) and Chlorine (page 362) for the common laboratory uses of this substance. It is also used in the glass industry.

Ozone

Ozone can be prepared from oxygen by the use of the silent electrical discharge. The apparatus is shown in Figure 133.

Dry oxygen is passed through the space between the glass tubes. Each tube is coated with tin-foil, which is connected to the terminals of an induction coil.

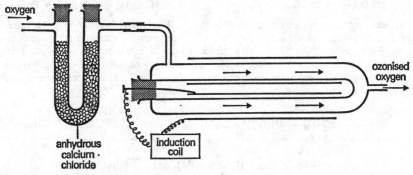

Figure 133

No actual sparking takes place between the layers of tin-foil, nor does the oxygen come into contact with them, but a state of electrical strain exists. The issuing gas may contain up to 5% of ozone. This ozonized oxygen should not be allowed to come into contact with rubber, which is attacked by ozone.

$$3O_2(g) \rightarrow 2O_3(g)$$

Other forms of apparatus may be used for making ozonized oxygen. Their principle is the same as above, but dilute sulphuric acid takes the place of the layers of tin-foil.

Tests for ozone. It possesses a smell which resembles that of very dilute chlorine. It is noticeable near an electrical machine in operation.

Ozone oxidizes mercury and makes the mercury 'tail', *i.e.*, leave a trail of mercury stuck to glass as the mercury flows across it. Quite small traces of ozone can be detected in this way.

Properties of ozone O₃

Ozone is a gas at ordinary temperature and pressure. It is extremely poisonous, and rooms containing operating electrical machinery must be well ventilated. It is obtained pure by liquefaction and is then a dark blue, explosive liquid, boiling at about −112 °C under ordinary pressures.

Ozone as an oxidizing agent

Ozone is a vigorous oxidizer. It oxidizes lead(II) sulphide to lead(II) sulphate

$$PbS(s) + 4O_3(g) \rightarrow PbSO_4(s) + 4O_2(g)$$

black   white
lead(II)   lead(II)
sulphide   sulphate

and hydrogen sulphide to sulphuric acid.

$$H_2S(g) + 4O_3(g) \rightarrow H_2SO_4(aq) + 4O_2(g)$$

It also liberates iodine from potassium iodide in acidic solution

$$2KI(aq) + H_2SO_4(aq) + O_3(g) \rightarrow I_2(aq) + O_2(g) + K_2SO_4(aq) + H_2O(l)$$

iodine
(brown
coloration)

This reaction is expressed in ionic terms as:

$$2I^-(aq) + 2H^+(aq) + O_3(g) \rightarrow I_2(aq) + O_2(g) + H_2O(l)$$

The iodide ion is oxidized by electron loss: $2I^- - 2e^- \rightarrow I_2$; the electrons are accepted by ozone acting as an oxidizing agent:

$$2H^+(aq) + O_3(g) + 2e^- \rightarrow O_2(g) + H_2O(l)$$

This powerful oxidizing action of ozone makes it useful in very high dilution for ventilating places to which fresh air has little access. It attacks the organic compounds which are responsible for the 'stuffy' smell. The gas becomes poisonous at concentrations exceeding about 1 in 50 000 of air, by volume.

Allotropy of oxygen

Oxygen and ozone are *allotropes* (page 91) of the same element; the difference between them is one of molecular complexity, oxygen having a diatomic molecule, O_2, and ozone a triatomic molecule, O_3. Their chemical identity is shown by the fact that each can be converted to the other (as shown in earlier pages) without change of mass.

The following table compares the two allotropes.

Oxygen O_2	Ozone O_3
Gas at s.t.p.	Gas at s.t.p.
Density 16 (H = 1).	Density 24 (H = 1).
Insoluble in turpentine.	Absorbed by turpentine.
Heat has no action.	Heat converts ozone into oxygen. $2O_3(g) \rightarrow 3O_2(g)$.
No effect on mercury at room temperature.	Makes mercury 'wet' glass ('tailing').
No effect on rubber.	Attacks rubber.
Has no effect on potassium iodide solution.	Liberates iodine from potassium iodide solution.
Oxidizing agent.	Vigorous oxidizing agent.

Hydrides of oxygen

Oxygen may be considered to form two principal hydrides. The commoner hydride is water H_2O (see Chapter 23). Oxygen also forms another hydride, hydrogen peroxide H_2O_2. This compound is a colourless liquid and has been known for more than a century. Its purification is difficult and it was not until 1894 that it was obtained in a pure state.

Experiment 93
Preparation of a solution of hydrogen peroxide H_2O_2

Hydrogen peroxide may be prepared by acting upon the peroxides of certain metals with acids.

The materials usually used are barium peroxide and dilute sulphuric acid, because the barium sulphate produced is insoluble and can be filtered off.

$$BaO_2(s) + H_2SO_4(aq) \longrightarrow$$
$$BaSO_4(s) + H_2O_2(aq)$$

To 200 cm³ of water add 20 cm³ of concentrated sulphuric acid. Place the beaker containing the dilute acid in a freezing-mixture of ice and salt and allow it to cool. Gradually add to it a quantity of previously moistened hydrated barium peroxide until the mixture only just reacts acid. Then allow the mixture to settle and filter it. Add to the filtrate a few drops of baryta water (barium hydroxide solution) until it is accurately neutral, and the resulting aqueous solution of hydrogen peroxide is ready for use.

Properties of hydrogen peroxide

The pure compound is a syrupy liquid. It is usually used in dilute solution in water.

Experiment 94
Action of heat on hydrogen peroxide

Warm hydrogen peroxide solution in a test-tube. Effervescence occurs. The gas given off is *oxygen*. The gas will not rekindle a glowing splint because of the presence of steam.

$$2H_2O_2(aq) \longrightarrow 2H_2O(l) + O_2(g)$$

Hydrogen peroxide as an oxidizing agent

In many of its reactions hydrogen peroxide acts as an oxidizing agent, as shown by the following experiments.

Experiment 95
Oxidation of lead(II) sulphide by hydrogen peroxide

Precipitate lead(II) sulphide by passing hydrogen sulphide into a solution of lead(II) nitrate in a boiling-tube. Allow the precipitate to settle, pour off the liquid, and add to the black lead(II) sulphide some hydrogen peroxide solution. Leave it to stand for some time, shaking occasionally. The precipitate gradually turns white, because it is slowly converted to lead(II) sulphate.

$$PbS(s) + 4H_2O_2(aq) \longrightarrow PbSO_4(s) + 4H_2O(l)$$

Hydrogen peroxide is here acting as an oxidizing agent, oxidizing lead(II) sulphide to lead(II) sulphate, and being itself reduced to water.

This reaction is used in restoring pictures. Hydrogen sulphide in the air reacts with the white lead paint (lead(II) carbonate) of the picture to produce lead(II) sulphide, which is brown and makes the picture dingy. Washing with hydrogen peroxide restores the white colour.

Experiment 96
Oxidation of acidified potassium iodide solution by hydrogen peroxide

Acidify a solution of potassium iodide with dilute sulphuric acid. Add hydrogen peroxide. A brown coloration is caused by the production of free iodine.

$$2KI(aq) + H_2SO_4(aq) + H_2O_2(aq) \longrightarrow$$
$$K_2SO_4(aq) + I_2(aq) + 2H_2O(l)$$

Here again, hydrogen peroxide is an oxidizing agent, oxidizing potassium iodide to iodine, and being reduced to water.

In ionic terms, this equation is:

$$2I^-(aq) + 2H^+(aq) + H_2O_2(aq) \rightarrow 2H_2O(l) + I_2(aq)$$

The iodine ion is oxidized by electron loss, as: $2I^- - 2e^- \rightarrow I_2$. The electrons are accepted by the oxidizing agent, as:

$$2H^+(aq) + H_2O_2(aq) + 2e^- \rightarrow 2H_2O(l)$$

That is, hydrogen peroxide is reduced by electron gain.

Hydrogen peroxide also oxidizes an iron(II) salt, *e.g.*, $FeSO_4$, in acidic solution to the iron(III) state, the solution turning from green to yellow.

$$2Fe^{2+}(aq) + H_2O_2(aq) + 2H^+(aq) \rightarrow 2Fe^{3+}(aq) + 2H_2O(l)$$

A solution of a soluble sulphite is similarly oxidized to sulphate.

$$SO_3^{2-}(aq) + H_2O_2(aq) \rightarrow SO_4^{2-}(aq) + H_2O(l)$$

Its powerful oxidizing action makes hydrogen peroxide useful as a bleaching agent. It bleaches hair to a blonde colour, and is used commercially for the bleaching of paper pulp, cotton, and other natural fibres.

Hydrogen peroxide as a reducing agent

Hydrogen peroxide is capable of reacting as a reducing agent towards certain other compounds; a confusing aspect of this property is that when it does so one of the products is usually gaseous oxygen! In these reactions, hydrogen peroxide appears to act as a reducing agent by losing electrons (in association with hydrogen ions), as:

$$H_2O_2(aq) \rightarrow 2H^+(aq) + O_2(g) + 2e^-$$

The electrons are accepted by the other reacting substance, acting as an oxidizing agent.

Experiment 97
Reduction of lead(IV) oxide by hydrogen peroxide

Suspend some lead(IV) oxide in dilute nitric acid. Add hydrogen peroxide. Effervescence occurs with evolution of oxygen (rekindles a glowing splint). A colourless solution remains. In this reaction, lead(IV) oxide is converted to lead(II) oxide PbO, which dissolves in the nitric acid. Hydrogen peroxide is converted to water.

$$PbO_2(s) + H_2O_2(aq) \rightarrow$$
$$PbO(s) + H_2O(l) + O_2(g)$$

$$PbO(s) + 2HNO_3(aq) \rightarrow$$
$$Pb(NO_3)_2(aq) + H_2O(l)$$

Similar reactions occur, with evolution of oxygen, when hydrogen peroxide reacts with *silver oxide, acidified potassium permanganate solution* (see page 283), *and ozone.*

$$Ag_2O(s) + H_2O_2(aq) \rightarrow 2Ag(s) + H_2O(l) + O_2(g)$$

Silver oxide is reduced to metallic silver (a black precipitate).

$$2MnO_4^-(aq) + 5H_2O_2(aq) + 6H^+(aq) \rightarrow 2Mn^{2+}(aq) + 8H_2O(l) + 5O_2(g)$$

The permanganate ion is reduced to a manganese salt.

$$O_3(g) + H_2O_2(aq) \rightarrow H_2O(l) + 2O_2(g)$$

Ozone is reduced to oxygen.

Decomposition of hydrogen peroxide

Hydrogen peroxide is decomposed catalytically by many substances, *e.g.*, manganese(IV) oxide, finely powdered gold, and platinum.

$$2H_2O_2(aq) \rightarrow 2H_2O(l) + O_2(g)$$

Add a pinch of manganese(IV) oxide to about 5 cm³ of hydrogen peroxide solution in a test-tube. Oxygen is rapidly evolved (see page 283).

Sale of hydrogen peroxide

Hydrogen peroxide is sold retail in '10 volume' and '20 volume' solutions, *i.e.*, 1 cm³ of the solution yields 10 cm³ or 20 cm³ of oxygen at s.t.p. when heated. To minimize loss by catalytic decomposition, the solutions should be as pure as possible and the containers free from roughness. Some additives, *e.g.*, acetanilide, can act as stabilizers (negative catalysts) but purity is the best safeguard.

Questions

1. Pure oxygen is given off when potassium permanganate is gently heated. Draw a labelled diagram of the apparatus you would use to measure, under room conditions of temperature and pressure, the volume of oxygen given off when a small quantity of potassium permanganate is heated.

Explain how you would (*a*) ensure that the volume of oxygen was measured under room conditions, (*b*) obtain the temperature and pressure of the room, (*c*) determine the mass of this volume of oxygen.

In an experiment, it was found that 228 cm³ of oxygen were produced, measured at 21 °C and 756 mm of mercury. The mass of the oxygen was 0.300 g. Calculate (i) the volume of the oxygen at s.t.p., (ii) the relative molecular mass of oxygen. (C.)

2. (*a*) Outline a method by which oxygen is manufactured from air. (*b*) State *two* methods by which this gas is conveyed to the consumer; and give *one* large scale use for the gas. (*c*) Name *two* compounds which, when heated alone, give off oxygen as the only gaseous product, and write equations for their decomposition. (*d*) Name *two* gases which are *compounds*, and which burn in air or oxygen. Write equations for the reactions which take place. In each case state how much oxygen is needed to burn 100 cm³ of the gas. (All measurements of gas volumes at s.t.p.) (S.)

3. Name the products formed when the following react with an excess of oxygen:

(*a*) carbon, (*b*) magnesium, (*c*) hydrogen, (*d*) zinc. Write equations for the reactions, if any, of these products with (i) dilute hydrochloric acid, (ii) sodium hydroxide solution. If no reaction occurs, write 'no reaction'. State the type of oxide formed by each of the four elements. (J.M.B.)

4. (*a*) Describe, giving a sketch of your apparatus, how you would attempt to measure the proportion of oxygen in air. (*b*) Given a supply of hydrogen peroxide solution, how would you use it to generate a supply of oxygen and how would you collect the gas? (*c*) Give the products of the combustion in oxygen of (i) a non-metal, (ii) a metal, each of your own choice. (O.)

5. Describe the preparation and properties of oxygen. If you had two vessels, one containing air and the other oxygen, how would you distinguish them by simple tests? (O. and C.)

6. Describe an accurate method of finding the percentage of oxygen in the air. How do you account for the percentage being so nearly constant when there are many causes which remove oxygen from the air? (L.)

7. Describe *two* experiments which you could perform in the laboratory, *one* in which oxygen justifies its name of acid-producer and *one* which gives an opposite result. What evidence is there that the gases of the atmosphere are not chemically combined with each other?

8. Describe the usual preparation and collection of oxygen in the laboratory. State how oxygen may be converted into (a) an acidic oxide; (b) an alkaline oxide; (c) an insoluble basic oxide; (d) an allotropic form. Give also a sketch illustrating (d). (L.)

9. Sulphur dioxide is called an acidic oxide and copper(II) oxide a basic oxide. What is meant by these terms? Give two other examples of each of these classes of oxides and describe how you would test an oxide in order to assign it to one of these classes. (O. and C.)

10. What chemical properties distinguish metals as a class? Describe in detail three methods by which a metal may be converted into its oxide. In each case name the metal which you would use. (L.)

11. What are the four chief classes of oxides? Give *one* example of each class.

How do oxides of metals differ chemically from those of non-metals?

State briefly how you would prepare from suitable oxides (a) a solution of copper(II) sulphate; (b) a dilute solution of hydrogen peroxide.

What happens when manganese(IV) oxide is added to a solution of hydrogen peroxide? (J.M.B.)

12. How is ozonized oxygen prepared in the laboratory? Describe *three* experiments to show how it differs from oxygen. If 100 cm³ of pure ozone were heated and then reduced to the original temperature and pressure, what would be the volume of the resulting gas? (J.M.B.)

13. How is ozonized oxygen obtained? In what respects does this gas differ from ordinary oxygen? How has it been shown (a) that ozone is composed of oxygen atoms only; (b) that its relative molecular mass is greater than that of oxygen? (O. and C.)

14. Describe the properties of ozone. On partly ozonizing 100 cm³ of oxygen a decrease of volume of 10 cm³ resulted. What volume of ozone had been produced? The resulting gas was treated with excess of a solution of potassium iodide when the following reaction took place:

$$O_3 + 2KI + H_2O \rightarrow 2KOH + I_2 + O_2$$

Calculate (a) what volume of gas would remain; (b) what weight of iodine would be liberated, assuming the volumes to have been measured at s.t.p. (C.)

15. How can oxygen be converted into its allotropic form? Compare the properties of oxygen with those of its allotropic form. (O. and C.)

16. Describe fully how you would prepare: (i) pure copper(II) oxide (CuO) from a mixture of copper(II) oxide and lead(II) oxide (PbO); (ii) pure lead sulphate (lead(II) sulphate) from a mixture of lead(II) oxide (PbO); and sand. (N.B. Lead(II) sulphate is insoluble in water.)

Give *two* different large-scale uses *each* for copper and lead, mentioning in *each* case the property of the metal on which the use depends. (C.)

17. Outline the preparation of the following oxides and describe their appearance: (i) copper(II) oxide starting with copper(II) sulphate solution; (ii) zinc oxide starting with powdered zinc carbonate; (iii) lead(II) oxide (lead monoxide) starting with crystals of lead(II) nitrate.

Analysis shows that a certain oxide of iron contains 30% oxygen. Derive its empirical formula. (O.)

18. Give the name and formula of a metallic oxide (different in each case) which (a) on heating yields oxygen and a lower oxide of the same metal, (b) is yellow when hot and white when cold, (c) is easily reduced by heating in a stream of hydrogen, (d) is formed by passing steam over the red-hot metal. (S.)

19. Suggest how you could determine experimentally the proportion of oxygen in a given oxide of iron. If the oxide was found to contain 30% oxygen, what would be its empirical formula?

Outline the industrial production of quicklime (calcium oxide) and slaked lime (calcium hydroxide) from limestone (calcium carbonate), pointing out any special features such as reversible or exothermic reactions. (O.)

20. Describe the reactions, if any, of each of the following oxides with (a) water, (b) dilute nitric acid, (c) sodium hydroxide solution: (i) calcium oxide, (ii) carbon monoxide, (iii) zinc oxide. Hence classify each of the oxides as acidic, amphoteric, basic, or neutral. (J.M.B.)

21. State what class of oxide each of the following belongs to: (a) copper(II) oxide (CuO), (b) sulphur dioxide, (c) aluminium oxide, (d) carbon monoxide. In each case give *one* fact to support your statement. (S.)

22. The observations given below relate to three oxides, A, B, and C. Classify the oxides, and in each case name *one* oxide which has the properties indicated.

A is a white crystalline solid which reacts vigorously with water, forming a solution which turns blue litmus paper red.

B is a white powder which is insoluble in water. It forms colourless solutions when separate portions are warmed with (i) dilute hydrochloric acid, (ii) concentrated sodium hydroxide solution.

C is a white solid which reacts vigorously with water forming a white suspension. When this is filtered the filtrate turns red litmus paper blue. (J.M.B.)

23. What would you observe if (*a*) a solution of hydrogen peroxide was added to manganese(IV) oxide, (*b*) an excess of hydrogen peroxide solution was added to (i) a solution of potassium permanganate acidified with dilute sulphuric acid, (ii) lead(II) sulphide. (J.M.B.)

24. (*a*) Write an equation for the decomposition of hydrogen peroxide in the presence of manganese(IV) oxide, and calculate the volume of oxygen liberated (measured at s.t.p.) from a solution containing one mole of hydrogen peroxide.

(*b*) When potassium permanganate solution is added to hydrogen peroxide solution in the presence of dilute sulphuric acid, the reaction can be represented by the equation

$$2MnO_4^- + 5H_2O_2 + 6H^+ \rightarrow$$
$$8H_2O + 5O_2 + 2Mn^{2+}$$

State what will be observed and calculate the volume of oxygen (measured at s.t.p.) released from a solution containing one mole of hydrogen peroxide.

(*c*) State what occurs when a solution of iron(II) sulphate acidified with dilute sulphuric acid is gently warmed with an excess of hydrogen peroxide solution and suggest an equation for the reaction. What reagent would you apply to the original and final solutions to confirm your statement and what would you expect to observe? (O.)

26 Hydrogen

Occurrence

Uncombined hydrogen does not occur in nature to any appreciable extent, but the element occurs in vast quantities in a combined state in such compounds as water, acids, and many organic substances.

Industrial preparation of hydrogen

Hydrogen has acquired much greater importance in recent years because of new uses given below. There are now two chief methods of manufacture – from hydrocarbons and by electrolysis.

(1) *From hydrocarbons*. In recent years, hydrogen has been made almost entirely from hydrocarbons. To use the simplest example, *methane* (natural gas) can be passed with *steam* over *nickel* catalyst at 800 °C and 30 atm.

$$CH_4(g) + H_2O(g) \rightarrow CO(g) + 3H_2(g)$$

The product is mixed with more steam and passed over *iron(III)oxide* (catalyst) at 450 °C. Carbon monoxide is converted to the dioxide with further yield of hydrogen.

$$CO(g) + 3H_2(g) + H_2O(g) \rightarrow CO_2(g) + 4H_2(g)$$

Carbon dioxide is dissolved out by water under 30 atm pressure.

$$H_2O(l) + CO_2(g) \rightleftharpoons H_2CO_3(aq)$$

Any remaining traces of carbon monoxide are absorbed under pressure, by copper(I) formate in ammonia.

(2) *By electrolysis*. Hydrogen is obtained as a by-product in the electrolytic manufacture of chlorine from common salt (page 365).

Where electrical power is cheap, hydrogen can be made by electrolysis of water containing sulphuric acid. The laboratory version of this method is fully discussed on page 155.

Uses of hydrogen

(1) *For filling balloons*. It is the least dense gas known but has the great disadvantage of flammability, and its use is therefore extremely limited.

(2) *In the 'hardening' of oils to make margarine*. Oils, *e.g.*, olive oil or whale-oil, are heated to 180 °C and finely divided nickel is added as catalyst. They are then treated with hydrogen at about 5 atm pressure. The oil combines with hydrogen and is converted to a *fat*, which is solid at ordinary temperature and is used in the manufacture of margarine. In this way, a liquid oil, unacceptable in our diet, is 'hardened' to an acceptable solid fat and used as a butter-substitute.

(3) *In the conversion of coal to synthetic 'petrol'*. A paste of coal-dust and oil,

containing iron oxide and alkali as catalyst, is sprayed into hydrogen at 200 atm pressure and 450 °C. The products are hydrocarbon gases and a liquid oil. On distillation, this oil yields a fraction suitable for petrol. (This process has been made obsolete by large, cheap supplies of natural petroleum.)

(4) *In the manufacture of ammonia.* This is Haber's Process, described on page 188.

(5) *In the synthesis of hydrochloric acid* $(H_2(g) + Cl_2(g) \rightarrow 2HCl(g))$ and of organic chemicals, *e.g.*, *methanol* CH_3OH.

(6) In the oxyhydrogen flame for cutting and welding steel.

Laboratory preparation of hydrogen

Hydrogen was first recognized by Cavendish (1766). It was called 'inflammable air', and the name hydrogen (*i.e.*, water-producer) was given to it by Lavoisier. It is most commonly prepared in the laboratory by the action of dilute mineral acids on certain metals, the method used by Cavendish himself.

Experiment 98
Preparation of hydrogen by the action of dilute acids on metals

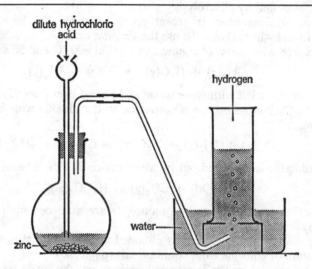

Figure 134

Note. Hydrogen explodes violently with air if a spark or flame reaches the mixture. For safety, always collect a *sample* of hydrogen and test as described on page 52, *before* lighting a jet or collecting the gas in bulk.

Into a flat-bottomed flask or bottle, put some pieces of zinc and add dilute hydrochloric acid by means of a thistle funnel. There is effervescence, and a gas is given off which is collected over water, as shown in Figure 134. Zinc chloride, which is formed, dissolves to form zinc chloride solution.

$$Zn(s) + 2HCl(aq) \rightarrow ZnCl_2(aq) + H_2(g)$$

or, $$Zn(s) + 2H_3O^+(aq) \rightarrow Zn^{2+}(aq) + 2H_2O(l) + H_2(g)$$

The extent to which a given metal will react with a dilute mineral acid to produce hydrogen depends, among other factors, on the position of the metal in the reactivity series. This is summarized as follows.

K
Na
Ca
Mg } Dilute sulphuric and hydrochloric acids attack these metals with
Al } the liberation of hydrogen.
Zn   (Al with hot concentrated HCl only)
Fe
Pb
(H)
Cu
Hg
Ag
Au

Dilute nitric acid, a common mineral acid, is *not* used for the preparation of hydrogen by reaction with a metal as the nitric acid is a strong oxidizing agent and will produce hydrogen only in the case of magnesium. Sodium and potassium are never used with a mineral acid to prepare hydrogen, as the reactions are extremely violent; these metals are so reactive they will displace hydrogen from cold water (see page 244).

Hydrogen may also be obtained by reaction between certain metals and caustic alkalis. Warm sodium (or potassium) hydroxide solution will react with zinc, aluminium, or silicon to liberate hydrogen and leave a solution of sodium (or potassium) zincate, aluminate, or silicate.

$$Zn(s) + 2NaOH(aq) + 2H_2O(l) \rightarrow Na_2Zn(OH)_4(aq) + H_2(g)$$
$$2Al(s) + 2NaOH(aq) + 6H_2O(l) \rightarrow 2NaAl(OH)_4(aq) + 3H_2(g)$$
$$Si(s) + 2NaOH(aq) + H_2O(l) \rightarrow Na_2SiO_3(aq) + 2H_2(g)$$

These methods are not usually used in the laboratory.

An alternative method (though rarely used nowadays) of preparing hydrogen is by the action of steam on heated iron, a reaction which is reversible, and which needs to be displaced from equilibrium in order to collect a high yield of hydrogen (see page 187).

Experiment 99
Hydrogen from water. Action of steam on heated iron

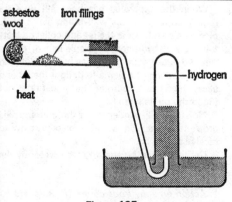

Figure 135

Into a hard glass test-tube place a plug of loosely packed asbestos wool soaked in water, as shown in Figure 135. With the test-tube in the horizontal position, insert a spatula measure of iron filings about half-way along the tube. Connect up the apparatus as shown.

Heat the tube under the iron filings, gently at first. By moving the flame of the Bunsen burner occasionally backwards towards the asbestos plug, steam will be generated and will pass over the hot metal. Keep the flow of steam constant, and collect the hydrogen by displacement of water.

$$3Fe(s) + 4H_2O(g) \rightleftharpoons Fe_3O_4(s) + 4H_2(g)$$

Zinc reacts in a similar manner with steam to produce hydrogen.

Test for hydrogen

A mixture of hydrogen and air explodes when a flame is applied.

Note that hydrogen/air explosions can be very dangerous if more than very small quantities are used, and therefore the test for hydrogen should always be carried out with very small volumes of the gas.

Properties of hydrogen

Hydrogen is an invisible gas, neutral to litmus, and, if pure, possesses no smell. It is much less dense than air, as can be shown using the simple apparatus shown in Figure 136.

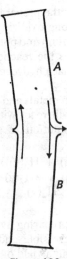

If a gas-jar full of hydrogen (*B*) is held under a gas-jar full of air (*A*), and the covers of both removed, after a short time (approximately fifteen seconds) the upper gas-jar may be removed and tested for hydrogen by applying a lighted splint. The contents of the gas-jar will explode (or burn), showing that hydrogen has passed upwards displacing the air in the top gas-jar to form a mixture. A similar test applied to the bottom gas-jar will produce no explosion or burning, showing that the hydrogen previously in the bottom jar has all been displaced by the air from the upper jar.

Although mixtures of hydrogen and oxygen will explode, it is possible to show that pure hydrogen burns steadily in air. A lighted splint applied to an inverted gas-jar of hydrogen will cause the gas to burn steadily around the edges of the jar, with a pale blue flame. A lighted splint pushed up into the gas will be seen to be extinguished, showing that hydrogen will burn in air but will not allow a splint to burn in it. The product of combustion of hydrogen is water,

Figure 136

$$2H_2(g) + O_2(g) \rightarrow 2H_2O(l)$$

Experiment 100
The burning of hydrogen in air to form water

Note: This experiment must not be carried out unless safety goggles are worn, and a safety screen is used.

This is the *synthesis* of water, *i.e.*, the building-up of water from its elements.

Fit up the apparatus as shown in Figure 137.

Hydrogen is generated by the action of fairly conc. hydrochloric acid on zinc.

$$Zn(s) + 2H^+(aq) \rightarrow Zn^{2+}(aq) + H_2(g)$$

The gas then passes through a U-tube containing anhydrous calcium chloride in order to dry the gas, the hydrogen is burnt at a jet, and the vapours are cooled by coming into contact with a can kept cool by water.

$$2H_2(g) + O_2(g) \rightarrow 2H_2O(l)$$
from
air

When the apparatus has been set up (use rubber stoppers or very well-fitting bark corks), place a test-tube over the jet and collect a test-tube full of hydrogen by displacement of air. When this test-tube full of gas burns quietly on the application of a flame to it, light the jet and allow the flame to burn so that it just does not touch the cooled can.

Moisture will condense on the can and will drop off into a dish which is placed below to receive the liquid.

The liquid can be shown to be water by the tests described below:

1. *Action on anhydrous copper(II) sulphate*

Allow a drop of the liquid to fall on to anhydrous copper(II) sulphate. A blue patch on the white solid (with hissing and development of heat)

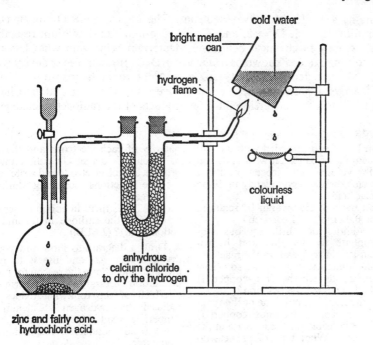

bright metal can

cold water

hydrogen flame

zinc and fairly conc. hydrochloric acid

anhydrous calcium chloride to dry the hydrogen

colourless liquid

Figure 137

indicates that water is present, but does **not** prove the liquid to be pure water.

$$CuSO_4(s) + 5H_2O(l) \longrightarrow CuSO_4.5H_2O(s)$$

2. *Boiling-point* (see Figure 138)

If the atmospheric pressure is 760 mm, the thermometer should register 100 °C.

These two tests together prove the liquid to be pure water. An additional, but less convenient test, is to find the freezing point, which should be 0 °C at 760 mm pressure.

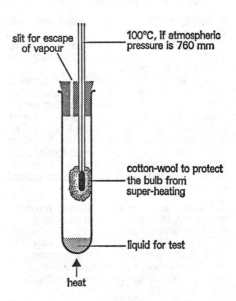

slit for escape of vapour

100°C, if atmospheric pressure is 760 mm

cotton-wool to protect the bulb from super-heating

liquid for test

heat

Figure 138

Isotopes of hydrogen

It may be briefly mentioned that hydrogen has at least three isotopes – ordinary hydrogen H, heavy hydrogen or *deuterium* D, and *tritium* T. All these have one electron per atom. In protium, the nucleus of the atom consists of a single proton; in deuterium, the nucleus contains a proton and a neutron; in tritium, it contains a proton and two neutrons. The approximate atomic masses are, therefore, H = 1, D = 2, and T = 3. Protium and deuterium resemble each other closely in chemical behaviour, deuterium being somewhat less reactive. Deuterium oxide is known as *heavy water* D_2O. It resembles ordinary water but has a higher density (about 1.10 g cm^{-3} at room temperature). Tritium is radioactive; it has a half-life of about twelve years – that is, in that time, half of the tritium is lost by conversion into the products of the radioactive change.

Questions

1. (a) You have been provided with a small weighed sample of zinc. How would you measure the volume of hydrogen produced when this zinc dissolves completely in dilute hydrochloric acid?

(b) What mass of zinc would be required to produce 0.01 mole of hydrogen?

(c) The addition of a little aqueous copper(II) sulphate to the zinc and hydrochloric acid is said to speed up the reaction. How would you test whether this is so? How would you decide whether the copper(II) sulphate had *catalysed* the reaction?

(d) It is thought that only one of the types of particle present in aqueous copper(II) sulphate is responsible for the increase in the rate of reaction. What type of particle do you think that this is? How would you test whether your suggestion is correct? (L.)

2. Briefly outline the apparatus and conditions that you would use to prepare and collect hydrogen by the action of water on (i) calcium, (ii) iron. Name the other products of the reactions.

State which of the following metals would give hydrogen when added to dilute hydrochloric acid, and give equations for the reactions: (iii) calcium, (iv) copper, (v) iron, (vi) aluminium. Describe how aluminium and iron(III) oxide (ferric oxide) react together. On the basis of the evidence from these reactions, arrange the four metals in order of increasing chemical reactivity. (J.M.B.)

3. (a) (i) Name the two reagents normally used in the laboratory preparation of hydrogen from a metal and an acid. (ii) Write an ionic equation for the reaction. (iii) The ionic equation indicates that the reaction is an electron transfer between the two reagents. Which of the reagents is oxidized and which reduced?

(b) Hydrogen is also obtained rapidly when steam is passed over heated magnesium. Describe what you would see in this reaction and give a reason why there is such an increase in the rate of reaction under these conditions compared to that with magnesium and cold water.

(c) How would you obtain hydrogen by electrolysis in the laboratory? Name the electrolyte used, and state the name and polarity of each electrode and the material of which it is made. Give the name of the product at each electrode and write equations for the reactions occurring during their discharge.

(d) What is there in common between the reaction at the cathode in (c) and the reaction in (a)? (J.M.B.)

4. Draw a diagram to show how you would dry hydrogen gas and use it to reduce a metallic oxide. State the reason for one safety precaution necessary in this experiment.

Which of the following oxides can be reduced by hydrogen? Calcium oxide, copper(II) oxide (cupric oxide), lead(II) oxide (lead monoxide), magnesium oxide, zinc oxide.

Choosing *one* of these oxides which can be reduced, (a) give the equation for the reaction, (b) state what you would see when the reduction is complete, (c) give *one* test to confirm that the solid product is a metal.

How is the ease of reduction of a metallic oxide related to the method chosen for the extraction of a metal from its ore? Illustrate your answer by briefly referring to the extraction of iron and aluminium. (J.M.B.)

5. Draw a fully labelled diagram to illustrate the usual laboratory preparation and collection of hydrogen by the action of a named metal on a named acid.

Hydrogen can be obtained by the action of iron on steam at 500 °C.

$$3Fe + 4H_2O \rightleftharpoons Fe_3O_4 + 4H_2$$

State what would be obtained if the following were left in a closed vessel at 500 °C: (i) iron and steam, (ii) tri-iron tetroxide (Fe_3O_4) and hydrogen.

Explain why the sign $\rightleftharpoons$ is used instead of an 'equals' sign (=) in the above equation.

Describe the action, if any, of water or steam on (a) zinc, (b) copper, (c) sodium. In each case state clearly whether water or steam would normally be used. If there is no reaction with either, state 'no reaction'.

Place the metals copper, iron, sodium, and zinc in order of their activity, placing the most reactive first. (J.M.B.)

6. With what results and under what conditions does hydrogen react with (i) nitrogen, (ii) sulphur, (iii) chlorine, and what is the effect of water upon each product?

Calculate the volume of hydrogen (measured at s.t.p.) released when 2 moles of water are (i) allowed to react with sodium, (ii) passed as steam over strongly heated magnesium, (iii) electrolysed after suitable acidification. (O.)

7. Sketch an apparatus by means of which you could measure the volume of gas liberated in a chemical reaction between a solid and a liquid. As a result of such an experiment it was found that 148.0 cm³ of hydrogen (corrected to s.t.p.) were released by dissolving 0.16 g of magnesium in hydrochloric acid. Given that one dm³ of hydrogen at s.t.p. weighs 0.09 g, calculate the mass of magnesium which displaces 2.0 g hydrogen. Give *three* chemical properties or reactions of hydrogen. (O.)

8. A method for manufacturing hydrogen involves the following steps:
(i) Methane and steam are passed over a nickel catalyst at 400 °C, producing a mixture of carbon monoxide and hydrogen.

(ii) The mixture of carbon monoxide and hydrogen, together with more steam, is passed over a hot iron oxide catalyst. This converts the carbon monoxide and steam into carbon dioxide and hydrogen.

(iii) The carbon dioxide is removed from the mixture.

(a) Write balanced equations for the reactions in steps (i) and (ii). (b) Both of the reactions in steps (i) and (ii) involve a catalyst. What is the function of a catalyst? (c) How does the composition of the gas mixture resulting from step (i) (1) resemble and (2) differ from that of water gas? (d) Suggest a method of removing the carbon dioxide in step (iii). (Remember that this is an industrial and not a laboratory process.) (e) If you had two gas-jars, one containing carbon monoxide and the other hydrogen, how would you find out which gas was in each jar? (Scottish.)

27 Carbon and its Compounds

Occurrence of carbon

Pure carbon is found in the form of diamond (India, South Africa) and impure carbon as graphite (Ceylon). Carbon is a constituent of numerous naturally occurring substances such as coal, mineral oils, carbonates, organic matter of all kinds, and occurs in the air to a small but very important extent (0.03–0.04% by volume) as carbon dioxide (see page 312).

Allotropes of carbon

Carbon exists in several allotropic modifications. Two of these, diamond and graphite, have already been mentioned (see page 91).

Diamond

The diamond is in the form of octahedral crystals of density 3.5 g cm^{-3} which have, when cut and polished, an amazing lustre which makes them valuable as jewellery. It is also the hardest substance known and has a commercial value for the manufacture of glass cutters and rock borers.

Graphite

Graphite exists as black, slippery, hexagonal crystals. It is found naturally as plumbago and is manufactured artificially by heating coke to a very high temperature in the electric furnace (Acheson process).

It is used extensively in the manufacture of lead pencils (which contain a long thin cylinder of graphite and clay) and also as 'black lead' as a protective coating for iron articles. It is used as a lubricant, particularly for small bearings (for example, those in dynamos and vacuum-cleaner motors) which require little, but regular, lubrication.

Amorphous carbon

Amorphous carbon (which is really made up of small crystals of graphite) exists in many forms:

Animal charcoal is made by heating animal refuse and bones with a limited supply of air. It also contains much calcium phosphate. It has the property of absorbing colouring matter (for example, litmus) and has a use in industry in removing the colouring matter from brown sugar.

Wood charcoal is made by heating wood with a limited supply of air. It is a light porous variety, and is a remarkably good absorbent for gases (1 cm^3 of wood charcoal will absorb nearly 100 cm^3 of ammonia gas at 0 °C).

Lampblack is made by burning oils (for example, turpentine) with a limited supply of air, and it is used for making printers' ink and shoe-polish.

Sugar charcoal is a very pure form of carbon, and is made by removing the elements of water from sugar.

Coke, gas carbon, and soot are other forms of impure amorphous carbon.

The crystal structures of diamond and graphite have already been described (see page 91) and the main differences in the properties of the two allotropes are summarized in the following table.

Graphite	Diamond
Density 2.3 g cm^{-3} average; variable	Density 3.5 g cm^{-3} average; variable
Black, opaque	Colourless, transparent; very high refractive index
Very soft, marks paper	Hardest known natural substance
Good electrical conductor	Electrical non-conductor
Attacked by potassium chlorate and nitric acid together	Not attacked by these reagents

Diamond is transparent to X-rays while glass is almost opaque.

Properties of carbon

Carbon is not a very reactive element. All forms of carbon can be made to burn in excess of oxygen to form carbon dioxide, although the temperature at which they begin to burn varies. As the carbon burns a great amount of heat is liberated.

$$C(s) + O_2(g) \rightarrow CO_2(g)$$

Coal is an impure form of carbon and, as such, is still used as a domestic source of heat. Its use is, however, uneconomic in an open grate, as a considerable amount of the heat of the combustion goes up the chimney and also many valuable products are lost. Further, much of the carbon escapes as soot, polluting the air in the larger industrial areas. Charcoal is also used in fires as a source of heat.

The reactions in a deep, brightly glowing coke or coal fire are those given below and illustrated in Figure 139.

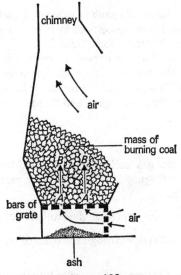

Figure 139

At A plenty of air is available. Carbon burns to carbon dioxide.

$$C(s) + O_2(g) \rightarrow CO_2(g)$$

At B ascending carbon dioxide is reduced by red-hot carbon to carbon monoxide.

$$CO_2(g) + C(s) \rightarrow 2CO(g)$$

At the surface the hot carbon monoxide burns in the air (to form carbon dioxide) with a flickering blue flame.

$$2CO(g) + O_2(g) \rightarrow 2CO_2(g)$$

Owing to the fact that carbon combines readily with oxygen, it acts as a reducing agent and is used in industrial practice in obtaining iron and zinc from their ores. (See page 462, and page 461.)

Experiment 101
Reducing property of carbon

Scrape a small hole in a charcoal block and place in the hole a mixture of lead(II) oxide and anhydrous sodium carbonate. (The carbonate melts and forms a protective coating, preventing the metal from being oxidized.) Turn the luminous Bunsen flame low, direct a jet of flame by means of a mouth blowpipe on to the mixture, and heat it for a few moments. Allow to cool and, on ejecting the substance from the hole, you will find a small grey globule of metallic lead which can be cut with a knife and which will mark paper.

$$PbO(s) + C(s) \rightarrow Pb(s) + CO(g)$$

In this reaction, carbon in association with the O^{2-} ion of the oxide makes electrons available and, therefore, acts as a reducing agent, as:

$$O^{2-}(s) + C(s) \rightarrow CO(g) + 2e^-$$

The ion, Pb^{2+}, of lead(II) oxide is reduced by accepting these electrons, as:

$$Pb^{2+}(s) + 2e^- \rightarrow Pb(s)$$

Carbon is insoluble in all common solvents, a fact which all motor-car owners know to their cost. Petrol consists of hydrocarbons similar to methane CH_4. In the cylinders these hydrocarbons are oxidized by the oxygen of the air. If the supply of air is insufficient a deposit of carbon is left inside the cylinders as a hard black solid. This would, in time, choke up the engine and hence the carbon has to be removed. Since carbon is not soluble in any common solvent, this has to be done by dismantling the engine and removing the carbon mechanically.

Producer gas

Producer gas is a fuel gas made by passing air through a thick layer of white-hot coke in a *producer* (Figure 140). In the lower part of the producer, with *air* in excess, the reaction is:

$$C(s) + O_2(g) \rightarrow CO_2(g) \text{ (strongly exothermic; } i.e., \text{ heat liberated)}$$

As the carbon dioxide rises through the mass of white-hot coke, the further reaction is:

$$CO_2(g) + C(s) \rightarrow 2CO(g) \quad \text{(endothermic; } i.e., \text{ heat absorbed)}$$

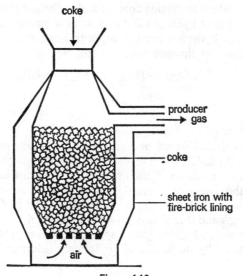

Figure 140

On the balance of these reactions, *net heat is liberated* and the reaction can continue indefinitely.

The gas emerging from the producer consists of carbon monoxide (one-third by volume) and unchanged nitrogen of the air (two thirds). This producer gas is usually distributed, while still hot from the producer, to points on the site where heating is required; it is mixed with air and burnt to produce more heat.

$$2CO(g) + O_2(g) \rightarrow 2CO_2(g) \quad \text{(exothermic; heat liberated)}$$

Since two-thirds (by volume) of the gas is non-combustible nitrogen, its calorific value is not high. It has been found very useful, however, in such processes as the firing of retorts and glass furnaces.

Water-gas

Water-gas is produced by the passage of steam through a mass of coke at a temperature (minimum 1000 °C) which allows the reaction:

$$C(s) + H_2O(g) \rightarrow CO(g) + H_2(g) \quad \text{(endothermic; heat absorbed)}$$

giving equal volumes of carbon monoxide and hydrogen, both of which are combustible, so the gas has a high calorific value.

As shown, the above reaction is endothermic, so the temperature of the coke falls. Below 1000 °C, much steam passes without reaction and, also, incombustible carbon dioxide is formed.

$$C(s) + 2H_2O(g) \rightarrow CO_2(g) + 2H_2(g)$$

To meet this situation, the plant similar to the producer above, is used in the following way. An air-blow of about two minutes raises the temperature of the coke to incandescence (perhaps 1400°–1500 °C), using the exothermic reaction of producer gas (above). Then a steam-blow of about four minutes produces water-gas, which is collected. This cools the coke and the air-blow is renewed and so on alternately.

Water-gas can be burnt as a fuel gas with air, both the following reactions being exothermic.

$$2H_2(g) + O_2(g) \rightarrow 2H_2O(g); \quad 2CO(g) + O_2(g) \rightarrow 2CO_2(g)$$

The high carbon monoxide content makes the gas poisonous. It has been an industrial source of hydrogen and of organic chemicals, *e.g.*, to make *methanol*. Water-gas and hydrogen are passed, with heating to 450 °C, over a catalyst (oxides of zinc and chromium) at 200 atm pressure.

$$\underbrace{CO(g) + H_2(g)}_{\text{water-gas}} + H_2(g) \rightarrow CH_3OH(l)$$

Carbon dioxide

This gas was first observed by Van Helmont towards the end of the sixteenth century, but Black (1728–99) first showed that the gas could be prepared by the action of dilute acids on calcium carbonate. It occurs in the air on the earth's surface to the extent of about 0.03% of its volume; it issues from rocks in volcanic regions, and occurs in mines as 'choke damp'. Certain mineral springs contain the gas and it is always present in natural drinking water because of its solubility in water. Its biological importance is dealt with on page 237.

It is prepared by the action of dilute hydrochloric acid on marble (calcium carbonate).

Experiment 102
Preparation of carbon dioxide

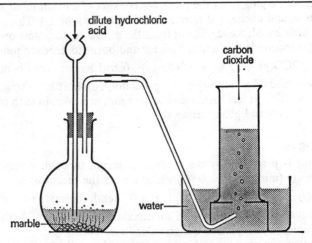

Figure 141

Place several pieces of marble in a flask (or bottle), as shown in Figure 141, and pour some dilute hydrochloric acid down the thistle funnel on to the marble. There is effervescence and a colourless gas is liberated which is collected over water* or by downward delivery, the gas being denser than air. Calcium chloride solution is left in the flask.

$$CaCO_3(s) + 2HCl(aq) \rightarrow$$
$$CaCl_2(aq) + H_2O(l) + CO_2(g)$$

or $\quad CO_3^{2-} + 2H^+ \rightarrow H_2O + CO_2$

If the gas is required pure and dry it can be passed through potassium hydrogencarbonate solution in a wash-bottle (to remove suspended hydrochloric acid spray), dried by passing it through a U-tube packed with anhydrous calcium chloride and collected by downward delivery (the gas being denser than air).

* There is some loss due to the solubility of the gas in water. Water, at ordinary temperature and pressure, absorbs its own volume of carbon dioxide.

After several gas-jars of carbon dioxide have been collected the following experiments may be performed to illustrate the general properties of the gas.

Experiment 103
Some properties of carbon dioxide

Effect of carbon dioxide on a lighted splint

Plunge a lighted splint into a gas-jar of the gas. It is extinguished. Carbon dioxide does not support combustion.

Action of carbon dioxide on lime-water

Pour lime-water into a gas-jar full of carbon dioxide. **The lime-water goes milky.** If the mixture is allowed to stand, you will see white solid particles separate out. These are particles of chalk. **The milkiness is due to a suspension of the insoluble substance, chalk, in water.**

$$Ca(OH)_2(aq) + CO_2(g) \longrightarrow CaCO_3(s) + H_2O(l)$$

calcium	carbon	calcium	water
hydroxide	dioxide	carbonate	
solution		(chalk)	

The above test serves to distinguish carbon dioxide from any other gas.

Effect of carbon dioxide on a lighted candle

Lower a candle on a deflagrating spoon into a gas-jar of air. The candle can be extinguished by 'pouring' carbon dioxide into the gas-jar in which the candle is burning (Figure 142). This shows carbon dioxide to be denser than air (density 22 relative to hydrogen).

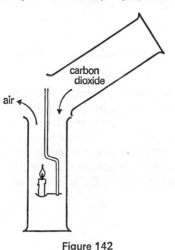

Figure 142

Effect of carbon dioxide on burning magnesium

Lower a piece of burning magnesium into carbon dioxide in a gas-jar. It continues to burn for a short time with a spluttering flame, and black specks of carbon can be seen on the sides of the gas-jar. The magnesium burns to magnesium oxide.

$$2Mg(s) + CO_2(g) \longrightarrow 2MgO(s) + C(s)$$

This clearly shows carbon dioxide to contain carbon and oxygen.

Solution of carbon dioxide in water

Invert a gas-jar of carbon dioxide in a trough of cold water and shake the gas-jar. The water rises slowly, showing that the gas is soluble. Put a glass plate over the mouth of the jar and remove it. To the liquid in it, add blue litmus solution and shake. The solution becomes claret-coloured but not red. This is because carbon dioxide reacts with water to produce *carbonic acid,* which is, however, too weak to turn litmus solution red.

$$H_2O(l) + CO_2(g) \rightleftharpoons H_2CO_3(aq) \rightleftharpoons$$

water   carbon   carbonic
dioxide   acid

$$2H^+(aq) + CO_3{}^{2-}(aq)$$
(very slight)

Carbon dioxide is an acidic oxide.

Action of carbon dioxide with sodium hydroxide solution

Repeat the above experiment, using sodium hydroxide solution instead of water. The rapid rise of the solution shows that the gas is quickly absorbed. The acidic carbon dioxide reacts with the alkaline solution producing sodium carbonate.

$$CO_2(g) + 2NaOH(aq) \longrightarrow Na_2CO_3(aq) + H_2O(l)$$

or
$$CO_2(g) + 2OH^-(aq) \longrightarrow CO_3{}^{2-}(aq) + H_2O(l)$$

This reaction is discussed more fully later (page 320).

Uses of carbon dioxide

Solutions of the gas in water have a pleasant taste (the taste of soda water), and hence the gas is used in the manufacture of the effervescing drinks called 'mineral waters'. The effervescence is caused by dissolving the gas in water at a pressure of several atmospheres; when the pressure is released (by opening the bottle) the gas is liberated.

Its use in the solid form is increasing. Carbon dioxide can be made into a white solid (carbon dioxide snow) by allowing liquid carbon dioxide to evaporate, the temperature falling to −78 °C as the solid forms. The solid evaporates when heated, leaving no residue, and it is, therefore, used as a refrigerating agent for perishable goods.

Owing to its non-flammable nature and its high density, carbon dioxide is used for extinguishing fires. One type of fire extinguisher contains a solution of sodium carbonate which can be made to come into contact with dilute sulphuric acid by striking a knob. The carbon dioxide liberated forces a stream of effervescing liquid on to the fire, and the carbon dioxide prevents the air from getting to the burning material and so helps to put out the fire. Another type of extinguisher contains carbon dioxide under high pressure; this type can be used intermittently, whereas the acid-carbonate type has to be recharged after each use. For solid CO_2 crystals, see page 87.

Test for carbon dioxide

Carbon can be tested for by passing the gas into lime water (a solution of calcium hydroxide) which turns milky due to the formation of calcium carbonate.

$$Ca(OH)_2(aq) + CO_2(g) \rightarrow CaCO_3(s) + H_2O(l)$$

If the gas is passed in excess, the lime water will turn clear again due to the formation of calcium hydrogencarbonate, which is soluble.

$$CaCO_3(s) + H_2O(l) + CO_2(g) \rightarrow Ca(HCO_3)_2(aq)$$

Carbon monoxide

Carbon monoxide is a poisonous, colourless gas with practically no smell. It is present in coal-gas and other gaseous fuels. It is formed by the partial combustion of carbon, and poisoning by the exhaust fumes of a motor-car in an enclosed space, for example a garage, is due to the presence of carbon monoxide. The blood of a person poisoned by the gas is a characteristic cherry-red in colour. An atmosphere containing as little as 0.5% carbon monoxide may cause death if breathed for some time, and an atmosphere containing 0.1% is injurious.

Experiment 104
Preparation of carbon monoxide from ethanedioic acid (oxalic acid)

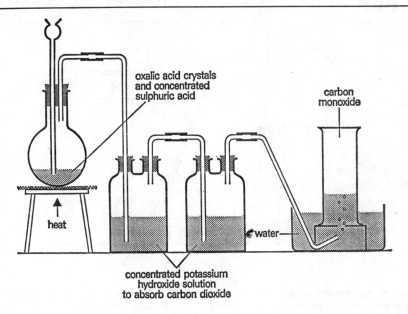

oxalic acid crystals
and concentrated
sulphuric acid

carbon
monoxide

heat

water

concentrated potassium
hydroxide solution
to absorb carbon dioxide

Figure 143

Fit up the apparatus as shown in Figure 143. Place some crystals ($H_2C_2O_4.2H_2O$) in the strong flat-bottomed flask and pour concentrated sulphuric acid down the thistle funnel. Warm the mixture gently (always have the greatest respect for hot concentrated sulphuric acid). The white crystals dissolve, effervescence is observed, and a mixture of carbon monoxide and carbon dioxide gases is evolved. By passing the mixture through a concentrated solution of caustic potash the carbon dioxide is absorbed, and the carbon monoxide passes on and is collected over water in which it is insoluble.

$$2KOH(aq) + CO_2(g) \longrightarrow$$
$$K_2CO_3(aq) + H_2O(l)$$

Chemistry of the action. Oxalic acid has the formula $H_2C_2O_4$ and the hot concentrated sulphuric acid removes **the elements of water** from the molecule of oxalic acid, leaving a mixture of equal volumes of carbon monoxide and carbon dioxide. It is because the carbon dioxide is there in quantity, and not merely as a small trace of impurity, that it is necessary to pass the gas through two wash-bottles containing potassium hydroxide solution.

$$H_2C_2O_4(s) \rightarrow CO(g) + CO_2(g) + H_2O(l)$$

oxalic   carbon   carbon   water
acid   monoxide   dioxide (removed
from acid)

Experiment 105
Preparation of carbon monoxide from sodium methanoate (sodium formate)

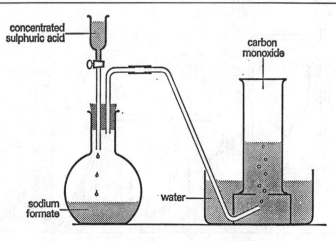

Figure 144

Fit up the apparatus shown in Figure 144. Place one or two spatula measures of sodium formate in the flat-bottomed flask and allow concentrated sulphuric acid to run in from the tap funnel. The reaction takes place in the cold, effervescence is observed, and carbon monoxide is collected over water in which it is insoluble.

Chemistry of the action. Sodium formate is converted into formic (methanoic) acid from which the elements of water are immediately removed by the concentrated sulphuric acid:

$$HCOONa(s) + H_2SO_4(aq) \rightarrow HCOOH(aq) + NaHSO_4(aq)$$
$$HCOOH(aq) \rightarrow CO(g) + \underset{\substack{\text{removed} \\ \text{by acid}}}{H_2O(l)}$$

The above actions take place simultaneously and can be represented by the equation:

$$HCOONa(s) + H_2SO_4(aq) \rightarrow NaHSO_4(aq) + CO(g) + H_2O(l)$$

Test. Carbon monoxide burns in air with a blue flame, forming carbon dioxide (the latter will turn lime water turbid, forming a precipitate of chalk).

Properties of carbon monoxide

Carbon monoxide is quite a strong reducing agent. For example, when the gas is passed over heated lead(II) oxide the formation of grey metallic lead can be observed. The carbon monoxide has reduced the lead(II) oxide to lead, being itself oxidized to carbon dioxide.

$$PbO(s) + CO(g) \rightarrow Pb(s) + CO_2(g)$$

The gas will similarly reduce copper(II) oxide and iron(III) oxide to the metal.

$$CuO(s) + CO(g) \rightarrow Cu(s) + CO_2(g)$$
$$Fe_2O_3(s) + 3CO(g) \rightarrow 2Fe(s) + 3CO_2(g)$$

In all these reactions, carbon monoxide, in association with the O^{2-} ion of the oxides, makes electrons available and so acts as a reducing agent, as:

$$O^{2-} + CO \rightarrow CO_2 + 2e^-$$

The metallic ion is reduced by accepting these electrons:

$$Cu^{2+} + 2e^- \rightarrow Cu$$
$$Pb^{2+} + 2e^- \rightarrow Pb$$
$$2Fe^{3+} + 6e^- \rightarrow 2Fe$$

As already noted (see page 314) the gas is extremely poisonous, since it combines with the haemoglobin in red blood corpuscles so preventing the carriage of oxygen by the haemoglobin to essential sites in body cells.

In contrast to carbon dioxide it is a **neutral oxide**; it is insoluble in water and does not react with either acids or alkalis under normal conditions.

Carbonates

Carbonates may be regarded as salts derived from carbonic acid H_2CO_3, which is formed when carbon dioxide is dissolved in water (page 313). The table below summarizes briefly the important properties of the common carbonates.

K \ Na /	Carbonates soluble.	Carbonates of these metals not decomposed by heat.	
Ca Mg Al Zn Fe Pb Cu	Carbonates of these metals insoluble in water.	Carbonates of these metals decomposed into oxide of the metal by heat (Al forms no carbonate)	Any carbonate with any acid liberates carbon dioxide.

Ammonium carbonate is also soluble in water.

Test for any carbonate

Put some of the suspected carbonate in a test-tube and add dilute nitric acid. If a carbonate is present there will be effervescence and the gas which comes off will turn lime-water milky:

$$CO_3{}^{2-}(s) + 2H^+(aq) \rightarrow H_2O(l) + CO_2(g)$$

Ammonium carbonate $(NH_4)_2CO_3$

This compound is prepared as a sublimate, by heating ammonium sulphate with limestone.

$$(NH_4)_2SO_4(s) + CaCO_3(s) \rightarrow (NH_4)_2CO_3(s) + CaSO_4(s)$$

It is used as a constituent of 'smelling salts' since it decomposes readily to ammonia, carbon dioxide, and water.

Potassium carbonate K_2CO_3

Potassium carbonate is very similar to sodium carbonate. (See page 318.) It cannot, however, be made by the Solvay process because the hydrogencarbonate of potassium is too soluble. It is made by the Leblanc process and in other ways.

Potassium carbonate differs chiefly from sodium carbonate in that it is very deliquescent, and is without water of crystallization.

It is used to make soft soap, hard glass, and potassium salts generally.

Sodium carbonate (soda ash) Na₂CO₃

Sodium carbonate is obtained in both the anhydrous and crystalline states by the Solvay process.

Solvay process

Very concentrated brine (28% NaCl) is saturated with ammonia gas in a tower and the ammoniacal brine is run down further Solvay towers up which carbon dioxide is forced. The towers are fitted with perforated mushroom-shaped baffles at intervals (Figure 145). These baffles delay the flow of liquid and present

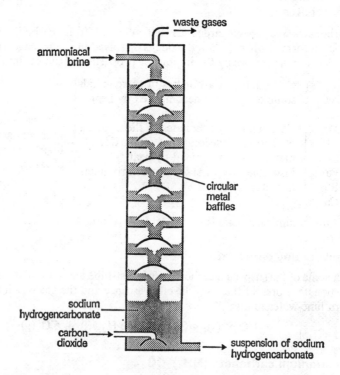

Figure 145

surfaces for reaction. Sodium hydrogencarbonate is formed. It is not very soluble in water, so it precipitates; precipitation is assisted by cooling the lowest third of the tower.

$$NaCl(aq) + NH_4OH(aq) + CO_2(g) \rightarrow NaHCO_3(s) + NH_4Cl(aq)$$

Sodium hydrogencarbonate is filtered from the white sludge and washed free from ammonium compounds. It is then heated to convert it to sodium carbonate, and the carbon dioxide evolved is used again.

$$2NaHCO_3(s) \rightarrow Na_2CO_3(s) + H_2O(g) + CO_2(g)$$

The substance formed is anhydrous sodium carbonate, which finds a wide market. If the crystalline form (washing soda) is required, the anhydrous solid is dissolved in such an amount of hot water that crystallization occurs on cooling. The crystals are removed and allowed to dry.

$$Na_2CO_3(s) + 10H_2O(l) \rightarrow Na_2CO_3.10H_2O(s)$$

Efficiency of the process

(1) The principal raw materials for the process are cheap and plentiful. They are *sodium chloride* and *limestone*. The common salt is extracted from deposits as brine; the limestone yields *quicklime* and *carbon dioxide* when heated.

$$CaCO_3(s) \rightleftharpoons CaO(s) + CO_2(g)$$

Carbon dioxide is passed to the carbonating tower and approximately half of it is recovered in the heating of sodium hydrogencarbonate above.

Ammonium chloride $NH_4{}^+Cl^-$ is left in the solution after precipitation of sodium hydrogencarbonate or is washed out of the precipitate. Ammonia is recovered by heating the solution and washings with quicklime (above) and returned to the ammoniating tower.

$$CaO(s) + H_2O(l) \rightarrow Ca(OH)_2(s)$$
$$2NH_4Cl(aq) + Ca(OH)_2(s) \rightarrow CaCl_2(aq) + 2H_2O(l) + 2NH_3(g)$$

Consequently, once the pipe-line is full, no further external supplies of ammonia are theoretically required. In practice, about 2% of the ammonium chloride in use is added per circuit to restore manipulative losses.

(2) The process is one of continuous flow and only a minimum labour cost.

(3) The principal weakness of the process is its failure to utilize the chlorine of the sodium chloride used. This is lost as calcium chloride (see equation in (1) above) for which there is only slight demand.

Uses of sodium carbonate

Three of the important uses of sodium carbonate are the following.

(*a*) *Manufacture of glass.* Ordinary bottle glass is made by fusing together sodium carbonate (or sulphate), calcium carbonate, silica, and a little carbon (reducing agent). Broken glass (cullet) is added to assist fusion.

$$Na_2CO_3(s) + SiO_2(s) \rightarrow Na_2SiO_3(s) + CO_2(g)$$
$$CaCO_3(s) + SiO_2(s) \rightarrow CaSiO_3(s) + CO_2(g)$$

The mixture of silicates (with some unchanged silica) constitutes the glass.

(*b*) *Manufacture of water-glass.* Sodium carbonate is fused with silica to produce sodium silicate.

$$Na_2CO_3(s) + SiO_2(s) \rightarrow Na_2SiO_3(s) + CO_2(g)$$

On cooling, a glassy solid is left, which is broken up, boiled with water and evaporated to a colourless, treacly material, known as water-glass. It is used in preserving eggs, in fire-proofing, and in producing cements.

(*c*) *In domestic water-softening.* Calcium ion, Ca^{2+}, which is the principal cause of hardness in water, is precipitated from the water as chalk $Ca^{2+}CO_3{}^{2-}$, by the addition of sodium carbonate.

Experiment 106
Preparation of sodium carbonate from sodium hydroxide solution

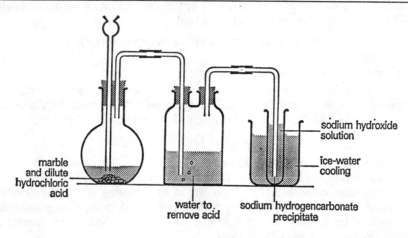

marble
and dilute
hydrochloric
acid

water to
remove acid

sodium hydroxide
solution

ice-water
cooling

sodium hydrogencarbonate
precipitate

Figure 146

Fit up the apparatus as shown in Figure 146. Pass carbon dioxide (free from hydrochloric acid) into a moderately concentrated solution of sodium hydroxide for some time until finally a white solid (sodium hydrogencarbonate) appears on the bottom of the boiling-tube.

$2NaOH(aq) + CO_2(g) \longrightarrow$
$\qquad Na_2CO_3(aq) + H_2O(l)$ (1st stage)

$Na_2CO_3(aq) + H_2O(l) + CO_2(g) \longrightarrow$
$\qquad 2NaHCO_3(s)$ (2nd stage)

In ionic terms:

$2OH^-(aq) + CO_2(g) \longrightarrow$
$\qquad CO_3{}^{2-}(aq) + H_2O(l)$ (1st stage)

$CO_3{}^{2-}(aq) + H_2O(l) + CO_2(g) \longrightarrow$
$\qquad 2HCO_3{}^-(s)$ (2nd stage)

Filter this off, wash the solid residue two or three times with a little cold water, and then transfer the solid to a dish and heat. Finally sodium carbonate will be obtained as a fine white powder, by heating the sodium hydrogencarbonate to constant mass.

$2NaHCO_3(s) \longrightarrow$
$\qquad Na_2CO_3(s) + H_2O(l) + CO_2(g)$

Carbon dioxide is evolved during the reaction.

Properties and uses of washing soda

Washing soda is sodium carbonate decahydrate $Na_2CO_3.10H_2O$, large translucent crystals.

Efflorescence. On exposure to air the crystals lose mass and become coated with a fine white powder which renders them opaque. Each molecule of washing soda has given up to the atmosphere 9 molecules of water of crystallization.

$$Na_2CO_3.10H_2(s) \longrightarrow Na_2CO_3.H_2O(s) + 9H_2O(g)$$
sodium carbonate   sodium carbonate
decahydrate   monohydrate

Such an action, that is, the giving up of water of crystallization to the atmosphere, is termed efflorescence.

The crystals are readily soluble in water.

Anhydrous sodium carbonate can be made by heating the hydrated sodium carbonate. It is a fine white powder and does not dissolve as readily in water as do the crystals.

Solutions of sodium carbonate in water are alkaline to litmus. This is due to the feebly acid properties of carbonic acid. This acid is expelled by almost every other acid, and hence, sodium carbonate acts like sodium hydroxide, although the former is a salt and the latter an alkali.

$$Na_2CO_3(aq) + 2HCl(aq) \rightarrow 2NaCl(aq) + H_2O(l) + CO_2(g)$$

Sodium carbonate can be used quantitatively in volumetric analysis, as if it were an alkali.

Notice that sodium and potassium carbonates are both soluble in water and are not decomposed at a red heat.

Sodium carbonate is used for the softening of water for domestic purposes, and in the manufacture of glass, borax, caustic soda, and water-glass. It is a constituent of many 'dry soap' powders.

Calcium carbonate CaCO₃

This occurs as limestone, marble, chalk, and in many other forms, and, since it is insoluble, can easily be made in the laboratory by double decomposition (see page 181). It is seen as a white precipitate when carbon dioxide is bubbled into lime water.

$$Ca(OH)_2(aq) + CO_2(g) \rightarrow CaCO_3(s) + H_2O(l)$$
$$\text{calcium}$$
$$\text{carbonate}$$
$$\text{(chalk)}$$

The chalk can be obtained by filtering off the precipitate, washing it a few times with hot water, and allowing it to dry.

It is attacked by dilute hydrochloric and nitric acids with the evolution of carbon dioxide, for example:

$$CaCO_3(s) + 2HCl(aq) \rightarrow CaCl_2(aq) + H_2O(l) + CO_2(g)$$

With dilute sulphuric acid, however, the action slows down and finally stops, particularly if the calcium carbonate is in lump form, for example, marble. The reason is that calcium sulphate, being only sparingly soluble, forms a protective layer on the outside preventing the sulphuric acid from acting upon the solid within.

Although practically insoluble in pure water, calcium carbonate is dissolved by water which contains dissolved carbon dioxide, because it forms soluble calcium hydrogencarbonate.

$$CaCO_3(s) + H_2O(l) + CO_2(g) \rightarrow Ca(HCO_3)_2(aq)$$
$$\text{calcium}$$
$$\text{hydrogencarbonate}$$

Zinc carbonate ZnCO₃

This is formed as a white precipitate when sodium hydrogencarbonate solution is added to a solution of zinc sulphate in water.

$$ZnSO_4(aq) + 2NaHCO_3(aq) \rightarrow ZnCO_3(s) + Na_2SO_4(aq) + H_2O(l) + CO_2(g)$$

(If sodium carbonate is used, *basic* carbonate of zinc is formed.)

The white precipitate is filtered off, washed with hot distilled water, and allowed to dry.

Zinc carbonate is attacked by dilute acids liberating carbon dioxide, for example:

$$ZnCO_3(s) + 2HCl(aq) \rightarrow ZnCl_2(aq) + H_2O(l) + CO_2(g)$$

On heating a little zinc carbonate in a test-tube, carbon dioxide is given off and zinc oxide (yellow when hot, white when cold) remains in the test-tube.

$$ZnCO_3(s) \rightarrow ZnO(s) + CO_2(g)$$

Lead(II) carbonate PbCO₃

This is made in the laboratory, as a white precipitate, by adding sodium hydrogencarbonate solution to a solution of lead(II) nitrate in water.

$$Pb(NO_3)_2(aq) + 2NaHCO_3(aq) \rightarrow PbCO_3(s) + 2NaNO_3(aq)$$
$$+ H_2O(l) + CO_2(g)$$

Sodium carbonate precipitates *basic* lead carbonate.

White lead, basic lead(II) carbonate $Pb(OH)_2.2PbCO_3$, is used extensively as a paint when mixed with oils. It is made by subjecting strips of lead to the action of acetic acid, water-vapour, carbon dioxide, and air. It is poisonous and blackens rapidly in industrial areas, where hydrogen sulphide occurs.

N.B. Lead(II) carbonate is not readily acted upon by either dilute hydrochloric or sulphuric acids. A layer of insoluble chloride or sulphate formed round the carbonate protects it from further action. Dilute nitric acid attacks it to liberate carbon dioxide in accordance with the general action of acids on carbonates:

$$CO_3{}^{2-}(s) + 2H^+(aq) \rightarrow H_2O(l) + CO_2(g)$$

Copper(II) carbonate CuCO₃

This is made by double decomposition and it is usually obtained as a basic salt, having the formula $CuCO_3.Cu(OH)_2$. This is a bright green powder which liberates carbon dioxide on being heated, and black copper(II) oxide is left.

$$CuCO_3(s) \rightarrow CuO(s) + CO_2(g); \quad Cu(OH)_2(s) \rightarrow CuO(s) + H_2O(g)$$

It dissolves in dilute acids with the liberation of carbon dioxide, for example:

$$CuCO_3(s) + H_2SO_4 \rightarrow CuSO_4(aq) + H_2O(l) + CO_2(g)$$

Hydrogencarbonates

Hydrogencarbonates may also be regarded as salts derived from carbonic acid H_2CO_3, formed by the partial replacement of the hydrogen by a metal or cationic radical.

We are only concerned with the hydrogencarbonates of sodium and calcium, all the others being unstable or unimportant.

Sodium hydrogencarbcnate (baking soda) NaHCO₃

Sodium hydrogencarbonate is manufactured by saturating a wet mush of Solvay sodium carbonate and water with carbon dioxide. The product is washed with cold water and dried.

$$CO_3{}^{2-}(aq) + H_2O(l) + CO_2(g) \rightarrow 2HCO_3{}^-(s)$$

Laboratory preparation of sodium hydrogencarbonate

Sodium hydrogencarbonate is made in the laboratory by bubbling carbon dioxide for some time through a concentrated solution in water of either sodium hydroxide or sodium carbonate (see Figure 146). In the latter case the reaction takes place much more quickly:

$$CO_2(g) + Na_2CO_3(aq) + H_2O(l) \rightarrow 2NaHCO_3(s)$$
sodium
carbonate

With sodium hydroxide

$$2NaOH(aq) + CO_2(g) \rightarrow Na_2CO_3(aq) + H_2O(l)$$
sodium
hydroxide

then $\quad Na_2CO_3(aq) + H_2O(l) + CO_2(g) \rightarrow 2NaHCO_3(s)$

In both cases the hydrogencarbonate is deposited as a white powder and this is filtered off, washed two or three times with a little cold distilled water, and allowed to dry.

With dilute acids, the hydrogencarbonate liberates carbon dioxide:

$$HCO_3^-(aq) + H^+(aq) \rightarrow H_2O(l) + CO_2(g)$$

Experiment 107
Action of heat on sodium hydrogencarbonate

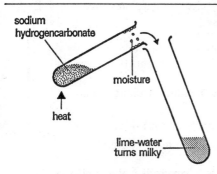

sodium hydrogencarbonate

moisture

heat

lime-water turns milky

Figure 147

Place a small amount of sodium hydrogen-carbonate in a dry test-tube and heat gently with the lip of the tube projecting into a boiling-tube containing lime water (see Figure 147). A gas is given off which turns the lime water milky and is, therefore, carbon dioxide. Water is seen to condense on the cooler parts of the tube.

$$2HCO_3^-(s) \rightarrow CO_3^{2-}(s) + H_2O(g) + CO_2(g)$$

The white residue is sodium carbonate. This reaction distinguishes sodium hydrogencarbonate from sodium carbonate, which is unaffected by heat.

Sodium hydrogencarbonate is used in the manufacture of baking powders. Under the action of heat it decomposes, as above, and gives off carbon dioxide which causes the cake to 'rise' and so be light. This is why it is commonly called 'baking soda'. Baking powders also contain rice powder as a diluent, and tartaric acid (or a similar compound) to react with the sodium carbonate, which would otherwise be left when the hydrogencarbonate decomposes.

Calcium hydrogencarbonate

See Hardness of Water, page 248.

Its method of preparation and its reactions are chemically similar to those of sodium hydrogencarbonate.

Calcium hydrogencarbonate cannot, however, be isolated as a solid, since it decomposes too easily, and all its reactions are carried out in solution.

Tetrachloromethane CCl_4 (carbon tetrachloride)

This compound is made by chlorinating carbon disulphide, boiling under reflux, with iodine as catalyst.

$$CS_2(l) + 3Cl_2(g) \rightarrow CCl_4(g) + S_2Cl_2(g)$$

The products can be separated by fractional distillation (boiling points, CCl_4 77 °C, S_2Cl_2 138 °C at standard pressure). The tetrachloromethane can be purified by shaking with dilute sodium hydroxide solution (to remove any chlorine) and then with water. This is done in a separating funnel (Figure 148). Tetrachloromethane is almost non-miscible with water and separates as the lower layer (density 1.63 g cm^{-3}) which is run off *through the tap*. The washing liquid (upper layer) is poured out *through the top* of the funnel. Anhydrous calcium chloride (bean size) is then added to the tetrachloromethane and the mixture is left in a stoppered flask for at least 12 hours. This will dry the tetrachloromethane as the calcium chloride takes up the water present as the solid hydrate $CaCl_2.6H_2O$. The liquid should then be decanted or filtered (through a dry filter paper) into a dry distillation flask and distilled (apparatus as page 11). If pure, the entire liquid should distil over at its boiling point, 77 °C at standard pressure.

Properties and uses

1. Tetrachloromethane is a good solvent for fats and greases and is used as a de-greasing and dry-cleaning agent. It has the advantage of being non-flammable.

2. By heating with iron and water, tetrachloromethane is converted to trichloromethane.

$$CCl_4(l) + Fe(s) + H^+(aq) \rightarrow Fe^{2+}(aq) + Cl^-(aq) + CHCl_3(l)$$

Figure 148

Otherwise, it is chemically rather inert; unlike most non-metallic chlorides it is unaffected by water.

The tetrachloromethane molecule is shaped like the methane molecule (page 87), substituting four chlorine atoms for the four hydrogen atoms of methane.

Questions

1. (*a*) Give three *contrasting* properties of diamond and graphite. (*b*) Explain what is meant by a *crystal lattice*. Give diagrams of the crystal lattices of diamond and graphite, and state how these lattices account for the differences in properties of the two allotropes. (*c*) If a hydrocarbon contains 80% carbon and 20% hydrogen, what is its empirical formula? If its relative molecular mass is 30, what is its molecular formula? (*d*) Give *two* tests which distinguish between this hydrocarbon and the corresponding hydrocarbon with two fewer hydrogen atoms in its molecule. (A.E.B.)

2. (*a*) Name *two* crystalline forms of carbon. (*b*) Describe experiments by which you could prepare good samples of *two* crystalline forms of sulphur from finely powdered sulphur.

(*c*) Weighed samples of two pure forms of carbon were heated in oxygen until combustion was complete. The carbon dioxide formed was absorbed in weighed bulbs containing potassium hydroxide solution. The experimental results were

Mass of sample	Mass of CO_2 formed
0.40 g	1.47 g
0.60 g	2.20 g

(i) Give the equation for *one* reaction between carbon dioxide and potassium hydroxide. (ii) Explain why potassium hydroxide solution was used rather than lime-water (calcium hydroxide solution). (iii) Show that the results are in agreement with the Law of Constant Composition. (C.)

3. By means of simple sketches, show the difference in crystalline structure between diamond and graphite. Give *one* physical property of each of these allotropes and explain it in terms of their crystalline structures.

Describe the preparation of carbon dioxide from calcium carbonate. No diagram is required, but you should say how a sample of the gas could be collected and give an equation for the reaction. Give *two* large-scale uses of carbon dioxide. (O. and C.)

4. By means of a labelled diagram and an equation, show how a sample of carbon dioxide can be made and collected in the laboratory. Without giving any details of the apparatus used, describe briefly how you could convert carbon dioxide into pure carbon monoxide.

A journalist in a motoring magazine wrote, 'On a busy roadway, the proportion of carbon monoxide has varied from 6 parts per million to 180 parts per million.' (*a*) At what time of day would you expect the concentration of carbon monoxide to be high? (*b*) By what reaction is the carbon monoxide formed? (*c*) What is the effect of carbon monoxide on blood, and why does this make the gas so poisonous? (J.M.B.)

5. Compare the reactions of carbon dioxide with carbon and with magnesium, mentioning the conditions under which each reaction can be carried out.

When carbon dioxide is bubbled into sodium hydroxide solution, the gas is absorbed but the products depend on the relative quantities of the gas and the alkali. What substances are formed when two solutions each containing 48.0 g sodium hydroxide absorb 26.4 g and 52.8 g of carbon dioxide respectively? Describe *one* test by means of which you could differentiate between the products.

Explain why lime-water is an effective reagent for the detection of small quantities of carbon dioxide, and give one reason why the proportion of carbon dioxide in the earth's atmosphere remains approximately constant. (O. and C.)

6. How would you show that the gas obtained by heating oxalic (ethanedioic) acid with concentrated sulphuric acid contains both carbon dioxide and carbon monoxide? (J.M.B.)

7. Carbon monoxide may be prepared by the action of sulphuric acid on *either* formic (methanoic) *or* oxalic (ethanedioic) acid. Choose *one* of these two methods, give a diagram of a suitable apparatus for carrying out the experiment, and write the equation for the reaction. State: (*a*) what concentration of sulphuric acid is used; (*b*) if it is necessary to warm the mixture; (*c*) the name of the product or products of the reaction other than carbon monoxide; (*d*) how the main impurity in the carbon monoxide can be removed.

Indicate briefly how (i) water gas and (ii) producer gas are manufactured. What are the constituents of these two gases? (A.E.B.)

8. A liquid organic compound of formula H_2CO_2 can be catalytically decomposed in two different ways, as shown in the two following equations:

$$H_2CO_2 \longrightarrow H_2O + CO$$
$$H_2CO_2 \longrightarrow H_2 + CO_2$$

Describe simple experiments you would carry out to test whether each reaction does produce the products indicated in the two equations. Describe one test for each of the four products.

If 23 g of the original compound were decomposed in each of these ways, what would be the volume of gases produced at s.t.p., in each case? (O. and C.)

9. (*a*) Give an account of the chemistry of respiration in mammals.

(*b*) Explain briefly why the proportion of carbon dioxide in the air remains approximately constant.

(*c*) Name *four* products that may be formed when a gaseous fuel of formula C_3H_8 burns in a limited supply of oxygen.

(*d*) What method is normally used in gas burners to ensure complete combustion? (C.)

10. When limestone (calcium carbonate) is strongly heated, it decomposes into calcium oxide and carbon dioxide. The change is reversible. (*a*) Write the equation for the reaction. (*b*) Explain what would happen if some limestone was strongly heated (i) in a closed vessel, (ii) in a current of air. (*c*) Describe the changes that occur when (i) water is slowly added to freshly made calcium oxide, (ii) a mixture of calcium oxide with excess coke is heated in an electric furnace. (*d*) State *one* commercial use for the principal products obtained in (*c*) (i) and (*c*) (ii).

How do you account for the presence of calcium hydrogencarbonate in some natural waters? Explain *one* reason why the presence of this substance may be disadvantageous and mention *one* reason why it may be beneficial. (A.E.B.)

11. (*a*) 'Calcite is pure calcium carbonate.' Is this statement borne out by the following experimental results?

A large crystal of calcite weighing 3.753 g was placed in 25.0 cm^3 of a 2.00M solution

Mass of MCO_3 (grammes)	0.025	0.050	0.100	0.150	0.200	0.300	0.400	0.500	0.600
Volume of CO_2 corrected to s.t.p. (cm³)	4.0	11.0	21.0	33.0	44.5	56.0	56.0	56.0	56.0

of hydrochloric acid and left there until action ceased. After washing and drying out, the crystal was found to weigh 1.253 g.

What volume of carbon dioxide at s.t.p. was formed during the action?

(b) Give the names and formulae of (i) *two* salts which contain water by crystallization, (ii) *two* salts, other than calcium carbonate, which form anhydrous crystals, (iii) two alkalis which contain different cations.

(c) What liquid would you use in each case to remove (i) rusty stains from a laboratory coat, (ii) sulphur from a crucible in which it had been melted? Explain briefly how the liquids selected perform their tasks. (C.)

12. Name the raw materials required for the manufacture of sodium carbonate by the ammonia-soda process, and outline the chemical changes by which sodium carbonate is obtained from them, paying particular attention to those features which make the process efficient. (Technical details and diagrams of the plant are not required.) How, and under what conditions, does sodium carbonate react with (a) ammonium sulphate, (b) slaked lime? (J.M.B.)

13. Azurite is a deep blue mineral which has been given the formula $Cu(OH)_2.2CuCO_3$. If you were presented with a sample of the mineral how would you collect experimental evidence that supported as fully as possible this formula? (N.I.)

14. What raw materials are required in the manufacture of sodium carbonate by the Solvay (ammonia-soda) process? Giving equations and reaction conditions, outline the chemistry of the process, saying how waste is kept to a minimum.

How may sodium carbonate be used to make a sample of (a) sodium hydrogen-carbonate (sodium bicarbonate), (b) glass? In (b) state *one* type of article for which this type of glass would be suitable, giving a reason. (J.M.B.)

15. How would you prepare dry crystals of sodium carbonate starting from sodium hydroxide solution? How, and under what conditions, does sodium carbonate react with (a) carbon dioxide, (b) calcium hydroxide, (c) nitric acid, and (d) lead(II) nitrate? (J.M.B.)

16. A pupil was given a bottle containing a carbonate which was known to have the formula MCO_3. He weighed out several samples of the carbonate and added excess hydrochloric acid solution to each. The volume of carbon dioxide liberated in each experiment was measured in a suitable apparatus and the results are given in the table above.

(a) Plot these points on a graph of volume of CO_2 (*y* axis) against mass of carbonate.

(b) Put on your graph the line that you would expect from these results.

(c) Do you think that the pupil always accurately recorded the mass of carbonate used? Explain your answer.

(d) Determine the molar mass of the carbonate.

(e) Before carrying out the investigations, the pupil added a little calcium carbonate to each portion of hydrochloric acid solution. Why do you think he did this?

(f) Would the carbonate MCO_3 be soluble in water? Give a reason for your answer. (S.)

17. A piece of marble (calcium carbonate) was placed in a beaker containing an excess of dilute hydrochloric acid and standing on a direct reading balance. The mass of the beaker and its contents was recorded every 2 minutes (as shown below).

(a) Why was there a loss of mass?

(b) Write an equation for the reaction.

(c) State *three* different ways in which the reaction could have been made more rapid.

(d) Why did the mass remain constant after 10 minutes?

(e) Write the name and formula of each of *two* ions remaining in the final solution.

(f) The solution was then evaporated to dryness in the same beaker and the mass of the beaker and remaining solid was 97.63 g. Next day, the mass was 98.63 g. Explain

Time (minutes)	0	2	4	6	8	10	12
Mass (g)	126.44	126.31	126.19	126.09	126.03	126.00	126.00

what had occurred to cause this change and name the phenomenon.

(*g*) Finally a little water was added to the contents of the beaker, and the resulting solution was added to solutions of (i) lead(II) nitrate, and (ii) copper(II) sulphate (cupric sulphate).

Describe and explain what would be observed in each case. (J.M.B.)

28 Organic Chemistry

Organic chemistry is the chemistry of the compounds of carbon; that is, all organic compounds contain carbon with, also, one or more other elements – hydrogen, oxygen, chlorine, nitrogen, etc. Carbon shows exceptional behaviour, in a chemical sense, by forming *chains* of its atoms (sometimes of very great length, e.g., about 2000 carbon atoms in polythene) and *rings* of its atoms. Sometimes, the chains and rings are included in the same molecule. Consequently, carbon produces a very great number of compounds, many of them very complex, and it has become necessary to make a separate study of these compounds as *organic chemistry*. Nominally, every compound of carbon is an organic compound. For historical and conventional reasons, however, a few of the simpler carbon compounds, such as carbon dioxide and sodium carbonate, are usually studied with non-carbon compounds in inorganic chemistry. Examples of chain and ring formation by carbon atoms will be found in the present chapter.

Some of the simplest organic compounds are the hydrocarbons, and this is the first class of compounds we shall study.

Hydrocarbons

Hydrocarbons are compounds containing *hydrogen* and *carbon* and no other element. That is, a hydrocarbon has the molecular formula C_xH_y, x and y being whole numbers. For example, *methane* CH_4, *ethene* C_2H_4, and *benzene* C_6H_6, are hydrocarbons.

The hydrocarbons are themselves classified into several types according to their structure. The main classes of hydrocarbon we shall study are the **alkanes,** the **alkenes,** and the **alkynes.** We shall mention only briefly the **arenes,** often called the aromatic hydrocarbons.

The Alkanes

The members of this group of hydrocarbons (also called simply *paraffins*) are distinguished by possessing the general molecular formula C_nH_{2n+2}, where n is $1, 2, 3$, etc., for successive members of the group. This general molecular formula will be justified after the properties of a typical member of the series have been studied. The first member of the series ($n = 1$) is *methane* CH_4, and the second ($n = 2$) is *ethane* C_2H_6. Both are gases at room temperature and pressure. The following table gives the molecular formula and name of the first few alkanes, plus an indication of their physical properties.

Name	Molecular formula	Melting point °C	Boiling point °C	Density g cm^{-3}
Methane	CH_4	-183	-162	gas
Ethane	C_2H_6	-172	-89	gas
Propane	C_3H_8	-188	-42	gas
Butane	C_4H_{10}	-135	-1	gas
Pentane	C_5H_{12}	-130	36	0.626
Hexane	C_6H_{14}	-95	69	0.659
Heptane	C_7H_{16}	-91	98	0.684
Octane	C_8H_{18}	-57	126	0.703
Nonane	C_9H_{20}	-54	151	0.718
Decane	$C_{10}H_{22}$	-30	174	0.730

Homologous series

A series of compounds related to each other as the alkanes are (above) is called a *homologous series*. Such a series has the following characteristics.

(1) All members conform to a general molecular formula, *e.g.*, for alkanes, C_nH_{2n+2}.

(2) Each member differs, in molecular formula, from the next by CH_2, *e.g.*, alkanes are CH_4, C_2H_6, C_3H_8, and so on.

(3) All members show similar chemical reactions, though varying in vigour. For example, all alkanes burn in air and give substitution reactions with chlorine (page 334).

(4) The physical properties of members change gradually in the same direction along the series, *e.g.*, in the alkanes, boiling points and freezing points rise (CH_4 – a gas; C_5H_{12} – a liquid; $C_{20}H_{42}$ – a solid, at ordinary temperature and pressure). Also, densities increase and solubility in water decreases as the number of carbon atoms per molecule increases (see table).

(5) General methods of preparation are known which can be applied to any member of the series.

Other homologous series are *alkene hydrocarbons* C_nH_{2n}, *alcohols* $C_nH_{2n+1}OH$, and *fatty acids* $C_nH_{2n+1}CO_2H$. These series are considered later in this chapter.

Structure of alkanes

In the alkane, all carbon atoms exercise a covalency of 4. The simplest alkane has, therefore, the molecular formula CH_4, and is *methane*. Each covalent bond

P

represents a shared pair of electrons, one each from the carbon atom and a hydrogen atom. In more complex alkanes, carbon atoms form chains by combining together by covalency. With chains of two carbon atoms (*ethane* C_2H_6) and three carbon atoms (*propane* C_3H_8), the molecular structures are shown below, together with *methane*.

$$
\begin{array}{ccc}
& \quad H & \quad\quad H \;\; H & \quad\quad\quad H \;\; H \;\; H \\
& | & \quad | \;\; | & \quad\quad | \;\; | \;\; | \\
H\!-\!\!& C\!-\!H & H\!-\!C\!-\!C\!-\!H & H\!-\!C\!-\!C\!-\!C\!-\!H \\
& | & \quad | \;\; | & \quad\quad | \;\; | \;\; | \\
& H & \quad H \;\; H & \quad\quad H \;\; H \;\; H
\end{array}
$$

methane   ethane   propane

It must be remembered that methane is in fact tetrahedral (see page 87) and so the chains – which consist of methane units joined together – cannot be in a true straight line. It is more convenient for the printer to print these structural formulae as if the molecules were flat; but it is *most important* that you should learn to think in three dimensions. Your teacher should be able to provide you with suitable models so that you can get the feel of the *actual shapes* of all the smaller molecules you meet in this chapter.

Much longer chains are also produced, such as the following (nonane):

$$
H\!-\!\underset{\displaystyle H}{\overset{\displaystyle H}{C}}\!-\!\underset{\displaystyle H}{\overset{\displaystyle H}{C}}\!-\!\underset{\displaystyle H}{\overset{\displaystyle H}{C}}\!-\!\underset{\displaystyle H}{\overset{\displaystyle H}{C}}\!-\!\underset{\displaystyle H}{\overset{\displaystyle H}{C}}\!-\!\underset{\displaystyle H}{\overset{\displaystyle H}{C}}\!-\!\underset{\displaystyle H}{\overset{\displaystyle H}{C}}\!-\!\underset{\displaystyle H}{\overset{\displaystyle H}{C}}\!-\!\underset{\displaystyle H}{\overset{\displaystyle H}{C}}\!-\!H
$$

In such structures, it will be seen that all the carbon atoms except the two at the ends of the chain use two units of valency to attach themselves to the carbon atoms on each side, so forming the carbon chain. Each carbon atom (except the end ones) then has two units of valency left to combine with two hydrogen atoms. That is, except for the end carbon atoms, the relation of carbon to hydrogen is C_nH_{2n}. But each of the end carbon atoms uses only one valency unit in forming the chain and so can combine with an extra hydrogen atom, so producing the general molecular formula C_nH_{2n+2}. Many alkanes have a branched carbon chain, such as the following.

$$
H\!-\!\underset{\displaystyle H}{\overset{\displaystyle H}{C}}\!-\!\underset{\displaystyle H}{\overset{\displaystyle H}{C}}\!-\!\underset{\displaystyle H}{\overset{\displaystyle H}{C}}\!-\!\underset{\displaystyle |}{C}\!-\!\underset{\displaystyle H}{\overset{\displaystyle H}{C}}\!-\!\underset{\displaystyle H}{\overset{\displaystyle H}{C}}\!-\!\underset{\displaystyle H}{\overset{\displaystyle H}{C}}\!-\!H
$$

$$
H\!-\!\underset{\displaystyle |}{\overset{\displaystyle |}{C}}\!-\!H
$$

$$
H\!-\!\underset{\displaystyle H}{\overset{\displaystyle |}{C}}\!-\!H
$$

In this case, a hydrogen atom is lost from the carbon atom on which the branching occurs. But it is restored at the open end of the branch, so the general molecular formula is unchanged.

Petroleum (crude oil) consists largely of a mixture of hydrocarbons from the simplest, an alkane, pentane C_5H_{12}, to complex hydrocarbons, with a total composition which depends on the geographical source of the petroleum. The four simplest alkanes are all gases in ordinary conditions, *i.e.*, CH_4–C_4H_{10}. Butane C_4H_{10} is supplied in pressurized containers for use when camping.

Isomerism

Isomerism occurs often among organic compounds because of their complexity, and the alkanes provide examples of isomerism due to the possibility of branching the carbon chain as seen in the last section.

Isomerism is the occurrence of two or more compounds with the same molecular formula but different molecular structures.

Isomers of the same molecular formula have different physical and chemical properties because of structural differences.

Isomers of molecular formula C_4H_{10}

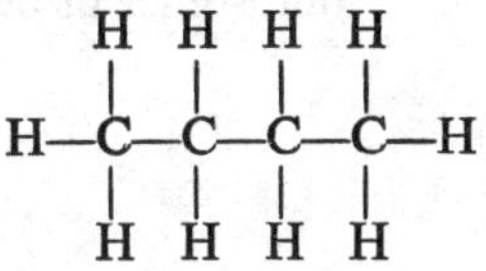

or, $CH_3.CH_2.CH_2.CH_3$
A

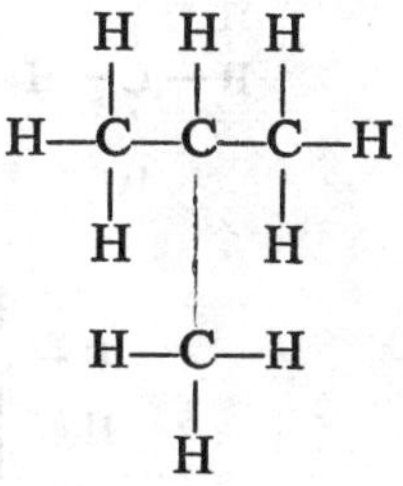

or, $CH_3.CH(CH_3).CH_3$
B

Isomer A is butane (see table on page 329), but by the definition of isomerism, isomer B also has the same molecular formula, and therefore we need to distinguish the two compounds in some way. The convention in naming such isomers is to write out the structural formula of the compound such that the longest possible straight chain of carbons is arranged horizontally. In the case of isomer B we should therefore write

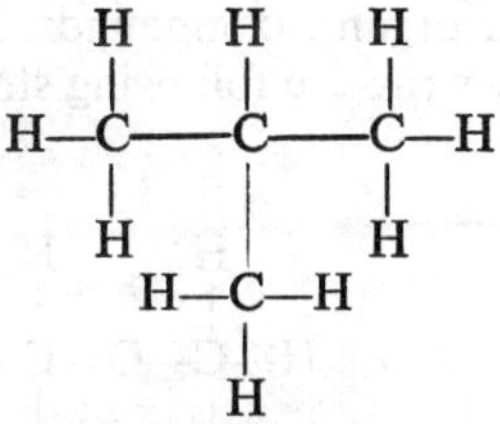

rather than

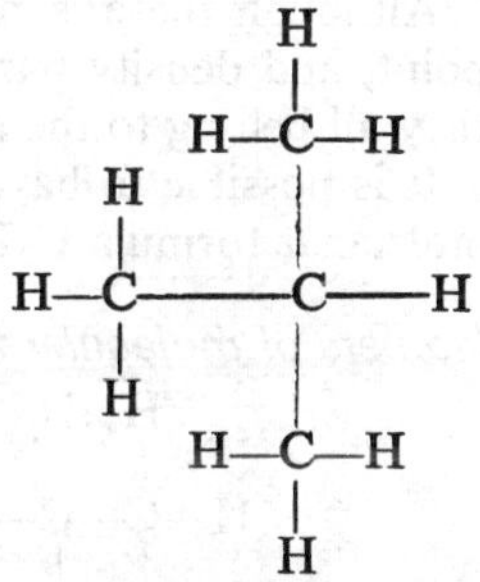

We then name the compound by the name of the longest chain hydrocarbon: thus in the case of isomer B since the largest straight chain of carbon atoms is three, it is a derivative of *propane*. However it is not propane as the two structures below indicate.

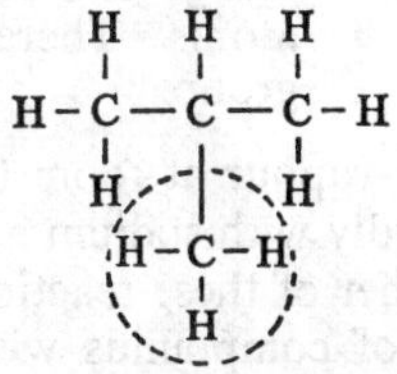

isomer B

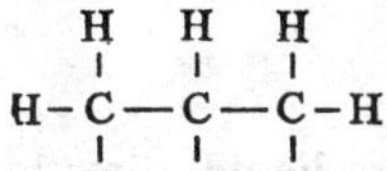

propane

The difference quite clearly is that isomer B has a —CH_3 group (dotted outline) in place of one of the hydrogen atoms on the second carbon atom of propane. The group —CH_3 is known as a **methyl** group, and the name of isomer B would therefore be 2-methylpropane (indicating that it is an alkane with a methyl group on the *second* carbon atom of a propane structure).

In a similar way the following hydrocarbons have the names indicated.

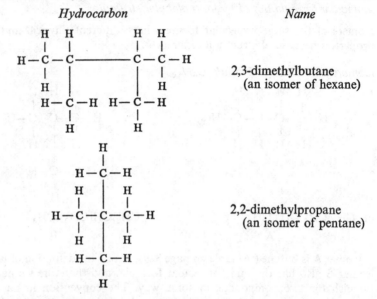

Hydrocarbon	Name
	2,3-dimethylbutane (an isomer of hexane)
	2,2-dimethylpropane (an isomer of pentane)

The name of the groups —CH_3, —C_2H_5, —C_3H_7, etc., are simply derived from the parent alkane, and they are called, respectively, methyl, ethyl, and propyl groups, and are known generally as *alkyl groups*.

Although these isomers differ in physical properties, boiling point, melting point, and density particularly, their chemical reactions are very similar since they all belong to the alkane series.

It is possible to have isomerism in other organic compounds. Hence for the molecular formula C_2H_6O it is possible to write the following structures.

Isomers of molecular formula C_2H_6O

or, $CH_3.CH_2.OH$

ethanol; an alcohol. The oxygen atom is part of a hydroxyl group, OH.

or, $CH_3.O.CH_3$

dimethyl ether (methoxymethane); the oxygen atom forms a bridge between the two carbon atoms. There is no hydroxyl group.

Ethanol is a liquid, dimethyl ether a vapour at room temperature and pressure. As an alcohol, ethanol reacts rapidly with sodium and phosphorus pentachloride (page 344). The ether has neither of these reactions.

Another example is the isomerism of compounds with molecular formula $C_2H_4Cl_2$.

Isomers of molecular formula C₂H₄Cl₂

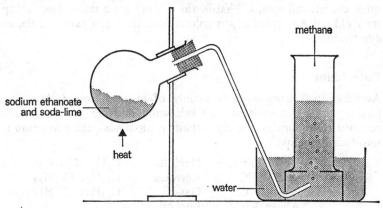

or, CH₂Cl.CH₂Cl

One chlorine atom is combined with each carbon atom.

or, CH₃.CHCl₂

Both chlorine atoms are combined with the same carbon atom.

These compounds differ in boiling point and freezing point. The compound on the left is 1,2-dichloroethane, and that on the right is 1,1-dichloroethane.

Laboratory preparation of methane

Since the method of preparing all alkanes is very similar, we shall only describe that for methane here.

Anhydrous sodium ethanoate (*sodium acetate*) is ground with an equal mass of *soda-lime* (a non-deliquescent form of sodium hydroxide). The mixture is heated in a hard glass flask (Figure 149). Methane is evolved and collected over water.

methane

sodium ethanoate and soda-lime

heat

water

Figure 149

When heating is finished, the delivery tube must be removed from the water *at once*; otherwise, water may be forced back into the hot flask and cause an explosion. Soda-lime acts as a source of sodium hydroxide.

$$CH_3COONa(s) + NaOH(s) \rightarrow Na_2CO_3(s) + CH_4(g)$$

Properties and reactions of methane

Physical properties

Methane is a colourless gas with no smell. It is almost insoluble in water and much less dense than air in the same conditions (vapour density of methane is 8, of air is 14.4).

Chemical reactions

(1) *Combustion*. Methane burns (or explodes) in air on the application of a flame (or electric spark). It produces carbon dioxide and steam.

$$CH_4(g) + 2O_2(g) \rightarrow CO_2(g) + 2H_2O(g); \quad \Delta H = -890 \text{ kJ}$$

This is an *exothermic* reaction, utilized as a means of industrial and domestic heating when methane is burnt as a constituent of town-gas. Similarly, other (liquid) alkanes are used as petrol or fuel oil in exothermic combustion with air. For example, *pentane*, a constituent of petrol, burns in the following way.

$$C_5H_{12}(g) + 8O_2(g) \rightarrow 5CO_2(g) + 6H_2O(g); \quad \Delta H = -3509 \text{ kJ}$$

Methane is the important constituent of *natural gas* which usually accompanies petroleum deposits. Natural gas is non-poisonous and has no smell.

(2) *With chlorine.* Methane reacts (slowly at ordinary temperature) with chlorine, the reaction being catalysed by light (*photocatalysis*). The first product is *monochloromethane* CH_3Cl (or *methyl chloride*, the group CH_3 – being known as *methyl*).

$$CH_4(g) + Cl_2(g) \rightarrow CH_3Cl(g) + HCl(g)$$

In a similar way, excess of chlorine may produce, with increasing difficulty, *dichloromethane* CH_2Cl_2, *trichloromethane* $CHCl_3$, and *tetrachloromethane* CCl_4.

These are called *substitution reactions* because chlorine successively replaces hydrogen in the methane molecule. Notice that the hydrogen is expelled in combination with chlorine as HCl, not as free hydrogen.

The above two reactions are the only ones given by methane or other alkanes in ordinary conditions, *i.e.*, without the use of high temperature and pressure or other exceptional means. That is, the alkanes are a rather inert group; in fact, their old name of **paraffin** was coined from the Latin *parum*, little, and *affinis*, affinity.

Petroleum

American petroleum consists mainly of alkanes from C_5H_{12}, pentane, to $C_{43}H_{88}$, usually associated with salt water and the gaseous paraffins. After removal of impurities (mainly sulphur compounds), the petroleum is distilled into four fractions:

Up to 200 °C	Naphtha	C_5H_{12}–$C_{12}H_{26}$
200–260 °C	Kerosene	$C_{12}H_{26}$–$C_{15}H_{32}$
260–300 °C	Gas-oil	$C_{15}H_{32}$–$C_{17}H_{36}$
Above 300 °C	Fuel oil	

Redistillation of naphtha yields *petrol*, b.p. 50–60 °C, mainly C_5H_{12} to C_9H_{20}. Other products of redistillation of the other fractions are *paraffin*, *lubricating oil*, *petroleum jelly*, and *paraffin wax*. Much of the gas-oil fraction is *cracked* to yield more petrol, *i.e.*, the complex molecules are broken down into simpler units. At about 12 atm pressure and 520 °C, gas-oil yields *petrol* hydrocarbons and valuable alkene by-products, *e.g.*, ethene C_2H_4, and propene C_3H_6. (For fractional distillation, see page 259.)

$$C_{15}H_{32}(g) \rightarrow C_8H_{18}(g) + 2C_2H_4(g) + C_3H_6(g)$$

Petroleum refining occurs usually on an estuary, to avoid land transport of bulky imported oil and to secure large flat areas for its storage, segregated because of fire risks.

The Alkenes

The alkenes are members of a homologous series of general molecular formula C_nH_{2n}. That is, each member of the series has two fewer hydrogen atoms per molecule than the corresponding alkane, *e.g.*, alkane C_3H_8, alkene C_3H_6. The reason for this will be mentioned later. The most important alkene is *ethene* (ethylene) C_2H_4, which is a gas at room temperature and pressure. For a discussion of the structure of ethene, see page 90.

Laboratory preparation of ethene (ethylene)

Ethene is usually prepared by the dehydration of ethanol (ethyl alcohol) by hot, concentrated sulphuric acid. To ethanol (50 cm³), slowly with shaking and cooling under the tap, is added concentrated sulphuric acid (100 cm³). The apparatus is set up as in Figure 150, and the mixture is heated with care to

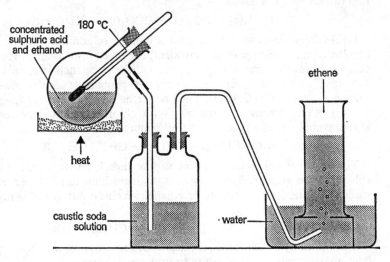

Figure 150

about 180 °C. Ethene is evolved and collected over water. The wash-bottle of alkali solution serves to remove sulphur dioxide, a by-product produced in small amount as the alcohol reduces the sulphuric acid slightly.

The alcohol and acid first produce ethyl hydrogen sulphate.

$$C_2H_5OH(l) + H_2SO_4(aq) \rightarrow C_2H_5HSO_4(aq) + H_2O(l)$$

When heated, ethyl hydrogen sulphate releases ethene.

$$C_2H_5.HSO_4(aq) \rightarrow H_2SO_4(aq) + C_2H_4(g)$$

These two reactions are equivalent to the dehydration of alcohol by the acid.

$$C_2H_5OH - H_2O \rightarrow C_2H_4$$

When the reaction is complete, the junction between the flask and the wash bottle in Figure 150 should be disconnected to avoid the possibility of caustic soda 'sucking back' into the hot concentrated sulphuric acid.

Industrial production of ethene

In industry, ethene is produced by cracking the gas-oil fraction from distillation of petroleum. At 12 atm pressure and 520 °C, alkanes in the petrol range (C_5H_{12}–C_9H_{20}) are produced, together with ethene and, possibly, other alkenes. For example,

$$C_{17}H_{36}(g) \rightarrow C_8H_{18}(g) + 3C_2H_4(g) + C_3H_6(g)$$

Properties and reactions of ethene

Physical properties

Ethene is gaseous at room temperature and pressure, colourless, almost insoluble in water and slightly less dense (vapour density = 14) than air (vapour density = 14.4).

Chemical reactions

(1) *Combustion.* Ethene burns (or explodes) in air if a light (or electric spark) is applied. The products of complete combustion are carbon dioxide and steam, but the flame tends to be smoky from unburnt carbon because of its high proportion (about 86%) in ethene.

$$C_2H_4(g) + 3O_2(g) \rightarrow 2CO_2(g) + 2H_2O(g)$$

(2) *Addition reactions.* Ethene gives a number of addition reactions in which two hydrogen atoms (or their equivalent) are taken into combination per molecule of ethene to form a single product. Ethene is, therefore, said to be *unsaturated.* Structurally, this means that it contains a *double bond* between carbon atoms, $H_2C{=}CH_2$. This will be further considered later (page 337).

(a) *With chlorine or bromine.* With chlorine at ordinary temperature, ethene combines rapidly to give an oily liquid, 1,2-dichloroethane.

$$H_2C{=}CH_2(g) + Cl_2(g) \rightarrow CH_2Cl.CH_2Cl(l)$$

The similar reaction with bromine vapour rapidly destroys the reddish-brown colour of the vapour. This acts as a distinguishing test between ethene and gaseous alkanes, *e.g.*, methane or ethane, which do not give this rapid colour change.

$$H_2C{=}CH_2(g) + Br_2(g) \rightarrow CH_2Br.CH_2Br(l)$$

(b) *With hydrogen iodide.* Ethene combines rapidly with hydrogen iodide (vapour) at ordinary temperature to produce *iodoethane*.

$$H_2C{=}CH_2(g) + HI(g) \rightarrow CH_3.CH_2I(l) \ (or C_2H_5I)$$

The group $C_2H_5{-}$ is called the *ethyl* group. The gases HBr and HCl combine similarly but more slowly.

(c) *With concentrated sulphuric acid.* Ethene is absorbed rapidly by this acid at room temperature to form *ethyl hydrogen sulphate*.

$$H_2C{=}CH_2(g) + H_2SO_4(aq) \rightarrow CH_3.CH_2HSO_4(aq) \ (or \ C_2H_5.HSO_4)$$

The action is reversed at about 180 °C, liberating ethene.

(d) *With hydrogen.* Ethene combines with hydrogen if the two are passed over finely divided *nickel* (catalyst) at about 200 °C. The product is the alkane, *ethane.*

$$H_2C{=}CH_2(g) + H_2(g) \rightarrow CH_3.CH_3(g) \ (or \ C_2H_6)$$

This (Sabatier's) reaction is important as representing the essential change in the conversion of oils into margarine. An oil possesses a long carbon chain (15–17 atoms) in which there is one ethylenic double bond, $-CH{=}CH-$. Pressure is used (5 atm) and a somewhat lower temperature (180 °C) and the oil is hydrogenated in the presence of nickel. This saturates the double bond, so converting the oil into a fat, which is then sold as margarine.

$$-CH{=}CH- + H_2 \rightarrow -CH_2-CH_2-$$

(e) *With potassium permanganate.* Ethene reacts rapidly if shaken with potassium permanganate solution at ordinary temperature. If acidic, the solution is decolorized, being reduced to a manganese salt; if alkaline, it turns green, being reduced to a manganate. In both cases, the ethene is converted to *ethane-1,2-diol* (*ethylene glycol*) by water and oxygen from the permanganate.

$$H_2C{=}CH_2 + H_2O + (O) \rightarrow C_2H_4(OH)_2$$

This reaction distinguishes ethene from all gaseous alkanes, *e.g.*, methane or ethane, which have no reaction with potassium permanganate. Ethane-1,2-diol is the material used in anti-freeze solutions for motor-car radiators and for the production of Terylene (page 348).

Polyethene (Polythene)

If ethene, with a trace of oxygen present, is pressurized to about 100 atmospheres and heated to start the action, it is polymerized to form polyethene. The polyethene produced by this method is a low-density material which softens at around 120 °C. If ethene is passed under pressure into an inert solvent containing a special catalyst (a Ziegler catalyst named after the inventor of the process) a high-density form of polyethene with a softening temperature of 130 °C is formed.

Polymerization is the combination of two or more molecules of the same compound to form one complex molecule with no gain or loss of material. The original material is called the *monomer*.

$$nA \rightarrow A_n$$

The product, A_n, is called a polymer of the original compound, A. In the case of ethene the polymerization is quite exothermic and, once the reaction has started, cooling is required.

Polyethene is very resistant to the common types of chemical action and can be moulded (while hot) into a great variety of domestic and scientific articles – buckets, bowls, bags, flexible containers, funnels, wash-bottles. The polymerization can be stated as:

$$3n(CH_2{=}CH_2) \rightarrow (-CH_2-CH_2-CH_2-CH_2-CH_2-CH_2-)_n$$

where n is about 300. From this, polyethene resembles a highly complex alkane and shares the chemical inertness of this group.

Polyethene is an *addition* polymer. There is only one monomer molecule, and only one product.

Unsaturation in ethene

Ethene is said to be unsaturated because it contains a double bond in its structure which means that one molecule of it can combine with two hydrogen atoms (or their equivalent) in addition reactions. In general,

$$H_2C{=}CH_2 + A.B \rightarrow AH_2C-CH_2B$$

where A and B are univalent atoms or groups. Several examples of these reactions were given in the recent sections of this chapter. The C to C double bond which represents this unsaturation indicates that, between the carbon atoms concerned, two pairs of electrons are shared in covalency.

When the addition reaction has taken place, *e.g.*, with Cl_2, only two electrons are shared between the carbon atoms; only single covalent bonds are present in the molecule, the compound is said to be *saturated* and is an alkane derivative. It gives no addition reactions but reacts by substitution. The addition of chlorine to ethene is shown by the following equation.

$$\underset{\underset{\textstyle H}{|}}{\overset{\overset{\textstyle H}{|}}{C}}=\underset{\underset{\textstyle H}{|}}{\overset{\overset{\textstyle H}{|}}{C}} + Cl_2 \rightarrow Cl-\underset{\underset{\textstyle H}{|}}{\overset{\overset{\textstyle H}{|}}{C}}-\underset{\underset{\textstyle H}{|}}{\overset{\overset{\textstyle H}{|}}{C}}-Cl$$

The Alkynes

Ethyne (acetylene) is really the first member of a homologous series of hydrocarbons called the alkynes. The general molecular formula of the series is C_nH_{2n-2}. For ethyne itself, $n = 2$, and the molecular formula is C_2H_2. So far, ethyne is the only member of the series which has acquired general importance. Alkynes are characterized by possessing a carbon-to-carbon triple bond at one point in the carbon chain. The triple bond involves the loss of *four* hydrogen atoms per molecule in an alkyne hydrocarbon when compared with the corresponding alkane.

$$H-C\equiv C-H \qquad\qquad H-\underset{\underset{\textstyle H}{|}}{\overset{\overset{\textstyle H}{|}}{C}}-\underset{\underset{\textstyle H}{|}}{\overset{\overset{\textstyle H}{|}}{C}}-H$$

  ethyne   ethane

The triple bond involves the sharing of *three* pairs of electrons in covalency between the two carbon atoms.

Preparation of ethyne

For apparatus, see Figure 151. Cold *water* is dripped on to *calcium carbide*.

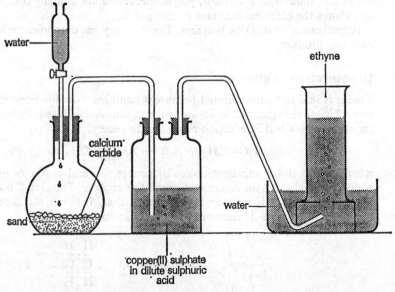

water

ethyne

calcium carbide

sand

water

copper(II) sulphate in dilute sulphuric acid

Figure 151

Much heat is evolved and sand protects the flask from breakage. The chief impurity, phosphine PH_3, is absorbed by the acidified copper(II) sulphate solution. Ethyne is collected over water.

$$CaC_2(s) + 2H_2O(l) \rightarrow Ca(OH)_2(aq) + C_2H_2(g)$$

Properties of ethyne

Physical

In ordinary conditions, ethyne is a colourless gas, almost insoluble in water, and having a sweet smell when pure. It is slightly less dense than air (vapour density is 13; vapour density of air is 14.4). Ethyne is strongly endothermic.

$$2C(s) + H_2(g) \rightarrow C_2H_2(g), \quad \Delta H = +227 \text{ kJ}$$

Like endothermic compounds in general, it is rather unstable and liable to explode if stored under compression alone. For industrial purposes, it is stored in steel cylinders in solution in acetone at about 12 atm pressure.

Chemical reactions

Having a carbon-to-carbon triple bond in its structure, ethyne is an *un-saturated* compound. It gives a number of **addition reactions**, combining with *four* hydrogen atoms per molecule (or their equivalent) as a maximum, then producing the saturated alkane, *ethane*, or one of its derivatives. These reactions are of the type:

$$\begin{matrix} CH \\ ||| \\ CH \end{matrix} + 2AB \rightarrow \begin{matrix} CHA_2 \\ | \\ CHB_2 \end{matrix} \quad \text{where A and B are univalent}$$

Notice that the two A groups usually combine with the same carbon atom, as do the two B groups. Compounds of the type CHAB.CHAB are rarely produced.

(a) *With bromine.* At ordinary temperature, ethyne combines rapidly with bromine, forming *tetrabromoethane*.

$$\begin{matrix} CH \\ ||| \\ CH \end{matrix} (g) + 2Br_2(l) \rightarrow \begin{matrix} CHBr_2 \\ | \\ CHBr_2 \end{matrix} (l)$$

Chlorine gives a corresponding reaction provided it is diluted with an inert gas, but if pure ethyne and chlorine are mixed, a violent explosion occurs, with formation of carbon and hydrogen chloride.

$$C_2H_2(g) + Cl_2(g) \rightarrow 2C(s) + 2HCl(g)$$

(b) *With hydrogen.* If passed with twice its own volume of hydrogen over *nickel* (catalyst) at about 200 °C, ethyne forms *ethane*.

$$C_2H_2(g) + 2H_2(g) \rightarrow C_2H_6(g)$$

(c) *With halogen acids (as gases).* Ethyne combines readily with hydrogen iodide at room temperature to form 1,1-*diiodoethane*.

$$\begin{matrix} CH \\ ||| \\ CH \end{matrix} (g) + 2HI(g) \rightarrow \begin{matrix} CHI_2 \\ | \\ CH_3 \end{matrix} (l)$$

A corresponding reaction is given by hydrogen bromide at 100 °C, but the reaction with hydrogen chloride is very slow.

(d) *With water.* Ethyne reacts additively with water if passed into dilute sulphuric acid at about 96 °C with *mercury(II) sulphate* present as a catalyst. The product is *ethanal (acetaldehyde)*.

$$C_2H_2(g) + H_2O(l) \rightarrow CH_3.CHO(l)$$

This is the first stage of the manufacture of ethanoic acid (acetic acid) from ethyne (see page 345).

(*e*) *With acidified potassium permanganate solution.* At room temperature, with shaking, ethyne quickly decolorizes this solution (*i.e.*, reduces it) with formation of *oxalic acid (ethanedioic acid)*.

$$8MnO_4^-(aq) + 5C_2H_2(g) + 24H^+(aq) \rightarrow$$
$$5H_2C_2O_4(aq) + 8Mn^{2+}(aq) + 12H_2O(l)$$

Notice that ethyne decolorizes bromine vapour (see (*a*) above) and decolorizes acidified potassium permanganate solution (see (*e*) above). These reactions distinguish it from alkanes such as methane and ethane which do not give these changes; they do not distinguish ethyne from ethene which also produces the same colour changes (see page 336).

Metallic derivatives of ethyne

Ethyne produces several metallic derivatives, copper(I) acetylide and silver acetylide being the most important.

Copper(I) acetylide. If ethyne is passed at room temperature into a solution of copper(I) chloride in ammonia, a reddish-brown precipitate of copper(I) acetylide is at once formed.

$$C_2H_2(g) + 2CuCl(aq) + NH_3(aq) \rightarrow Cu_2C_2(s) + 2NH_4Cl(aq)$$

Silver acetylide. If ethyne is passed at room temperature into a solution of silver oxide in ammonia, a yellow precipitate is produced, turning rapidly grey. It is *silver acetylide*.

$$C_2H_2(g) + Ag_2O(aq) \rightarrow Ag_2C_2(s) + H_2O(l)$$

If allowed to dry and heated, both acetylides are explosive, the silver salt much more violently. Both give off ethyne if warmed with dilute acid.

$$Ag_2C_2(s) + H_2SO_4(aq) \rightarrow Ag_2SO_4(s) + C_2H_2(g)$$

The two reactions above distinguish ethyne from ethene, which forms no metallic derivatives.

Polymerization. Ethyne polymerizes into the cyclic hydrocarbon, *benzene*, if passed through a tube at red heat.

$$3C_2H_2(g) \rightarrow C_6H_6(g)$$

Combustion. Ethyne burns in air if ignited by a flame or electric spark and mixtures of the two can explode very violently. The flame is usually very sooty; ethyne contains over 90% of carbon and much of it remains unburnt unless special means are used to provide a good air supply.

$$2C_2H_2(g) + 5O_2(g) \rightarrow 4CO_2(g) + 2H_2O(g)$$

A flame of ethyne burning in oxygen (the oxyacetylene flame) is very hot and can give a temperature of 2200 °C. It is used in welding and in cutting up steel scrap.

Manufacturing uses of ethyne

Ethyne has recently become very important as a source of organic chemicals on the large scale, though some of the chemicals made from it are very complex. The manufacture of ethanal and hence ethanoic acid from ethyne is given on page 345. Ethanal is also dimerized to ethyl ethanoate using aluminium ethoxide as catalyst at 0 °C.

$$2CH_3.CHO(l) \rightarrow CH_3.COOC_2H_5(l)$$

Ethyne can also be converted to *ethanoic anhydride*, which, together with ethanoic acid (also an ethyne product), converts cellulose to cellulose ethanoate which is used as *celanese* rayon and as *cellophane*. Ethyne is also the starting material for making *polyvinyl chloride*, which is used in electrical insulation and waterproofing, and *polyvinyl ethanoate*, which is used in sheets as the middle layer of Triplex safety glass. The uses of the oxyacetylene flame have been mentioned above.

Ethyne is important as a source of *polyvinyl chloride* (PVC) plastic. Hydrogen chloride reacts additively with ethyne to give *vinyl chloride* (chloroethene).

$$HC{\equiv}CH(g) + HCl(g) \rightarrow H_2C{=}CHCl(g)$$
(the group $H_2C{=}CH{-}$ is called *vinyl*)

Vinyl chloride can be polymerized (in aqueous emulsion and initiated by peroxides) to PVC, which consists of long molecular chains of the type:

... $-CH_2-CHCl-CH_2-CHCl-CH_2-CHCl-CH_2-CHCl-CH_2-$
... $-CHCl-CH_2-CHCl-CH_2-CHCl-CH_2-CHCl-$ ...

(Compare the polymerization of ethene to polyethene, page 337.) PVC is a thermoplastic much used in electrical insulation and waterproofing. Notice that the double bonds (unsaturation) are lost in polymerization and PVC is a saturated compound.

Cyclic hydrocarbons

In addition to producing open chains of atoms, as in the alkanes, carbon can produce closed rings of atoms. Compounds possessing such carbon rings are said to be *cyclic compounds*. The commonest rings are those composed of five or six carbon atoms. This is so because the four valencies of the carbon atom are directional and distributed symmetrically in three-dimensional space (Figure 152). This gives a natural angle of about 109° between any pair of valency

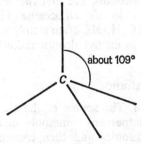

Figure 152

directions. The angles in a regular five-membered ring are 108° and in a regular six-membered ring are 120°. That is, the natural angle between the carbon valency directions needs very little strain to form 5- or 6-membered rings.

The most important is the 6-membered ring of *benzene* C_6H_6 and its derivatives. This is usually represented with alternate single and double carbon-to-carbon valency bonds (Kekulé's formula), though this representation is known to be unsatisfactory in some respects, and a circle is increasingly used. It is usual to omit the symbol for the carbon atoms in the ring, and for the carbon

atoms joined directly to them. Derivatives of benzene are very important chemicals, *e.g.*, *aniline* (phenylamine) $C_6H_5NH_2$, and *phenol* (benzenol) C_6H_5OH, both of which can be obtained from coal-tar as source. More complex benzene derivatives include the M. and B. drugs, aspirin, and a wide range of dyestuffs. Cyclic compounds may also contain straight carbon chains, too, which are known as *side-chains* in this case. An example is the following.

It is obvious that such combinations may become very complex. This is why carbon appears to be the only element which can supply the very varied and complex compounds needed by living organisms to carry on their vital processes.

The Alcohols

The alcohols form a homologous series of general molecular formula $C_nH_{2n+1}OH$ (or $C_nH_{2n+2}O$). The first of these formulae is usually preferred because it shows the presence of the hydroxyl group, OH, which is the characteristic group of the alcohols. The first two members of the series are *methanol* CH_3OH, also known as *wood spirit* because of its early production by distillation of wood, and *ethanol* C_2H_5OH, often simply known as *alcohol*. The two are also known (respectively) as methyl alcohol and ethyl alcohol.

Manufacture of ethanol

(1) *By fermentation.* The source of the alcohol is some form of starch; in western Europe, potatoes are commonly used. They are pressure-cooked to release the starch granules and then treated with malt (partially sprouted barley) for an hour at 60 °C. Malt supplies an enzyme, *i.e.*, an organic catalyst, called *diastase*. In these conditions, the starch is hydrolysed to a sugar, *maltose*. The empirical formula of starch is $C_6H_{10}O_5$.

$$2C_6H_{10}O_5(aq) + H_2O(l) \rightarrow C_{12}H_{22}O_{11}(aq) \text{ (maltose)}$$

At room temperature, yeast is added. One of its enzymes, *maltase*, catalyses the hydrolysis of maltose to *glucose*.

$$C_{12}H_{22}O_{11}(aq) + H_2O(l) \rightarrow 2C_6H_{12}O_6(aq)$$

Another enzyme of yeast, *zymase*, catalyses the decomposition of glucose to *ethanol* and carbon dioxide.

$$C_6H_{12}O_6(aq) \rightarrow 2C_2H_5OH(aq) + 2CO_2(g)$$

The 'wash', containing less than 11% of ethanol, is converted by distillation in a very efficient still to a liquid containing about 95% of ethanol and this is the purest form (surgical spirit) in which the alcohol is usually sold. The material known as *methylated spirit* contains about 85% of ethanol (with water) and has pyridine and colouring matter added to discourage its being drunk.

Fermentation of starch (or the sugar of grapes) produces many alcoholic beverages, *e.g.*, varieties of beer and wines with 8–20% of alcohol; fermented and distilled products are known as *spirits* – whisky, gin, brandy, rum – and contain about 30% of alcohol.

The effect of alcohol on human capabilities is, of course, attracting much contemporary attention, especially in its relation to motoring. Alcohol tends to impair skill and cloud judgement, though the extent of these effects depends on the individual and his alcoholic tolerance. Quantities of alcohol lead to lack of muscular control, *e.g.*, the drunken stagger, and, ultimately, to coma, the state of dead-drunk. A sufficiently large alcoholic consumption taken over a long period causes liver deterioration and accelerates death. Methanol is more poisonous still: even moderate amounts may affect the optic nerve and produce blindness for several days, while greater amounts may produce permanent blindness or prove fatal.

The breaking down of a complex material such as starch into much simpler substances, *e.g.*, by fermentation into alcohol and carbon dioxide, is known as *degradation* and is a common way of utilizing natural products. The distillation of coal in air-free retorts to yield coal-gas, tar, coke, and ammonia liquor is another example of degradation.

(2) *Manufacture of alcohol from petroleum products.* The cracking of petroleum oils to produce ethene was described on page 335. If ethene is absorbed at 80 °C under slight pressure in concentrated sulphuric acid, ethyl hydrogen-sulphate is formed. If this product is diluted with water and distilled, ethanol is obtained.

$$C_2H_4(g) + H_2SO_4(aq) \rightarrow C_2H_5.HSO_4(aq)$$
$$C_2H_5.HSO_4(aq) + H_2O(l) \rightarrow C_2H_5OH(l) + H_2SO_4(aq)$$

In a newer process, ethene and steam are passed over *phosphoric acid* (catalyst) at 600 °C and 60 atm pressure. The greater part of the world's industrial ethanol is now produced by these two processes. The alcohol is used as a solvent in stains and polishes and in the manufacture of essences, perfumes, and drugs. Its use in beverages and methylated spirit was mentioned above.

Properties of alcohols

Physical

The simpler alcohols, *e.g.*, methanol and ethanol, are liquids at room temperature and pressure, volatile, colourless, and possessed of a characteristic smell. They mix with water in all proportions. Some alcohols are known with a carbon chain of about 20 atoms. These are solids resembling paraffin wax in appearance.

Chemical reactions

(1) *Reactions of the hydroxyl group.* All alcohols contain the hydroxyl group, OH, and show corresponding reactions.

(a) *With sodium.* If a small piece of sodium is added to 1–2 cm^3 of ethanol at room temperature, effervescence will occur with liberation of *hydrogen* (explosion on the application of a flame). *Sodium ethoxide* will eventually remain as a white, deliquescent solid. Methanol behaves similarly.

$$2C_2H_5OH(l) + 2Na(s) \rightarrow H_2(g) + 2C_2H_5ONa(s) \text{ (sodium ethoxide)}$$
$$2CH_3OH(l) + 2Na(s) \rightarrow H_2(g) + 2CH_3ONa(s) \text{ (sodium methoxide)}$$

(b) *With a chloride of phosphorus.* If a little phosphorus(V) chloride is added to 1–2 cm^3 of ethanol at room temperature, a vigorous reaction occurs with the liberation of 'steamy' fumes of hydrogen chloride. The organic product is *chloroethane*, which escapes as vapour. This is the recognized test for the hydroxyl group, OH, in a straight-chain organic compound. Its effect is to replace the OH group by Cl.

$$C_2H_5OH(l) + PCl_5(l) \rightarrow C_2H_5Cl(g) + POCl_3(l) + HCl(g)$$

Methanol CH$_3$OH gives a similar reaction to produce *chloromethane* CH$_3$Cl, which is also a vapour.

Phosphorus(III) chloride gives a similar reaction, but usually less vigorously.

$$3C_2H_5OH(l) + 2PCl_3(l) \rightarrow 3C_2H_5Cl(g) + P_2O_3(s) + 3HCl(g)$$

All members of the alcohol homologous series give the reactions (a) and (b) above, though with decreasing vigour as the number of carbon atoms increases.

(2) *Reaction with sulphuric acid.* Ethanol reacts with sulphuric acid in two distinct ways. If the acid is in considerable excess and the temperature is about 180 °C, the product is the gaseous alkene *ethene* (page 335).

$$C_2H_5OH(l) + H_2SO_4(aq) \rightarrow C_2H_5.HSO_4(aq) + H_2O(l)$$
$$C_2H_5.HSO_4(aq) \rightarrow C_2H_4(g) + H_2SO_4(aq)$$

With a smaller proportion of the concentrated acid and a lower temperature (145 °C), the chief product is *ether* (diethyl ether, ethoxyethane). This is a very volatile, sweet-smelling liquid, formerly much used as an anaesthetic.

$$C_2H_5OH(l) + H_2SO_4(aq) \rightarrow C_2H_5.HSO_4(aq) + H_2O(l)$$

Then, with excess alcohol available,

$$C_2H_5.HSO_4(aq) + HOC_2H_5(l) \rightarrow C_2H_5.O.C_2H_5(g) + H_2SO_4(aq)$$

These are both *dehydrating actions* of sulphuric acid. The severer conditions (excess of the acid and a higher temperature) extracted H$_2$O from one molecule of the alcohol.

$$C_2H_5OH - H_2O \rightarrow C_2H_4$$

Milder conditions (relatively less acid and a lower temperature) extract H$_2$O between two molecules of the alcohol.

$$2C_2H_5OH - H_2O \rightarrow C_2H_5.O.C_2H_5$$

(3) *Oxidation.* Ethanol can be oxidized in two stages. If 1–2 cm^3 of ethanol are heated with potassium dichromate solution and dilute sulphuric acid, the liquid turns green (by reduction of the dichromate) and *ethanal* is given off as a very acrid vapour.

$$CH_3.CH_2OH(aq) \rightarrow CH_3.CHO(g) + 2H^+(aq) + 2e^-$$

Ethanal can be further oxidized by more heated potassium dichromate in dilute sulphuric acid; product is ethanoic acid (acetic acid), the acid of vinegar.

$$CH_3.CHO(g) + H_2O(l) \rightarrow CH_3.COOH(aq) + 2H^+(aq) + 2e^-$$

No further oxidation is possible except by combustion of ethanoic acid vapour to carbon dioxide and water. Because of its oxidation product (acetic acid), ethanal is also known as *acetaldehyde*.

Both these stages of oxidation can also be brought about by passing the alcohol or aldehyde, as vapour, with air over a heated catalyst. For the first stage, a copper coil is a suitable catalyst.

$$CH_3.CH_2OH(g) + \tfrac{1}{2}O_2(g) \rightarrow CH_3.CHO(g) + H_2O(g)$$

For the second stage, manganese(II) ethanoate is catalytic.

$$CH_3.CHO(g) + \tfrac{1}{2}O_2(g) \rightarrow CH_3.COOH(l)$$

(4) *Ester formation.* See Experiment 108.

The Carboxylic Acids

Ethanoic acid is the second member of the homologous series of *fatty acids*, so called because some members of the series, *e.g.*, stearic acid $C_{17}H_{35}.COOH$, can be prepared from fats which are their esters. The general molecular formula of the acids is $C_nH_{2n+1}.COOH$, or $C_nH_{2n}O_2$. The first formula is usually preferred because it expresses the presence of the *carboxyl* group, COOH, which is the characteristic group of the series. The first two members of the series are *methanoic acid* (commonly known as *formic acid*) H.COOH or CH_2O_2, and *ethanoic acid* (commonly known as *acetic acid*) $CH_3.COOH$ or $C_2H_4O_2$. Notice that, as in all homologous series, successive members have molecular formulae which differ by CH_2.

Laboratory preparation of ethanoic acid (acetic acid)

Ethanoic acid is the final organic oxidation product of ethanol. The oxidation occurs in two stages – the first to *ethanal*, the second to *ethanoic acid*. Both stages can be carried out in the laboratory by heating with chromic acid (*i.e.*, potassium dichromate in dilute sulphuric acid).

$$CH_3.CH_2OH(aq) \rightarrow CH_3.CHO(aq) + 2H^+(aq) + 2e^-$$
$$CH_3.CHO(aq) + H_2O(l) \rightarrow CH_3.COOH(aq) + 2H^+(aq) + 2e^-$$

In practice, this method gives a dilute solution of the acid from which a pure product is not easily obtained. An easier preparation is the distillation of anhydrous *sodium ethanoate* with *concentrated sulphuric acid*. Hydrochloric acid is *not* suitable because it is volatile and would distil with the ethanoic acid.

$$CH_3.COONa(s) + H_2SO_4(aq) \rightarrow NaHSO_4(aq) + CH_3.COOH(g)$$

The apparatus must be all-glass because hot ethanoic acid vapour attacks cork or rubber stoppers. A retort is suitable (see nitric acid, page 436) or a modern glass-jointed distillation apparatus. Ethanoic acid distils as a colourless liquid.

Large-scale production of ethanoic acid (acetic acid)

Pure ethanoic acid is now made on the large scale from *ethyne* C_2H_2, in two stages. In the first stage, ethyne is passed into dilute sulphuric acid at about 96 °C, with *mercury(II) sulphate* present as *catalyst*.

$$C_2H_2(g) + H_2O(l) \rightarrow CH_3.CHO(g)$$
$$\text{ethanal}$$

The ethanal vapour produced is then passed over a heated *catalyst, manganese(II) ethanoate*, with air as oxidizing agent.

$$CH_3.CHO(g) + \tfrac{1}{2}O_2(g) \rightarrow CH_3.COOH(g)$$

Ethanoic acid is also manufactured in the form of *vinegar*, which contains about 4% of the acid. A dilute aqueous solution of ethanol, *e.g.*, a wine of poor quality, is run slowly over coke in a well-aerated vat. The coke is smeared with a white solid material containing the bacterium, *Mycoderma aceti*. This acts as a catalyst to oxidation of ethanol by oxygen of the air. The liquid may have to pass through the vat several times to secure full oxidation.

$$CH_3.CH_2OH(aq) + O_2(g) \rightarrow CH_3.COOH(aq) + H_2O(l)$$

Properties and reactions of ethanoic acid

Physical properties

Ethanoic acid is usually a colourless liquid with a strong, very irritating smell. It mixes with water in all proportions and a dilute solution has the usual sour taste of an acid. The melting point of the acid (17 °C) is near room temperature. In winter in cold countries the pure acid may freeze to crystals resembling ice; hence the name *glacial ethanoic acid* which is usually used for the pure material.

Chemical reactions

(1) *Acidic behaviour.* Ethanoic acid is a *weak* acid, that is, it is slightly ionized in dilute solution. For example, at 0.1M, or 6 g dm^{-3}, the acid has a degree of dissociation of 1.4%; at M, it is 0.4%.

$$CH_3.COOH(aq) \rightleftharpoons CH_3.COO^-(aq) + H^+(aq)$$

The presence of H^+ gives the solution the usual acidic behaviour, as below.

It turns blue litmus to red.

It forms a salt with an alkali or base.

$$CH_3.COO^-(aq) + H^+(aq) + Na^+(aq) + OH^-(aq) \rightarrow$$
$$CH_3.COO^-Na^+(aq) + H_2O(l)$$

It liberates carbon dioxide from a carbonate or hydrogencarbonate.

$$CO_3{}^{2-}(s) + 2H^+(aq) \rightarrow H_2O(l) + CO_2(g)$$
$$HCO_3{}^-(s) + H^+(aq) \rightarrow H_2O(l) + CO_2(g)$$

It liberates hydrogen on contact with a strongly electropositive metal, *e.g.*, Mg.

$$Mg(s) + 2H^+(aq) \rightarrow Mg^{2+}(aq) + H_2(g)$$

All compounds containing the *carboxyl group*, COOH, show these weakly acidic reactions in aqueous solution. This includes all the fatty acids and also compounds which do not belong to this series, *e.g.*, oxalic acid and tartaric acid. These are both *dicarboxylic* acids, containing two carboxyl groups per molecule.

Oxalic (Ethanedioic) acid	Tartaric acid
$H_2C_2O_4$ or $(COOH)_2$	$C_2H_4O_2(COOH)_2$

Both acids can neutralize alkali in two stages, giving an acid salt and a normal salt. For oxalic acid with sodium hydroxide solution, the stages are:

$$H_2C_2O_4(aq) + NaOH(aq) \rightarrow NaHC_2O_4(aq) + H_2O(l)$$
$$H_2C_2O_4(aq) + 2NaOH(aq) \rightarrow Na_2C_2O_4(aq) + 2H_2O(l)$$

The other acidic properties given for acetic acid are also shown by these acids, but the liberation of hydrogen with metal may be negligible for very weak organic acids.

(2) *Ester formation*. A mixture of glacial ethanoic acid and moderate excess of a simple alcohol, *e.g.*, CH_3OH or C_2H_5OH, boiled under reflux with a little concentrated sulphuric acid or dry hydrogen chloride present as catalyst, produces an *ester*. For example, with ethanol:

$$CH_3.COOH(l) + C_2H_5OH(l) \rightleftharpoons CH_3.COOC_2H_5(l) + H_2O(l)$$

For more details, see Experiment 108, page 348.

(3) *With chlorides of phosphorus*. Glacial ethanoic acid reacts rapidly in the cold with phosphorus(V) chloride. 'Steamy' fumes of hydrogen chloride are given off and the organic product is *ethanoyl chloride*, a colourless liquid which fumes in air.

$$CH_3.COOH(l) + PCl_5(l) \rightarrow CH_3.COCl(l) + HCl(g) + POCl_3(l)$$

This reaction shows the presence of the hydroxyl group, OH, in the acid; it is part of the carboxyl group, COOH.

Phosphorus(III) chloride reacts similarly but more slowly: slight heating is required.

$$3CH_3.COOH(l) + 2PCl_3(l) \rightarrow 3CH_3.COCl(l) + P_2O_3(s) + 3HCl(g)$$

Notice that the chlorides of phosphorus attack the COOH group of the acid but leave the CH_3 group unchanged.

(4) *With chlorine*. If glacial ethanoic acid is boiled under reflux while chlorine is passed in, *monochloroethanoic acid* is produced. Light catalyses this reaction (photocatalysis).

$$Cl_2(g) + CH_3.COOH(l) \rightarrow CH_2Cl.COOH(l) + HCl(g)$$

Further passage of chlorine will give, in succession, *di-* and *trichloroethanoic acid* $CHCl_2.COOH$ and $CCl_3.COOH$ respectively. Notice that chlorine attacks only the CH_3 group, leaving the COOH group intact.

If monochloroethanoic acid is allowed to react with aqueous ammonia, it is converted to *aminoethanoic acid*, also called *glycine*, $CH_2NH_2.COOH$. NH_2 is called the *amino* group. Excess of ammonia would produce the ammonium salt of glycine, $CH_2NH_2.COONH_4$. Glycine and other, more complex, aminoacids are very important biological materials. For example, a protein, boiled with dilute acid, *e.g.*, H_2SO_4, is hydrolysed into aminoacids, such as glycine and *alanine* $CH_3.CH(NH_2).COOH$. This occurs because the protein contains *peptide* (or *amide*) linkages,—CO.NH—, and they break down in hydrolysis by using the elements of a molecule of water.

$$-CO.NH- + H_2O \rightarrow -COOH + NH_2-$$

The COOH and NH_2 groups are then constituents of different aminoacids. Quite complex mixtures of aminoacids and other organic mixtures of compounds can now be investigated very efficiently by the technique of *chromatography*, about which something will be written in the last section of this chapter.

Complex compounds closely resembling natural proteins can be built up in the laboratory by condensing aminoacids together, *i.e.*, causing them to combine by eliminating water molecules between the NH_2 group of one aminoacid and the COOH group of another. This forms a peptide linkage in the course of creating a more complex molecule.

$$R.COOH + NH_2.R_1 \rightarrow R.CO.NH.R_1 + H_2O$$

R_1 and R_2 represent the rest of the aminoacid molecules. Using ingenious techniques like anchoring one end of the growing polypeptide chain on a resin bead, the condensation can be continued to give a quite large, protein-like molecule; insulin and other naturally occurring proteins have been synthesized.

The Esters

Experiment 108
Formation of an ester

Into a dry test-tube, put about 2 cm³ of glacial ethanoic acid. Add a few *drops* of concentrated sulphuric acid (catalyst) and shake. Add 4 cm³ of ethanol, shake and place the test-tube in a beaker of cold water. Heat it slowly and keep the temperature as high as is possible without allowing the mixture in the test-tube to boil. After 20–25 minutes, pour the contents of the test-tube into an evaporating dish about half-full of a concentrated common salt solution. Oily drops of liquid should rise to the surface (the *ester, ethyl ethanoate*), having a pleasant smell of pear-drops.

$$CH_3.COOH(l) + HOC_2H_5(l) \rightleftharpoons$$
$$CH_3.COOC_2H_5(l) + H_2O(l)$$

Notice that an ester is formed from the acid by replacing ionizable hydrogen (of the COOH group) by the organic *ethyl* group from the ethanol. All the common alcohols produce esters with organic acids in this way; the esters usually have a pleasant fruity smell and find uses as flavouring materials and in perfumes. The production of a more complex product like the ester from simpler materials like the acid and alcohol is called *synthesis* and is a very frequent organic process.

It can be reversed, to recover the acid and alcohol again, by *hydrolysis*, theoretically by water but, in practice, by boiling with dilute mineral acid (HCl or H_2SO_4) or with aqueous caustic alkali as catalysts. In the acidic case, the products are the alcohol and acid, from which the ester is derived.

$$CH_3.COOC_2H_5(l) + H_2O(l) \rightleftharpoons CH_3.COOH(aq) + C_2H_5OH(aq)$$

In the alkaline case, the above reaction occurs and is then followed by the production of the sodium (or potassium) salt of the acid, in this instance, *sodium ethanoate.*

$$CH_3.COOH(aq) + NaOH(aq) \rightarrow CH_3.COONa(aq) + H_2O(l)$$

The production of soap by boiling a fat or oil with aqueous caustic soda is alkaline hydrolysis of an ester. (The hydrolysis of *any* ester used to be called *saponification*.) The ester is the fat or oil, the alcohol produced is glycerol (propane-1,2,3-triol) and the sodium salt is the soap.

$$\text{Fat or oil} + \text{caustic soda} \rightarrow \text{soap} + \text{glycerol}$$

The chemical formulae are very complex. For a discussion of the manufacture of soap, see page 249.

Synthesis of Terylene

Terylene is a very complex ester (a *polyester*) synthesized from the alcohol, *ethane-1,2-diol (glycol)* $C_2H_4(OH)_2$, and *benzene-1,4-dicarboxylic acid (terephthalic acid)* $C_6H_4(COOH)_2$. These materials condense together when heated, eliminating molecules of water between them. This is exactly similar in principle to the production of ethyl ethanoate (above) from ethanoic acid and ethanol, though more complex. The first stage is:

$$2C_2H_4(OH)_2 + C_6H_4(COOH)_2 \rightarrow C_6H_4(C_3H_4O_2)_2(OH)_2 + 2H_2O$$

The organic product is still an alcohol (notice the two OH groups) and it can condense with more acid, a process which continues until the relative molecular mass of the product, *terylene*, exceeds 50 000. Terylene can be extruded as fibres and woven into fabrics, sometimes in combination with other textiles.

Unlike the *addition* polymers (polyethene, PVC, etc., see page 341), Terylene is a *condensation* polymer. Formation of such polymers requires two types of monomer molecule, each with two reactive groups:

Every time a new bond is formed, a small molecule such as water is eliminated; but reacting groups are still left available for further reaction on each end of the chain.

Terylene (also known as *Dacron* and *Fortrel*) is a typical *polyester*.

The Amines

Amines are members of the homologous series which has a general molecular formula of $C_nH_{2n+1}NH_2$ for its constituents. The first two members (both gases in ordinary conditions) are *methylamine* CH_3NH_2, and *ethylamine* $C_2H_5NH_2$. The amino group, NH_2, is the characteristic group of all members of the series. Like all such amines, these two compounds are derived from ammonia by replacing one hydrogen atom of the NH_3 molecule by an organic group such as CH_3 or C_2H_5.

$$\overset{\displaystyle H}{\underset{}{H-N-H}}$$
ammonia

$$\overset{\displaystyle C_2H_5}{\underset{}{H-N-H}} \quad \text{or} \quad C_2H_5NH_2$$
ethylamine

Properties of simples amines

Methylamine and ethylamine are colourless gases at room temperature and pressure; they have a strong smell of fish and ammonia and, like ammonia, are very soluble in cold water. They burn in air and this distinguishes them from ammonia, which does not.

Basic nature of amines

Like ammonia, the amines give an *alkaline* reaction in aqueous solution by producing a considerable concentration of hydroxide ion, OH^-.

$$C_2H_5NH_2(g) + H_2O(l) \rightleftharpoons C_2H_5NH_3^+(aq) + OH^-(aq)$$

Compare: $NH_3(g) + H_2O(l) \rightleftharpoons NH_4^+(aq) + OH^-(aq)$

Both methylamine and ethylamine turn damp red litmus paper blue and are, in fact, stronger bases than ammonia at corresponding dilutions.

Just as ammonia forms salts with mineral acids, the amines produce corresponding salts, which can be obtained as similar white crystals.

$$C_2H_5NH_2(g) + HCl(g) \rightarrow C_2H_5NH_3^+Cl^-(s) \text{ (ethylammonium chloride)}$$

Compare:

$$NH_3(g) + HCl(g) \rightarrow NH_4^+Cl^-(s) \text{ (ammonium chloride)}$$

These amine salts liberate the free amine when warmed with aqueous caustic alkali, just as an ammonium salt liberates ammonia in the same conditions.

$$C_2H_5NH_3{}^+Cl^-(s) + Na^+OH^-(aq) \rightarrow Na^+Cl^-(aq) + H_2O(l) + C_2H_5NH_2(aq)$$

Compare:

$$NH_4{}^+Cl^-(s) + Na^+OH^-(aq) \rightarrow Na^+Cl^-(aq) + H_2O(l) + NH_3(aq)$$

Like aqueous ammonia, a solution of a simple amine supplies a large enough concentration of hydroxide ion to precipitate the insoluble hydroxides of several metals from solutions of their salts, *e.g.*, iron(III), zinc, and aluminium hydroxides.

$$Fe^{3+}(aq) + 3OH^-(aq) \rightarrow Fe(OH)_3(s) \quad \text{(Al similar)}$$
$$Zn^{2+}(aq) + 2OH^-(aq) \rightarrow Zn(OH)_2(s)$$

Carbohydrates

The carbohydrates are a family of compounds containing carbon, hydrogen, and oxygen only. They are so called because they were found to have the general empirical formula $C_n(H_2O)_m$, *i.e.*, they appeared to be 'hydrates of carbon'. It would be misleading to think of them in this way, however, and in reality the carbohydrates are classified in two principal ways. One class is the so-called *sugars*, which includes such compounds as glucose and fructose (formula $C_6H_{12}O_6$) and sucrose ($C_{12}H_{22}O_{11}$), and the other class is the *polysaccharides* which contain polymeric material such as starch and cellulose [general formula $(C_6H_{10}O_5)_n$].

Starch has many uses. Wheat starch is the main constituent of bread in Europe and America, and maize starch of cornflour. Potato starch is widely consumed as food and other edible forms are sago, tapioca, rice, and arrowroot. Potato starch is used in the manufacture of glucose, dextrins, and starch syrup. Until recently starch was the main source of ethanol (alcohol) by fermentation and a great quantity of alcoholic beverages is still produced this way, but industrial alcohol is now largely made from petroleum products such as ethene (page 343).

When used as human food, starch is hydrolysed by enzyme action to glucose which is utilized in three principal ways: for production of energy by oxidation to carbon dioxide, for storage as fat, and for storage as glycogen, a compound which readily breaks down to glucose as the body needs it.

Starch can be chemically hydrolysed in several successive stages. Cold, dilute mineral acid converts it to *soluble starch*, which dissolves in hot water and does not set to a paste on cooling. Dry heat (as in baking bread) produces *dextrins* of the kind which colour and flavour bread crust or toast and more complete hydrolysis produces *glucose* (a sugar).

Test for starch

Starch itself, and soluble starch, give a *dark blue coloration* with *iodine*, which disappears on heating but reappears on cooling. (The more complex dextrins produce a red colour with iodine but simpler dextrins and glucose give no colour reaction with iodine.)

Conversion of starch to glucose

Starch is mixed with about $2\frac{1}{2}$ times its mass of water and about 1% of sulphuric acid is added. This mixture is heated under pressure for about 90 min, after which glucose solution is left.

$$(C_6H_{10}O_5)_n + nH_2O \rightarrow nC_6H_{12}O_6$$

A slight excess of chalk is added to destroy the remaining acid and calcium sulphate is filtered off. After decolorization with charcoal and evaporation under reduced pressure, glucose crystallizes out after cooling and can be separated.

Structure of the starch molecule

A starch molecule is produced by the combination of many molecules of glucose $C_6H_{12}O_6$, with elimination of molecules of water. Since this splitting out of H_2O molecules is classed as *condensation* and *many* molecules of glucose are involved, starch can be called a *condensation polymer* of glucose. For this purpose, glucose can be written as $HO.C_6H_{10}O_4.OH$. Hydroxyl groups from a pair of different glucose molecules eliminate a water molecule between them and this happens repeatedly to form a very large linear molecule in the following way. Letting $C_6H_{10}O_4 = X$,

$$HO-X-O\boxed{H+HO}-X-O\boxed{H+HO}-X-O\boxed{H+HO}-X-OH \longrightarrow$$

$$HO-X-O-X-O-X-O-X-OH + 3 H_2O$$

Between 200 and 500 glucose molecules can group themselves together in an actual starch molecule, giving a possible molecular mass in the region of 60 000. If the chain-molecule formed remains straight, the material is called *amylose*, but if a branched structure is formed with many offshoots (usually about 20 glucose units long), the product is *amylopectin*. Various samples of starch have varying proportions of the two. Hydrolysis with hot, dilute, sulphuric acid puts back the H_2O molecules and breaks up the chain to single glucose molecules again. The two components of starch differ somewhat in behaviour with iodine. Amylose gives a *blue* colour and amylopectin *reddish purple*.

Synthetic polymers

Starch is an example of a polymeric substance which occurs in nature. However, we are all very much aware of the importance which synthetic (or 'artificial') polymers play in our lives, and the manner in which three of the most common polymers are manufactured will be discussed.

Polystyrene (*Polyphenylethene*)

Styrene is *phenylethene* $C_6H_5.CH{=}CH_2$. It is a colourless liquid prepared (in two stages) from benzene C_6H_6, and ethene. It polymerizes slowly at ordinary temperature and rapidly if heated to 100–150 °C, preferably in nitrogen to prevent oxidation. It produces polymers of average molecular mass about 120 000 at 100 °C and about 23 000 at 150 °C. The polymers have a linear structure resembling that of polyethene (page 337), but with a phenyl (C_6H_5) group attached to each alternate carbon atom.

$$-CH_2-CH-CH_2-CH-CH_2-CH-CH_2-CH-CH_2-CH-$$
$$C_6H_5C_6H_5C_6H_5C_6H_5C_6H_5$$

The polymer is a white solid and can be depolymerized (back to styrene) at about 350 °C.

Polystyrene is a very good electrical insulator (especially for high-frequency work) and, being water-resistant, can be moulded into battery cases and refrigerator parts. Expanded polystyrene (in the form of blocks and mouldings) can be used in thermal insulation and also in packaging and display. Great quantities of phenylethene (25%) are co-polymerized with butadiene (75%) in making a synthetic rubber.

Perspex

An acidic derivative of ethene is *propenoic acid* (*acrylic acid*) $CH_2{=}CH.COOH$. 2-Methylpropenoic acid is $CH_2{=}C(CH_3).COOH$ and its methyl ester is $CH_2{=}C(CH_3).COOCH_3$. This ester, having a double bond like ethene, polymerizes in a corresponding way. The polymerization is carried out at 80–90 °C in an emulsion of the ester in dilute aqueous soap solution with hydrogen peroxide present. After cooling, the mixture is poured into very dilute hydrochloric acid when the polymer coagulates and is separated, washed, and dried. The polymer is linear, like polyethene, but has CH_3 and $COOCH_3$ groups attached to alternate carbon atoms.

$$-CH_2{-}\underset{\underset{\displaystyle COOCH_3}{|}}{\overset{\overset{\displaystyle CH_3}{|}}{C}}{-}CH_2{-}\underset{\underset{\displaystyle COOCH_3}{|}}{\overset{\overset{\displaystyle CH_3}{|}}{C}}{-}CH_2{-}\underset{\underset{\displaystyle COOCH_3}{|}}{\overset{\overset{\displaystyle CH_3}{|}}{C}}{-}CH_2{-}\underset{\underset{\displaystyle COOCH_3}{|}}{\overset{\overset{\displaystyle CH_3}{|}}{C}}{-}CH_2{-}\underset{\underset{\displaystyle COOCH_3}{|}}{\overset{\overset{\displaystyle CH_3}{|}}{C}}{-}CH_2{-}$$

The polymer is colourless and transparent. When softened by heat, it can be moulded into objects with optical uses, *e.g.*, lenses and reflectors, and, in sheet form as 'Perspex' can replace the much denser glass in illuminated signs and windows. It is also moulded into artificial dentures and small household articles. (Notice how, in both polystyrene and the Perspex polymer, the double bonds of the simple compound are lost and the polymer is a *saturated* material.)

Nylon

Nylon is called a condensation polymer (page 349) because water is split out between pairs of molecules as it forms. These molecules are of the types HO—X—OH and H—Y—H and the essential reaction is:

$$...-\boxed{H + HO}-X-\boxed{OH + H}-Y-\boxed{H + HO}-X-\boxed{OH + H}-Y-\boxed{H + HO}-X-\boxed{OH + H}-...\longrightarrow$$

$$-X-Y-X-Y-X-Y-X-Y-+ nH_2O$$

This produces a greatly extended linear molecule which is *nylon*. The process is done in two stages, the second with heat and pressure in an autoclave and, after cooling, the nylon appears as chips. By melting the chips and forcing the liquid through tiny holes in a metal disc (a *spinneret*), filaments are formed, stretched between rollers, and then gathered into nylon yarn. This yarn is woven into a variety of garments (underwear, stockings, shirts) and into tow ropes, tyre-cord, nets, and racquet strings, which require strength and durability. Nylon powder can be moulded, while hot, into various plastic articles, *e.g.*, gears and bearings in small machines such as electric mixers and razors, where toughness is required.

The actual compounds used in making nylon, both petroleum derivatives, are:

$$HO{-}CO(CH_2)_4CO{-}OH \quad \text{and} \quad H{-}NH(CH_2)_6NH{-}H$$

hexane-1,6-dioic acid 1,6-diaminohexane

From this, X is $-CO(CH_2)_4CO-$ and Y is $-NH(CH_2)_6NH-$ and the linear nylon molecule is:

$$... -CO(CH_2)_4CO{-}NH(CH_2)_6NH{-}CO(CH_2)_4{-}CO{-}NH(CH_2)_6NH{-} ...$$

Notice the amide linkages, —CO—NH—, in this chain. It is found that, in 'condensation polymers' like nylon, the *total* number of CH_2 groups in the two constituents must be at least eight, and preferably nine. Otherwise, cyclic compounds of small molecular mass will form, and not very large linear molecules.)

The separation of organic compounds by chromatography

Chromatography is a means of separating the constituents (usually organic but by no means exclusively so) of a mixture by taking advantage of their different rates of movement (in a solvent) over an adsorbent medium. For each constituent, the rate of movement depends on the relative affinities of the constituent for the solvent and the adsorbent medium. In paper chromatography, which is one of its most useful forms, the adsorbent medium is filter paper and a variety of solvents can be used, *e.g.*, ethyl ethanoate, methanol, aqueous phenol. The following experiment illustrates the process of chromatography in a simple way. A pinhole is made at the centre of a Whatman No. 1 filter paper of 12.5 cm diameter. String is threaded through the hole and cut to 0.5 cm length on one side and 5 cm on the other. A drop of BDH universal indicator is then placed in the centre of the filter paper and allowed to dry out and turn brown in air. A Petri dish (or shallow crystallizing dish) of 10 cm diameter is about one-third filled with the solvent, *ethyl ethanoate*, and the filter paper is placed over it so that the longer end of string dips into the solvent. Another similar dish is placed on top (Figure 153), to make a closed system. Ethyl ethanoate will ascend the string and spread over the filter paper so developing a chromatogram of the constituents of the indicator. That is, after a suitable time (up to an hour), bands of the different dyes present will be seen – a purple colour ahead, a yellow colour about half-way, and a red-brown dye remaining close to the centre. If *methanol* is the solvent, an outer yellow band occurs, with an inner red band and no residue at the centre.

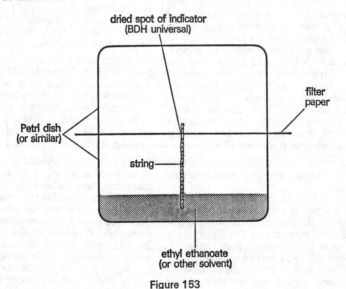

Figure 153

Questions

1. Select *one* member of each of the following homologous series: (i) the alkenes, (ii) the alcohols, and (iii) the mono-carboxylic acids. In each case give its name, molecular formula and structural formula, two chemical reactions, and two other uses in everyday life. (N.I.)

2. Outline how *five* of the following substances can be obtained: (*a*) ethene (ethylene)

Q

from petroleum; (*b*) ethanol (ethyl alcohol) from ethene; (*c*) ethene from ethanol; (*d*) polyethene (polyethylene) from ethene; (*e*) ethanoic acid from ethanol; (*f*) ethyl ethanoate from ethanoic acid; (*g*) soap from a fat. (A.E.B.)

3. When a mixture of solid sodium ethanoate and solid sodium hydroxide is heated, the following reaction takes place:

$$CH_3.COONa + NaOH \rightarrow$$
$$CH_4 + Na_2CO_3$$

(*a*) Assuming that the sodium salts of other organic acids react with sodium hydroxide in a similar way, each forming sodium carbonate and one other product, write equations for the reactions you would expect when the following salts are heated with sodium hydroxide: (i) sodium methanoate H.COONa; (ii) sodium propanoate $CH_3.CH_2.COONa$; (iii) sodium propenoate $CH_2=CH.COONa$. Name the products that you would expect to be formed in each reaction.

(*b*) Draw a diagram of the apparatus you would use to try out these reactions and to collect small samples of the products.

(*c*) Describe simple tests you could carry out with the products of *two* of the reactions you have suggested in (*a*) in order to confirm your predictions. (C.)

4. What do you understand by the terms (*a*) *unsaturated*, (*b*) *polymerization*, and (*c*) *alkane*? Illustrate your answers by reference to simple hydrocarbons.

What is the reaction of methane, ethene, and ethyne with chlorine? Draw attention to any characteristic types of reactions involved. Ethyne reacts more vigorously than ethene with chlorine while ethane reacts less vigorously. With the aid of three-dimensional sketches showing the molecular structures of these hydrocarbons, account for this variation in reactivity. (S.)

5. (*a*) Describe briefly how you would obtain some dilute aqueous ethanol, starting from sugar.

(*b*) What is the gas produced at the same time as the aqueous ethanol?

(*c*) How would you obtain a small specimen of almost pure ethanol from aqueous ethanol? Give brief experimental details.

(*d*) How would you make a small quantity of ethyl ethanoate (on a test-tube scale) from ethanol? How would you be able to tell whether you had any ethyl ethanoate in your reaction?

(*e*) Much ethanol is manufactured from ethene. (i) What is the main source of ethene and how is ethene obtained from this source? (ii) Name one other important product that is manufactured from ethene and give a use of this product.

(*f*) Why are many centres of the petroleum industry situated near the coast? (L.)

6. Distinguish between an *addition* and a *substitution* reaction. Give one example of each. (S.)

7. Olive oil is composed mainly of an *ester* of an *unsaturated* organic acid. Explain the meaning of the terms in italics.

One of the largest users of vegetable oils is the soap industry. State briefly how soap is produced from vegetable oils. Explain how soap acts in removing grease from dishes. Explain why the efficiency of soap is reduced if the tap water contains Ca^{2+} or Mg^{2+} ions. (Scottish.)

8. Give *two* chemical reactions of each of the following, naming the products formed and giving equations in each case: (*a*) methane; (*b*) ethene (ethylene); (*c*) ethanol; (*d*) ethanoic acid. State very briefly how either methane or ethene can be prepared. (A.E.B.)

9. Describe what is seen when ethene (ethylene) is bubbled through bromine water. Give an equation for the reaction and explain what can be deduced about the structure of ethene.

Describe what is seen when ethanol is warmed with acidified potassium dichromate solution. Give the name and structural formula of *one* organic compound formed in this reaction.

Write down the structural formula of ethyl ethanoate and describe in outline how it can be prepared from ethanol. (O.)

10. Write down one possible structural formula for aminoacetic acid (glycine). How does this compound react with (*a*) sodium hydroxide solution, (*b*) hydrochloric acid?

Describe what happens when a protein, such as egg-white, is boiled for a long time with dilute hydrochloric acid. What class of product is obtained? (O. and C.)

11. Under what conditions does ethanol react with sulphuric acid to produce the colourless gas ethene (ethylene) C_2H_4? Write an equation for this reaction.

What is the action of ethene upon bromine and upon hydrogen chloride? Give the names and formulae of the products. What are the reactions (if any) of ethane upon bromine and upon hydrogen chloride? How is the difference in behaviour between ethene and ethane accounted for?

Ethene will burn in a plentiful supply of air to form carbon dioxide and water. Assuming that the air contains 20% by volume of oxygen and that all volumes are measured at s.t.p., what volume of air will be required to burn completely 100 cm^3 of ethene and what will be the volume of the carbon dioxide produced? (O.)

12. Ethanol can be produced either by fermentation or from ethene. (*a*) Explain, with reference to the formation of ethanol, the meaning of the term *fermentation*. (*b*) Describe briefly how ethene is obtained on a large scale. (*c*) How is ethene converted to ethanol? (*d*) Give *one* other large-scale use of ethene. (*e*) Name the products formed when ethanol is burnt in excess air. Give the equation from this reaction. (*f*) Give the name and formula of an ester which can be prepared from ethanol. (O. and C.)

13. (*a*) Ethanol can be produced by *fermentation*. Explain the term *fermentation* and give the full structural formula of ethanol.

(*b*) Under certain conditions the gas ethene can be *polymerized* to give a solid. Explain the meaning of the word *polymerized* and indicate how the reaction takes place.

(*c*) An ester can be prepared from ethanol by reaction with ethanoic acid in the presence of concentrated sulphuric acid. Give the *name* and *structural formula* of this ester and explain why concentrated sulphuric acid is needed.

(*d*) *Photosynthesis* is a natural process. What is *photosynthesis*, and under what conditions does it take place? (O. and C.)

14. The gas ethene C_2H_4 is obtained by the *cracking* of oils and natural gas. With the aid of heat, pressure, and *catalysts*, polyethene is obtained from ethene.

(*a*) Write down the structural formula of ethene. (*b*) Explain what is meant by the term *cracking*. (*c*) Define the term *catalyst*. (*d*) Explain the process which takes place in the manufacture of polyethene. (*e*) Describe what happens when ethene is passed into bromine water. (O. and C.)

15. Give the names of each of the following compounds:

A $CH_3.OH$;   D $CH_2=CH_2$;
B $CH_3.CH_2.CH_3$;   E $CH_3.COOCH_3$.
C $CH_3.CH_2.COOH$;

Which of these compounds are gases at room temperature?

Which *two* of these compounds have identical molecular masses? Describe, with an equation, a reaction of *one* of these compounds that would distinguish it from the other.

What reagents would you use to carry out (i) a substitution reaction with compound **B**, (ii) an addition reaction with compound **D**? Write equations for these two reactions. (C.)

16. State *one* chemical and *one* physical property of ethane. Suppose that, in the molecule of ethane, (*a*) one H is replaced by OH, (*b*) one H is replaced by NH_2, (*c*) two H atoms are removed. State, in each case, the name, *one* chemical property, and *one*

physical property of the new compound formed.

If two hydrogen atoms in the molecule of ethane are replaced by chlorine atoms, it is possible to obtain two different compounds as products. Explain this. (C.)

17. Name the process used to obtain an aqueous solution of ethanol from a sugar, for example, glucose $C_6H_{12}O_6$. Describe how you would carry out this preparation.

State, without further description, the method you would use to obtain a sample of reasonably pure ethanol from the aqueous solution and indicate the physical bases of this separation.

Write down the *structural* formula of ethanol.

Explain the following, stating the chemical reactions which take place: (*a*) ethanol burns in air with a pale blue flame, (*b*) when sodium is added to pure ethanol a colourless gas is evolved, (*c*) when a mixture of equal volumes of ethanol and ethanoic acid is warmed with a few drops of concentrated sulphuric acid, a vapour with a pleasant smell is obtained. (A.E.B.)

18. Explain the basic nature of ethylamine $C_2H_5NH_2$, and compare the behaviour of this compound with that of ammonia towards hydrochloric acid. How, by *one* chemical test, would you distinguish ethylamine from ammonia (both gaseous)? If ethylamine is the second member, write the molecular formula of the fifth member of the homologous series. Mention *two* differences of physical properties you would expect between the two compounds.

19. Explain what is meant by *polymerization*. Outline the manufacture of Polythene. Write a short account of its uses and why it is suited to them. Explain why Terylene is called a *polyester*. Why is it called a condensation polymer? What is the chief use of Terylene in ordinary life?

20. There are two isomers of molecular formula C_2H_6O, *i.e.*, an alcohol and a compound of another type. Write their full structural formulae. Give *one* chemical test to distinguish between them. Write full structural formulae for the two isomers of butane C_4H_{10}. Pentane C_5H_{12} has three isomers. Write full structural formulae for *two* of them.

21. (*a*) Draw diagrams to show the structural formulae of (i) ethane, (ii) ethene, and (iii) ethyne.

(*b*) Under what conditions does ethene react with (i) hydrogen, and (ii) hydrogen chloride? In each case, name the product and write an equation.

(*c*) Under what conditions does ethyne

react with an excess of (i) hydrogen, and (ii) bromine? In each case, name the product and write an equation.

(*d*) When ethyne reacts with hydrogen chloride, a compound is formed which will undergo polymerization. Draw a diagram to illustrate the structure of a section of the polymer which is formed. (J.M.B.)

29 Silicon and its Compounds

Occurrence

Silicon is the second most abundant element in the earth's crust, the most abundant of all being oxygen. Silicon is found in the following forms:

1. *As metallic silicates.* Igneous rocks, such as granite and basalt, consist largely of mixtures of silicates, those of magnesium, aluminium, potassium, and iron being most common. China clay, kaolin, is a hydrated silicate of aluminium; ordinary clay is a mixture of particles of quartz, mica, and other substances bound together by a sticky material which is a hydrated silicate of aluminium of approximate formula $Al_2Si_2O_7.2H_2O$.

2. *As silica (silicon dioxide) SiO_2.* The purest form of silica is rock crystal or transparent quartz. Sand usually consists of silica, with various impurities, and opal, hornstone, and jasper are forms of this oxide. It also exists in a less pure form as flint.

The porous material 'kieselguhr' is a form of silica made up of the fossil shells of small plants called diatoms. It is used to absorb the explosive, 'nitroglycerine', forming the product called 'dynamite', which is safer to transport and handle than the explosive itself.

Preparation of silicon

The element may be prepared in the amorphous state by heating magnesium powder with *dry* silica SiO_2.

$$SiO_2(s) + 2Mg(s) \rightarrow 2MgO(s) + Si(s)$$

The reaction is very violent, and great care needs to be taken in obtaining the element by this method. In the industrial production of silicon the silica is heated in an electric furnace with carbon.

$$SiO_2(s) + C(s) \rightarrow Si(s) + CO_2(g)$$

Reactions of silicon

Silicon burns in air forming a white solid, silica.

$$Si(s) + O_2(g) \rightarrow SiO_2(s)$$

Amorphous silicon combines with chlorine at low red heat and also decomposes steam.

$$Si(s) + 2Cl_2(g) \rightarrow SiCl_4(l)$$
$$Si(s) + 2H_2O(g) \rightarrow SiO_2(s) + 2H_2(g)$$

Silicon is used in the manufacture of certain types of steel and bronze, and in the production of silicones, transistors, and microcircuits.

Silicon dioxide (silica) SiO_2

Experiment 109
Preparation of silica from sodium silicate

Add concentrated hydrochloric acid to a solution of sodium silicate and warm if necessary. White hydrated silica (silicic acid) will come down as a gelatinous precipitate.

$$Na_2SiO_3(aq) + 2HCl(aq) \longrightarrow$$
$$2NaCl(aq) + SiO_2.H_2O(s)$$

Show that the precipitate is readily soluble in dilute caustic soda solution.

$$2NaOH(aq) + SiO_2(s) \longrightarrow$$
$$Na_2SiO_3(aq) + H_2O(l)$$

Uses and properties of silica

In its naturally occurring form, sand, silica finds great use as a constituent of mortar and cement, for the manufacture of glass (see page 360) and for filtration of water in bulk.

Fused silica is used for the manufacture of certain types of laboratory apparatus. Its coefficient of expansion is very low (about one-fiftieth of that of glass) and consequently the strains set up in it by irregular contraction under sudden reduction of temperature are small and insufficient to break it. A red-hot silica tube or crucible can be plunged into cold water without damage. Fused quartz threads are used in the construction of physical apparatus. They can be made so thin as to be invisible to the naked eye and yet strong enough to support a mass of 2 g. Their tenuity and elasticity make them invaluable for light, delicate suspensions. The difficulty in working fused silica is that the very high temperature of at least 1500 °C is required.

Silicon dioxide (silica) is almost insoluble in water. It acts as the acidic oxide of a non-metal and forms *silicates* with caustic alkalis and metallic oxides (all heated).

$$SiO_2(s) + 2OH^-(aq) \rightarrow SiO_3{}^{2-}(aq) + H_2O(l)$$
or
$$SiO_2(s) + O^{2-}(s) \rightarrow SiO_3{}^{2-}(s)$$

Silicon dioxide forms three-dimensional crystalline systems of great complexity. In its simplest form (*cristobalite*), the silicon atoms are arranged like the carbon atoms of diamond (page 91) but with oxygen atoms midway between them. This giant covalent structure accounts for the high melting point of silica in contrast with the gaseous state of carbon dioxide at s.t.p. This gas contains only separate molecules of carbon dioxide (page 87).

Experiment 110
Sodium silicate from silica

Wash some of the white gelatinous precipitate of hydrated silica obtained above and ignite it in a crucible. The product is amorphous silica and is no longer easily soluble in caustic soda solution even if hot and concentrated. That silica is an acidic oxide can, however, be shown as follows: put a small piece of solid sodium hydroxide about half an inch long into a crucible, add a spatula measure of amorphous silica and heat the mass strongly for about ten minutes. Take great care when doing this that none of the molten mass is spilled or comes into contact with the flesh. Allow the melt to cool, fill the crucible two-thirds full of water and warm. The solution obtained is sodium silicate and from it gelatinous silica can be obtained by the action of hydrochloric acid.

Silica is non-volatile and can, at high temperatures, displace volatile acidic oxides from combination. They pass off as vapour, leaving a silicate, *e.g.*,

$$Na_2SO_4(s) + SiO_2(s) \longrightarrow$$
$$Na_2SiO_3(s) + SO_3(g)$$

Silicates

Silicates are salts of the silicic acids. The chemistry of these acids is complex, and little is known for certain about them.

The gelatinous substance precipitated by acidifying a hot solution of 'water-glass' has the empirical formula H_2SiO_3, and is called meta-silicic acid, and an acid of approximate formula H_4SiO_4, ortho-silicic acid, is obtained by the action of water on silicon tetrachloride.

$$Na_2SiO_3(aq) + 2HCl(aq) \rightarrow H_2SiO_3(aq) + 2NaCl(aq)$$
$$SiCl_4(l) + 4H_2O(l) \rightarrow H_4SiO_4(aq) + 4HCl(aq)$$

The salts of these acids are called silicates. The meta-silicates are the most important.

Sodium silicate (water-glass) Na_2SiO_3

Sodium silicate is prepared by heating two parts by mass of silica with one part of sodium carbonate.

$$Na_2CO_3(s) + SiO_2(s) \rightarrow Na_2SiO_3(s) + CO_2(g)$$

The product is a glassy solid. It is heated with water (under pressure) to dissolve it, and is sold in tins in the form of a concentrated solution, similar in consistency to 'golden syrup', but colourless. This is called 'water-glass'.

Water-glass is chiefly used for preserving eggs. The shell of an egg consists largely of calcium carbonate. This, when the egg is immersed in a suitable solution of 'water-glass', reacts with the sodium silicate to produce a precipitate of calcium silicate which seals the pores of the shell, excluding bacteria and so preserving the egg from putrefaction.

$$CaCO_3(s) + Na_2SiO_3(aq) \rightarrow CaSiO_3(s) + Na_2CO_3(aq)$$

Experiment 111
Formation of a silica garden

An interesting chemical phenomenon, called a 'silica garden', can be shown in the following way. Dilute some water-glass until it has a density of 1.1 g cm^{-3} (test with a hydrometer), then filter it and put the liquid into a tall, rather narrow vessel. Drop into it crystals of manganese(II) chloride, cobalt(II) nitrate, iron(II) sulphate, and copper(II) sulphate. From these crystals there will shoot fantastic coloured growths. Those from the cobalt salt usually appear within a few seconds and grow rapidly. They are dark blue. Growths from the other crystals appear more slowly. Those from the manganese salt are pale pink, from copper sulphate light blue, and from the iron(II) salt green. The growths are tubes of the silicates of the metals.

Silica gel

If hydrochloric acid is added to a concentrated solution of sodium silicate at about 100 °C, a product known as *silica gel* is precipitated. It contains 5–7% of water and is regarded as a partially hydrated silica.

$$Na_2SiO_3(aq) + 2HCl(aq) \rightarrow 2NaCl(aq) + SiO_2.H_2O(s)$$

Silica gel absorbs water-vapour very readily and is used for drying gases on the industrial scale, *e.g.*, in drying air for the blast of a blast-furnace. When spent, the gel can be re-activated by heating to a suitable temperature.

Silica gel can also absorb volatile solvents, *e.g.*, carbon disulphide or propanone, which would otherwise be lost as effluent. The solvents can be recovered by suitable heating of the gel. It has also been used successfully for freeing petroleum oils from sulphur compounds.

Glass

The substance to which the term 'glass' is usually applied consists of a mixture of two or more silicates. Common soft glass, of which bottles are made, is prepared by heating together silica in the form of sand, sodium carbonate or sodium sulphate, and chalk or limestone (calcium carbonate). (Some broken glass and a little coke are usually added.) The glass so prepared consists of a mixture of sodium silicate and calcium silicate.

$$Na_2CO_3(s) + SiO_2(s) \rightarrow Na_2SiO_3(s) + CO_2(g)$$
$$\text{(or } Na_2SO_4) \qquad\qquad\qquad \text{(or } SO_3)$$
$$CaCO_3(s) + SiO_2(s) \rightarrow CaSiO_3(s) + CO_2(g)$$

If potassium carbonate is used instead of sodium carbonate, a 'hard' glass, that is, a glass needing a higher temperature to melt it, will be produced, and will consist of a mixture of calcium silicate and potassium silicate.

Glass containing lead silicate has a brilliant appearance, and is made by adding to the ordinary glass mixture some red lead oxide Pb_3O_4, or massicot PbO. Such a glass ('flint glass') has a high refractive index and is used for prisms and lenses, but it is soft and should be wiped only with silk, to avoid scratching the surface.

Coloured glass is obtained by addition of metallic oxides, for example, cobalt produces blue glass, chromium produces green glass. Opalescent glass may be produced by addition of calcium phosphate.

Physical nature of glass

Glass is not really a solid. This is shown by the fact that, when heated, it does not suddenly pass from the solid to the liquid state at a definite temperature, but softens slowly as the temperature rises and gradually becomes liquid. A true solid would melt suddenly over a small temperature range. Glass is a supercooled liquid. By this we mean that its molecules have not taken up a definite formation to produce crystals, but are arranged at random as they were in liquid glass.

The softening of glass before it melts is a very valuable property. It enables a skilled worker or a machine to blow it into various shapes – bottles, flasks, beakers – by using suitable moulds. While soft, glass can also be moulded and joined to produce elaborate scientific apparatus. Large sheets of glass have been produced by rolling out a mass of hot glass on a long, flat table or, more recently, by floating glass on molten tin while heating the top surface.

Articles made of glass, particularly those needed for optical purposes, usually have to be annealed, that is, heated to a suitable temperature and then allowed to cool very slowly and uniformly so that no stresses are set up within the glass. The annealing of considerable masses of glass may require several weeks, or even months. Working with glass is a very ancient art. It was practised by the Egyptians at least 2300 years ago, while some of the finest specimens of glass-ware are the work of Venetian craftsmen of the Middle Ages.

Silicon tetrachloride

Silicon combines with chlorine when heated to form silicon tetrachloride, but a more convenient method of producing this compound (because it does not require the element, silicon) is to heat an intimate mixture of silica and carbon in a current of *dry* chlorine.

$$SiO_2(s) + 2C(s) + 2Cl_2(g) \rightarrow SiCl_4(l) + 2CO(g)$$

Silicon tetrachloride distils off as vapour and can be condensed to a colourless liquid of boiling point 58 °C at 760 mm. It may contain a small proportion of other chlorides, *e.g.*, Si_2Cl_6.

Properties

(1) Silicon tetrachloride reacts rapidly with cold water, forming hydrogen chloride and silicic acid.

$$SiCl_4(l) + 3H_2O(l) \rightarrow 4HCl(g) + H_2SiO_3(aq)$$

This reaction (with atmospheric moisture) causes the tetrachloride to fume in moist air. Carbon tetrachloride CCl_4, however, has no reaction with water, which is unusual for the chloride of a non-metal.

(2) If heated with an alkali metal, silicon tetrachloride is decomposed to form silicon.

$$SiCl_4(l) + 4K(s) \rightarrow Si(s) + 4KCl(s)$$

Questions

1. What is the chemical nature of glass? How is it made? Why is glass stated to be a super-cooled liquid?

2. Silica is an acid-forming oxide. Justify this statement by reference to the chemical properties of silica.

3. In what forms does silica occur in nature? How may the element silicon be obtained from silica?

4. What is silica? Starting with the naturally occurring substance describe how you would obtain (*a*) water-glass; (*b*) a solution of silicic acid; and (*c*) pure silica. (L.)

5. Describe in outline *two* methods by which silicon may be prepared in the laboratory from silica SiO_2. How, and in what con-ditions, does silicon react with (*a*) sodium hydroxide, (*b*) chlorine, (*c*) oxygen? Compare the behaviour of carbon with that of silicon towards these three reagents.

6. Describe a laboratory preparation of a concentrated solution of sodium silicate from silica SiO_2. How may the product *silica gel* be obtained from this solution? Briefly discuss the large-scale uses of this product.

7. Describe in brief outline a preparation of silicon tetrachloride from silica SiO_2. Compare and contrast the elements *carbon* and *silicon* with reference to the chemical nature of their dioxides and the behaviour of their tetrachlorides with water. Mention the experimental evidence for your statements.

30 The Halogens

The four non-metals, fluorine, chlorine, bromine, and iodine, make up a family of related elements, the chemical properties of which form an interesting study. The name given to this family of elements is the Halogens, meaning literally 'salt-producer', since all the halogen elements react with most metals to form electrovalent, salt-like, compounds. The first of the series is the very active element fluorine, which is so active that it evaded efforts to isolate it for many years, because it reacted with almost every element or substance with which it came into contact – even the glass of the apparatus in which the reaction took place! Because fluorine is so difficult to isolate safely, its properties have no practical application in the school laboratory. We shall see that bromine behaves, in reactivity, between the very reactive chlorine and the not-so-reactive iodine.

Chlorine

Chlorine was first isolated by Scheele in 1774. Scheele was a Swedish apothecary who carried out, during the short time which he lived (he died when only 44), a vast number of illuminating experiments, performed in an old shed attached to his house. It was whilst he was investigating the properties of pyrolusite (impure manganese(IV) oxide) that he heated it with concentrated hydrochloric acid in a retort and collected the chlorine gas which came off into a pig's bladder.

It is usually made by the oxidation of concentrated hydrochloric acid. The oxidation can be brought about by many oxidizing agents, for example, lead(IV) oxide, manganese(IV) oxide, red lead oxide, or potassium permanganate.

$$2Cl^-(aq) \rightarrow Cl_2(aq) + 2e^-$$

from
hydrochloric
acid

to
oxidizing agent

Laboratory preparation of chlorine

Experiment 112
Preparation of chlorine from concentrated hydrochloric acid by oxidation with manganese(IV) oxide

Fit up the apparatus as shown in Figure 154. Put some manganese(IV) oxide into the flask, pour concentrated hydrochloric acid down the funnel and shake well before connecting up the flask with the rest of the apparatus. (Note: the use of a gas-ring and gauze keeps the flask low and makes the apparatus more stable.)

Heat the mixture in the flask and effervescence is observed. A greenish-yellow gas is evolved which, together with a certain amount of hydrogen chloride (misty fumes), passes over into the first bottle which contains water. This removes the hydrochloric acid gas (which is very soluble in water), and the concentrated sulphuric acid in the second bottle dries the gas which is collected by downward delivery, the gas being denser than air.

$$MnO_2(s) + 4HCl(aq) \rightarrow$$
$$MnCl_2(aq) + 2H_2O(l) + Cl_2(g)$$

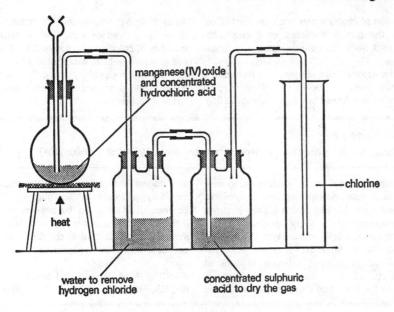

manganese (IV) oxide
and concentrated
hydrochloric acid

chlorine

heat

water to remove
hydrogen chloride

concentrated sulphuric
acid to dry the gas

Figure 154

The above experiment should be carried out in a fume-chamber, as should any preparation of chlorine in which it is collected by displacement of air.

Experiment 113
Preparation of chlorine from concentrated hydrochloric acid by oxidation with potassium permanganate

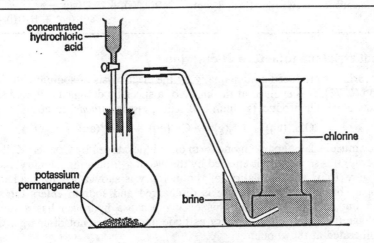

concentrated
hydrochloric
acid

chlorine

potassium
permanganate

brine

Figure 155

[If the chlorine is required pure and dry, insert wash-bottles containing (*a*) water, and (*b*) concentrated sulphuric acid in Figure 155 and collect by displacement of air or in a gas syringe.]

$$2KMnO_4(aq) + 16HCl(aq) \longrightarrow$$
$$2KCl(aq) + 2MnCl_2(aq) + 8H_2O(l) + 5Cl_2(g)$$

This is a very convenient laboratory method because it takes place in the cold and the rate of production of chlorine can easily be controlled, and, if the gas is collected over brine, the experiment need not be conducted in a fume-chamber.

Solid potassium permanganate is placed in a flask and concentrated hydrochloric acid is dropped on to it from a tap-funnel (Figure 155).

As each drop of acid reaches the permanganate, there is evolved at once the corresponding quantity of chlorine. The apparatus is filled with the greenish-yellow fumes of the gas. Several gas-jars of the gas should be collected and the experiments described later performed to illustrate its properties.

Experiment 114
Preparation of chlorine from common salt (sodium chloride)

Chemistry of the action. Concentrated sulphuric acid acts upon the sodium chloride to form hydrogen chloride, which is then oxidized to chlorine by manganese(IV) oxide.

The apparatus is identical with Figure 154 and the experiment is performed in a similar manner except that an intimate mixture of sodium chloride and manganese(IV) oxide is placed in the flask and concentrated sulphuric

acid added. In this experiment the presence of hydrogen chloride as an impurity is more obvious. It is removed by passing the gases through water, and the chlorine is dried by means of concentrated sulphuric acid and collected as shown in Figure 154.

$$2NaCl(s) + MnO_2(s) + 2H_2SO_4(aq) \rightarrow$$
$$Na_2SO_4(aq) + MnSO_4(aq) + 2H_2O(l) + Cl_2(g)$$

Experiment 115
Preparation of chlorine from bleaching powder

In this case the chlorine is not prepared by the oxidation of concentrated hydrochloric acid. The apparatus used is identical with Figure 155, bleaching powder is placed in the flask and a dilute acid, *e.g.,* nitric acid, is dropped on to the powder. Effervescence occurs and the greenish-yellow gas can be collected by either of the methods mentioned previously. Heat is not required.

$$\text{'CaOCl}_2\text{'(s)} + 2HNO_3(aq) \rightarrow$$
bleaching
powder

$$Ca(NO_3)_2(aq) + H_2O(l) + Cl_2(g)$$

Similarly:

$$\text{'CaOCl}_2\text{'(s)} + 2HCl(aq) \rightarrow$$
$$CaCl_2(aq) + H_2O(l) + Cl_2(g)$$

Industrial manufacture of chlorine

Chlorine is produced commercially by the electrolysis of sodium chloride solution (*brine*). It is evolved at the *anode* of a specially designed cell, and since the other electrode product (sodium hydroxide, at the *cathode*) reacts with chlorine,

$$2OH^-(aq) + Cl_2(g) \rightarrow Cl^-(aq) + OCl^-(aq) + H_2O(l)$$

they must be kept apart. In one form of cell (developed by Castner, Kellner, and Solvay) this separation is effected by the use of a circulating mercury diaphragm (Figure 156). The circulation of the mercury is shown heavily shaded. In the upper (brine) cell, sodium ion is discharged and sodium enters the mercury forming sodium amalgam. This occurs because hydrogen has a high over-voltage (0.78 volt) at a mercury cathode and so does not discharge. Chlorine is liberated at the anode.

At cathode	*At anode*
$Na^+ + e^- \rightarrow Na$	$Cl^- - e^- \rightarrow \frac{1}{2}Cl_2$

Sodium amalgam, flowing into the lower (soda) cell, encounters a flow of distilled water in contact with steel grids on which hydrogen has only a very low over-voltage. Sodium hydroxide is formed and hydrogen is liberated.

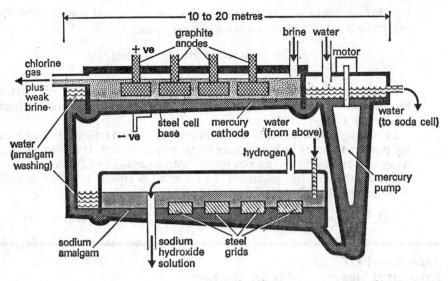

Figure 156

$$2\text{Na(s)} + 2(\text{H}^+\text{OH}^-)(\text{l}) \rightarrow 2(\text{Na}^+\text{OH}^-)(\text{aq}) + \text{H}_2(\text{g})$$

The high capital charge on the mercury used adds considerably to the cost of this process and rival cells using asbestos separating diaphragms, *e.g.*, the Hooker cell, have been tried. Here, the advantage of cheapness is offset by the fact that sodium chloride penetrates the diaphragm and pollutes the product.

Properties of chlorine

Chlorine is a greenish-yellow gas with a choking, unpleasant, irritating smell. It is very poisonous if inhaled to even a small extent (1 part of chlorine in 50 000 of air may be injurious). It was used extensively during the war of 1914–18 and, being about $2\frac{1}{2}$ times as dense as air, it would roll along the ground when propelled by a very gentle wind without a great deal of it escaping upwards.

It bleaches damp litmus and is a very reactive gas indeed. The following experiments illustrate its properties and they are classified according to the various ways in which the gas can act.

Experiment 116
Chlorine as a bleaching agent

Pour a little litmus solution into a gas-jar of the gas. The litmus immediately turns colourless. Chlorine will bleach the colour from most dyes and will remove writing ink (but not printer's ink, which consists mainly of carbon, which chlorine does not attack).

Bleaching action. The chlorine reacts with the water, forming hypochlorous acid.

$$\underset{\text{chlorine}}{\text{Cl}_2(\text{g})} + \underset{\text{water}}{\text{H}_2\text{O(l)}} \rightarrow \underset{\substack{\text{hypochlorous} \\ \text{acid}}}{\text{HOCl(aq)}} + \underset{\substack{\text{hydrochloric} \\ \text{acid}}}{\text{HCl(aq)}}$$

This hypochlorous acid is a very reactive compound and readily gives up its oxygen to the dye, to form a colourless compound.

$$\underset{\text{coloured}}{\text{dye}} + \text{HOCl} \rightarrow \text{HCl} + \underset{\text{colourless}}{(\text{dye} + \text{O})}$$

Notice that hydrochloric acid is produced whenever chlorine bleaches, and hence an article must be thoroughly washed after bleaching or it will be attacked by the free acid.

In industry, the article to be bleached is dipped into a tank containing bleaching powder in water, then into very dilute sulphuric acid. It is then washed by water to remove acid and may be treated with an *anti-chlor* to remove remaining chlorine, which, if left, might rot the material. A typical anti-chlor is sodium thiosulphate ('hypo'), which reacts with chlorine:

$$S_2O_3{}^{2-}(aq) + 4Cl_2(g) + 5H_2O(l) \rightarrow 2SO_4{}^{2-}(aq) + 8Cl^-(aq) + 10H^+(aq)$$

Experiment 117
Attempted bleaching with dry chlorine

Pour about 20 cm³ of concentrated sulphuric acid into a gas-jar of the gas as soon as it is collected. After a time suspend a piece of dry coloured cloth into the gas-jar. At the same time put a piece of damp coloured cloth into a gas-jar of chlorine; the latter is immediately bleached, whereas the former remains unattacked. (If the chlorine and cloth are perfectly dry, the cloth remains unbleached indefinitely.)

From the equation above, it is seen that water is necessary for the formation of hypochlorous acid, the compound which liberates the oxygen and which performs the bleaching. Hence, if no water is present no bleaching can occur.

Test for chlorine. Chlorine is a greenish-yellow gas which rapidly bleaches damp litmus paper. Since the gas is acidic, if damp blue litmus paper is used a red colour is often seen before it is bleached.

The formation of hydrogen chloride from chlorine

In many of the reactions of chlorine with other substances, hydrogen chloride is produced. The following experiments illustrate the readiness with which chlorine will combine with hydrogen, even when the hydrogen is present in a compound.

Experiment 118
Effect of sunlight on chlorine water

Pass chlorine gas into water in a beaker for some time until the water becomes quite yellowish-green in colour. Fill a long tube with this chlorine water, invert it in a beaker containing some of the water and expose to bright sunlight (Figure 157). After some time, a gas collects in the tube and on applying a glowing splint, the latter is rekindled, showing the gas to be oxygen.

$$2Cl_2(g) + 2H_2O(l) \rightarrow 4HCl(aq) + O_2(g)$$

The chlorine has combined with the hydrogen of the water to form hydrogen chloride and oxygen is liberated.

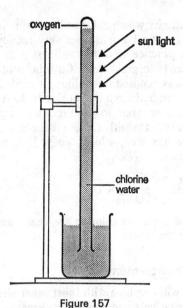

oxygen
sun light
chlorine water

Figure 157

The above reaction probably occurs in two stages, as indicated by the equations:

$$H_2O(l) + Cl_2(g) \longrightarrow$$
water   chlorine

$$HOCl(aq) + HCl(aq) \quad \text{(instantaneous)}$$
hypochlorous   hydrochloric
acid   acid

$$2HOCl(aq) \longrightarrow 2HCl(aq) + O_2(g) \text{ (slow)}$$
oxygen

Experiment 119
Reaction between chlorine and warm turpentine

Care should be taken in carrying out this experiment.

Warm a little turpentine in a dish, dip into it a filter paper, and then drop this into a gas-jar of chlorine. There is a red flash accompanied by a violent action whilst a black cloud of solid particles of carbon is also formed. Hydrogen chloride can be shown to be present by blowing the fumes from an ammonia bottle across the top of the jar, when dense white fumes of ammonium chloride are observed.

$$NH_3(g) + HCl(g) \longrightarrow NH_4^+Cl^-(s)$$

Turpentine (a hydrocarbon) consists of hydrogen and carbon combined together. The chlorine combines with the hydrogen and leaves the black carbon behind.

$$C_{10}H_{16}(l) + 8Cl_2(g) \longrightarrow 10C(s) + 16HCl(g)$$

Experiment 120
Effect of chlorine on a burning taper

Lower a burning taper into a gas-jar of chlorine. It burns with a small, red, and sooty flame. Wax consists mainly of hydrocarbons and, as with the turpentine, the hydrogen forms hydrogen chloride and leaves the carbon.

Experiment 121
Reaction between chlorine and hydrogen sulphide

Invert a gas-jar of hydrogen sulphide over a gas-jar of chlorine and remove the plates. You will observe a yellow precipitate of sulphur, and hydrogen chloride will be formed.

$$H_2S(g) + Cl_2(g) \longrightarrow 2HCl(g) + S(s)$$
hydrogen   chlorine   hydrogen   sulphur
sulphide   chloride

This can also be regarded as a case of oxidation – reduction, in which chlorine is reduced to its ions by electron gain and the sulphide ion (from H_2S) is oxidized to sulphur by electron loss. Chlorine is the oxidizing agent, hydrogen sulphide the reducing agent.

$$\begin{cases} H_2S \rightleftharpoons 2H^+ + S^{2-} \\ S^{2-} \longrightarrow S + 2e^- \quad \text{(oxidation)} \end{cases}$$
$$Cl_2 + 2e^- \longrightarrow 2Cl^- \quad \text{(reduction)}$$

The above experiments illustrate the affinity which chlorine has for hydrogen, and it is to be expected that hydrogen gas itself would react very readily with chlorine. This is found to be the case in practice, and the readiness with which hydrogen and chlorine combine together is so great that, if a tube containing equal volumes of chlorine and hydrogen is exposed to sunlight, it explodes. In the absence of sunlight it is possible to burn hydrogen in chlorine by lowering a jet of burning hydrogen (from a cylinder, or laboratory-source) into a gas-jar full of chlorine. The hydrogen continues to burn with a white flame and clouds of steamy fumes of hydrogen chloride are seen, whilst the yellowish-green colour of the chlorine gradually disappears.

$$\underset{\text{hydrogen}}{H_2} + \underset{\text{chlorine}}{Cl_2} \rightarrow \underset{\substack{\text{hydrogen} \\ \text{chloride}}}{2HCl}$$

Both of these experiments between hydrogen and chlorine are dangerous, and should not be attempted by the student in the laboratory.

Reaction between chlorine and other elements

Chlorine is a very reactive element and will combine with most other elements to form chlorides. In many cases chlorine will combine with elements spontaneously, *i.e.*, without applying a flame or in any way inducing the reaction to take place.

Experiment 122
Action of chlorine on phosphorus

Lower a piece of dry yellow phosphorus into a gas-jar of chlorine. It burns spontaneously, giving off white fumes of chlorides of phosphorus.

$$P_4(s) + 6Cl_2(g) \rightarrow \underset{\substack{\text{phosphorus(III)} \\ \text{chloride}}}{4PCl_3(g)}$$

$$P_4(s) + 10Cl_2(g) \rightarrow \underset{\substack{\text{phosphorus(V)} \\ \text{chloride}}}{4PCl_5(g)}$$

Experiment 123
Action of chlorine on copper

Drop a piece of Dutch metal (a very thin sheet of an alloy of copper and zinc, mainly copper) into a gas-jar of chlorine. It burns spontaneously with a green flame to form copper(II) chloride and a little zinc chloride.

$$Cu(s) + Cl_2(g) \rightarrow \underset{\substack{\text{copper(II)} \\ \text{chloride}}}{CuCl_2(s)}$$

$$Zn(s) + Cl_2(g) \rightarrow \underset{\substack{\text{zinc} \\ \text{chloride}}}{ZnCl_2(s)}$$

Experiment 124
Action of chlorine on iron

Place a coil of iron wire in the hard glass tube in the apparatus shown in Figure 158 and pass a stream of pure dry chlorine over it. On heating the wire by means of a burner, the wire glows and the reaction continues without application of the flame, black crystals of ferric chloride, or iron(III) chloride, collecting in the small bottle, which acts as a condenser.

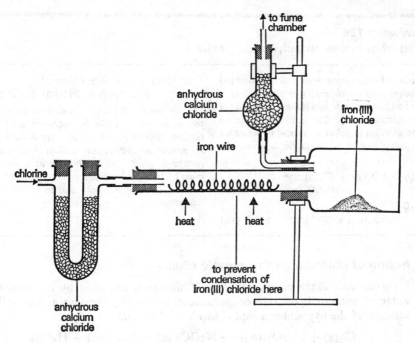

Figure 158

$$2Fe(s) + 3Cl_2(g) \longrightarrow 2FeCl_3(s)$$

iron   chlorine   iron(III)

chloride

Note that the *iron(III)* salt is formed – an

indication that chlorine is an oxidizing agent.

The black crystals of anhydrous iron(III) chloride should be removed and placed in a desiccator, as they are very deliquescent.

Chlorine as an oxidizing agent

In its oxidizing behaviour, chlorine, like other oxidizing agents, acts as an acceptor of electrons. It is converted to chlorine ions, as:

$$Cl_2 + 2e^- \longrightarrow 2Cl^-$$

The electrons for this purpose are supplied by the reducing agents with which the chlorine is reacting. In the two cases given below, these are:

(*a*) The iron(II) ion, which is oxidized to the iron(III) ion by electron loss,

$$Fe^{2+} - e^- \longrightarrow Fe^{3+}$$

(*b*) The sulphite ion, which, in association with water, is oxidized to the sulphate ion,

$$SO_3{}^{2-}(aq) + H_2O(l) - 2e^- \longrightarrow SO_4{}^{2-}(aq) + 2H^+(aq)$$

also with electron loss.

Experiment 125
Action of chlorine on iron(II) chloride solution

Bubble a stream of chlorine through a solution of iron(II) chloride (which is pale green in colour). The colour changes to yellow, and on adding a little caustic alkali solution there is obtained a reddish-brown precipitate of iron(III)

hydroxide, showing that the iron(II) chloride has been oxidized to iron(III) chloride.

$$2FeCl_2(aq) + Cl_2(g) \longrightarrow 2FeCl_3(aq)$$

or $2Fe^{2+}(aq) + Cl_2(g) \longrightarrow$

$$2Fe^{3+}(aq) + 2Cl^-(aq)$$

Experiment 126
Action of chlorine on sulphurous acid

A solution of sulphurous acid is first generated by passing sulphur dioxide gas slowly into a beaker half full of water, in a fume-cupboard (or well-ventilated area). On bubbling chlorine into the solution of sulphurous acid in water for a few minutes, dilute sulphuric acid is obtained.

$$2H_2O(l) + SO_2(g) + Cl_2(g) \longrightarrow$$
$$H_2SO_4(aq) + 2HCl(aq)$$

or $H_2O(l) + SO_2(g) \rightleftharpoons$
$$H_2SO_3(aq) \rightleftharpoons 2H^+(aq) + SO_3^{2-}(aq)$$

$$SO_3^{2-}(aq) + H_2O(l) + Cl_2(g) \longrightarrow$$
$$SO_4^{2-}(aq) + 2H^+(aq) + 2Cl^-(aq)$$

The presence of the sulphuric acid can be shown by testing with dilute hydrochloric acid and barium chloride before and after the experiment. A white precipitate of barium sulphate is obtained. [*N.B.* The sulphurous acid solution must be fresh, otherwise it will contain a certain amount of sulphuric acid due to atmospheric oxidation.]

Action of chlorine on the caustic alkalis

On the cold dilute aqueous solution. Chlorine is absorbed by a solution in water of sodium hydroxide or potassium hydroxide, forming a pale yellow solution of the hypochlorite and chloride of the metal.

$$Cl_2(g) + 2NaOH(aq) \rightarrow NaOCl(aq) + NaCl(aq) + H_2O(l)$$
$$Cl_2(g) + 2KOH(aq) \rightarrow KOCl(aq) + KCl(aq) + H_2O(l)$$
or $\quad Cl_2(g) + 2OH^-(aq) \rightarrow OCl^-(aq) + Cl^-(aq) + H_2O(l)$

On the hot concentrated aqueous solution. If chlorine is passed into a hot concentrated solution of potassium hydroxide for some time, a mixture of potassium chloride and potassium chlorate is formed, and the latter can be obtained by crystallizing the mixture when crystals of potassium chlorate separate first. (These can be purified by recrystallization.)

$$\underset{\substack{\text{potassium} \\ \text{hydroxide}}}{6KOH(aq)} + 3Cl_2(g) \rightarrow \underset{\substack{\text{potassium} \\ \text{chlorate}}}{KClO_3(aq)} + \underset{\substack{\text{potassium} \\ \text{chloride}}}{5KCl(aq)} + 3H_2O(l)$$

or $\quad 6OH^-(aq) + 3Cl_2(g) \rightarrow ClO_3^-(aq) + 5Cl^-(aq) + 3H_2O(l)$

A similar action is observed if hot concentrated sodium hydroxide solution or milk of lime is substituted for the potassium hydroxide solution.

Notice that in the above actions the alkalis are dissolved (or suspended) in water.

Bleaching powder

If chlorine is passed for a considerable time over **solid slaked lime** the product is bleaching powder, formerly used as an easily transported substitute for free chlorine. Now, however, liquid chlorine is regularly transported in bulk and bleaching powder is less important. It is prepared in the laboratory by taking a gas-jar full of chlorine and shaking a spatula-measure of freshly prepared slaked lime into it. The colour of the chlorine disappears immediately. The product may be used to absorb the chlorine from several more gas-jars before absorption is complete.

$$\underset{\substack{\text{calcium} \\ \text{hydroxide}}}{Ca(OH)_2(s)} + \underset{\text{chlorine}}{Cl_2(g)} \rightarrow \underset{\substack{\text{bleaching} \\ \text{powder}}}{\text{`}CaOCl_2.H_2O\text{'}(s)}$$

In the manufacture of bleaching powder, slaked lime is moved forward by Archimedean screws through a series of pipes against a counter-current of chlorine until the requisite mass of chlorine has been absorbed. The solid is removed and packed.

It contains about 36% of available chlorine, *i.e.*, chlorine which can be removed by dilute acids and even by the carbonic acid of the atmosphere. Hence bleaching powder usually smells of chlorine and deteriorates if in contact with air. It has an extensive use in dye works, and in laundries.

$$CaOCl_2(s) + CO_2(g) \rightarrow CaCO_3(s) + Cl_2(g)$$

bleaching   carbon   chlorine
powder   dioxide

Chlorine in chemical industry

(*a*) Chlorine is extensively used as a bleaching agent and in the manufacture of bleaching agents and domestic antiseptic solutions such as sodium hypochlorite. It is also used in the manufacture of chlorates, used, for example, as weed-killers, and in the manufacture of hydrogen chloride. Hydrogen chloride is used in the production of PVC (page 341).

(*b*) Many organic chemicals are manufactured with the help of chlorine, *e.g.*, tetrachloromethane CCl_4, 1,1,2-trichloroethene C_2HCl_3, chloral $CCl_3.CHO$, and many others. These compounds are useful as degreasing agents, dry-cleaning fluids, and as sources for other products, *e.g.*, chloral for DDT.

(*c*) Chlorine is used to sterilize water for domestic and industrial use and for swimming-baths.

Chlorine occurs as an important by-product of electrolytic cells which produce sodium hydroxide from the raw material, sodium chloride. For this, see page 364. The products appear in the mass proportion of NaOH : Cl, *i.e.*, 40 : 35.5.

In the older *Deacon's Process*, hydrogen chloride (produced from sodium chloride and concentrated sulphuric acid) was oxidized by air to chlorine at about 450 °C in the presence of the *catalyst*, copper(II) chloride $CuCl_2$, distributed on broken brick to increase the available surface.

$$4HCl(g) + O_2(g) \rightarrow 2H_2O(g) + 2Cl_2(g)$$

Bromine

Bromine was discovered by Balard in 1826. He passed chlorine through the mother liquor obtained after crystallizing common salt from sea-water. The liquor turned red and from it he was able to isolate bromine and to show that it was an element.

Liebig, some years previously, had received the dark red liquid with a request to examine it, but thinking that it was merely a compound of iodine and chlorine, he did not pay it much attention.

It occurs chiefly as the bromides of potassium, sodium, and magnesium, usually in association with larger proportions of the chlorides of those metals. Since the bromides are much more soluble in water than the chlorides, a liquid rich in bromides is left by crystallizing out the chlorides. Treated in this way, the mother liquors from the Stassfurt deposits in Germany (which consist mainly of carnallite, $KCl.MgCl_2.6H_2O$), after the removal of a large proportion of the potassium chloride, contain about $\frac{1}{4}$% of bromide and the bromine is obtained by allowing this solution to come into contact with chlorine, which displaces the bromine:

$$MgBr_2(aq) + Cl_2(g) \rightarrow MgCl_2(aq) + Br_2(aq)$$

magnesium   bromine
bromide

Laboratory preparation of bromine

Bromine can be prepared in a way exactly analogous to one of the methods for making chlorine. An intimate mixture of potassium bromide and manganese(IV) oxide is placed in a retort (Figure 159); concentrated sulphuric acid is added

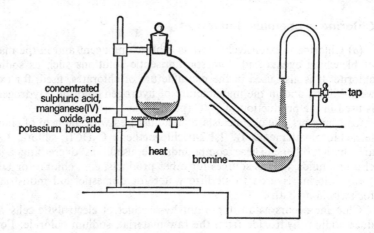

concentrated
sulphuric acid,
manganese(IV)
oxide, and
potassium bromide

tap

heat

bromine

Figure 159

and the mixture warmed. A red gas is given off (together with some misty fumes of hydrogen bromide) which condenses to a red liquid in the cooled receiver. This is bromine.

$$2Br^-(aq) + MnO_2(s) + 4H^+(aq) \rightarrow Mn^{2+}(aq) + 2H_2O(l) + Br_2(g)$$

Properties of bromine

Physical

(1) It is a heavy (density $= 3.2$ g cm^{-3}) red, volatile liquid (boiling point 59 °C).

(2) It has a choking, irritating smell. (Its name means 'a stench'.) The liquid causes sores on the flesh, which heal with difficulty.

(3) It is slightly soluble in water, forming a red solution containing about 3% of bromine at ordinary temperatures.

As a bleaching agent

Bromine is a bleaching agent, not so rapid as chlorine. A piece of damp litmus paper is bleached when placed in the vapour of bromine.

With hydrogen

Bromine combines with hydrogen, but not as readily as does chlorine. A mixture of chlorine and hydrogen will explode when merely exposed to sunlight, but a mixture of bromine and hydrogen needs the application of heat to induce combination, and the compound formed (hydrogen bromide) is not as stable as hydrogen chloride.

$$H_2(g) + Br_2(g) \rightarrow 2HBr(g)$$

Reaction between bromine and other elements

Bromine combines readily with most metals and non-metals to form bromides, for example, copper, iron, sodium, sulphur. It explodes when mixed with yellow phosphorus, so vigorous is the action. Phosphorus tribromide is made by gradually adding a solution of bromine in carbon tetrachloride to red phosphorus. The solution is used in order to moderate the action.

As an oxidizing agent

Bromine is an oxidizing agent, $Br_2 + 2e^- \rightarrow 2Br^-$, but not quite as vigorous as chlorine. It will perform the majority of the oxidations attributed to chlorine. Thus, on shaking an acidified solution of iron(II) sulphate with a few drops of bromine in a test-tube, the bromine colour soon disappears and the iron(II) sulphate has been converted into iron(III) sulphate.

$$2Fe^{2+}(aq) + Br_2(g) \rightarrow 2Fe^{3+}(aq) + 2Br^-(aq)$$

Action on the alkalis

The action of bromine on an alkaline solution is exactly analogous to that of chlorine. Thus:

Cold potassium hydroxide solution

$$2KOH(aq) + Br_2(g) \rightarrow KBr(aq) + KBrO(aq) + H_2O(l)$$
$$\text{potassium} \quad \text{potassium}$$
$$\text{bromide} \quad \text{hypobromite}$$

Hot potassium hydroxide solution

$$6KOH(aq) + 3Br_2(g) \rightarrow 5KBr(aq) + KBrO_3(aq) + 3H_2O(l)$$
$$\text{potassium}$$
$$\text{bromate}$$

In all these reactions bromine displays a great similarity to chlorine, but differs from it principally by its lower reactivity.

Iodine

Iodine was discovered in 1812 by Courtois. He treated with concentrated sulphuric acid the mother liquors obtained after extracting sodium carbonate from the ash obtained by burning seaweed (kelp). The ash contains a small percentage of iodides and the concentrated sulphuric acid formed hydrogen iodide and oxidized it to iodine. Gay-Lussac and Davy investigated the properties of the black solid, which was called iodine by Gay-Lussac.

Most of the iodine used today occurs as calcium iodate $Ca(IO_3)_2$, in the sodium nitrate deposits in Chile. The amount is very small (about 0.1%) but after the removal of the sodium nitrate by crystallization the proportion is much higher in the residues. The iodine is obtained by treatment with sodium hydrogensulphite.

Laboratory preparation of iodine

Experiment 127
Preparation of iodine from potassium iodide

Grind together some potassium iodide and manganese(IV) oxide in a mortar and place the mixture in a dish. Add concentrated sulphuric acid and place an inverted funnel over the dish as shown in Figure 160. Warm the mixture carefully and the violet vapour of iodine will be seen to condense on the cooler parts of the funnel to black shining plates. The chemistry of the action is similar to the formation of chlorine from common salt. The hydrogen iodide is, however, much more easily oxidized than even hydrogen bromide.

$$MnO_2(s) + 2I^-(aq) + 4H^+(aq) \longrightarrow$$
$$Mn^{2+}(aq) + 2H_2O(l) + I_2(s)$$

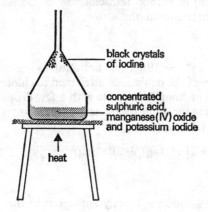

black crystals of iodine

concentrated sulphuric acid, manganese(IV) oxide and potassium iodide

heat

Figure 160

Properties of iodine

(i) It is a black shiny solid. Density 4.9 g cm^{-3}.

(ii) It sublimes when heated rapidly, forming a violet vapour from which the black solid can again be obtained by cooling.

(iii) It is almost insoluble in water but readily soluble in potassium iodide solution. This is due to the formation of a compound of potassium iodide and iodine, KI_3, which readily dissolves. This solution is brown. It also dissolves in ethanol and ether, forming brown solutions, and in carbon disulphide and trichloromethane, forming violet solutions.

Iodine does not bleach, and has little affinity for hydrogen. The effect of heating hydrogen iodide is to decompose the compound into its elements.

Iodine is a fairly active element and will combine with many metals to form iodides, but it does so much less readily than either chlorine or bromine.

Iodine is a mild oxidizing agent. It will not perform many of the ordinary oxidizing actions attributed to chlorine and bromine. It will, however, oxidize hydrogen sulphide to form hydrogen iodide and liberate sulphur.

$$H_2S(g) + I_2(g) \rightarrow 2HI(g) + S(s)$$

The action of iodine with alkalis is similar to the reactions of chlorine and bromine with alkalis. Hypoiodites, iodides, and iodates are produced (page 370).

Experiment 128
Action of iodine with starch solution

Place a 400 cm³ beaker full of water on a tripod and gauze and heat to boiling. Make a paste of a small amount of starch (about 1 g) and a little water, and pour this into the *boiling* water and stir. Allow to cool, or if the starch paste is required immediately, pour some of the paste into a boiling-tube and cool under the tap.

Add the smallest possible quantity of a solu-

tion of iodine (the test is sensitive to one part in one million) and immediately you will observe a blue coloration. Warm the mixture and the blue colour will disappear, but will return on cooling.

This test is given only by free iodine and is not given by, say, a solution of potassium iodide in water.

Uses of iodine

The antiseptic properties of iodine, due to its oxidizing nature, have been used in the treatment of small cuts. It is inadvisable to use it on larger wounds, because some people react violently to contact with it, with occasionally fatal results. It is sold as 'tincture of iodine' – a dilute solution of iodine in alcohol. It is used as iodine and iodides in medicine to treat cases of goitre, a disease which is sometimes due to a deficiency of iodine intake into the body. Small amounts of iodine have in fact been shown to be essential to the human body and all other forms of vertebrate life.

Summary comparison of the halogen elements

It has already been noted that the halogen elements show very marked similarity in chemical properties (page 97). This arises because each of the elements has a similar outer shell electronic structure; each element has seven electrons in the outermost shell.

$$F \quad 2, 7$$

$$Cl \quad 2, 8, 7$$

$$Br \quad 2, 8, 18, 7$$

$$I \quad 2, 8, 18, 18, 7$$

They are all univalent, forming an ion by gain of one electron per atom. This completes the external octet.

$$X_2 + 2e^- \rightarrow 2X^-, \quad \text{where X is Cl, Br, or I}$$

Being electron acceptors, they are all non-metals and oxidizing agents; fluorine is the most powerful oxidizing agent and iodine the least. Typical oxidations are:

$$2Fe^{2+} + Cl_2 \rightarrow 2Fe^{3+} + 2Cl^- \text{ (iron(II) ion to iron(III) ion)}$$
$$S^{2-} + Br_2 \rightarrow S + 2Br^- \quad \text{(H}_2\text{S to sulphur)}$$

Because of their very marked oxidizing action, chlorine and bromine are common bleaching agents; iodine is not.

In the order: Cl → Br → I, each halogen displaces an element to the right of it from simple salts. To do this, the more powerful oxidizing agent oxidizes the ion of the other halogen, so liberating the element.

$$Cl_2(g) + 2Br^-(aq) \rightarrow 2Cl^-(aq) + Br_2(l)$$
$$Br_2(l) + 2I^-(aq) \rightarrow 2Br^-(aq) + I_2(s)$$

Experiment 129
Displacement reactions of the halogens

(a) *Chlorine*

Displacement of bromine. Bubble chlorine into a solution of potassium bromide in water. The clear solution immediately turns red (due to formation of bromine water) and finally a drop of a red liquid (bromine) is observed at the bottom of the boiling-tube.

$$2KBr(aq) + Cl_2 \rightarrow 2KCl(aq) + Br_2(l)$$

Displacement of iodine. The above experiment is repeated with potassium iodide solution. The clear solution turns to the characteristic dark brown 'iodine' colour and finally a black solid (iodine) is deposited. On warming the solution the characteristic violet vapour of iodine is seen.

$$2KI(aq) + Cl_2(g) \rightarrow 2KCl(aq) + I_2(g)$$

In these two reactions, chlorine oxidizes the ions, Br^- and I^-, by attracting the electrons

from them, and chlorine is reduced to its ions by electron gain.

$$2Br^- \text{ (or } 2I^-) + Cl_2 \rightarrow Br_2 \text{ (or } I_2) + 2Cl^-$$

(b) *Bromine*

Bromine can displace iodine from iodides but cannot displace chlorine from chlorides. Thus, on adding a few drops of bromine to a solution of potassium iodide in water, the characteristic brown colour of the solution of iodine in potassium iodide is seen. On boiling the solution the violet vapour of iodine may be observed.

$$2KI(aq) + Br_2(l) \rightarrow 2KBr(aq) + I_2(aq)$$
$$\text{or } 2I^-(aq) + Br_2(l) \rightarrow 2Br^-(aq) + I_2(aq)$$

(c) *Iodine*

Iodine cannot displace chlorine or bromine from chlorides or bromides.

All three halogens behave in a similar way with aqueous caustic alkali solution.

$$\text{Cold dilute: } X_2 + 2OH^- \rightarrow X^- + XO^- + H_2O$$
$$\text{Hot conc.: } 3X_2 + 6OH^- \rightarrow 5X^- + XO_3^- + 3H_2O$$

where X is Cl, Br, or I. For names of products, see page 370.

The silver salts of the three halogens, AgCl, AgBr, AgI, are all insoluble in water and in dilute nitric acid. In the order given, they show gradation of colour: white → pale yellow → yellow, and gradation of solubility in ammonia: very soluble → slightly soluble → insoluble. All the silver salts are blackened by exposure to light, with reduction to metallic silver.

All these halogens combine directly with hydrogen (chlorine most readily, iodine least) and the halides, HCl, HBr, HI, are gaseous in ordinary conditions, very soluble in water and strongly acidic. The gases show a gradation of stability, being stable up to 1500 °C, 800 °C, and 180 °C respectively.

Chlorine can exhibit a covalency of seven, *i.e.*, the number of electrons in the outer shell, forming the acidic oxide Cl_2O_7. Neither of the other halogens shows this maximum valency towards oxygen; iodine forms the acidic oxide I_2O_5, but bromine forms no stable oxide.

Fluorine differs from the other halogens in that the number of electrons in its outer shell can never exceed the octet; so it always possesses a covalency of one.

It is the most electronegative element known, and in combination its oxidation number is always -1 (see page 173).

Questions

1. Describe the essential chemistry involved in the manufacture of chlorine from brine. Given calcium carbonate, potassium permanganate, sodium chloride, concentrated sulphuric acid and water, outline how you would prepare a sample of bleaching powder. What would you see when bleaching powder is added to: (*a*) purple litmus solution; (*b*) concentrated hydrochloric acid? (S.)

2. When dry chlorine is passed through hot iron wool in a combustion tube, the iron glows strongly and black crystals are deposited in the cooler parts of the tube. (i) Name the product of, and write an equation for, this reaction. (ii) Name *three* members of the halogen family other than chlorine. (iii) State whether each of the halogens named in (ii) would react with iron more vigorously or less vigorously than would chlorine.

Describe the reaction of chlorine with hydrogen under any one stated set of conditions. Compare the vigour of the reaction of hydrogen with chlorine with that of the reactions of hydrogen with the other halogens you have named. Name *two* halogens which react with potassium bromide solution. Write an equation for one of the reactions which occur. (J.M.B.)

3. How would you prepare and collect chlorine? Describe what you would observe and say what is formed when (i) a lighted taper *or* candle is placed in a jar of chlorine; (ii) dry chlorine is passed over heated iron wire; (iii) a solution of chlorine in water is left in sunlight. (C.)

4. Describe a method of manufacturing chlorine by an electrolytic process. Name the electrolyte and give the name, sign, and material of each electrode. Name also the products of the electrolysis and write ionic equations for the reactions occurring at the electrodes.

When dry hydrogen burns at a jet in a gas-jar of dry chlorine the following changes are observed: the colour of the chlorine disappears and a gas is formed which fumes in moist air and turns moist blue litmus red. Explain these observations.

Give the name and the structural formula for the product formed when one g-molecule (mole) of hydrogen chloride reacts with one g-molecule (mole) of acetylene (ethyne). Give *one* use of this product. (J.M.B.)

5. Describe how chlorine is manufactured from brine. Indicate the material, name, and polarity of each electrode, and write the ionic equation for the reaction at each electrode. Details of the industrial plant are not required. What other important substances are formed in this process? Using only the products of this process, what other two commercially important compounds can be made? Name each compound and write an equation for its formation. Give one use for each of the two products that you have named. (J.M.B.)

6. Chlorine is liberated when a concentrated solution of hydrogen chloride is oxidized. Name *two* reagents which will bring about this oxidation and write an equation for the reaction of *one* of them with concentrated hydrochloric acid.

Name the products and write equations for the reactions of chlorine with (i) a metal, (ii) a hydrocarbon, (iii) an aqueous solution of sulphur dioxide, (iv) an aqueous solution of potassium iodide. Explain why you would classify chlorine as an oxidizing agent, selecting *one* of these reactions to illustrate your argument.

Name *two* elements which are (together with chlorine) members of Group 7 of the Periodic Table, and mention *one* physical or chemical property in which lead chloride and lead iodide resemble each other and *one* in which they differ. (O.)

7. Sea-water contains about 2.5% by mass of sodium chloride and traces of potassium bromide. Suggest practical methods for obtaining from sea-water: (*a*) a sample of chlorine, (*b*) reasonably pure sodium chloride, (*c*) water, free from sodium and potassium ions. (O. and C.)

8. This is a question about the element fluorine, symbol F. It is a member of the halogen family, and is more reactive than chlorine. You will not be familiar with the chemistry of fluorine, but you should be able to predict some of its important properties from what you know about the properties of the other halogens.

(*a*) What physical properties would you expect fluorine to have?

(*b*) How would you expect fluorine to occur in nature?

(*c*) Suggest a method which might be used to obtain fluorine as the free element.

(*d*) What would you expect to see if fluorine was reacted with aqueous solutions of potassium chloride, potassium bromide, and potassium iodide? Write an equation for any *one* of these reactions.

(e) How would you expect fluorine to react with water? Predict the approximate pH of the resulting liquid.

(f) How would you expect fluorine to react with hydrogen? Name the product of this reaction and write an equation for its formation. What properties would you expect this product to have?

(g) Fluorine reacts with sodium to form fluoride. What properties would you expect this compound to have? (L.)

9. The halogens have the following atomic numbers: fluorine 9, chlorine 17, bromine 35, iodine 53.

(a) Give the electronic configuration for (i) an atom of fluorine, (ii) an atom of chlorine.

(b) How many valency electrons has an iodine atom?

(c) Write down the names of the four halogens and after each state whether it is a solid, a liquid, or a gas under ordinary laboratory conditions.

(d) How, and under what conditions, does chlorine react with (i) hydrogen, (ii) iron, (iii) potassium bromide?

(e) State whether iodine is more or less reactive than chlorine. Give an example to illustrate your answer. (J.M.B.)

10. Iodine is a solid element which resembles chlorine in chemical properties and in the formulae of corresponding compounds, though it is less electronegative than chlorine (*i.e.*, 'less reactive').

(a) Write down the formulae of hydriodic acid, potassium iodate, calcium iodide, and potassium iodide.

(b) Chlorine gas is bubbled through an aqueous solution of sodium iodide in one test-tube; in another iodine is shaken with an aqueous solution of sodium chloride. Describe what you would expect to *observe* in each test-tube and write equations for any reactions you think might occur.

(c) Iodine forms two oxides, A and B. When 1.590 g of the oxide A is heated to 180 °C it decomposes to 0.254 g of iodine and 1.336 g of oxide B. When B is heated to 300 °C it decomposes to iodine and oxygen; 0.668 g of B gives 0.508 g of iodine. Find the formulae of A and of B and write the equation for the reaction which occurs when A is heated to 180 °C.

(d) What would you expect to happen if a solution of silver nitrate is added to one of sodium iodide? (L.)

11. The halogens have the following boiling points: fluorine (-188 °C), chlorine (-34 °C), bromine ($+59$ °C), and iodine ($+184$ °C). The heats of dissociation of the molecules are:

$$F_2 \rightarrow 2F \quad \Delta H = +158 \text{ kJ mol}^{-1}$$
$$Cl_2 \rightarrow 2Cl \quad \Delta H = +242 \text{ kJ mol}^{-1}$$
$$Br_2 \rightarrow 2Br \quad \Delta H = +193 \text{ kJ mol}^{-1}$$
$$I_2 \rightarrow 2I \quad \Delta H = +151 \text{ kJ mol}^{-1}$$

State how the stabilities of the halogen molecules vary. At 1700 °C one halogen only is completely dissociated into atoms. Name this element.

Both the boiling points and the heats of dissociation show a progressive change through the family of halogens. Show that this pattern of change is also followed in their reactivity with (i) hydrogen, (ii) metals, and (iii) in the displacement of halogens from halides. How does chlorine react with sodium hydroxide solution? (J.M.B.)

12. Chlorine is one of the products of the electrolysis of molten sodium chloride under suitable conditions; write an equation for the electrode reaction by which the chlorine is released, stating at which electrode this occurs. What quantity of electricity (in Faradays) is required to release one mole of chlorine gas? How, and under what conditions, does chlorine react with (i) iron, (ii) potassium iodide, (iii) hydrogen sulphide?

Knowing that a close similarity exists between chlorine, bromine, and iodine, suggest what you would expect to be the effect of water on silver bromide and of hydrogen sulphide on iodine. (O.)

31 Compounds of the Halogens

The principal compounds to be considered in this chapter are the hydrogen halides (of general formula HX, where X = Cl, Br, or I) and the ionic halides (of general formula MX_n, where M is usually a metallic ion).

The Hydrogen Halides

Hydrogen Chloride

The gas is usually called hydrogen chloride, or hydrochloric acid gas, whereas the solution of the gas in water (which when saturated contains about 36% by mass of the gas) is termed hydrochloric acid.

Laboratory preparation of hydrogen chloride

Experiment 130
Preparation of hydrogen chloride from common salt

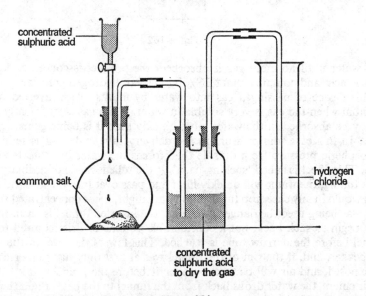

concentrated
sulphuric acid

hydrogen
chloride

common salt

concentrated
sulphuric acid
to dry the gas

Figure 161

Common salt (sodium chloride) is placed in a flask fitted with a dropping funnel and delivery-tube, and concentrated sulphuric acid is added (Figure 161). There is effervescence. and misty fumes are observed. The gas is passed through a wash-bottle containing concentrated sulphuric acid to dry it, and collected as shown by downward delivery, the gas being denser than air.

$$H_2SO_4(l) + NaCl(s) \longrightarrow$$
$$NaHSO_4(s) + HCl(g)$$

or $\quad H_2SO_4 + Cl^- \longrightarrow$
$$HSO_4^- + HCl$$

The reaction proceeds in the cold, although a further yield of the gas was obtained in the industrial process by heating to a red heat. The sulphate is not obtainable under laboratory conditions. Notice that here is another case of a volatile acid (HCl) being driven off by a comparatively involatile acid.

$$NaCl(s) + NaHSO_4(s) \longrightarrow Na_2SO_4(s) + HCl(g)$$

sodium   sodium   sodium   hydrogen
chloride   hydrogen-   sulphate   chloride
sulphate

(The above indicates clearly the acid nature of sodium hydrogensulphate.)

A solution of the gas in water can be made by means of the funnel arrangement as in Figure 162. This solution is hydrochloric acid. If the gas is passed

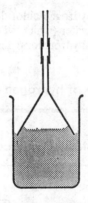

Figure 162

into water until no more gas is absorbed, the product is concentrated hydrochloric acid and contains about 36% by mass of hydrogen chloride.

This device of passing a gas into water by means of an inverted funnel is essential when the gas is very soluble in water. If the gas is sufficiently soluble, it may be absorbed in the water more quickly than it is being generated in the flask. In this case, the pressure in the delivery tube and flask is reduced and atmospheric pressure from outside then forces the water back up the delivery tube. This effect is called 'sucking-back'. If the tube is made of ordinary narrow glass tubing, the water will quickly fill it and pass over into the generating flask. This would in any case stop the reaction and might, if hot concentrated sulphuric acid was being used, be dangerous. When the inverted funnel is used, the water may begin to suck back, but a considerable volume of it is required to fill the funnel before the narrow tube is reached. This lowers the level of the water in the beaker, and, if the rim of the funnel was at first only just immersed, it will be exposed, and air will be forced under it before the funnel is filled. As soon as air enters, the water drops back from the funnel to the beaker and the process begins again. Notice that, to be effective, the funnel must be arranged with its rim only just immersed. An additional advantage is that the funnel offers a large water surface for absorption of the gas. This device is also useful when solutions of sulphur dioxide or ammonia are being prepared (page 408).

Test for hydrogen chloride

It is a clear gas (although in damp air it appears misty), acid to litmus, **and produces a white precipitate of silver chloride in a drop of a solution of silver nitrate and nitric acid which is held on a glass rod in the gas.**

$$Ag^+(aq) + Cl^-(aq) \rightarrow AgCl(s)$$

It also produces dense white fumes in the presence of ammonia.

Properties of hydrogen chloride

It has a choking, irritating smell, and is an acid gas, which is very soluble in water. These latter two properties are neatly shown by the *fountain experiment* (Figure 180). In this case, blue litmus is placed in the trough and turns red in the flask, showing the acidity of hydrogen chloride in aqueous solution.

The aqueous solution is known as hydrochloric acid, and being almost completely ionized in aqueous solution, this acid is very strong:

$$HCl(g) + H_2O(l) \rightleftharpoons H_3O^+(aq) + Cl^-(aq)$$

the position of equilibrium being far over to the right. The solution shows the usual acidic properties:

(i) *liberation of hydrogen with certain metals* (Mg, Zn, Fe):

$$Zn(s) + 2H^+(aq) \rightarrow Zn^{2+}(aq) + H_2(g)$$

(ii) *neutralization of bases to form salts and water*

$$Na^+OH^-(aq) + H^+Cl^-(aq) \rightarrow Na^+Cl^-(aq) + H_2O(l)$$

(iii) *liberation of carbon dioxide from carbonates*

$$2H^+(aq) + CO_3{}^{2-}(aq) \rightarrow H_2O(l) + CO_2(g)$$

In (i) notice, however, that **hydrochloric acid does not react with copper, or with other metals below hydrogen in the reactivity series.**

If a concentrated hydrochloric acid solution is heated, hydrogen chloride escapes into the air. If a very dilute solution is heated, water is lost, making the acid more concentrated. In both cases a mixture is finally obtained containing 20.24% of hydrochloric acid, which distils over unchanged at 760 mm mercury pressure. This is termed a 'constant boiling mixture'.

Since chlorine is the product of oxidation of hydrochloric acid (page 362), most oxidizing agents will liberate chlorine from it, and it may therefore *not* be used for the acidification of a solution of a reducing agent which is to be titrated against an oxidizing agent (for example, the titration of iron(II) in solution against potassium permanganate solution).

Manufacture of hydrogen chloride

For many years hydrogen chloride was manufactured exclusively by heating common salt with concentrated sulphuric acid. This was the first stage of the Leblanc process.

$$2NaCl(s) + H_2SO_4(l) \rightarrow 2HCl(g) + Na_2SO_4(s)$$

The gas was absorbed in water to form hydrochloric acid.

Recently, improved methods of manufacture of hydrogen have made it possible to produce hydrogen chloride in quantity by direct combination of hydrogen with chlorine, which is obtained by electrolysis of brine (page 364).

$$H_2(g) + Cl_2(g) \rightarrow 2HCl(g)$$

Hydrogen Bromide

Hydrogen bromide is not as stable as hydrogen chloride and is decomposed to some extent by the action of heat. This, together with the fact that hot concentrated sulphuric acid is an oxidizing agent, makes it impossible to prepare pure hydrogen bromide by the action of heat on a mixture of concentrated sulphuric acid and potassium bromide. The products of such an attempt would be hydrogen bromide, bromine, and sulphur dioxide.

It may be made by the action of bromine on a mixture of red phosphorus and water. The chemistry of the action is that bromides of phosphorus are formed which are decomposed by the water.

Laboratory preparation of hydrogen bromide

Note: It is not intended that this experiment be carried out by the student.

A paste of red phosphorus and water is placed in a round-bottomed flask (sand may be added to 'dilute' the mixture). Bromine is dropped in gradually from a tap funnel and the reaction proceeds at ordinary temperature (Figure 163).

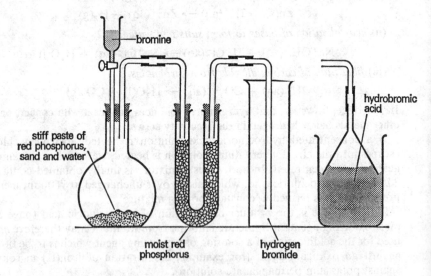

Figure 163

Heat is evolved during the process and a considerable amount of bromine may be volatilized and would, if not removed, contaminate the hydrogen bromide. The bromine is removed by passing it through a U-tube containing beads smeared with red phosphorus and water. **This U-tube is, in fact, a secondary generating apparatus** so arranged as to offer a large area of phosphorus to the bromine so that it is removed as completely as possible. The misty gas, very similar in appearance to hydrogen chloride, is collected by displacement of air as shown in Figure 163, the gas being denser than air. It can also be dissolved in water to form hydrobromic acid by the apparatus shown on page 382.

$$4P(s) \ + 6Br_2(l) \rightarrow 4PBr_3(g)$$
phosphorus   bromine   phosphorus
tribromide

$$PBr_3(g) + 3H_2O(l) \rightarrow H_3PO_3(aq) + 3HBr(g)$$
phosphorous   hydrogen
acid   bromide

(Alternatively hydrogen bromide may be prepared by heating potassium bromide with conc. sulphuric acid diluted with half its own volume of water.)

Test for hydrogen bromide

Hydrogen bromide turns damp blue litmus paper red, and gives a pale yellow precipitate of silver bromide with a mixture of silver nitrate solution and nitric acid. The precipitate is only slightly soluble in aqueous ammonia.

$$Ag^+(aq) + Br^-(aq) \rightarrow AgBr(s)$$

Properties of hydrogen bromide

(i) It is a fuming gas with a choking smell and its density is almost three times that of air.

(ii) It is very soluble in water, forming a strongly acid solution. A saturated solution of hydrogen bromide contains about 70% by mass of hydrogen bromide at ordinary temperatures.

(iii) It is less stable than hydrogen chloride, being more easily decomposed into its elements.

(iv) It is, in its general chemical properties, similar to hydrogen chloride.

Hydrogen Iodide

This gas is much less stable than even hydrogen bromide. A solution of hydrogen iodide in water quickly darkens because of the formation of iodine.

Laboratory preparation of hydrogen iodide

It is usually prepared by the action of water on a mixture of red phosphorus and iodine. Since iodine is a solid and does not volatilize to an appreciable extent during the reaction, there is no need for a U-tube as in the case of the similar preparation of hydrogen bromide (Figure 163).

For class purposes, the reactions of hydrogen iodide can easily be shown by the very simple apparatus shown in Figure 164. Grind a little red phosphorus with iodine in a mortar and introduce the mixture into a dry boiling-tube. Add about four drops of water, insert a cork fitted with delivery tube, and allow the gas to fall (hydrogen iodide is four times as dense as air) into test-tubes, one containing silver nitrate solution and the other containing a few drops of ammonia. If the tube is heated, the decomposition of the gas is shown by the violet vapour of iodine which then appears.

$$4P(s) + 6I_2(s) \rightarrow 4PI_3(s)$$
$$PI_3(s) \ + 3H_2O(l) \rightarrow 3HI(g) \ + H_3PO_3(aq)$$
phosphorus   hydrogen   phosphorous
tri-iodide   iodide   acid

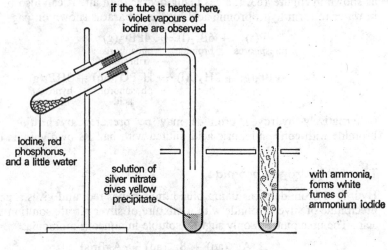

if the tube is heated here, violet vapours of iodine are observed

iodine, red phosphorus, and a little water

solution of silver nitrate gives yellow precipitate

with ammonia, forms white fumes of ammonium iodide

Figure 164

A solution of hydriodic acid can be more simply obtained by bubbling hydrogen sulphide into a suspension of iodine in water. The end of the reaction is reached when all the iodine is seen to have disappeared. The precipitated sulphur is filtered off.

$$H_2S(g) + I_2(aq) \rightarrow 2HI(aq) + S(s)$$

Test for hydrogen iodide

Add a little chlorine-water to a gas-jar of the gas, and pour a few drops of the liquid into starch paste. A blue colour is observed.

Properties of hydrogen iodide

Hydrogen iodide is a fuming, choking gas, very soluble in water, forming *hydriodic acid*. The gas readily dissociates (reversibly) above 180 °C.

$$2HI(g) \rightleftharpoons H_2(g) + I_2(g)$$

The acid, usually used as *acidified potassium iodide* solution, is a vigorous reducing agent $(2I^- - 2e^- \rightarrow I_2)$. It is oxidized with liberation of *iodine* (brown) by exposure to air and by:

hydrogen peroxide

$$H_2O_2(aq) + 2I^-(aq) + 2H^+(aq) \rightarrow 2H_2O(l) + I_2(aq)$$

potassium permanganate

$$2MnO_4^-(aq) + 10I^-(aq) + 16H^+(aq) \rightarrow 2Mn^{2+}(aq) + 8H_2O(l) + 5I_2(aq)$$

The Halides

It has already been seen (page 375) that the halogens usually react by gaining an electron for each halogen atom, so completing the outermost octet of electrons.

$$X_2 + 2e^- \rightarrow 2X^-$$

where X is F, Cl, Br, or I.

Thus, each halogen gives rise to a well-defined series of salts, usually by reaction with metallic elements from which the electron may be transferred. These salts are known as the *halides*, and since the most common element amongst the halogens is chlorine, it is the properties of the *chlorides* which will be examined in what follows, with brief comparisons for the *bromides* and *iodides*.

Chlorides

Preparation of metallic chlorides

All metals are attacked by chlorine to form chlorides. The methods of preparation are summarized below.

The chlorides of these metals are made by the action of:

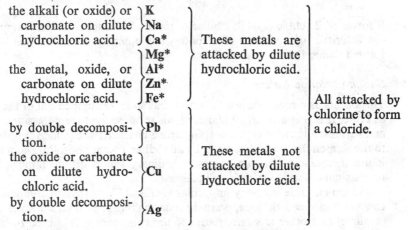

N.B. Iron(III) chloride is made by the action of chlorine on the metal.

* The chlorides of these metals are very deliquescent (see page 226). The above is a list of common metals in the order of the reacting series.

Characteristics of selected chlorides

Aluminium chloride AlCl₃

This is a pale yellow solid. Being readily hydrolysed by water, it fumes in damp air with evolution of hydrogen chloride. If required anhydrous, it must be prepared by *heating* aluminium foil in *dry* chlorine or *dry* hydrogen chloride. The apparatus is the same as for iron(III) chloride (page 369).

$$2Al(s) + 3Cl_2(g) \rightarrow 2AlCl_3(s)$$
$$2Al(s) + 6HCl(g) \rightarrow 2AlCl_3(s) + 3H_2(g)$$

The anhydrous solid reacts rather violently with water. It forms the hydrate, $AlCl_3.6H_2O$, and, with excess water, dissolves and hydrolyses considerably.

$$AlCl_3(s) + 3H_2O(l) \rightleftharpoons Al(OH)_3(s) + 3HCl(aq)$$

On evaporation to dryness, the solution leaves hydroxide as residue.

Ammonium chloride (sal-ammoniac) NH₄Cl

This compound has many uses, two of the chief being as a constituent of the Leclanché electric battery, and as a flux in soldering. It may be prepared by boiling ammonium sulphate solution with common salt.

$$(NH_4)_2SO_4(aq) + 2NaCl(aq) \rightleftharpoons 2NH_4Cl(aq) + Na_2SO_4(aq)$$

On cooling, sodium sulphate crystallizes first and may be removed by filtration, after which ammonium chloride may be obtained from the filtrate as a white solid. It sublimes on being heated. For the action of heat on ammonium chloride see page 431.

Potassium chloride KCl

This occurs as carnallite ($KCl.MgCl_2.6H_2O$). Potassium chloride can be prepared in the laboratory by the neutralization of potassium hydroxide solution with dilute hydrochloric acid.

$$KOH(aq) + HCl(aq) \rightarrow KCl(aq) + H_2O(l)$$

potassium   dilute   potassium   water
hydroxide   hydrochloric chloride
             acid

It forms white cubic crystals similar to those of sodium chloride. It imparts the characteristic *lilac* colour of a potassium compound to the non-luminous Bunsen flame. It is not deliquescent.

Sodium chloride NaCl

This occurs as a rock salt, which is mined as solid or pumped out of the ground as brine. It is prepared by similar methods to those used for potassium chloride and forms white cubic crystals. It imparts a characteristic golden-yellow colour to the Bunsen flame which is given by all sodium compounds. Pure common salt is not deliquescent (the dampness of ordinary salt is due to impurities, *e.g.*, magnesium chloride which is deliquescent).

Sodium chloride is a very important chemical. As a sodium compound, it is converted into caustic soda, washing soda, baking soda, and salt-cake (sodium sulphate) and other less important sodium compounds. It is used to 'salt-out' soap. As a chloride, it yields hydrochloric acid and chlorine, used as a bleaching agent, and in manufacture of hypochlorite solutions for home use, as well as in fine chemical manufacture.

Sodium chloride crystallizes as a face-centred cube (see page 80, and Figure 34). When heated the ions in the solid lattice vibrate increasingly; at 801 °C the lattice collapses, *i.e.*, the salt melts and the ions become mobile. The fused salt conducts electricity by migration of Na^+ to cathode and Cl^- to anode where each is discharged to give *sodium* and *chlorine*.

$$2Na^+ + 2e^- \rightarrow 2Na; \quad 2Cl^- \rightarrow Cl_2 + 2e^-$$

The lattice arrangement in the solid is determined by X-ray diffraction. A narrow beam of X-rays is passed through a crystal of the compound and on to a photographic plate. After development, the plate shows a central spot, produced by X-rays which have passed straight through the crystal, and, round it, a diffraction pattern of spots from which the lattice arrangement can be deduced.

Calcium chloride CaCl₂

This is very deliquescent and can be used as a drying agent for most gases (but not for ammonia, with which it forms a compound). The anhydrous salt is prepared by evaporating a solution until the solid formed fuses. The solution is most easily prepared by adding marble or limestone to dilute hydrochloric acid until a little of the marble remains. The mixture is then filtered.

$$CaCO_3(s) + 2HCl(aq) \rightarrow CaCl_2(aq) + H_2O(l) + CO_2(g)$$

Anhydrous iron(III) chloride FeCl₃

This is made by the action of iron on chlorine by the method described on page 368.

The anhydrous salt cannot be made by the evaporation of the solution because the chloride is attacked by water when a concentrated solution is evaporated. This type of action is termed *hydrolysis*.

$$FeCl_3(aq) + 3H_2O(l) \rightarrow Fe(OH)_3(s) + 3HCl(g)$$

$$2Fe(OH)_3(s) \xrightarrow{heat} Fe_2O_3(s) + 3H_2O(g)$$

On heating the solution in air the final product is iron(III) oxide. This action is the reverse of neutralization.

Iron(III) chloride is a black solid in the anhydrous state, but forms a brown solution if concentrated, and a yellow solution if dilute. It can be reduced by reducing agents (*e.g.*, zinc and dilute hydrochloric acid) to iron(II) chloride.

$$Zn(s) + 2Fe^{3+}(aq) \rightarrow Zn^{2+}(aq) + 2Fe^{2+}(aq)$$

With alkaline solutions, it gives (as do all iron(III) salts dissolved in water) a reddish-brown gelatinous precipitate of iron(III) hydroxide.

$$FeCl_3(aq) + 3NaOH(aq) \rightarrow Fe(OH)_3(s) + 3NaCl(aq)$$

ionically: $Fe^{3+}(aq) + 3OH^-(aq) \rightarrow Fe(OH)_3(s)$

Iron(II) chloride FeCl₂

Anhydrous iron(II) chloride, a white solid, is made by heating iron wire strongly in a stream of dry hydrogen chloride.

$$\underset{\text{iron}}{Fe(s)} + \underset{\substack{\text{hydrogen} \\ \text{chloride}}}{2HCl(g)} \rightarrow \underset{\substack{\text{iron(II)} \\ \text{chloride}}}{FeCl_2(s)} + \underset{\text{hydrogen}}{H_2(g)}$$

It forms a pale green solution which gives, with alkaline solutions, a dirty green precipitate of iron(II) hydroxide (as will any iron(II) salt dissolved in water).

$$\underset{\substack{\text{iron(II)} \\ \text{chloride}}}{FeCl_2(aq)} + 2NaOH(aq) \rightarrow 2NaCl(aq) + \underset{\substack{\text{iron(II)} \\ \text{hydroxide}}}{Fe(OH)_2(s)}$$

ionically: $Fe^{2+}(aq) + 2OH^-(aq) \rightarrow Fe(OH)_2(s)$

Lead(II) chloride PbCl₂

This is a white insoluble substance made by the interaction of a solution of *any* soluble lead(II) salt with a solution of *any* soluble chloride.

Into a beaker put some dilute hydrochloric acid and add lead(II) nitrate solution. There is a white precipitate.

$$Pb(NO_3)_2(aq) + 2HCl(aq) \rightarrow PbCl_2(s) + 2HNO_3(aq)$$

ionically: $Pb^{2+}(aq) + 2Cl^-(aq) \rightarrow PbCl_2(s)$

Filter off the white precipitate, wash it two or three times with a little *cold* distilled water, and put it on a porous plate to dry.

Lead(II) chloride is almost insoluble in cold water yet fairly soluble in hot water.

Silver chloride AgCl

This is a white insoluble compound made by adding a solution of *any* soluble silver salt to a solution of *any* soluble chloride.

$$\underset{\substack{\text{silver} \\ \text{nitrate}}}{AgNO_3(aq)} + \underset{\substack{\text{sodium} \\ \text{chloride}}}{NaCl(aq)} \rightarrow \underset{\substack{\text{silver} \\ \text{chloride}}}{AgCl(s)} + \underset{\substack{\text{sodium} \\ \text{nitrate}}}{NaNO_3(aq)}$$

ionically: $Ag^+(aq) + Cl^-(aq) \rightarrow AgCl(s)$

The white solid is filtered off, washed two or three times with hot distilled water, and dried on a porous plate. (The whole action should be performed in the absence of light since silver chloride turns violet on exposure to light.)

All the silver halides, AgCl, AgBr, and AgI, emit electrons when exposed to light (a *photoelectric* effect) and are slowly reduced, turning violet, and, eventually, black. The chief reduction product is metallic silver.

$$Ag^+ + e^- \rightarrow Ag$$

The halides show decreasing photoelectric activity in the order: AgBr → AgCl → AgI. This is why silver bromide is used as the principal silver compound suspended in the gelatine of a photographic plate. The other silver halides are less sensitive to the action of light. Silver chloride dissolves readily in ammonia solution, which distinguishes it from silver bromide and silver iodide, and therefore serves as one test for the chloride ion (see page 389).

General properties of chlorides

(1) *Action with concentrated sulphuric acid.* On being treated with concentrated sulphuric acid, a chloride evolves hydrogen chloride, *e.g.,*

$$NaCl(s) + H_2SO_4(l) \rightarrow NaHSO_4(s) + HCl(g)$$

(2) *Volatility.* Chlorides are more volatile than most salts. This makes them suitable for use in the 'flame-test' in which certain metals can be detected by the colour their vapour imparts to the Bunsen flame. To perform the flame-test, the substance under consideration is moistened with concentrated hydrochloric acid and a nichrome or platinum wire is dipped into the mixture and applied to the non-luminous Bunsen flame.

Metal chloride	Colour
Sodium	Persistent golden yellow (invisible through blue glass)
Potassium	Lilac flame (visible through blue glass)
Lithium	Carmine ('rich red')
Copper	Green (blue zone)
Calcium	Red ('brick red')

(3) *Hydrolysis of chlorides.* Several chlorides are readily hydrolysed by water, *e.g.,* magnesium, zinc, and iron chlorides. If solutions of the chlorides are evaporated a basic salt or the oxide of the metal remains.

(4) *Action of concentrated sulphuric acid on mixture of chloride and oxidizing agent.* Mix together a little common salt and manganese(IV) oxide (many other oxidizing agents would be suitable). Put this into a test-tube, add a few drops of concentrated sulphuric acid, and warm. A green gas, chlorine, is evolved.

$$2NaCl(s) + 2H_2SO_4(aq) + MnO_2(s) \rightarrow MnSO_4(aq) +$$
sodium   concentrated   manganese(IV)   manganese
chloride   sulphuric   oxide   sulphate
    acid

$$Na_2SO_4(aq) + 2H_2O(l) + Cl_2(g)$$
sodium   water   chlorine
sulphate

Test for a soluble chloride

Dissolve a suspected chloride in *distilled water* and add a little nitric acid and then silver nitrate solution. If a chloride is present a white precipitate of silver chloride will be seen.

$$Ag^+(aq) + Cl^-(aq) \rightarrow AgCl(s)$$

Divide the precipitate into two parts. Add dilute ammonium hydroxide to one and observe that the precipitate dissolves. Allow the other to be exposed to the light for a few minutes. The precipitate will turn violet.

Silver chloride is insoluble in nitric acid but soluble in ammonium hydroxide. The only two common insoluble chlorides are lead chloride and silver chloride.

Bromides

The bromides are prepared, generally speaking, by the same methods as the chlorides and possess similar properties. They can readily be distinguished from the chlorides by the action of chlorine gas which has no effect on the chlorides but displaces bromine from bromides (see page 376 for experimental details).

Iodides

Iodides are similar to chlorides and bromides but can be readily distinguished by the action of chlorine or bromine, which liberate iodine.

$$2KI(aq) + Cl_2(g) \rightarrow 2KCl(aq) + I_2(aq)$$

Addition of silver nitrate in dilute nitric acid to an iodide solution precipitates silver iodide, a yellow salt which is insoluble in ammonia solution.

$$NaI(aq) + AgNO_3(aq) \rightarrow AgI(s) + NaNO_3(aq)$$

Questions

1. Give the formulae of the chlorides of the following elements and the physical state in which each compound exists at room temperature: (i) carbon, (ii) hydrogen, (iii) magnesium.

Give an equation to show what change occurs when hydrogen chloride is added to water. (J.M.B.)

2. You have been provided with a mixture of copper(II) oxide and common salt. How would you obtain from this (a) pure dry sodium chloride, (b) pure copper, and (c) chlorine gas? (L.)

3. Hydrogen chloride can be prepared from sodium chloride and sulphuric acid. (a) State the conditions necessary to obtain a steady supply of the gas in the laboratory. (b) Give the equation for the reaction which occurs under these conditions. (c) How is the gas collected? (d) Give a diagram of the apparatus you would use. (e) What precaution would you take when dissolving the gas in water? (f) What reaction occurs when hydrogen chloride molecules dissolve in water?

Describe reactions (*one* in each case) in which hydrochloric acid (i) is oxidized; (ii) reacts to form a gaseous compound; (iii) reacts to give a precipitate of an insoluble salt. (A.E.B.)

4. When dilute hydrochloric acid is warmed with each of the following compounds, a gas is evolved. Name this gas and give a chemical test to establish its identity. (a) Sodium carbonate, (b) sodium sulphide, (c) sodium sulphate. (J.M.B.)

5. Describe how you would prepare reasonably pure samples of (a) lead(II) chloride given lead, concentrated nitric acid, and sodium chloride, (b) anhydrous iron(III) chloride given iron, concentrated sulphuric acid, sodium chloride, and manganese(IV) oxide. (O. and C.)

6. Outline experiments by which you could prepare (a) hydrogen chloride from chlorine, (b) chlorine from hydrogen chloride. Explain the bleaching action of chlorine on a dye such as litmus. Describe what happens in *two* reactions (other than bleaching) in

which chlorine and bromine behave similarly. Write the equations for these reactions. (S.)

7. Given ammonium chloride, and any other necessary reagents, describe how you would obtain from it (*a*) ammonia, (*b*) hydrogen chloride, (*c*) nitrogen, (*d*) chlorine. In each case the necessary conditions for the reaction and the method of collection of the gas should be stated, but diagrams of the apparatus are *not* required. (J.M.B.)

8. Describe how you would obtain a sample of hydrogen chloride gas from sodium chloride. You should give the *name* of the other chemical required and the equation for the reaction. No diagram is required.

With the aid of a diagram, explain how you would obtain an aqueous solution of this gas.

Describe, giving an equation in each case, the reaction of hydrochloric acid with (i) silver nitrate solution, (ii) granulated zinc. (O. and C.)

9. You are supplied with a solution of sodium hydroxide, crystalline lead nitrate, iron filings, dilute hydrochloric acid, and a source of chlorine. Describe how you would prepare samples of the anhydrous chlorides of sodium, iron(III), and lead(II). (O. and C.)

32 Sulphur

Occurrence

Sulphur occurs:

1. In Louisiana and Texas, U.S.A., as free sulphur.
2. In petroleum gases, *e.g.*, at Lacq, as hydrogen sulphide.

Extraction of sulphur

In America the deposits lie at a depth of about 160 m with deposits of limestone, clay, and sand between the ground level and the sulphur (Figure 165). It is not

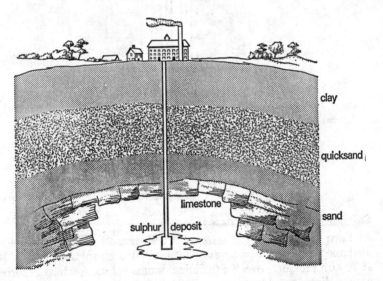

clay

quicksand

limestone

sand

sulphur deposit

Figure 165

necessary to mine the sulphur by sinking shafts as in the case of coal, for sulphur differs from coal in having a fairly low melting point (115 °C). By utilizing this property the sulphur can be extracted, by a method invented by Frasch, cheaply, rapidly, and in a high state of purity.

A hole about 30 cm in diameter is bored down through the clay, sand, and limestone to the sulphur beds. This boring is lined with an iron pipe and, inside the pipe, is sunk a device called the sulphur pump. It consists of three concentric tubes which terminate in a reservoir of larger diameter (see Figure 166). Down the outermost of the three tubes is forced a stream of water at about 170 °C. This water must be kept at a pressure of about 10 atm to maintain it in the liquid state, *i.e.*, it is super-heated water, and it is hot enough to melt the sulphur.

The molten sulphur flows into the reservoir at the base of the pump and is forced up to the surface through the second of the three tubes by means of a blast of hot compressed air at a pressure of about 15 atm, which is forced down the narrowest tube. The sulphur is run into large tanks, where it solidifies and can be separated from the water. Sulphur more than 99% pure is produced by this operation and a single pump may produce up to 500 tonnes of this high-grade sulphur daily.

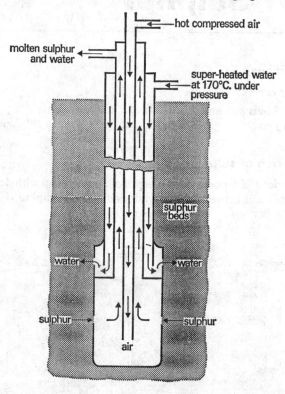

Figure 166

Sulphur from petroleum gases

At Lacq (S. France), gas associated with petroleum deposits contains mainly methane CH_4, with carbon dioxide and hydrogen sulphide (15%). It is passed, at 70 atm pressure, over an (alkaline) amine solution, which absorbs the acidic carbon dioxide and hydrogen sulphide. These gases are then released by heating at atmospheric pressure. Controlled supplies of air are added and the hydrogen sulphide is oxidized to sulphur in three stages, the last two in contact with a heated catalyst, *bauxite*.

$$2H_2S(g) + O_2(g) \rightarrow 2H_2O(l) + 2S(s)$$

The sulphur vapour is condensed and then cooled to solid. The product is 99.9% pure. The deposits have been worked since 1957 and present yield is about 1.4 million tonnes per annum.

Uses of sulphur

The element, sulphur, is a yellow solid. It is usually sold as either 'flowers of sulphur', a powder, or 'roll sulphur', cylindrical sticks. The output of sulphur in the world today exceeds 32 million tonnes annually, about two-thirds of it being produced by the United States of America. This vast amount is used in the following ways:

1. For the manufacture of sulphuric acid (see page 411).
2. For dusting vines to prevent the growth of certain kinds of fungus.
3. In making calcium hydrogensulphite $Ca(HSO_3)_2$, which is used as a bleacher of wood-pulp in the manufacture of paper.
4. For the vulcanization of rubber, a process which converts the soft pliable rubber into the hard, tough substance of which motor tyres and similar products are made.
5. In smaller quantities for the manufacture of dyes, fireworks, sulphur compounds such as carbon disulphide CS_2, and medicinally in ointments.

Effect of heat on sulphur

Experiment 131
The action of heat on sulphur when air is excluded

Place some powdered roll sulphur in a narrow test-tube and warm it gently, shaking well. Try to avoid local overheating by rotating the test-tube. The sulphur passes through the following stages as the temperature rises:

1. It melts at about 115° C to an amber-coloured, mobile liquid.
2. It becomes much darker in colour and, suddenly, at 160 °C, very viscous. So

viscous does it become that the test-tube may be inverted without loss of sulphur.

3. The sulphur gradually becomes more mobile again and very dark reddish brown in colour.
4. The sulphur boils at 444 °C, giving off light brown sulphur vapour.

These changes occur in the reverse order as the sulphur cools.

Experiment 132
The action of heat on sulphur with a plentiful supply of air

Plunge a deflagrating spoon containing burning sulphur into a gas-jar of air. The sulphur burns with a blue flame and leaves a misty gas. (The mist is due to traces of sulphur trioxide formed simultaneously.)

Treat several gas-jars in this way and use them for the following tests:

Add blue litmus solution.
It is turned red. The gas is an acidic oxide.
Add a dilute (pink) solution of potassium permanganate.

It is turned colourless.
Add a dilute (golden yellow) solution of potassium dichromate.
It is turned green.

The results of these tests prove that the gas is sulphur dioxide (see page 405).
Sulphur burns in air, forming sulphur dioxide.

$$S(s) + O_2(g) \longrightarrow SO_2(g)$$

Formation of sulphides from sulphur

Sulphur will combine directly with many elements forming sulphides. For example, if a finely ground mixture of iron filings and sulphur, in the proportions of 56 to 32 by mass (Fe = 56; S = 32), is heated, the two elements will combine vigorously and the whole mass will glow spontaneously when once the combination has been started at one point. A black, or dark grey, residue of *iron(II) sulphide* is left.

$$Fe(s) + S(s) \rightarrow \underset{\substack{\text{(iron(II)}\\\text{sulphide)}}}{FeS(s)}$$

Hot copper foil or wire will similarly glow in sulphur vapour, forming *copper(I) sulphide* Cu_2S.

$$2Cu(s) + S(s) \rightarrow Cu_2S(s)$$

Carbon combines directly with sulphur to form the important liquid, carbon disulphide CS_2.

$$C(s) + 2S(s) \rightarrow CS_2(l)$$

A very high temperature is required to bring about the combination, and this is secured by means of the electric furnace, in which an electric arc is struck between carbon electrodes and raises coke to white heat. Sulphur is also fed into the furnace. It vaporizes and combines with the white-hot coke. Carbon disulphide vapour passes off and is condensed.

Carbon disulphide is poisonous and may be used to destroy low and harmful forms of life, such as grain weevils (which feed on stored grain) or cockroaches. It is very flammable, and must be used with care. It is also an excellent solvent. (See also CCl_4, page 324.)

Action of acids on sulphur

Dilute acids do not act upon sulphur. It is oxidized by hot concentrated sulphuric acid with formation of sulphur dioxide.

$$S(s) + 2H_2SO_4(aq) \rightarrow 3SO_2(g) + 2H_2O(l)$$

In this reaction the sulphur is oxidized by the acid to sulphur dioxide and the acid is reduced to the same substance. Of the three molecules of sulphur dioxide in the equation, one is the product of oxidation of sulphur and two are the products of reduction of the sulphuric acid. The action is too slow to have practical value.

Sulphur is oxidized by hot concentrated nitric acid, with bromine as the best catalyst, to sulphuric acid.

$$S(s) + 6HNO_3(aq) \rightarrow H_2SO_4(aq) + 6NO_2(g) + 2H_2O(l)$$

Allotropy of sulphur

The following experiments show that sulphur exists in several different forms, called 'allotropes'. The meaning of this term will be considered more fully after the experiments have been described.

Experiment 133
Preparation of rhombic or octahedral sulphur (α-sulphur)

Shake some powdered sulphur with carbon disulphide for some time in a test-tube. (Take care to extinguish all flames in the vicinity.)

Filter the contents of the test-tube into a dry beaker through a dry filter paper and funnel. Fasten a filter paper over the mouth of the

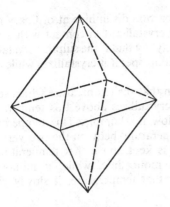

beaker, pierce a few pinholes in it, and set the beaker aside. The carbon disulphide will slowly evaporate, depositing crystals of sulphur, which, because of the slow evaporation, will be large enough for their shape to be seen. They will have the shape shown in Figure 167.

This variety of sulphur is called *rhombic sulphur* or *octahedral sulphur* or *α-sulphur*.

Note especially that the formation of the crystals takes place at ordinary room temperature.

Figure 167

Experiment 134
Preparation of monoclinic or prismatic sulphur (β-sulphur)

Place powdered sulphur in a very large crucible or an evaporating dish. Heat it and stir gradually adding more sulphur until the crucible or dish is almost brim-full of molten sulphur. Use a small flame for the heating or the sulphur may begin to burn. Then allow the sulphur to cool. After a time, a solid crust will begin to form on the surface. When the crust is continuous, pierce it at two widely separated points with a glass rod and rapidly pour out the liquid sulphur from inside. With a pen-knife, cut through the solid crust all the way round the crucible or dish, near the rim, and lift it out. Underneath will be seen long 'needle-shaped' crystals or sulphur whose shape is shown in Figure 168. They are crystals of *monoclinic sulphur* or *prismatic sulphur* or *β-sulphur*.

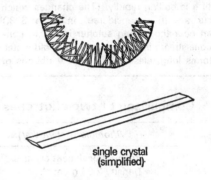

single crystal
(simplified)

Figure 168

Note that this variety crystallizes in close contact with hot, molten sulphur.

Amorphous sulphur (δ-sulphur)

This variety of sulphur may be prepared in several ways. One is to saturate distilled water with hydrogen sulphide and then expose the solution to the air. Sulphur is deposited as an almost white powder, *amorphous sulphur* or *δ-sulphur*.

$$2H_2S(g) + O_2(g) \rightarrow 2H_2O(l) + 2S(s)$$

In the experiments just described, three different varieties of the element sulphur were prepared. They have different properties, *e.g.*, their densities (g cm^{-3}) differ (rhombic, 2.08; monoclinic, 1.98), but they all consist of pure sulphur and nothing else. When an element can exist in several different forms in the same state it is said to show allotropy. (See page 91.)

Relation between monoclinic and rhombic sulphur

The factor determining which of these two allotropes will be obtained in an experiment is temperature. In our experiments, rhombic sulphur was crystallized by evaporation of a solution of sulphur in carbon disulphide at ordinary room temperature, while monoclinic sulphur was crystallized in contact with a mass of hot, molten sulphur. Roughly, then, we may say that if the sulphur crystallizes while still hot, it does so as the monoclinic allotrope; if it crystallizes while cold, the rhombic allotrope is formed.

We can go further. Experiment has shown that the temperature which separates the two varieties is 96 °C. If sulphur crystallizes above this temperature, monoclinic crystals are formed, and if below it, rhombic. This temperature, 96 °C, is therefore called the 'transition temperature' between the two varieties.

If rhombic sulphur, stable below 96 °C, is kept above that temperature, it changes its crystalline form and becomes monoclinic, while, if monoclinic sulphur, stable above 96 °C, is kept below that temperature, it slowly yields rhombic sulphur.

Experiment 135
Formation of plastic sulphur

Heat some powdered roll sulphur in a test-tube until it is boiling rapidly. (The changes which occur are fully considered on page 393.) Then pour the boiling sulphur in a thin continuous stream into a beaker full of cold water. It forms long, elastic, light-yellow ribbons of 'plastic sulphur', which are insoluble in carbon disulphide. This variety is not a true allotrope of sulphur. If kept for a few days, plastic sulphur becomes hard. This hard variety of sulphur is insoluble in carbon disulphide.

Comparison of two allotropes of sulphur

Rhombic (*octahedral*)	Monoclinic (*prismatic*)
Yellow translucent crystals. Density 2.08 g cm^{-3}. Melting point 114 °C. Stable at temperatures below 96 °C.	Transparent amber crystals. Density 1.98 g cm^{-3}. Melting point 119 °C. Unstable at temperatures below 96 °C, reverting to rhombic variety.

Experimental evidence of the *chemical* identity of these allotropes is given by the fact that each is convertible into the other (by temperature change) without change of mass and, if equal masses of the allotropes are converted into a given compound (*e.g.*, sulphur dioxide), identical masses of product are given.

Questions

1. You are provided with a finely powdered mixture of potassium nitrate and sulphur. Describe how you would obtain pure crystals of potassium nitrate from this mixture. Starting from powdered sulphur, describe how you would prepare samples of *two* crystalline allotropes of this element. In each case give a sketch showing the crystalline form.

Describe the changes which occur when powdered sulphur is gently heated in a test-tube up to its boiling point. (O. and C.)

2. Explain the following observations: (*a*) If molten sulphur is heated in air a blue flame sometimes appears above its surface. (*b*) When melted sulphur is allowed to re-solidify it has a crystalline appearance different from that of the original solid. (*c*) Changes are observed when hydrogen sulphide is bubbled through a solution of iodine. (O. and C.)

3. Name *three* different chemical sources of sulphur. Give *two* reasons for regarding sul-

phur as a non-metal. Describe how you could obtain a sample of pure sulphur from a mixture of sulphur and sand. Calculate the mass of sulphuric acid that could be made from one kilogramme of sulphur. (O. and C.)

4. Describe how you could prepare *two* different crystalline allotropes of sulphur from crushed roll sulphur. With the aid of simple sketches, show the difference in shape of the crystals of these two allotropes.

A mixture of approximately equal masses of sulphur and pure iron filings is carefully heated in a hard glass test-tube until no further reaction occurs. Describe what happens during this heating. Explain how you could distinguish, by means of *two* tests, between the original mixture and the final solid residue. (O. and C.)

5. What do you understand by the term 'allotropy'? Give the names of three sub-stances to which this term is applicable. If you were provided with a supply of rhombic sulphur, describe how you would obtain from it (*a*) monoclinic sulphur; (*b*) plastic sulphur; (*c*) several gas-jars of hydrogen sulphide (no diagram of the apparatus is required); (*d*) a sample of lead(II) sulphide. (L.)

6. 'Rhombic and monoclinic sulphur are allotropes.' Briefly explain this statement. (Details of the preparations of these forms of sulphur are not required.)

Describe the contact process for the preparation of sulphur(VI) oxide, paying particular attention to the effects of the catalyst and changes of temperature and pressure on (i) the equilibrium concentration of sulphur(VI) oxide, and (ii) the rate of the reaction. (The formation of sulphur(VI) oxide is exothermic.) (J.M.B.)

33 Compounds of Sulphur

Hydrogen sulphide H₂S

Preparation

Hydrogen sulphide was obtained in the experiment described on page 7 by the action of dilute hydrochloric acid on iron(II) sulphide. This is the most convenient method of preparation, using the apparatus shown in Figure 169. The preparation must be carried out in a fume-cupboard, since hydrogen sulphide is very poisonous.

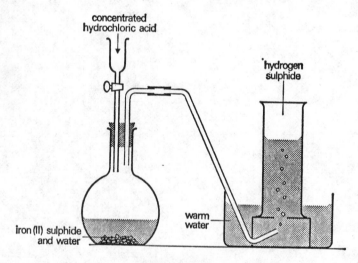

Figure 169

If it is required to prepare the gas starting from sulphur, the best way is to prepare iron(II) sulphide first by the method of page 393 and then to use it in the way about to be described.

As the acid reaches the iron(II) sulphide, effervescence begins and the hydrogen sulphide is collected over water. It is rather soluble in cold water (about three volumes of the gas in one volume of water), but like all gases, it is less soluble in hot water.

$$FeS(s) + 2HCl(aq) \rightarrow FeCl_2(aq) + H_2S(g)$$

Dilute sulphuric acid may also be used.

$$FeS(s) + H_2SO_4(aq) \rightarrow FeSO_4(aq) + H_2S(g)$$

Experiment 136
Characteristic test for hydrogen sulphide

Soak a strip of filter paper in lead(II) ethanoate solution and drop it into a gas-jar of hydrogen sulphide. The paper turns dark brown or black. This colour change is caused by precipitation of black lead(II) sulphide.

$$Pb(C_2H_3O_2)_2(aq) + H_2S(g) \rightarrow PbS(s) + 2C_2H_4O_2(aq)$$

A fairly pure specimen of hydrogen sulphide may be obtained by warming antimony(III) sulphide with concentrated hydrochloric acid.

$$Sb_2S_3(s) + 6HCl(aq) \rightarrow 2SbCl_3(aq) + 3H_2S(g)$$

If required dry, the gas may be dried by passing it over calcium chloride and collected by downward delivery, as the gas is somewhat denser than the air.

Kipp's apparatus

Kipp's apparatus is a device for obtaining intermittent supplies of a frequently used gas such as hydrogen, carbon dioxide or hydrogen sulphide (see Figure 170).

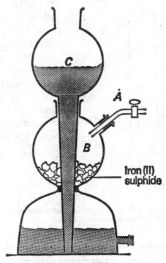

Figure 170

When the tap A is opened, the acid rises into the bulb B and attacks the iron(II) sulphide, producing hydrogen sulphide, which is delivered through A. When the gas is no longer required, A is turned off. The gas is still being generated, which raises the pressure in B. The acid is therefore forced out of B and up into C. The generation of hydrogen sulphide now stops because acid and iron(II) sulphide are no longer in contact and the apparatus will remain inactive until tap A is again opened to obtain gas.

Properties of hydrogen sulphide

The gas is colourless, and has a repulsive, rather sweet smell similar to that of a rotten egg. It is, in fact, given off from putrefying eggs and also from decaying cabbages, both of which contain sulphur. It is also fairly soluble in water giving a weakly acidic solution; at ordinary temperatures one volume of water can dissolve about three volumes of hydrogen sulphide. It is one of the weakest acids known.

Hydrogen sulphide is contained in the water of many sulphur springs and these 'waters' are said to have curative properties. They certainly possess, as a consequence of their hydrogen sulphide content, all the unpleasant taste usually associated with medicines. The gas has a density of 17 compared with that of hydrogen, and is somewhat denser than air, which is 14.4 times denser than hydrogen.

Warning. The gas is extremely poisonous; it is often found in sewers and has caused many fatalities among workmen. It should never be handled (except in minute quantities) except in an efficient fume-cupboard.

Although *very poisonous* under normal conditions, it is not so *dangerous* as the much less poisonous carbon monoxide. Can you say why this is?

Experiment 137
Combustion of hydrogen sulphide with a plentiful supply of air

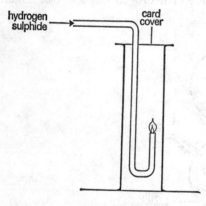

Figure 171

When several gas-jars of hydrogen sulphide have been collected remove the delivery tube and fix the tube as shown below (Figure 171). Apply a lighted taper. The hydrogen sulphide burns with a blue flame similar to that of sulphur. Lower the tube into a wide gas-jar, closing the mouth with a square of cardboard and, when the flame is extinguished, remove the tube and add a dilute pink solution of acidified potassium permanganate. On shaking, the solution becomes colourless and remains clear. This test proves the presence of *sulphur dioxide*.

$$2H_2S(g) + 3O_2(g) \longrightarrow 2H_2O(l) + 2SO_2(g)$$

Experiment 138
Combustion of hydrogen sulphide with a limited supply of air

Cut down the air supply to the flame obtained in the last section by putting into it a crucible lid.

After a few seconds, a yellow deposit of sulphur will be seen on the lid. The reduced oxygen supply cannot oxidize the gas completely and free sulphur is deposited.

$$2H_2S(g) + O_2(g) \longrightarrow 2S(s) + 2H_2O(l)$$

Hydrogen sulphide as a reducing agent

Hydrogen sulphide is a powerful reducing agent as the following experiments show. Like all reducing agents, it operates as a supplier of electrons. The usual product is a precipitate of sulphur, arising from the changes:

$$\begin{cases} H_2S \rightleftharpoons 2H^+ + S^{2-} \\ S^{2-} \rightarrow S + 2e^- \end{cases}$$

The electrons are accepted by the oxidizing agent with which the hydrogen sulphide is reacting. Very powerful oxidation, *e.g.*, by concentrated nitric acid, may convert hydrogen sulphide to sulphuric acid.

Experiment 139
Action of hydrogen sulphide with nitric acid

Dilute some concentrated nitric acid with about one-third of its volume of water in a boiling-tube and pass hydrogen sulphide into it. Brown fumes of nitrogen dioxide are given off, a pale yellow deposit of sulphur appears, and the liquid becomes hot. The hydrogen sulphide has reduced the nitric acid to nitrogen dioxide and has itself been oxidized to sulphur.

$$2HNO_3(aq) + H_2S(g) \longrightarrow$$
$$2H_2O(l) + 2NO_2(g) + S(s)$$

The solution also contains sulphuric acid, produced by the reaction:

$$H_2S(g) + 8HNO_3(aq) \longrightarrow$$
$$H_2SO_4(aq) + 8NO_2(g) + 4H_2O(l)$$

Experiment 140
Action of hydrogen sulphide with iron(III) chloride solution

Perform the experiment as above, using iron(III) chloride solution.

A yellow deposit of sulphur appears and, on heating to coagulate the sulphur and filtering, a pale green solution of iron(II) chloride is obtained. The hydrogen sulphide has reduced the yellow iron(III) chloride to green iron(II) chloride, being itself oxidized to hydrogen chloride, which dissolves in the water, and sulphur.

$$2FeCl_3(aq) + H_2S(g) \longrightarrow$$
$$2FeCl_2(aq) + 2HCl(aq) + S(s)$$

$$\text{or } 2Fe^{3+}(aq) + S^{2-}(aq) \longrightarrow 2Fe^{2+}(aq) + S(s)$$

Experiment 141
Action of air on hydrogen sulphide

Pass a stream of hydrogen sulphide into distilled water in a beaker for about half an hour. Leave the solution exposed to air. After a few days a white deposit of amorphous sulphur will have appeared. The oxygen of the air has oxidized the hydrogen sulphide to sulphur and water.

$$2H_2S(g) + O_2(g) \longrightarrow 2H_2O(l) + 2S(s)$$

Hydrogen sulphide will reduce concentrated sulphuric acid, depositing sulphur. For this reason, the acid cannot be used to dry it.

$$3H_2S(g) + H_2SO_4(aq) \longrightarrow 4H_2O(l) + 4S(s)$$

Acidified potassium permanganate and dichromate solutions are reduced by the gas. The effect produced differs from that produced by sulphur dioxide because, while either gas decolorizes the permanganate and turns the dichromate from yellow to green, hydrogen sulphide also leaves a precipitate of sulphur while sulphur dioxide does not.

$$2MnO_4^-(aq) + 5H_2S(g) + 6H^+(aq) \longrightarrow 2Mn^{2+}(aq) + 8H_2O(l) + 5S(s)$$
$$Cr_2O_7^{2-}(aq) + 3H_2S(g) + 8H^+(aq) \longrightarrow 2Cr^{3+}(aq) + 7H_2O(l) + 3S(s)$$

For the action of hydrogen sulphide with sulphur dioxide see page 407, and for its action with the halogen elements see page 375.

Experiment 142
The action of hydrogen sulphide on salts of metals

Copper(II) sulphate.

Heat a solution of copper(II) sulphate in a boiling-tube and pass hydrogen sulphide into it. A dark brown precipitate appears, copper(II) sulphide. Filter the mixture. If sufficient hydrogen sulphide has been passed, the filtrate will be colourless because all the copper, which formerly coloured it, is now precipitated as copper(II) sulphide. The filtrate is dilute sulphuric acid.

$$CuSO_4(aq) + H_2S(g) \longrightarrow$$
$$CuS(s) + H_2SO_4(aq)$$
$$\text{copper(II)}$$
$$\text{sulphide}$$

Lead(II) nitrate.

Experiment as above. Here, a black precipitate of lead(II) sulphide is produced and the filtrate is dilute nitric acid.

$$Pb(NO_3)_2(aq) + H_2S(g) \longrightarrow$$
$$PbS(s) + 2HNO_3(aq)$$

Hydrogen sulphide as an acid

Hydrogen sulphide acts as a weak dibasic acid. It forms with sodium hydroxide two salts, normal sodium sulphide Na_2S,

$$2NaOH(aq) + H_2S(g) \rightarrow Na_2S(aq) + 2H_2O(l)$$

or, with excess of hydrogen sulphide, the acid salt, sodium hydrogensulphide NaHS.

$$NaOH(aq) + H_2S(g) \rightarrow NaHS(aq) + H_2O(l)$$

Potassium hydroxide reacts similarly.

Since the possible reactions of hydrogen sulphide with sodium hydroxide solution are:

$$2NaOH(aq) + H_2S(g) \rightarrow Na_2S(aq) + 2H_2O(l)$$
$$2NaOH(aq) + 2H_2S(g) \rightarrow 2NaHS(aq) + 2H_2O(l)$$

it is clear, from the equations, that the volume of hydrogen sulphide needed to convert a given mass of sodium hydroxide into sodium hydrogensulphide NaHS, is twice that required to convert it to sodium sulphide Na_2S. It is impossible in practice to determine when just enough hydrogen sulphide has been used to convert the alkali into sodium sulphide, so the best way of carrying out the preparation is to convert half of the sodium hydroxide into sodium hydrogensulphide by saturation with hydrogen sulphide, and then to form the normal salt from the acid salt by addition of the other half of the sodium hydroxide.

$$NaOH(aq) + H_2S(g) \rightarrow NaHS(aq) + H_2O(l)$$
$$NaHS(aq) + NaOH(aq) \rightarrow Na_2S(aq) + H_2O(l)$$

Experiment 143
Preparation of sodium hydrogensulphide and sodium sulphide

Measure out 50 cm³ of bench (about 2M) sodium hydroxide solution, divide it into two equal parts and, into one of them, pass hydrogen sulphide until no more is absorbed and the liquid smells strongly of hydrogen sulphide. Add the other half of the sodium hydroxide solution and obtain crystals of sodium sulphide by the method described on page 271.

Sulphides

K	} Sulphides of these metals	
Na	are soluble in water.	Sulphides of these metals will not precipitate from acidified solutions.
Ca		
Mg		
Zn	Sulphides of these metals	
Fe	are insoluble in water.	
Pb		Sulphides of these metals will precipitate from acidified solutions.
Cu		

Potassium sulphide K₂S

This is similar to sodium sulphide and is similarly prepared.

Sodium sulphide Na₂S

The preparation of this compound by neutralization of sodium hydroxide by hydrogen sulphide is described above. Its aqueous solution is alkaline and smells of hydrogen sulphide. When heated with sulphur it forms 'polysulphides' of sodium. For example:

$$Na_2S(s) + 4S(s) \rightarrow Na_2S_5(s)$$

In industry, sodium sulphide is prepared by heating sodium sulphate with coke (page 416).

Uses. (1) For preparing a class of very 'fast' dyes.
(2) For stripping the hair from hides.

Calcium sulphide CaS

This compound was chiefly important in the form of the 'alkali waste' of the Leblanc process. Sulphur was recovered from it.

If it contains traces of certain metals, for example, 0.01% bismuth, it is 'phosphorescent', that is, after exposure to light, it will emit a violet glow whose intensity gradually diminishes. The glow fades out after some hours.

Zinc sulphide ZnS

This compound occurs as the mineral 'zinc blende'. It may be precipitated by hydrogen sulphide from a *neutral* (or alkaline) solution of a zinc salt.

$$Zn^{2+}(aq) + S^{2-}(aq) \rightarrow ZnS(s)$$

Like calcium sulphide, and under similar conditions, zinc sulphide is phosphorescent. The luminous paint on watches is usually zinc sulphide, containing about 1 part of a radium salt in 100 000 000 of the sulphide.

Iron(II) sulphide FeS

This black, insoluble compound is usually employed for the preparation of hydrogen sulphide (page 398). It is prepared by heating iron with sulphur in the calculated quantities (page 6).

Iron(II) disulphide FeS₂

This occurs as a hard, brassy mineral. Because of its appearance it is often known as fool's gold. There are deposits of it in various parts of the world including Spain. It is the cheapest source of sulphur dioxide which it gives off when burnt in air.

Lead(II) sulphide PbS

Lead(II) sulphide occurs as the mineral galena, and is precipitated from solutions of lead salts by hydrogen sulphide (page 402). The most satisfactory test for hydrogen sulphide (page 399) is the production of a dark brown (almost black) stain of lead(II) sulphide on a filter paper soaked in lead(II) ethanoate solution.

$$Pb^{2+}(aq) + S^{2-}(aq) \rightarrow PbS(s)$$

Copper(II) sulphide CuS

This is a black insoluble compound precipitated from a solution of a copper(II) salt by hydrogen sulphide (page 402).

$$Cu^{2+}(aq) + S^{2-}(aq) \rightarrow CuS(s)$$

Sulphur dioxide SO₂

Preparation of sulphur dioxide

This compound, which is a gas under ordinary conditions, is conveniently prepared in the laboratory by the apparatus of Figure 172.

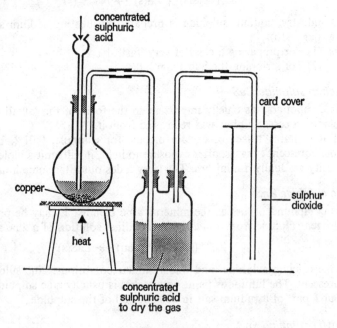

Figure 172

There is no action until the mixture in the flask becomes hot. Then rapid effervescence occurs and the sulphur dioxide, being very soluble in water and denser than air, is usually collected as shown:

$$Cu(s) + 2H_2SO_4(l) \rightarrow CuSO_4(s) + 2H_2O(l) + SO_2(g)$$

A dark brown mixture is left in the flask. It contains anhydrous copper(II) sulphate and certain impurities. Crystals of copper(II) sulphate may be obtained from it by the method described on page 275.

Tests. 1. The gas has a very irritating smell and a metallic taste.

2. *Action on potassium permanganate solution.* The solution is turned from purple to colourless by sulphur dioxide. (No precipitate is left as in the case of reduction of the permanganate by hydrogen sulphide.)

$$5SO_2(g) + 2MnO_4^-(aq) + 2H_2O(l) \rightarrow$$
$$5SO_4^{2-}(aq) + 2Mn^{2+}(aq) + 4H^+(aq)$$

The explanation of the change of colour is that the potassium permanganate is decomposed and all the products of the reaction give colourless solutions. (The concentration of the manganese sulphate is too small for its very pale pink colour to be observed.)

3. *Action on acidified potassium dichromate solution.* The solution is turned from golden yellow to green by sulphur dioxide (see Experiment 146).

$$3SO_2(g) + Cr_2O_7^{2-}(aq) + 2H^+(aq) \rightarrow$$
$$3SO_4^{2-}(aq) + 2Cr^{3+}(aq) + H_2O(l)$$

Properties of sulphur dioxide

The gas is colourless, and has an irritating smell and a rather sweet taste. It is fairly poisonous and is used for fumigation. It is also readily soluble in water, and litmus added to the resultant solution turns red, indicating that *the solution is acidic*. The sulphur dioxide reacts chemically with the water to produce *sulphurous acid*.

$$H_2O(l) + SO_2(g) \rightleftharpoons H_2SO_3(aq) \rightleftharpoons 2H^+(aq) + SO_3^{2-}(aq)$$

This acid will be considered more fully later.

Sulphur dioxide as a reducing agent

Sulphur dioxide, in the presence of water, is a powerful reducing agent. It reacts with water to form sulphurous acid and the sulphite ion, SO_3^{2-}, and this ion, like reducing agents in general, acts as a supplier of electrons. This occurs in association with water.

$$H_2O(l) + SO_2(g) \rightleftharpoons H_2SO_3(aq) \rightleftharpoons 2H^+(aq) + SO_3^{2-}(aq)$$
$$SO_3^{2-}(aq) + H_2O(l) \rightarrow SO_4^{2-}(aq) + 2H^+(aq) + 2e^-$$

The electrons are accepted by the oxidizing agent with which the SO_2–water system is reacting, *e.g.*,

iron(III) ion, which is reduced to iron(II) ion:

$$Fe^{3+} + e^- \rightarrow Fe^{2+}$$

chlorine, which is reduced to its ions:

$$Cl_2 + 2e^- \rightarrow 2Cl^-$$

acidified potassium permanganate, which is reduced to a manganese(II) salt:

$$MnO_4^-(aq) + 8H^+(aq) + 5e^- \rightarrow Mn^{2+}(aq) + 4H_2O(l)$$

The following are important examples of the reducing action of sulphur dioxide in aqueous solution.

Experiment 144
Action of sulphur dioxide with concentrated nitric acid

Put some concentrated nitric acid into a boiling-tube and pass into it a current of sulphur dioxide from a siphon of liquid sulphur dioxide. Brown fumes are evolved (nitrogen dioxide) and the liquid becomes warm. Dilute some of the liquid and add dilute hydrochloric acid and barium chloride solution (the recognized test for a soluble sulphate). The white precipitate of barium sulphate proves the presence of sulphuric acid.

$$BaCl_2(aq) + H_2SO_4(aq) \longrightarrow$$
$$BaSO_4(s) + 2HCl(aq)$$

The concentrated nitric acid has oxidized the sulphur dioxide in the presence of water to sulphuric acid and has been itself reduced to nitrogen dioxide.

$$SO_2(g) + 2HNO_3(aq) \longrightarrow$$
$$H_2SO_4(aq) + 2NO_2(g)$$

Experiment 145
Action of sulphur dioxide on iron(III) sulphate solution

Make a solution of iron(III) sulphate (or iron ammonium alum) in water in a boiling-tube and pass into it sulphur dioxide as above. The brownish colour of the solution is rapidly converted to pale green. The sulphur dioxide has reduced the brown iron(III) sulphate to light green iron(II) sulphate and has itself been oxidized to sulphuric acid.

$$2Fe^{3+}(aq) + 2H_2O(l) + SO_2(g) \longrightarrow$$
$$2Fe^{2+}(aq) + SO_4^{2-}(aq) + 4H^+(aq)$$

The red solution which may be formed is a complex sulphite which decomposes on heating, leaving the products as indicated by the equation above.

Experiment 146
Action of sulphur dioxide on potassium dichromate

See test for sulphur dioxide on page 405.
Acidify a solution of potassium dichromate in a boiling-tube with dilute sulphuric acid, and pass through it a stream of sulphur dioxide from a siphon of liquid sulphur dioxide. There is a rapid colour change from golden yellow to green, but no precipitate appears (compare the action of hydrogen sulphide, page 401).

$$3SO_2(g) + Cr_2O_7^{2-}(aq) + 2H^+(aq) \longrightarrow$$
$$3SO_4^{2-}(aq) + 2Cr^{3+}(aq) + H_2O(l)$$

The potassium dichromate has oxidized the sulphur dioxide in the presence of water to sulphuric acid, being itself reduced to green chromium(III) sulphate.

Experiment 147
Action of sulphur dioxide on potassium permanganate

See test for sulphur dioxide on page 405.
The potassium permanganate oxidized the sulphur dioxide in the presence of water to sulphuric acid, and was itself reduced to man-ganese(II) sulphate, the colour of the solution being almost totally discharged.

$$5SO_2(g) + 2MnO_4^-(aq) + 2H_2O(l) \longrightarrow$$
$$5SO_4^{2-}(aq) + 2Mn^{2+}(aq) + 4H^+(aq)$$

Experiment 148
Bleaching action of sulphur dioxide

Sulphurous acid is a bleaching agent. This may easily be shown by dropping into a gas-jar of the gas (containing some water) a few blue flowers, *e.g.*, blue crocus, iris, or bluebells. After a few minutes, the flowers will have lost their blue colour.

This bleaching is also a reducing action. The sulphurous acid takes up oxygen from the colouring matter of the flowers and forms sulphuric acid; the removal of oxygen from the dye converts it to a colourless compound. Sulphur dioxide is used industrially for bleaching sponges and straw for straw hats. The oxygen of the air may oxidize the reduced colourless compound back to the original coloured compound, which explains why straw hats gradually become yellow with usage.

Experiment 149
Action of sulphur dioxide with hydrogen sulphide

Add to a gas-jar of sulphur dioxide a little water, invert over it a gas-jar of hydrogen sulphide and allow the gases to mix. A yellow deposit of sulphur will be produced at once. The **dry** gases do not react.

$$2H_2S(g) + SO_2(g) \rightarrow 2H_2O(l) + 3S(s)$$

Note that here, the sulphur dioxide is **actually acting as an oxidizing agent supplying oxygen to the hydrogen sulphide.** As we have seen above, however, sulphur dioxide usually shows reducing properties. Here it has encountered in hydrogen sulphide a more powerful reducer than itself, which takes up its oxygen and causes it to act as an oxidizer.

Liquefaction of sulphur dioxide

Sulphur dioxide can readily be liquefied by being dried by concentrated sulphuric acid and passed through a freezing mixture of ice and salt. It liquefies under ordinary atmospheric pressure at about $-10\,°C$. It can be kept liquid at ordinary room temperature if under slight pressure, and it is sold in siphons under pressure.

Sulphur dioxide in chemical industry

Sulphur dioxide is very important as an intermediate compound in the manufacture of sulphuric acid (page 411). It is prepared by burning sulphur in air,

$$S(s) + O_2(g) \rightarrow SO_2(g)$$

or by burning iron pyrites in air.

$$4FeS_2(s) + 11O_2(g) \rightarrow 2Fe_2O_3(s) + 8SO_2(g)$$

Sulphurous acid H_2SO_3

This acid has never been obtained free from water. Any attempt to prepare the pure acid always results in its decomposition into sulphur dioxide and water.

It is prepared by passing sulphur dioxide into water. The gas is readily soluble and it is advisable to prevent 'sucking back' by the use of a funnel just touching the water surface (Figure 173).

The reaction in the flask is the same as described under the preparation of sulphur dioxide (page 404). It is, of course, not necessary here to dry the sulphur dioxide.

$$Cu(s) + 2H_2SO_4(aq) \rightarrow CuSO_4(aq) + 2H_2O(l) + SO_2(g)$$
$$SO_2(g) + H_2O(l) \rightarrow H_2SO_3(aq)$$

Sulphur dioxide is the anhydride of sulphurous acid and may be called 'sulphurous anhydride'.

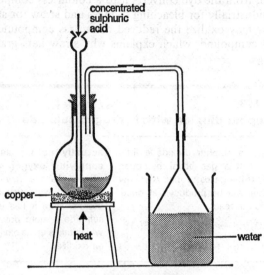

concentrated sulphuric acid

copper

heat

water

Figure 173

Definition. An anhydride is the oxide of a non-metal, which, which combined with water, forms an acid.

$$CO_2(g) + H_2O(l) \rightleftharpoons H_2CO_3(aq) \rightleftharpoons 2H^+(aq) + CO_3^{2-}(aq)$$
carbonic anhydride   carbonic acid

$$SO_2(g) + H_2O(l) \rightleftharpoons H_2SO_3(aq) \rightleftharpoons 2H^+(aq) + SO_3^{2-}(aq)$$
sulphurous anhydride   sulphurous acid

$$SO_3(g) + H_2O(l) \rightleftharpoons H_2SO_4(aq) \rightleftharpoons 2H^+(aq) + SO_4^{2-}(aq)$$
sulphuric anhydride   sulphuric acid

An anhydride will not always combine directly with water to give the corresponding acid, *e.g.*, silicon dioxide SiO_2 is the anhydride of silicic acid H_2SiO_3, though the acid cannot be prepared by direct combination of its anhydride with water. The acid is prepared from one of its salts and, when heated, loses water, leaving silicon dioxide as the residue.

Properties of sulphurous acid

Sulphurous acid is a colourless liquid which smells strongly of sulphur dioxide. The acid has all the reducing actions described previously as those of sulphur dioxide in the presence of water (pages 405–6).

Experiment 150
Effect of exposure to air on sulphurous acid

Leave a beaker of sulphurous acid exposed to air for a few days. Then add to it hydrochloric acid and barium chloride solution. The white precipitate of barium sulphate proves that the oxygen of the air has oxidized the sulphurous acid to sulphuric acid.

$$2H_2SO_3(aq) + O_2(g) \longrightarrow 2H_2SO_4(aq)$$
$$Ba^{2+}(aq) + SO_4^{2-}(aq) \longrightarrow BaSO_4(s)$$

Action of sulphurous acid with alkalis

Sulphurous acid is a dibasic acid and with sodium hydroxide forms two sodium salts, the acid salt, sodium hydrogensulphite $NaHSO_3$, and the normal salt, sodium sulphite Na_2SO_3.

$$NaOH(aq) + H_2SO_3(aq) \rightarrow NaHSO_3(aq) + H_2O(l)$$
$$2NaOH(aq) + H_2SO_3(aq) \rightarrow Na_2SO_3(aq) + 2H_2O(l)$$

Potassium hydroxide solution behaves similarly.

Laboratory preparation of sodium sulphite

This is similar to the preparation of sodium sulphide, described on page 402, using sulphur dioxide instead of hydrogen sulphide.

$$NaOH(aq) + H_2SO_3(aq) \rightarrow NaHSO_3(aq) + H_2O(l)$$
$$NaHSO_3(aq) + NaOH(aq) \rightarrow Na_2SO_3(aq) + H_2O(l)$$

Sulphites give off sulphur dioxide when warmed with dilute hydrochloric acid or dilute sulphuric acid (test for SO_2, page 405), *e.g.*,

$$Na_2SO_3(aq) + H_2SO_4(aq) \rightarrow Na_2SO_4(aq) + H_2O(l) + SO_2(g)$$

This is occasionally used as a method of preparing sulphur dioxide.

Used in dilute acidified solutions, sulphites have all the reducing actions of sulphur dioxide and water, or sulphurous acid (see page 408).

Preparation of sulphurous acid H_2SO_3 from sulphur

To convert sulphur into sulphurous acid it is necessary first to oxidize the sulphur to sulphur dioxide and then absorb this gas in water.

This can be done using the apparatus of Figure 174.

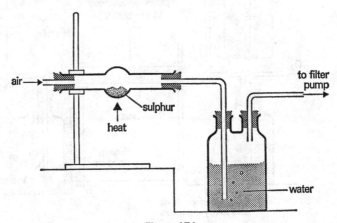

Figure 174

Heat the sulphur and, by means of a filter-pump, draw over it a rapid stream of air. The sulphur burns and the sulphur dioxide produced is absorbed as it passes through the water in the Woulff's bottle. The liquid left is sulphurous acid. Sulphur vapour may be carried over unburnt and appear as a yellow precipitate in the bottle. Remove it by filtration.

$$S(s) + O_2(g) \rightarrow SO_2(g)$$
$$H_2O(l) + SO_2(g) \rightarrow H_2SO_3(aq)$$

The method of producing sulphurous acid given on page 408 is much more convenient in the laboratory, but the sulphurous acid prepared on the large scale is made by modification of the above method or by burning iron pyrites FeS_2.

$$4FeS_2(s) + 11O_2(g) \rightarrow 2Fe_2O_3(s) + 8SO_2(g)$$

Sulphur trioxide SO₃

This compound is a white hygroscopic solid. A sample of it is usually kept in a sealed glass bulb as a laboratory exhibit.

Preparation of sulphur trioxide

It is prepared by passing a mixture of dry sulphur dioxide and dry air, or oxygen, over heated platinized asbestos (or vanadium(V) oxide). Platinized asbestos is made by soaking asbestos in platinum chloride solution and then igniting it, when platinum is left in a very finely divided form.

$$PtCl_4(aq) \rightarrow 2Cl_2(g) + Pt(s)$$

The platinum is a catalyst and the best temperature is 450–500 °C. The sulphur trioxide is seen as dense white fumes and may be solidified in a freezing mixture of ice and a little common salt (Figure 175).

$$2SO_2(g) + O_2(g) \rightarrow 2SO_3(g)$$

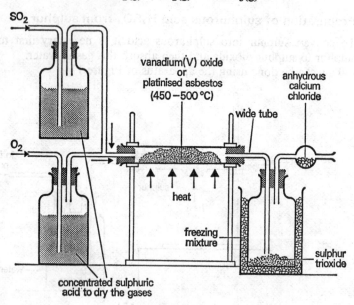

Figure 175

The sulphur trioxide container is protected from atmospheric moisture by a calcium chloride tube. Sulphur trioxide is important because it combines vigorously with water, giving sulphuric acid.

$$H_2O(l) + SO_3(g) \rightarrow H_2SO_4(aq)$$

It is the anhydride of this acid and sulphur trioxide may be termed 'sulphuric anhydride'. (See also page 408.)

Sulphuric acid H_2SO_4

Lead chamber process for the manufacture of sulphuric acid

To offset lack of the raw materials, sulphur, and pyrites, a process is now producing *sulphur dioxide* by strongly heating anhydrite, $CaSO_4$, with coke. The main reaction is:

$$2CaSO_4(s) + C(s) \rightarrow 2CaO(s) + 2SO_2(g) + CO_2(g)$$

By including sand and ashes containing alumina, the quicklime is converted to a valuable by-product, cement (calcium silicate(IV) and aluminate). Pyrites is also burnt in air to produce *sulphur dioxide*.

$$4FeS_2(s) + 11O_2(g) \rightarrow 2Fe_2O_3(s) + 8SO_2(g)$$

It is converted by the oxygen of the air, in the presence of steam, into sulphuric acid. Nitrogen oxide is used as a catalyst or oxygen-carrier.

$$2NO(g) + O_2(g) \rightarrow 2NO_2(g)$$
$$\text{from}$$
$$\text{air}$$
$$NO_2(g) + H_2O(l) + SO_2(g) \rightarrow NO(g) + H_2SO_4(aq)$$

The nitrogen oxide is usually supplied in a modern plant by oxidizing **ammonia** by oxygen of the air. The two gases are passed over heated platinum.

$$4NH_3(g) + 5O_2(g) \rightarrow 4NO(g) + 6H_2O(l)$$

The main oxidation of the sulphur dioxide is carried out in large lead chambers, on the floors of which 'chamber-acid' (65% sulphuric acid) accumulates. It is not very pure but finds a ready sale where an acid of high purity is not needed. The plant used is made more elaborate by devices for recovery of the nitrogen oxide, which would otherwise escape, and for its restoration into the reacting gases. This method, in its original form, is now almost obsolete.

Contact process for manufacture of sulphuric acid

Sulphur dioxide (prepared by burning sulphur) and air are passed over a catalyst, heated to 450–500 °C. About 98% of the possible yield of sulphur trioxide is obtained.

$$2SO_2(g) + O_2(g) \rightleftharpoons 2SO_3(g)$$

Originally, platinized asbestos was used as catalyst but platinum is very expensive and easily 'poisoned' by impurity which made elaborate purification of the gases necessary (especially from arsenical compounds). Vanadium(V) oxide V_2O_5 has replaced platinum as the usual catalyst employed.

The sulphur trioxide cannot be satisfactorily absorbed by water. A mist of fine drops of dilute sulphuric acid fills the factory if direct absorption in water is tried. It is dissolved in concentrated sulphuric acid, forming a fuming liquid called 'oleum' for which there is some demand. Most of the 'oleum' is carefully diluted with the correct amount of water to give ordinary concentrated sulphuric acid.

$$H_2SO_4(aq) + SO_3(g) \rightarrow H_2S_2O_7(aq)$$
$$H_2S_2O_7(aq) + H_2O(l) \rightarrow 2H_2SO_4(aq)$$

Properties of sulphuric acid

Sulphuric acid is a dense oily liquid, 'Oil of Vitriol'. It has several very important properties.

Dilute sulphuric acid – as an acid

Sulphuric acid is *dibasic*, ionizing in two stages to produce first a hydrogen ion and a *hydrogensulphate* ion, HSO_4^-, from a molecule of the acid, after which the hydrogensulphate ion may ionize further to produce a hydrogen ion and the *sulphate* ion, SO_4^{2-}

$$H_2SO_4(aq) \rightleftharpoons H^+(aq) + HSO_4^-(aq) \rightleftharpoons 2H^+(aq) + SO_4^{2-}(aq)$$

In dilute solution, the acid is almost completely ionized and so is a *strong* acid.

Because of its dibasic character, this acid forms two sodium salts, sodium sulphate $(Na^+)_2SO_4^{2-}$, and sodium hydrogensulphate $Na^+HSO_4^-$.

$$2(Na^+OH^-)(aq) + (H^+)_2SO_4^{2-}(aq) \rightarrow (Na^+)_2SO_4^{2-}(aq) + 2H_2O(l)$$
$$Na^+OH^-(aq) + (H^+)_2SO_4^{2-}(aq) \rightarrow Na^+HSO_4^-(aq) + H_2O(l)$$

Experiment 151
Preparation of sodium sulphate and sodium hydrogensulphate

It is evident from the previous equations that the amount of sodium hydroxide needed to convert a given amount of sulphuric acid into sodium hydrogensulphate is half that required to convert it to sodium sulphate.

Measure out, say, 100 cm³ of bench (2M) sodium hydroxide solution into a flask, add litmus, and then run in carefully, from a burette, bench (2M) dilute sulphuric acid, until the solution is neutral (purple). Note the volume of dilute sulphuric acid needed (say x cm³). This solution now contains sodium sulphate.

$$2NaOH(aq) + H_2SO_4(aq) \rightarrow$$
$$Na_2SO_4(aq) + 2H_2O(l)$$

Then measure out a further 100 cm³ of the same sodium hydroxide solution and add to it, from the burette, $2x$ cm³ of the same acid. This solution now contains sodium hydrogensulphate.

$$2NaOH(aq) + 2H_2SO_4(aq) \rightarrow$$
$$2NaHSO_4(aq) + 2H_2O(l)$$

Obtain crystals in the usual way from both solutions (see page 271).

Similarly, two potassium salts, potassium sulphate K_2SO_4, and potassium hydrogensulphate $KHSO_4$, can be made.

Dilute sulphuric acid also neutralizes basic oxides or hydroxides to form salts and water, *e.g.*,

$$CuO(s) + H_2SO_4(aq) \rightarrow CuSO_4(aq) + H_2O(l)$$
$$ZnO(s) + H_2SO_4(aq) \rightarrow ZnSO_4(aq) + H_2O(l)$$
$$Cu(OH)_2(s) + H_2SO_4(aq) \rightarrow CuSO_4(aq) + 2H_2O(l)$$
$$Zn(OH)_2(s) + H_2SO_4(aq) \rightarrow ZnSO_4(aq) + 2H_2O(l)$$

Action of dilute sulphuric acid with metals

Some of the common metals displace hydrogen from dilute sulphuric acid, *e.g.*,

$$Zn(s) + 2H^+(aq) \rightarrow Zn^{2+}(aq) + H_2(g)$$
$$Fe(s) + 2H^+(aq) \rightarrow Fe^{2+}(aq) + H_2(g)$$
$$Mg(s) + 2H^+(aq) \rightarrow Mg^{2+}(aq) + H_2(g)$$

Copper is, however, without action on this acid. Note that cold, concentrated sulphuric acid, in the complete absence of water, is not attacked by any metal.

Action of sulphuric acid with carbonates

If the sulphate of a metal is soluble, dilute sulphuric acid readily attacks its carbonate with evolution of carbon dioxide, *e.g.*,

$$Na_2CO_3(aq) + H_2SO_4(aq) \rightarrow Na_2SO_4(aq) + H_2O(l) + CO_2(g)$$
$$MgCO_3(aq) + H_2SO_4(aq) \rightarrow MgSO_4(aq) + H_2O(l) + CO_2(g)$$
or $$CO_3^{2-}(aq) + 2H^+(aq) \rightarrow H_2O(l) + CO_2(g)$$

If dilute sulphuric acid is added to marble $CaCO_3$, however, the effervescence is checked after a few seconds. This is because the calcium sulphate which is formed is only sparingly soluble in water and soon forms a deposit on the surface of the marble, separating it from the acid and checking the action.

Concentrated sulphuric acid as an oxidizing agent

Like all oxidizing agents, sulphuric acid acts as an acceptor of electrons. When *hot and concentrated*, the acid shows, as its principal reaction:

$$2H_2SO_4(aq) + 2e^- \rightarrow SO_4{}^{2-}(aq) + 2H_2O(l) + SO_2(g)$$

The electrons are supplied by the reducing agent concerned in the reaction. This may be a metal, such as copper or zinc:

$$Cu(s) \text{ (or Zn)} \rightarrow Cu^{2+}(aq) \text{ (or } Zn^{2+}) + 2e^-$$

The metallic ion is left associated with the $SO_4{}^{2-}$ ion as the corresponding metallic sulphate, and the reaction is usually written in a single equation, as:

$$Cu(s) \text{ (or Zn)} + 2H_2SO_4(aq) \rightarrow CuSO_4(aq) \text{ (or } ZnSO_4) + 2H_2O(l) + SO_2(g)$$

The non-metals, *carbon* and *sulphur*, are also oxidized by the hot, concentrated acid to give *sulphur dioxide* or *carbon dioxide*.

$$S(s) + 2H_2SO_4(aq) \rightarrow 2H_2O(l) + 3SO_2(g)$$
$$C(s) + 2H_2SO_4(aq) \rightarrow 2H_2O(l) + 2SO_2(g) + CO_2(g)$$

The sulphur dioxide given off may be detected using a strip of filter paper moistened with potassium dichromate which turns green.

Concentrated sulphuric acid possesses an affinity for water

The acid has a great affinity for water. The two, when mixed in equal volumes at room temperature, may give a liquid whose temperature is as high as 120 °C. This indicates chemical reaction, and the heat is due to the hydration of the ions which result from the H_2SO_4 molecule. **It is very important when mixing the acid with water to add the acid to the water and NEVER the water to the acid.**

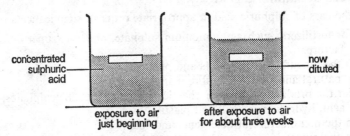

concentrated
sulphuric
acid

now
dituted

exposure to air
just beginning

after exposure to air
for about three weeks

Figure 176

It is necessary to stir the liquid as the acid enters to prevent formation of a *lower* layer of acid.

Concentrated sulphuric acid is *hygroscopic, i.e.,* it absorbs water-vapour out of the air, increasing in bulk and becoming dilute. This can be shown by the following experiment for which Figure 176 is sufficient explanation. The concentrated acid is used for drying gases, *e.g.,* sulphur dioxide, chlorine, hydrogen chloride. It cannot be used to dry a *reducing* gas like hydrogen sulphide, or an *alkaline* gas like ammonia.

So great is the affinity of concentrated sulphuric acid for water that it can decompose many compounds by removing from them the hydrogen and oxygen necessary to form water, with which it then combines. This is called a *dehydrating* action.

Experiment 152
Dehydration of sugar by concentrated sulphuric acid

Place about two or three spatula-measures of sugar in a 450 cm³ beaker and cover it with water. Place the beaker in a trough, for safety, and pour in a steady stream of concentrated sulphuric acid. The sugar is charred and a spongy black mass of charcoal rises, filling the beaker. Steam is given off and the whole mass becomes very hot.

The acid has taken out the *elements of water* from the sugar leaving a black mass of carbon.

$$C_{12}H_{22}O_{11}(s) (+ nH_2SO_4(l)) \longrightarrow$$
$$12C(s) + (11H_2O(l) + nH_2SO_4(aq))$$

A similar action is the explanation of the very marked corrosive action of the acid on cloth, *e.g.,* cotton. Cotton is largely cellulose, whose simplest formula is $(C_6H_{10}O_5)_n$.

$$C_6H_{10}O_5 (s)(+ nH_2SO_4(l)) \longrightarrow$$
$$6C(s) + (5H_2O(l) + nH_2SO_4(aq))$$

and a hole appears in the cloth. Similar reactions account for its rapid and serious burning of the skin.

Experiment 153
Dehydration of oxalic (ethanedioic) acid by concentrated sulphuric acid

Place a little oxalic acid in a test-tube, add a little concentrated sulphuric acid and warm gently. Effervescence occurs. Apply a lighted splint to the test-tube. The gas burns with a blue flame, showing that carbon monoxide is given off. Extinguish the flame and pass the gas into lime-water held in a boiling-tube. The turbidity shows that carbon dioxide is also present. The reaction is of the same type as those above, and it is used for the laboratory preparation of carbon monoxide (see page 315).

$$H_2C_2O_4(s) + H_2SO_4(l) \longrightarrow$$
$$CO(g) + CO_2(g) + (H_2O + H_2SO_4(aq))$$

Uses of sulphuric acid

The uses of sulphuric acid in appropriate order of significance are given below.

For fertilizers, such as ammonium sulphate, and for 'superphosphate' manufacture.

In the manufacture of paints and pigments.

In natural and man-made fibre manufacture.

For the production of other chemicals such as metallic sulphates, hydrochloric acid, hydrofluoric acid, and plastics.

In the manufacture of detergents and soap.

In the extraction of metals, and metal manufacturing including 'pickling' to clean metallic surfaces.

Experiment 154
Test for sulphuric acid and soluble sulphates

To a little dilute sulphuric acid in a boiling-tube add dilute hydrochloric acid and barium chloride solution. (Barium nitrate solution and dilute nitric acid are also often used.). A white precipitate of barium sulphate is formed which is insoluble in excess acid. This is the characteristic test of any soluble sulphate.

$$BaCl_2(aq) + H_2SO_4(aq) \longrightarrow$$
$$2HCl(aq) + BaSO_4(s)$$

Methods of preparation of sulphates

These are fully dealt with in Chapter 24, pages 274–5. Briefly summarized they are:

(1) *By the action of sulphuric acid on a metal*

Dilute acid

$$Fe(s) + H_2SO_4(aq) \rightarrow FeSO_4(aq) + H_2(g)$$
iron(II)
sulphate

$$Mg(s) + H_2SO_4(aq) \rightarrow MgSO_4(aq) + H_2(g)$$
magnesium(II)
sulphate

$$Zn(s) + H_2SO_4(aq) \rightarrow ZnSO_4(aq) + H_2(g)$$
zinc(II)
sulphate

Hot concentrated acid

$$Cu(s) + 2H_2SO_4(l) \rightarrow CuSO_4(aq) + 2H_2O(l) + SO_2(g)$$

(2) *By the action of dilute sulphuric acid on the oxide, hydroxide, or carbonate of the metal*

For example:
$$CuO(s) + H_2SO_4(aq) \rightarrow CuSO_4(aq) + H_2O(l)$$
$$Zn(OH)_2(s) + H_2SO_4(aq) \rightarrow ZnSO_4(aq) + 2H_2O(l)$$
$$Na_2CO_3(s) + H_2SO_4(aq) \rightarrow Na_2SO_4(aq) + H_2O(l) + CO_2(g)$$
sodium
sulphate

(3) *By double decomposition*

This is limited in application to the preparation of insoluble sulphates. Only two common sulphates are insoluble, barium(II) sulphate and lead(II) sulphate. (Calcium(II) sulphate is sparingly soluble.)

$$Pb^{2+}(aq) + SO_4^{2-}(aq) \rightarrow PbSO_4(s)$$
$$Ba^{2+}(aq) + SO_4^{2-}(aq) \rightarrow BaSO_4(s)$$

Aluminium(III) sulphate $Al_2(SO_4)_3.18H_2O$

This is a white solid. It is conveniently prepared by dissolving the oxide or hydroxide of the metal in dilute sulphuric acid, leaving the acid in slight excess to counter hydrolysis. The sulphate can be obtained by evaporation to small bulk and cooling, but it does not crystallize well.

$$2Al(OH)_3(s) + 3H_2SO_4(aq) \rightarrow Al_2(SO_4)_3(aq) + 6H_2O(l)$$

It is most commonly encountered in the form of *potash alum,* one of an important group of salts called *the alums.*

The alums

These are double salts of general formula

$$X_2SO_4.Y_2(SO_4)_3.24H_2O \quad \text{or} \quad X^+Y^{3+}(SO_4^{2-})_2.12H_2O$$

where   X is Na, K, or NH_4
and   Y is Fe(III), Al, or Cr.

(Note that X is a *mono*valent and Y a *tri*valent metal.)

The alums crystallize well from water. They all have similar crystalline shape, consequently crystalline layers of different alums may be deposited on one another to produce large, composite crystals with layers of varying colours.

The two commonest alums are:

Potash alum $K_2SO_4.Al_2(SO_4)_3.24H_2O$ (colourless)
Iron(III) alum $(NH_4)_2SO_4.Fe_2(SO_4)_3.24H_2O$ (purple)

Preparation of potash alum

(This is commonly called simply 'alum'.) Potassium sulphate and aluminium sulphate are weighed out approximately in the proportions of their formula masses

$$K_2SO_4 : Al_2(SO_4)_3.18H_2O$$
$$174 \text{ g} \qquad 666 \text{ g}$$

Use, say, one-twentieth of these figures, *i.e.*, 8.7 g and 33 g. These amounts are dissolved, with heat, in as little water as possible. In the case of the aluminium salt, the water should be slightly acidified with dilute sulphuric acid. The hot solutions are then mixed and stirred. On cooling, colourless alum crystals separate out and are filtered, washed with cold distilled water, and dried.

Ammonium sulphate (sulphate of ammonia) $(NH_4)_2SO_4$

This compound is very widely used as a nitrogenous fertilizer. It may be made in the laboratory by neutralization of dilute sulphuric acid with ammonia (page 274). In industry, it is produced by the action of ammonia and carbon dioxide on the mineral 'anhydrite', calcium sulphate, in the presence of hot water.

$$CaSO_4(s) + 2NH_3(aq) + CO_2(g) + H_2O(l) \rightarrow CaCO_3(s) + (NH_4)_2SO_4(aq)$$

The calcium carbonate is filtered off and the ammonium sulphate crystallized.

Potassium sulphate K_2SO_4

This compound may be prepared in the laboratory by neutralization of potassium hydroxide solution by dilute sulphuric acid (page 274). In industry, it is usually prepared by heating potassium chloride with concentrated sulphuric acid at a very high temperature. This reaction is not possible in normal laboratory apparatus.

$$2KCl(s) + H_2SO_4(l) \rightarrow K_2SO_4(s) + 2HCl(g)$$

Unlike most soluble sulphates, it crystallizes without water of crystallization.

Sodium sulphate Na_2SO_4

This salt is usually met with in the form of transparent crystals of the decahydrate $Na_2SO_4.10H_2O$, Glauber's salt. In the laboratory, it may be made by neutralizing sodium hydroxide solution by dilute sulphuric acid (page 274). In industry, it is prepared by heating sodium chloride with concentrated sulphuric acid at a very high temperature.

It is used in medicine, in the manufacture of glass, and, by heating with coke, for the manufacture of sodium sulphide.

$$Na_2SO_4(s) + 4C(s) \rightarrow Na_2S(s) + 4CO(g)$$

Calcium sulphate CaSO₄

This salt occurs naturally as anhydrite $CaSO_4$, and gypsum $CaSO_4.2H_2O$. For the use of anhydrite in making sulphuric acid and ammonium sulphate see pages 411 and 416.

Gypsum is chiefly employed for the manufacture of plaster of Paris.

Plaster of Paris (CaSO₄)₂.H₂O

Calcium sulphate hemihydrate. This compound is made by heating gypsum in large steel vessels of several tons capacity. The gypsum is stirred mechanically and the temperature is maintained between 100 °C and 200 °C.

$$2(CaSO_4.2H_2O)(s) \rightarrow (CaSO_4)_2.H_2O(s) + 3H_2O(g)$$
$$\text{plaster of Paris}$$

When mixed with water, plaster of Paris sets to a hard interlacing mass of fine needles of gypsum, expanding at the same time. It is used for making casts for statuary (the expansion during setting ensures a fine impression), in surgery to maintain joints in a fixed position, and in cements and wall-plasters.

Magnesium sulphate MgSO₄

Magnesium sulphate heptahydrate $MgSO_4.7H_2O$ is the familiar substance, 'Epsom salt', which occurs naturally in springs at Epsom and elsewhere and is usually prepared from the mineral, kieserite $MgSO_4.H_2O$, found at Stassfurt. It acts as a mild purgative.

In the laboratory it may be prepared by the method described on page 271.

Zinc sulphate ZnSO₄

This salt is usually encountered as the heptahydrate $ZnSO_4.7H_2O$, 'white vitriol'. It can be prepared in the laboratory from zinc, zinc oxide, or zinc carbonate, and its preparation is fully described on page 271.

Its transparent crystals are very soluble in water (138 g in 100 g water at 10 °C) and the salt is used as an emetic and for the treatment of certain skin diseases.

Iron(II) sulphate FeSO₄

Iron(II) sulphate heptahydrate $FeSO_4.7H_2O$ is known as 'green vitriol'. It is usually prepared in the laboratory by the action of iron (wire, filings, or borings) on dilute sulphuric acid (page 271).

In industry, it is obtained by the action of air and water on the mineral, iron pyrites (iron(II) disulphide) FeS_2.

$$2FeS_2(s) + 7O_2(g) + 2H_2O(l) \rightarrow 2FeSO_4(aq) + 2H_2SO_4(aq)$$

The sulphuric acid is neutralized by scrap iron and the iron(II) sulphate is crystallized.

$$Fe(s) + H_2SO_4(aq) \rightarrow FeSO_4(aq) + H_2(g)$$

Action of heat

On heating, iron(II) sulphate first loses its water of crystallization, the original green crystals being converted into a dirty-yellow anhydrous solid.

$$FeSO_4.7H_2O(s) \rightarrow FeSO_4(s) + 7H_2O(g)$$

When more strongly heated, it gives off sulphur dioxide (test – paper dipped in potassium chromate solution turns green) in addition to white fumes of sulphur trioxide, and leaves a reddish-brown solid, iron(III) oxide Fe_2O_3, 'jewellers' rouge'. This is used in pigments (venetian red, red ochre) and as a polishing powder.

$$2FeSO_4(s) \rightarrow Fe_2O_3(s) + SO_3(g) + SO_2(g)$$

Sulphuric acid was prepared by Glauber (1648) by distilling iron(II) sulphate crystals. The sulphur trioxide given off in the second stage reacted with the water driven off in the first.

$$H_2O(l) + SO_3(g) \rightarrow H_2SO_4(aq)$$

Iron(II) sulphate is used in the brown ring test for nitrates (page 440) and it gives a similar colour with nitrogen oxide (page 434).

Like all iron(II) salts, iron(II) sulphate is a reducing agent, operating by electron loss, which converts iron(II) ions, Fe^{2+}, into iron(III) ions, Fe^{3+}. For example, it reduces nitric acid to nitrogen oxide and chlorine to its ions. The two oxidizing agents accept the electrons lost by the iron(II) ions.

$$6Fe^{2+}(aq) + 6H^+(aq) + 2HNO_3(aq) \rightarrow 6Fe^{3+}(aq) + 4H_2O(l) + 2NO(g)$$
$$2Fe^{2+}(aq) + Cl_2(g) \rightarrow 2Fe^{3+}(aq) + 2Cl^-(aq)$$

The iron(II) sulphate is usually used in solution in dilute sulphuric acid to prevent hydrolysis.

When exposed to air, iron(II) sulphate crystals become covered with a brownish deposit of a basic iron(III) sulphate, by a reaction of the type:

$$12FeSO_4(s) + 6H_2O(l) + 3O_2(g) \rightarrow 4\{Fe_2(SO_4)_3.Fe(OH)_3\}(s)$$
$$\text{(from the air)}$$

Large quantities of iron(II) sulphate are used with gallic acid in the manufacture of ink. This recipe has been known for more than 2000 years.

Iron(III) sulphate $Fe_2(SO_4)_3$

This salt may be prepared by oxidizing iron(II) sulphate by nitric acid in the presence of sulphuric acid (equation above).

It forms alums, for example $K_2SO_4.Fe_2(SO_4)_3.24H_2O$, which are more important than iron(III) sulphate itself, because they can be more readily purified by crystallization.

Copper(II) sulphate (cupric sulphate) $CuSO_4$

'Blue vitriol' $CuSO_4.5H_2O$ is copper(II) sulphate pentahydrate. The preparation of the salt from copper is fully described on page 271; it may also be prepared from the oxide or carbonate of the metal and dilute sulphuric acid (page 275).

On the large scale, it is made by first heating scrap copper with sulphur,

$$Cu(s) + S(s) \rightarrow CuS(s)$$

and then oxidizing the sulphide by heating it with access of air.

$$CuS(s) + 2O_2(g) \rightarrow CuSO_4(s)$$

The sulphate is then crystallized.

When heated, the pentahydrate loses water of crystallization and leaves white anhydrous copper(II) sulphate (page 254).

$$CuSO_4.5H_2O(s) \rightarrow CuSO_4(s) + 5H_2O(g)$$

The formation of the blue pentahydrate from the anhydrous salt is used as a test for the presence of water.

Use. Copper(II) sulphate is used in making washes such as 'Bordeaux mixture' (11 parts of lime and 16 parts of copper sulphate in 1000 parts of water), used in spraying vines and potatoes to kill moulds which would injure the plants. It is also used in the manufacture of certain green pigments.

Sulphites

The general properties of sulphites are discussed on page 405.

Calcium hydrogensulphite, $Ca(HSO_3)_2$

This is prepared by passing sulphur dioxide into milk of lime (a paste of slaked lime and water) and is used for bleaching the pulp in paper-making.

$$Ca(OH)_2(s) + 2SO_2(g) \rightarrow Ca(HSO_3)_2(aq)$$

Questions

1. Give an account of the preparation and properties of hydrogen sulphide.
   If a specimen of hydrogen sulphide were contaminated with hydrogen, how could you obtain the hydrogen sulphide free from hydrogen? (C.)

2. Starting from sulphur, describe how you could prepare specimens of (*a*) plastic sulphur; (*b*) sulphur dioxide; (*c*) sulphur trioxide; and (*d*) hydrogen sulphide. (J.M.B.)

3. Starting with roll sulphur, how would you prepare: (*a*) Rhombic crystals of sulphur? (*b*) Monoclinic (prismatic) crystals of sulphur? (*c*) Plastic sulphur?
   Mention two other elements which, like sulphur, exist in more than one variety. (C.)

4. Describe *one* laboratory method of preparing and collecting hydrogen sulphide, and mention a suitable drying agent. What is the effect of passing hydrogen sulphide into solutions of (*a*) copper(II) sulphate; (*b*) iron(III) chloride; (*c*) chlorine; (*d*) ammonia; (*e*) litmus?
   Briefly describe what happens when hydrogen sulphide burns in (*a*) excess of air; (*b*) a deficit of air.

5. Give an account of the important properties of hydrogen sulphide. A specimen of this gas prepared from iron(II) sulphide is found on analysis to contain 10% by volume of free hydrogen. Assuming that the iron(II) sulphide contained no other impurity than metallic iron, calculate the percentage of free iron present. (L.)

6. From what sources is sulphur obtained? How can sulphur be used for the preparation of (*a*) sulphuric acid; (*b*) sulphurous acid? (O. and C.)

7. How would you prepare in the laboratory hydrogen sulphide gas? Sketch the apparatus. What is the effect of the gas on (*a*) a solution of lead nitrate; (*b*) sulphur dioxide; (*c*) bromine water? (J.M.B.)

8. Give a short account of the chemical reactions which take place in the manufacture of sulphuric acid.
   Describe experiments illustrating the action of this acid on metals. (O. and C.)

9. Describe the properties of sulphuric acid.
   Why is this compound regarded as (*a*) an acid; (*b*) a dibasic acid? (O. and C.)

10. How would you prepare a quantity of dry sulphur dioxide? How may it be shown that the formula for this gas is SO_2? (J.M.B.)

11. Describe briefly two distinct methods which could be used for the preparation of sulphur dioxide.
    How is sulphur dioxide converted into sulphuric acid in the 'contact' process?
    What simple experiment shows that sulphur dioxide contains its own volume of oxygen?

12. Describe the preparation and collection of dry sulphur dioxide in the laboratory. Mention, without giving details of the manufacturing plants, how it is prepared on the industrial scale. What is the action of sulphur dioxide on (*a*) water; (*b*) oxygen; (*c*) chlorine; (*d*) hydrogen sulphide; (*e*) nitrogen dioxide?

13. Describe the preparation and collection of sulphur dioxide. Describe an experiment to show that sulphur dioxide contains its own volume of oxygen. What additional information would you require in order to determine the molecular formula of sulphur

dioxide? Show clearly how you would use the results of the experiment, and the additional information in determining this formula. (J.M.B.)

14. Describe the reaction which takes place when copper is heated with concentrated sulphuric acid. The resulting gas is passed into (a) litmus solution; (b) chlorine water; (c) a solution of hydrogen sulphide. What would be observed in each case and what explanations would you give of the results obtained? (L.)

15. Describe fully how to prepare and collect in the laboratory sulphur dioxide from sulphuric acid. How would you show the action of this gas on (a) chlorine water; (b) moist hydrogen sulphide?

What takes place when a solution of the gas is allowed to stand in contact with air?

Explain the above reactions by equations or otherwise. (J.M.B.)

16. What is meant by the term allotropy? Describe the preparation of two allotropic forms of sulphur.

Starting from sulphur, how would you obtain fairly pure samples of (a) sulphur dioxide; (b) sulphur trioxide? (W.)

17. Describe, with a diagram, how you would prepare and collect hydrogen sulphide in the laboratory.

Describe how hydrogen sulphide reacts with (a) sulphur dioxide; (b) iron(III) chloride solution.

When electric sparks from an induction coil are passed for some time through a volume of hydrogen sulphide the gas is decomposed, sulphur is deposited on the sides of the vessel and on cooling to the original conditions hydrogen remains, the volume of which is equal to that of the hydrogen sulphide. The vapour density of the hydrogen sulphide being 17 calculate *from these facts* the formula of hydrogen sulphide. (H, 1; S, 32.) (J.M.B.)

18. Explain the construction and the working of a Kipp's apparatus for generating hydrogen sulphide. Describe and explain the effect of passing the gas through aqueous solutions of (a) copper(II) sulphate; (b) blue litmus; (c) chlorine. What happens if the resulting solution from (b) is boiled? (L.)

19. How would you prepare a sample of pure dry sodium sulphate starting with solutions of sodium hydroxide and sulphuric acid? How, and under what conditions, does hydrogen sulphide react with (a) sulphur dioxide, (b) lead nitrate, (c) chlorine? (S.)

20. Write equations for the reaction occurring when: (a) hydrogen sulphide is bubbled through a solution of sulphur dioxide; (b) sodium hydrogencarbonate is heated; (c) concentrated nitric acid is added to copper. (S.)

21. Outline how you would obtain a sample of hydrogen sulphide from a mixture of iron filings and powdered sulphur. (No details of collection are required.)

State and explain the reactions of hydrogen sulphide with *three* of the following reagents in aqueous solution: (i) sulphur dioxide; (ii) iron(III) chloride; (iii) copper(II) sulphate; (iv) sodium hydroxide.

Molybdenum(IV) sulphide (MoS_2) contains 60% of molybdenum. Calculate the atomic mass of molybdenum. (A.E.B.)

22. Give a reaction by which hydrogen sulphide can be conveniently prepared in the laboratory. Neither diagrams nor method of collection are required.

Describe the reactions of hydrogen sulphide with *three* of the following: (i) air *or* oxygen, (ii) chlorine, (iii) zinc sulphate solution, (iv) concentrated nitric acid. Why is it important to remove hydrogen sulphide from town gas or coal gas?

The compound sulphur dichloride dioxide (formula SO_2Cl_2) reacts readily with cold water to give a solution containing sulphuric acid and hydrochloric acid. Write the equation for this reaction and calculate the volume of 1M sodium hydroxide needed to neutralize the solution formed when one tenth of a mole of sulphur dichloride dioxide reacts completely with water. (C.)

23. 0.625 grammes of a mineral yield 175 cm^3 of hydrogen sulphide measured at s.t.p. after suitable treatment. Calculate the percentage of sulphur in the mineral. (S.)

24. Describe what you would see and give equations for the reactions which occur when hydrogen sulphide reacts with (a) an excess of air, (b) chlorine water, (c) lead nitrate solution, (d) iron(III) chloride solution.

In each case state, giving a reason, whether the hydrogen sulphide is acting as an oxidizing agent, a reducing agent, or as neither of these. (J.M.B.)

25. Hydrogen sulphide is usually prepared in the laboratory by the action of hydrochloric acid on iron(II) sulphide.

(a) Give a diagram of an apparatus for preparing hydrogen sulphide by use of which a supply of the gas is available whenever required. Explain briefly how the apparatus works.

(b) What is the approximate concentration of the acid used?

(c) Give the equation for the reaction.

(d) Indicate how one main impurity can be removed from the gas.

(e) Calculate the volume at s.t.p. of hydrogen sulphide which can be obtained by using 11 g of iron(II) sulphide.

Give *two* reactions in which hydrogen sulphide acts as a reducing agent, explaining why hydrogen sulphide is considered to be a reducing agent in each reaction. (A.E.B.)

26. Write an equation for the reaction and state the colour of the precipitate when hydrogen sulphide is passed into solutions of (a) copper(II) sulphate, (b) zinc sulphate. (S.)

27. Describe *in outline* the chemistry of the industrial preparation of sulphuric acid from sulphur. Give *one* example to illustrate each of the following uses of sulphuric acid: (a) as a source of hydrogen; (b) as an oxidizing agent; (c) as a dehydrating agent. (S.)

28. Sulphuric acid is now made from sulphur dioxide, oxygen, and water by the 'contact process'. (a) Why must the gases be purified and dried first? (b) How is the purification done? (c) State the conditions under which sulphur dioxide and oxygen are made to combine and name the product formed. (d) To convert the product to sulphuric acid, it must be made to react with water. How is this done commercially?

Describe (i) *two* reactions in which sulphuric acid behaves as an acid, (ii) *one* reaction in which it behaves as a dehydrating agent, (iii) *one* reaction in which it acts as an oxidizing agent, explaining why you regard this reaction as oxidation. (A.E.B.)

29. Draw a fully labelled diagram and write an equation to illustrate the usual laboratory preparation and collection of sulphur dioxide from sulphuric acid. How, and under what conditions, does sulphur dioxide react with (a) magnesium, (b) nitric acid, (c) hydrogen sulphide? (J.M.B.)

30. A pupil added concentrated sulphuric acid to a substance and noted that a colourless gas was given off which decolorized bromine water. He concluded that (i) the gas was sulphur dioxide, and (ii) that the sulphur dioxide had been produced by the reduction of the sulphuric acid.

(a) He may have been wrong in his first conclusion. Why?

(b) Name *one* substance which would produce sulphur dioxide when acted upon by concentrated sulphuric acid *without involving the reduction of the sulphuric acid*. Write a balanced equation for the reaction, and use it to show that reduction of the acid has not occurred.

(c) (i) A solution of barium chloride was added to a solution of sodium sulphite and a white precipitate was obtained. What is this precipitate likely to be? (ii) When dilute hydrochloric acid was added to the precipitate it reacted and

a pungent smelling gas was given off. What is this gas? (iii) A solution of sulphur dioxide in water was warmed with a little nitric acid. Barium chloride solution was added and a white precipitate was formed which did *not* dissolve when dilute hydrochloric acid was added. What is the precipitate? What happened to the solution of sulphur dioxide when it was acted upon by the nitric acid?

(d) Name *two* substances which are used for the manufacture of sulphur dioxide in industry. (Scottish.)

31. By means of a labelled diagram together with an equation for the action, show how you would prepare sulphur dioxide.

Describe, including any changes you would see, the reactions that occur (a) when a freshly prepared solution of sulphur dioxide in water is: (i) allowed to stand in an open beaker for a few days, (ii) added in excess to a solution of iron(III) sulphate, (b) when a gas-jar of sulphur dioxide is shaken with a few cubic centimetres of bromine water. (C.)

32. Write an equation for the production of sulphur dioxide by roasting iron pyrites (FeS$_2$) in an excess of air, and outline the subsequent conversion of sulphur dioxide into sulphuric acid.

Explain carefully, the following: (i) calcium sulphate causes water to be hard and that hardness to be permanent; (ii) iron(II) sulphate is a reducing agent; (iii) sodium sulphate decahydrate is efflorescent. (O.)

33. Under what conditions is sulphur trioxide formed on an industrial scale and how is the product converted into sulphuric acid?

Describe the action of sulphuric acid upon (i) any one metal of your own choice, (ii) any hydroxide of your own choice, (iii) sucrose C$_{12}$H$_{22}$O$_{11}$, (iv) sodium chloride. State clearly the conditions under which each reaction would be carried out and the concentration of the acid to be used.

How would you use (i) hydrochloric acid, (ii) barium chloride, (iii) hydrochloric acid with barium chloride, to distinguish between sodium sulphite and sodium sulphate? (O.)

34. A specimen of sulphur trioxide SO$_3$ can be made in the laboratory by passing a mixture of oxygen and sulphur dioxide (obtained from cylinders) over heated platinized asbestos. The sulphur trioxide is condensed to a solid by cooling. The sulphur trioxide can be converted to sulphuric acid by the addition of water.

(a) Sketch the apparatus you would use to make sulphur trioxide by the method outlined above, and write the equation for the reaction taking place.

(b) It is necessary to make certain that the apparatus and gases are completely dry during the preparation of sulphur trioxide. Suggest one reason why this is so.

(c) Platinized asbestos is finely divided platinum supported by asbestos wool. What is the function of the platinum and why is it 'finely divided'?

(d) In what ratio by volume would you arrange for the oxygen and sulphur dioxide to mix during the preparation of sulphur trioxide? Explain your reason for choosing this volume ratio.

(e) Sulphur trioxide dissolves in water to produce dilute sulphuric acid. Write the equation for this reaction. What tests would you carry out on the resulting liquid to show that dilute sulphuric acid had been formed?

(f) What is the maximum number of moles of sulphuric acid that could be obtained from 0.1 mole of oxygen and an unlimited supply of sulphur dioxide and water? (L.)

35. How may sulphur be converted into sulphuric acid? Describe *two* reactions in *each* case which show that sulphuric acid can react as (a) an acid, (b) an oxidizing agent, (c) a dehydrating agent. (S.)

36. Describe, with equations, *one* method for the manufacture of sulphuric acid from sulphur. No description of the industrial plant is required.

Describe how you would obtain in the laboratory dry crystals of sodium sulphate $Na_2SO_4.10H_2O$. You are provided with dilute aqueous solutions of sulphuric acid and sodium hydroxide. (O. and C.)

34 Nitrogen and its Compounds

Nitrogen

Occurrence

About four-fifths of the atmosphere is free nitrogen. The element also occurs combined in the form of sodium nitrate, Chile saltpetre $NaNO_3$, as a mineral deposit in Chile, and distributed everywhere in the soil in minute quantities as ammonium sulphate $(NH_4)_2SO_4$, and sodium nitrate $NaNO_3$, potassium nitrate KNO_3, and calcium nitrate $Ca(NO_3)_2$. (The very great importance of these compounds of nitrogen in maintaining the fertility of the soil is discussed on page 443.)

Combined nitrogen is always found as a constituent of the living matter of plants and animals.

Preparation of nitrogen from the atmosphere

The most important gases present in dry air are oxygen, about 21% by volume, carbon dioxide, about 0.03% by volume, and 'atmospheric nitrogen', about 79% by volume. The first two of these gases can be removed and the nitrogen collected by the apparatus of Figure 177.

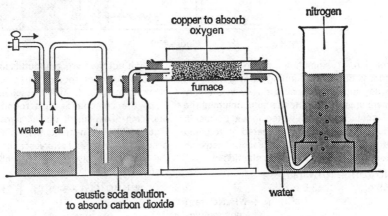

Figure 177

The equations are:

Absorption of carbon dioxide

$$2NaOH(aq) + CO_2(g) \rightarrow Na_2CO_3(aq) + H_2O(l)$$

Absorption of oxygen

$$2Cu(s) + O_2(g) \rightarrow 2CuO(s)$$

If the nitrogen is required dry, it may, after leaving the heated copper, be passed through a U-tube containing glass beads wetted with concentrated sulphuric acid to dry it and then collected in a syringe.

The product of this experiment is not pure nitrogen. It contains about 1% by volume of the 'noble gases', chiefly argon, the removal of which is not possible by chemical methods. The presence of these gases makes 'atmospheric nitrogen' denser than the pure gas.

Another method of preparing 'atmospheric' nitrogen is to absorb both carbon dioxide and oxygen together by shaking air with a solution of pyrogallol in sodium hydroxide solution. The sodium hydroxide absorbs the carbon dioxide and the pyrogallol absorbs the oxygen to form an oxidation product of itself.

This method is, however, only suitable for the preparation of small samples of 'atmospheric nitrogen'.

Experiment 155
Preparation of nitrogen by the action of heat on ammonium nitrite (ammonium nitrate(III))

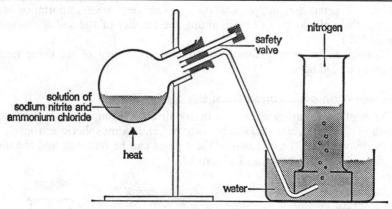

Figure 178

A solution of ammonium nitrite readily decomposes on slight warming to give nitrogen. This decomposition occurs slowly at ordinary temperatures, so that neither ammonium nitrite itself, nor its solution in water, should be kept in stock. The compound is prepared as required by a double decomposition reaction between sodium nitrite and ammonium chloride.

$$NaNO_2(aq) + NH_4Cl(aq) \longrightarrow$$
$$\quad 69\ g \qquad\quad 53.5\ g$$
$$\quad\quad NaCl(aq) + NH_4NO_2(aq)$$
$$\quad\quad\quad\quad\quad\quad ammonium$$
$$\quad\quad\quad\quad\quad\quad nitrite$$

Weigh out the two compounds in these proportions, 14 g of sodium nitrite and 11 g of ammonium chloride will be suitable masses. Place the compounds in a round flask, add 350 cm³ of water, fit up the apparatus as in Figure 178, and heat gently. As the solution becomes warm, rapid effervescence occurs and the nitrogen evolved may be collected over water.

$$NH_4NO_2(aq) \longrightarrow N_2(g) + 2H_2O(l)$$

Other chemical methods of preparation of nitrogen
The action of chlorine on excess ammonia.

$$3Cl_2(g) + 8NH_3(g) \longrightarrow N_2(g) + 6NH_4Cl(s)$$

(See page 124.)

Passing ammonia gas over heated copper(II) oxide (see page 429).

$$2NH_3(g) + 3CuO(s) \rightarrow 3Cu(s) + N_2(g) + 3H_2O(g)$$

Reduction of oxides of nitrogen by heated copper, *e.g.*,

$$2Cu(s) + 2NO(g) \rightarrow 2CuO(s) + N_2(g)$$

These methods are all much less convenient than the heating of ammonium nitrite.

Industrial manufacture of nitrogen

Nitrogen is obtained in industry by the fractional distillation of liquid air. The air, liquefied by the method described on page 284 for the industrial preparation of oxygen, is distilled under controlled conditions when nitrogen (boiling point 77 K at standard atmospheric pressure) is evolved. Since the boiling point of oxygen is higher than this (it is 90 K at standard atmospheric pressure) a complete separation of the nitrogen and oxygen is readily achieved. The separated nitrogen is reliquefied and stored in specially designed containers ready for use. It is also sold as a compressed gas.

Tests for nitrogen

At ordinary temperatures, nitrogen is so inert that no positive tests can be applied. We can only show a given gas to be nitrogen by elimination of other possibilities.

Lighted splint. Place a lighted splint into a gas-jar of the gas. It is extinguished and the gas does not burn. It cannot, therefore, be any gas which supports combustion, *e.g.*, oxygen, dinitrogen oxide, or any combustible gas, *e.g.*, hydrogen sulphide, carbon monoxide, hydrogen.

Smell. The gas has no smell. This distinguishes it from gases such as sulphur dioxide, ammonia, hydrochloric acid gas.

Action of lime-water. After the above tests the only gas with which nitrogen may be confused is carbon dioxide. To distinguish it from this, add lime-water and shake. Nitrogen leaves the lime-water unchanged; with carbon dioxide, the lime-water is turned milky.

Properties of nitrogen

Nitrogen is colourless and odourless. It is slightly less dense than air and only slightly soluble in water (about 2 volumes of the gas dissolve in 100 volumes of water at ordinary temperature).

Under ordinary conditions the gas is very inert, but, by applying the results of much research, it has been made to combine with hydrogen to produce ammonia.

$$N_2(g) + 3H_2(g) \rightleftharpoons 2NH_3(g)$$

For this reaction applied in Haber's Process, see page 188.

Nitrogen will combine directly with many metals forming nitrides, *e.g.*,

$$3Mg(s) + N_2(g) \rightarrow Mg_3N_2(s)$$
$$\text{magnesium}$$
$$\text{nitride}$$

To illustrate this, burn some magnesium ribbon in a crucible and allow the product to cool. Add a few drops of water and smell the mixture. The choking smell is that of ammonia. It is evolved by the action of water on the magnesium nitride which was formed, in small amount, by combination of the magnesium with nitrogen of the air.

$$Mg_3N_2(s) + 6H_2O(l) \rightarrow 2NH_3(g) + 3Mg(OH)_2(s)$$

magnesium   water   ammonia   magnesium
nitride   hydroxide

Ammonia

This hydride of nitrogen, NH_3, can be made in very small amounts by heating nitrogenous organic materials such as hoofs and horns of animals. Its old name was, in fact, 'spirit of hartshorn'.

Preparation of ammonia

Ammonia may be prepared in the laboratory by heating any ammonium salt with an alkali. Usually a mixture of ammonium chloride and calcium hydroxide (slaked lime, the cheapest alkali) is used. Both are solids so they must be thoroughly ground first to give a very fine mixture in which the reaction can occur satisfactorily.

$$Ca(OH)_2(s) + 2NH_4Cl(s) \rightarrow CaCl_2(s) + 2H_2O(l) + 2NH_3(g)$$

74 g   2 × 53.5 g
               107 g

Experiment 156
Preparation of ammonia from ammonium chloride

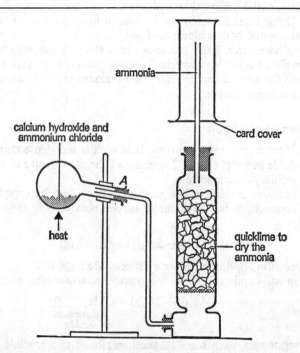

Figure 179

An excess of the slaked lime is preferable. Weigh out 25 g of slaked lime and 16 g of ammonium chloride. Grind the mixture well in a mortar, place it in a round flask of resistance glass and set up apparatus as in Figure 179. The neck of the flask should slope towards *A* as shown, because water will condense and, if allowed to run back on to the hot flask, might break it. Heat the flask. Ammonia gas is evolved. **It is dried by a rather unusual drying agent, quicklime CaO,** because it reacts with all the usual drying agents. Concentrated sulphuric acid is acidic and would absorb the gas forming a salt, *e.g.,*

$$2NH_3(g) + H_2SO_4(l) \longrightarrow (NH_4)_2SO_4(s)$$

while it reacts with calcium chloride, forming solid complex compounds, *e.g.,*

$$CaCl_2(s) + 4NH_3(g) \longrightarrow CaCl_2.4NH_3(s)$$

Ammonia is less dense than air and very soluble in water, so it is collected as shown by upward delivery.

Instead of calcium hydroxide, sodium hydroxide (or potassium hydroxide) solution may be used, in which case the flask would be placed in the vertical position and heated on a tripod and gauze.

$$NaOH(aq) + NH_4Cl(s) \longrightarrow$$
$$NH_3(g) + H_2O(l) + NaCl(aq)$$
$$\text{or } OH^-(aq) + NH_4^+(s) \longrightarrow NH_3(g) + H_2O(l)$$

Ammonium sulphate may be used instead of ammonium chloride.

$$Ca(OH)_2(s) + (NH_4)_2SO_4(s) \longrightarrow$$
$$CaSO_4(g) + 2H_2O(l) + 2NH_3(g)$$
$$2NaOH(aq) + (NH_4)_2SO_4(s) \longrightarrow$$
$$Na_2SO_4(aq) + 2H_2O(l) + 2NH_3(g)$$

Ammonia gas in quantity

If several gas-jars of ammonia are required, it is very convenient to fill these by heating a concentrated solution of the gas in water. In this case the flask containing the ammonium chloride and slaked lime in the previous experiment is replaced by a vertical flask containing a concentrated solution of ammonia. *N.B. This is not, strictly speaking, a preparation of ammonia gas but merely obtaining the gas from its solution in water.*

Manufacture of ammonia

Ammonia is obtained on an industrial scale by the Haber Process, which is based on the direct combination of nitrogen and hydrogen:

$$N_2(g) + 3H_2(g) \rightleftharpoons 2NH_3(g)$$

This reaction, which is a reversible one, requires special conditions for an optimum yield of ammonia to be obtained at an economically desirable rate. The factors involved in determining these conditions have already been fully discussed (see page 188), and it is sufficient simply to restate them here. The nitrogen and hydrogen are dried, and mixed in volume proportions of 1 : 3 respectively at a relatively high pressure (usually between 200 and 500 atmospheres) and are passed over a catalyst (of finely divided reduced iron impregnated with alumina) at a temperature of 450 °C (723 K). The ammonia produced is liquefied by refrigeration and stored for further use.

Tests for ammonia

Smell. A characteristic choking smell. The choking smell is due to the fact that the gas temporarily paralyses the respiratory muscles and breathing is checked. In large quantities, the gas causes asphyxiation and it should therefore always be treated with respect; it should be sniffed cautiously, wafting the gas towards the nose with the hand.

Action with litmus. Expose damp red litmus paper to the gas. It is turned blue. **Ammonia is the only common alkaline gas.**

Properties of ammonia

Appearance. A colourless gas.
Density. Less dense than air.
  Density relative to hydrogen, 8.5.
  Density of air relative to hydrogen, 14.4.

Experiment 157
Solubility of ammonia in water. The fountain experiment

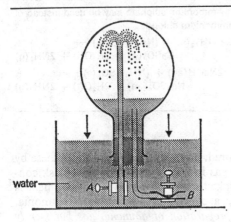

Figure 180

The very great solubility of ammonia in water is illustrated in the fountain experiment (Figure 180). Replace the gas-jar of Figure 179 by a dry, thick-walled flask of about 1000 cm³ capacity and pass ammonia into it for some time. (It is better to supply another flask with fresh reaction mixture, to give off a satisfactory stream of ammonia.) Fit the flask with a rubber stopper carrying tubes and clips as shown. Place the tubes and clips under water, open clip *B* for a moment, close it and allow the few drops of water which have entered to run down into the round part of the flask. Then replace the tubes and clips under water and open clip *A*. A fountain will at once play, as in sketch, and **will continue until the flask is as full of water as it was formerly full of ammonia.**

The alkaline nature of ammonia can be shown in this experiment by adding a little litmus solution to the water in the trough and making it turn red by the addition of a drop of acid. When the ammonia dissolves in the litmus solution, the latter is turned blue.

Explanation. Ammonia has the highest solubility of all known gases (about 800 vol. of gas in 1 vol. of water at 15 °C). The first few drops of water, which entered when clip *B* was opened, dissolved nearly 800 times their own volume of ammonia. This reduced the gas pressure inside the flask to only a fraction of its former value, atmospheric pressure. As soon as the clip *A* was opened, the water was forced into the flask because the atmospheric pressure from outside overcame the resistance of the reduced gas pressure inside the flask. The water, entering in a fountain, dissolved the remaining ammonia, maintaining the fountain until no air was left in the flask. (A thin-walled flat-bottomed flask must not be used for this experiment; the reduction of pressure inside would almost certainly cause it to collapse inwards.)

Experiment 158
Action of ammonia with hydrogen chloride

Place a gas-jar of ammonia over a gas-jar of hydrogen chloride, remove the covers and allow the gases to mix. Dense white fumes will be seen which will settle to a white solid, sal-ammoniac or ammonium chloride, on the sides of the gas-jar.

$$NH_3(g) + HCl(g) \longrightarrow NH_4Cl(s)$$

Combustion of ammonia

Ammonia will burn in an atmosphere of air slightly enriched by oxygen but not in air alone. The chief products are nitrogen and water.

$$4NH_3(g) + 3O_2(g) \rightarrow 2N_2(g) + 6H_2O(l)$$

Ammonia as a reducing agent

Experiment 159
Action of ammonia on copper(II) oxide

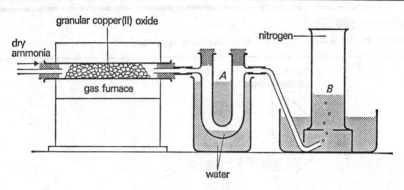

Figure 181

Set up the apparatus as shown in Figure 181. The colourless liquid collecting at A is water (for tests, see page 254) and the colourless gas at B is nitrogen (for tests see page 425). The ammonia has **reduced** the copper(II) oxide to copper and has itself been **oxidized** to nitrogen and water.

$$3CuO(s) + 2NH_3(g) \rightarrow$$
$$3Cu(s) + 3H_2O(l) + N_2(g)$$

The combustion of ammonia (see above) is also an example of a reaction in which the ammonia may be considered to be a reducing agent.

Liquefaction of ammonia gas

Ammonia gas can be liquefied at ordinary temperatures by compression. The colourless liquid boils at about $-33\,°C$ under ordinary atmospheric pressures. It has the peculiar property of dissolving alkali metals (without the evolution of hydrogen) to give blue, strongly conducting solutions. Liquid ammonia can behave as a poor solvent, similar in many ways to water.

Uses of ammonia

(1) Ammonia solution is used in laundry work. It removes temporary hardness by precipitating the calcium ion of calcium hydrogencarbonate as chalk.

$$Ca^{2+}(aq) + 2HCO_3^-(aq) + 2OH^-(aq) \rightarrow CaCO_3(s) + 2H_2O(l) + CO_3^{2-}(aq)$$

It also dissolves out acids left by evaporation of perspiration from underclothing.

(2) Ammonia is converted to nitric acid. (See page 437.)

(3) Ammonium sulphate is made from ammonia to be used as a nitrogenous

fertilizer. This may be done by direct neutralization with sulphuric acid or by passing carbon dioxide and ammonia through a steam-heated mixture of calcium sulphate (anhydrite) and water.

$$CaSO_4(s) + 2NH_3(aq) + CO_2(g) + H_2O(l) \rightarrow CaCO_3(s) + (NH_4)_2SO_4(aq)$$

Calcium carbonate is filtered off and ammonium sulphate can be crystallized.

(4) Liquid ammonia is used in large-scale refrigerating plant, such as in ships and warehouses. It used to be used in domestic refrigerators, but has now been replaced by non-toxic, non-corrosive chlorofluorohydrocarbons.

Ammonium salts

Experiment 160
Preparation of a solution of ammonium hydroxide

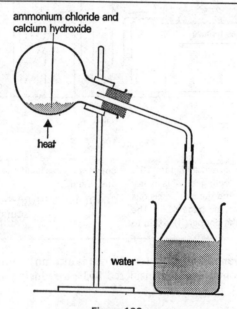

ammonium chloride and calcium hydroxide

heat

water

Figure 182

Set up apparatus as in Figure 182 and heat the flask gently as in the preparation of ammonia gas. The rim of the inverted funnel should just touch the surface of the water. This is a device to prevent the water from 'sucking back' into the flask. After a time the water in the beaker will be found to have acquired the smell of ammonia gas which has dissolved in it. The solution is known as *ammonia solution*.

At 0 °C and ordinary pressure, one volume of water dissolves about 1000 volumes of the gas. Some of this gas reacts with protons from the water; this disturbs the ionic equilibrium in water (page 277) and gives rise to a higher concentration of hydroxyl ions than is present in pure water.

$$H_2O(l) \rightleftharpoons H^+(aq) + OH^-(aq)$$
$$NH_3(aq) + H^+(aq) \rightleftharpoons NH_4^+(aq)$$

The excess of hydroxyl ions, OH^-, gives the solution its alkaline reaction towards litmus and many properties resembling those of the caustic alkalis

Na^+OH^- and K^+OH^-. Like them, it will precipitate insoluble metallic hydroxides when mixed with solutions of salts of the metals, *e.g.*,

$$3KOH(aq) + FeCl_3(aq) \rightarrow Fe(OH)_3(s) + 3KCl(aq)$$
$$3NH_4OH(aq) + FeCl_3(aq) \rightarrow Fe(OH)_3(s) + 3NH_4Cl(aq)$$

Zinc and copper hydroxides will dissolve in excess ammonia solution to give solutions of complex ammines.

Ammonia solution will also neutralize acids to form *ammonium salts*, which can be crystallized out and are generally similar to ordinary metallic salts, *e.g.*,

$$NaOH(aq) + HCl(aq) \rightarrow NaCl(aq) + H_2O(l) \ \Big\}$$
$$NH_4OH(aq) + HCl(aq) \rightarrow NH_4Cl(aq) + H_2O(l) \Big\}$$
$$2NaOH(aq) + H_2SO_4(aq) \rightarrow Na_2SO_4(aq) + 2H_2O(l) \ \Big\}$$
$$2NH_4OH(aq) + H_2SO_4(aq) \rightarrow (NH_4)_2SO_4(aq) + 2H_2O(l) \Big\}$$

Notice that ammonium salts are electrovalent compounds, containing the *ammonium ion*, NH_4^+, in combination with a corresponding amount of an acidic ion, such as Cl^- or NO_3^-. Since the ammonia molecule NH_3 can combine with a hydrogen ion (proton) by donation of its lone pair of electrons, it must be regarded as a base.

$$NH_3(aq) + H^+(aq) \rightleftharpoons NH_4^+(aq) \quad \text{(See page 430).}$$

It is a much weaker base than the hydroxide ion so ammonium salts of strong acids are considerably hydrolysed in their aqueous solutions and these solutions show appreciable acidity.

Action of alkalis on ammonium salts

Any ammonium salt, if heated with sodium hydroxide (or potassium hydroxide) solution or with calcium hydroxide and a little water, gives off ammonia gas (which turns red litmus paper blue). This distinguishes ammonium salts from those of any metal.

$$NH_4^+(aq) + OH^-(aq) \rightarrow NH_3(g) + H_2O(l)$$

Experiment 161
Action of heat on ammonium salts

Place a little ammonium chloride in a dry test-tube, and heat the tube gently below the solid.

Carefully observe the changes which take place along the whole length of the tube.

Usually when a vapour is cooled, it condenses first to a liquid, and, later, on further cooling, solidifies, *e.g.*, steam $\rightarrow$ water $\rightarrow$ ice.

Ammonium salts are always decomposed by heat and, sometimes, sublime. The best example of *sublimation* is provided by ammonium chloride. The characteristic feature of sublimation is that, on cooling, the vapour condenses to the solid without the intermediate liquid state. Usually, the converse is also true, that the subliming solid is converted directly to vapour on heating without an intermediate liquid stage, but in some cases (for example, that of iodine) melting may occur. In the case of ammonium chloride above, a white *sublimate* of ammonium chloride could be observed on the upper part of the tube, a region which is relatively cool (Figure 183).

Very few substances sublime. Among those which sublime are a few ammonium salts (especially the chloride), iodine, and naphthalene. Sublimation is a very effective means of purifying them, because their impurities are very unlikely to sublime. The white sublimate of ammonium chloride in Figure 183 will be a

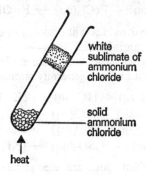

white
sublimate of
ammonium
chloride

solid
ammonium
chloride

heat

Figure 183

purer sample of the compound than the original material used for the experiment. For thermal dissociation, see page 190.

Decomposition of ammonium salts by heat. Ammonium salts of acids having a high proportion of oxygen are usually decomposed by heat, *e.g.*

(i) *Ammonium nitrite*

$$NH_4NO_2(s) \rightarrow N_2(g) + 2H_2O(g)$$

This reaction is fully dealt with on page 424.

(ii) *Ammonium nitrate.*

$$NH_4NO_3(s) \rightarrow \underset{\substack{\text{dinitrogen}\\\text{oxide}}}{N_2O(g)} + 2H_2O(g)$$

This reaction is fully considered below.

Oxides of nitrogen

There are three common oxides of nitrogen: dinitrogen oxide (formerly called *nitrous oxide*) N_2O, nitrogen oxide (formerly called *nitric oxide*) NO, and nitrogen dioxide or dinitrogen tetroxide NO_2 or N_2O_4.

Dinitrogen oxide

The gas may be prepared by heating any mixture of salts which, by double decomposition, will yield ammonium nitrate – *e.g.*, a mixture of potassium nitrate and ammonium sulphate, finely ground (Figure 184).

$$(NH_4)_2SO_4(s) + 2KNO_3(s) \rightarrow 2NH_4NO_3(s) + K_2SO_4(s)$$

On heating, the ammonium nitrate melts and effervesces. Dinitrogen oxide and steam are given off. Most of the steam is condensed to water as the bubbles pass up the gas-jar.

$$NH_4NO_3(l) \rightarrow N_2O(g) + 2H_2O(l)$$

Ammonium nitrate itself will explode, on heating, especially if the amount of material in the vessel becomes very small, but the reaction is quite safe if the ammonium nitrate is prepared by double decomposition during the preparation, as above.

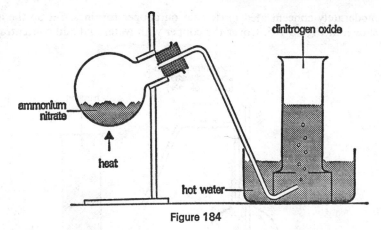

Figure 184

Test for dinitrogen oxide

The gas rekindles a brightly glowing splint.

It may be distinguished from oxygen by the following tests:

Dinitrogen oxide has a faint sweet and sickly smell (oxygen has no smell).

Invert a gas-jar of dinitrogen oxide over cold water and shake it. The level of water in the gas-jar will rise, showing the gas to be fairly soluble in water. Oxygen is almost insoluble in water and no rise would be observed.

Dinitrogen oxide does not give brown fumes with nitrogen monoxide. If a little oxygen is bubbled up into a volume of nitrogen monoxide enclosed over water, brown fumes of nitrogen dioxide are formed.

Properties of dinitrogen oxide

The gas is colourless, neutral to litmus, and has a faint, rather sweet, sickly smell. It can produce insensibility for short periods and is used as an anaesthetic for minor surgical operations, such as are required in dentistry. Insensibility lasts for a minute or two only. The period of insensibility can be prolonged if the gas is mixed with about 20% oxygen and inhaled, the oxygen being necessary to keep the patient alive. A trace of carbon dioxide must also be present to stimulate the respiratory centres and maintain breathing. Patients recovering from the effects of dinitrogen oxide may become hysterical; hence its common name – 'laughing gas'.

Density. Dinitrogen oxide has a density of 22 compared with hydrogen. (Relative molecular mass of dinitrogen oxide = 44.)

Solubility. The gas is fairly soluble in cold water.

Action on a glowing splint. If the splint is glowing brightly, it will be rekindled by dinitrogen oxide, but, if only feebly glowing, it will be extinguished.

To be rekindled, the glowing portion of the splint must be hot enough to decompose some dinitrogen oxide into nitrogen and oxygen. The mixture will then be rich enough in oxygen to stimulate the combustion of the splint and cause it to burst into flame. A feebly glowing splint will not be hot enough to decompose the dinitrogen oxide, and so will be extinguished, having no free oxygen with which to burn.

Nitrogen oxide

Nitrogen oxide is produced, mixed always with other oxides of nitrogen, by the action of nitric acid on most metals. The commonest reaction used is that of moderately concentrated nitric acid on copper turnings. Set up the apparatus shown in Figure 185. Cover the copper with water and add concentrated nitric

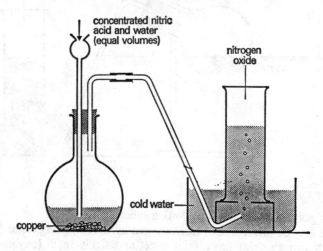

Figure 185

acid, about equal in volume to the water. Vigorous effervescence occurs and the flask is filled with brown fumes. These fumes are nitrogen dioxide, produced partly by the action of the acid upon the copper and partly by the oxidation of the main product, nitric oxide, by the oxygen of the air in the flask.

$$2NO(g) + O_2(g) \longrightarrow 2NO_2(g)$$
$$\text{nitrogen}$$
$$\text{dioxide}$$

The brown fumes dissolve in the water over which the nitrogen oxide is collected as a colourless gas. A green solution of copper(II) nitrate is left in the flask.

$$3Cu(s) + 8HNO_3(aq) \longrightarrow 3Cu(NO_3)_2(aq) + 4H_2O(l) + 2NO(s)$$

Tests for nitrogen oxide

Exposure to air. Remove the cover from a gas-jar of nitrogen oxide. **Reddish-brown fumes are at once produced by oxidation of the gas by oxygen of the air.**

$$2NO(g) + O_2(g) \longrightarrow 2NO_2(g)$$

Nitrogen oxide is the only gas to give this action.

Action on iron(II) sulphate solution. Prepare a cold solution of iron(II) sulphate in dilute sulphuric acid. Pour it into a gas-jar of nitrogen oxide. The dark brown or black coloration is caused by formation of a black compound, (FeSO₄).NO.

$$FeSO_4(aq) + NO(g) \longrightarrow (FeSO_4).NO(aq)$$

(The gas can be obtained in a pure state by heating this compound.)

Properties of nitrogen oxide

Nitrogen oxide is colourless and almost insoluble in water. Its density relative to that of hydrogen is 15; it is slightly denser than air (density 14.4). It is neutral to litmus.

The smell of the gas is unknown because it combines with the oxygen of the air (see tests preceding); it can never be collected by displacement of air. In the lungs it would react with oxygen to produce poisonous nitrogen dioxide.

Nitrogen oxide will support the combustion of those burning materials whose flames are hot enough to decompose it and so liberate free oxygen with which the material may combine. A splint, candle, sulphur, and glowing charcoal are all extinguished by the gas, but it supports the combustion of strongly burning phosphorus or magnesium.

$$P_4(s) + 10NO(g) \rightarrow P_4O_{10}(s) + 5N_2(g)$$
$$2Mg(s) + 2NO(g) \rightarrow 2MgO(s) + N_2(g)$$

Nitrogen dioxide

Nitrogen dioxide is given off, together with oxygen, when nitrates of heavy metals are heated. The most suitable nitrate to use is lead(II) nitrate, because it crystallizes without water of crystallization, which is found in crystals of most nitrates and which would interfere with the preparation. The experiment should be done in a fume-cupboard.

The lead(II) nitrate decrepitates and melts on heating. It effervesces, giving a brown gas, nitrogen dioxide, and oxygen. The nitrogen dioxide is liquefied in the freezing-mixture (Figure 186) and collects in the U-tube as a green liquid

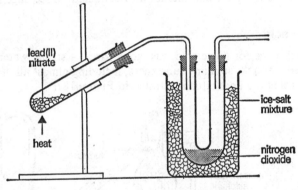

Figure 186

(yellow if pure), oxygen passing on as gas and escaping or being collected over water. Lead(II) oxide remains in the tube as a yellow solid fused into the glass.

$$2Pb(NO_3)_2(s) \rightarrow 2PbO(s) + 4NO_2(g) + O_2(g)$$

Properties of nitrogen dioxide

Nitrogen dioxide is usually seen as a reddish-brown gas, though the boiling point of liquid nitrogen dioxide (22 °C) is above the usual atmospheric temperatures. It has a pungent, irritating smell and is a dangerous gas on account of its tendency to set up septic pneumonia if inhaled. It should not be allowed to escape in quantity into the open laboratory.

Action with water. When the brown gas is passed into water a pale blue solution results. On testing this solution with indicator it is found to be acidic, and does, in fact, contain the compounds nitrous and nitric acids.

$$2NO_2(g) + H_2O(l) \rightarrow \underset{\substack{\text{nitric} \\ \text{acid}}}{HNO_3(aq)} + \underset{\substack{\text{nitrous} \\ \text{acid}}}{HNO_2(aq)}$$

The gas is a *mixed anhydride*. Similarly with sodium hydroxide, it gives a mixture of the corresponding sodium salts of the acids.

$$2NaOH(aq) + 2NO_2(g) \rightarrow NaNO_3(aq) + NaNO_2(aq) + H_2O(l)$$

Like the other two oxides of nitrogen we have considered, nitrogen dioxide will support the combustion of a burning substance whose flame is hot enough to decompose it and supply free oxygen.

Dissociation of nitrogen dioxide. At 150 °C, nitrogen dioxide is very dark brown in colour. Its vapour density is 23, and its relative molecular mass is therefore 46, corresponding to the formula NO_2 ($N = 14$; $O = 16$). As the temperature falls, the vapour density of the gas gradually increases until at 22 °C it approaches 46, corresponding to a relative molecular mass of 92 and a formula of N_2O_4. At the same time it becomes lighter in colour.

This must mean that, on heating, nitrogen dioxide dissociates, and, at any temperature between 22 °C and 150 °C, contains both N_2O_4 and NO_2 molecules, the proportion of the latter increasing as the temperature rises to 150 °C, when only NO_2 molecules are present.

$$\underset{\text{light yellow}}{N_2O_4(g)} \underset{\text{cool}}{\overset{\text{heat}}{\rightleftharpoons}} \underset{\text{dark brown}}{2NO_2(g)}$$

For a more complete consideration of dissociation, see page 190.

Nitric acid

The old name for nitric acid was 'aqua fortis' – strong water. It was so called because it attacks so many substances, including almost all the metals. It is prepared using the apparatus illustrated in Figure 187.

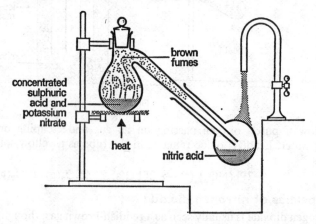

Figure 187

The potassium nitrate crystals are placed in the bulb of the retort and concentrated sulphuric acid added. The retort is heated gently, the potassium nitrate gradually dissolves, and effervescence occurs.

$$\underset{\substack{\text{potassium} \\ \text{hydrogensulphate}}}{KNO_3(s) + H_2SO_4(aq) \rightarrow KHSO_4(s)} + HNO_3(g)$$

The nitric acid distils and collects in the cooled receiver as a yellow liquid, while drops of the acid can be seen running down the bulb and neck of the retort. The brown fumes are nitrogen dioxide formed by slight decomposition of the nitric acid by heat,

$$4HNO_3(l) \rightarrow 2H_2O(l) + 4NO_2(g) + O_2(g)$$
nitrogen
dioxide

and they impart a yellow colour to the acid by dissolving in it. The nitrogen dioxide impurity can be removed from the acid by bubbling air through it; the pure acid is colourless.

This reaction is a general one. Any metallic nitrate, when heated with concentrated sulphuric acid, gives off nitric acid, *e.g.*,

$$NaNO_3(s) + H_2SO_4(aq) \rightarrow NaHSO_4(aq) + HNO_3(aq)$$
sodium nitrate

Again, as in the preparation of hydrogen chloride (page 380), a less volatile acid (concentrated sulphuric acid, b.pt. 360 °C) has displaced a more volatile one (nitric acid, b.pt. 85 °C). Unlike the gas hydrogen chloride however, the nitric acid has to be gently distilled off.

Manufacture of nitric acid

Nitric acid is manufactured by the oxidation of ammonia.

Fixation of atmospheric nitrogen by Haber's process has already been considered (page 188). The product, ammonia, may be converted to *ammonium sulphate* or oxidized by passing the ammonia with excess of air over a platinum (90%)/rhodium (10%) gauze catalyst. The catalyst is heated to red heat to start the reaction. The reaction is exothermic and, once started, maintains the temperature of the gauze.

$$4NH_3(g) + 5O_2(g) \rightarrow 4NO(g) + 6H_2O(l)$$

The nitrogen oxide so formed is rapidly cooled and combines with the oxygen from excess of air to form nitrogen dioxide.

$$2NO(g) + O_2(g) \rightarrow 2NO_2(g)$$

The nitrogen dioxide, in the presence of more air, is then absorbed in hot water, the conditions being chosen to yield nitric acid by the equation:

$$2H_2O(l) + 4NO_2(g) + O_2(g) \dashrightarrow 4HNO_3(aq)$$

Nitric acid is used mainly for the manufacture of explosives and dyes. A valuable fertilizer can be obtained by neutralizing the acid with lime and the calcium nitrate, mixed with excess of lime to form a non-deliquescent basic salt, is applied to the soil.

$$Ca(OH)_2(s) + 2HNO_3(aq) \rightarrow Ca(NO_3)_2(aq) + 2H_2O(l)$$

Properties of nitric acid

Nitric acid is a colourless, fuming liquid of density 1.5 g cm^{-3} and boiling point 85 °C at ordinary atmospheric pressure. The ordinary concentrated nitric acid, as sold, contains about 70% by mass of the pure acid and 30% of water. The pure acid is corrosive and destroys organic matter very readily. The skin is stained yellow by it and, if the acid is left in contact with it for even a very short time, the skin is destroyed.

Chemical properties of nitric acid. Nitric acid can behave chemically in two ways:

It is (1) a very strong acid,
      (2) a powerful oxidizing agent.

Nitric acid acting as an acid

Nitric acid is a very strong acid, being almost completely ionized in dilute solution with the production of the hydrogen ion and the nitrate ion, NO_3^-.

$$HNO_3(aq) \rightleftharpoons H^+(aq) + NO_3^-(aq)$$

This ionization confers on it the usual acidic properties, modified to some extent by the powerful oxidizing action of the acid.

(a) Nitric acid neutralizes bases, forming metallic nitrates, *e.g.*,

$$K^+OH^-(aq) + H^+NO_3^-(aq) \rightarrow K^+NO_3^-(aq) + H_2O(l)$$
$$Cu^{2+}O^{2-}(s) + 2(H^+NO_3^-)(aq) \rightarrow Cu^{2+}(NO_3^-)_2(aq) + H_2O(l)$$

A full account of reactions of this type is given in Chapter 24, page 274.

(b) Nitric acid liberates carbon dioxide in reaction with metallic carbonates, as:

$$CO_3^{2-}(s) + 2H^+(aq) \rightarrow H_2O(l) + CO_2(g)$$

It is useful to remember that **all metallic nitrates are soluble in water.**

(c) It is characteristic of strong acids that, when dilute, they react with the more electropositive metals, liberating hydrogen, as:

$$Zn(s) \text{ (or Mg)} + 2H^+(aq) \rightarrow Zn^{2+} \text{ (aq) (or Mg}^{2+}) + H_2(g)$$

This reaction cannot occur in this simple form with nitric acid; it is complicated by the fact that nitric acid is a powerful oxidizing agent. Any hydrogen initially produced by the action of a metal on nitric acid is at once oxidized by more of the acid to water, and the reduction products of the acid are liberated. These may include *nitrogen dioxide, nitrogen oxide,* or even ammonia, which is at once converted, by excess of the acid, to *ammonium nitrate.* Exact equations cannot be written because the reactions are always complex. If, however, copper is used with the concentrated acid, the principal reaction is:

$$Cu(s) + 4HNO_3(aq) \rightarrow Cu(NO_3)_2(aq) + 2H_2O(l) + 2NO_2(g)$$

If the concentrated acid is diluted with its own volume of water, copper gives the principal reaction:

$$3Cu(s) + 8HNO_3(aq) \rightarrow 3Cu(NO_3)_2(aq) + 4H_2O(l) + 2NO(g)$$

This is the usual laboratory preparation of nitrogen oxide. The important point is that, if nitric acid acts with a metal *under ordinary experimental conditions,* the product is never hydrogen but rather a reduction product of the acid, such as nitrogen dioxide, nitrogen oxide, or ammonium nitrate. If, however, *very dilute* nitric acid is used (about 1%) with magnesium or manganese, some hydrogen will be produced, escaping oxidation because of the very dilute condition of the acid. The hydrogen is not pure, being accompanied to some extent by reduction products of the acid.

Nitric acid as an oxidizing agent

Like oxidizing agents in general, nitric acid acts as an acceptor of electrons. It can do this in several different ways. Two of the more important ones are:

$$4HNO_3(aq) + 2e^- \rightarrow 2NO_3^-(aq) + 2H_2O(l) + 2NO_2(g)$$
$$8HNO_3(aq) + 6e^- \rightarrow 6NO_3^-(aq) + 4H_2O(l) + 2NO(g)$$

giving nitrogen dioxide and nitrogen oxide. The electrons are supplied by the reducing agent which takes part in the reaction. This is often a metal, *e.g.*, copper. This forms ions, as: $Cu \rightarrow Cu^{2+} + 2e^-$. By combining this ionization with the two equations given above, the reactions shown between copper and nitric acid in the last paragraph are obtained. Other metals give similar behaviour, varying in detail with the nature of the metal, the concentration of the acid, and the temperature employed.

Nitric acid also oxidizes certain non-metallic elements and certain compounds; examples are given below.

Experiment 162
The oxidizing action of nitric acid on an iron(II) salt

Dissolve a few crystals of iron(II) sulphate in dilute sulphuric acid in a test-tube. Add a little concentrated nitric acid and heat. Brown fumes of nitrogen dioxide are seen and a brown or yellow solution is left, instead of the original pale green solution.

The nitric acid has oxidized the green iron(II) sulphate to brown or yellow iron(III) sulphate and has itself been reduced to nitrogen oxide which, in the air, forms nitrogen dioxide (page 434).

$$6FeSO_4(aq) + 3H_2SO_4(aq) + 2HNO_3(aq) \rightarrow$$
$$3Fe_2(SO_4)_3(aq) + 4H_2O(l) + 2NO(g)$$
$$2NO(g) + O_2(g) \rightarrow 2NO_2(g)$$
$$\text{of the}$$
$$\text{air}$$

An important point to notice about nitric acid as an oxidizer is that it introduces no solid into a mixture. The acid itself is volatile and its reduction products, oxides of nitrogen, are gaseous, while the other product, water, can also be evaporated off. Thus, it leaves no solids to complicate purification of the product of oxidation.

Other oxidizing actions of hot, concentrated nitric acid are:
Carbon to carbon dioxide.

$$C(s) + 4HNO_3(aq) \rightarrow 2H_2O(l) + 4NO_2(g) + CO_2(g)$$

Red phosphorus to phosphoric(V) acid.

$$P(s) + 5HNO_3(aq) \rightarrow H_3PO_4(aq) + 5NO_2(g) + H_2O(l)$$
$$\text{phosphoric(V)}$$
$$\text{acid}$$

Nitrates

The properties of the nitrates vary according to the position of the metal in the reactivity series. This is summarized in the table below.

K ⎱	Nitrates of these metals are decomposed by	⎱
Na ⎰	heat to the *nitrite* and *oxygen*.	
Ca ⎱		All
Mg		nitrates
Al	Nitrates of these metals are decomposed on	are
Zn	heating to the *oxide of the metal, nitrogen*	soluble
Fe	*dioxide*, and *oxygen*.	in
Pb		water.
Cu ⎰		
Hg ⎱	Nitrates of these metals are decomposed on	
Ag ⎰	heating to the *metal, nitrogen dioxide*, and *oxygen*, because the metal oxides are unstable to heat.	

All nitrates, however, irrespective of the position of the metal in the reactivity series, undergo the same reaction with iron(II) sulphate and concentrated sulphuric acid, and this reaction has become a test for soluble nitrates.

Experiment 163
Test for soluble nitrates

Crush a few potassium nitrate crystals in a mortar and put them into a test-tube and add water to a depth of about 2 cm. Shake to dissolve the potassium nitrate. Add a little sulphuric acid and then two or three crystals of iron(II) sulphate, which have also been crushed. Shake to dissolve them. Hold the test-tube in a slanting position and pour a slow continuous stream of concentrated sulphuric acid down the side. (Care!) It will form a separate layer underneath the aqueous layer and, at the junction of the two, a brown ring will be seen. **This brown ring is the characteristic test for a soluble nitrate.**

Explanation. The concentrated sulphuric acid and the nitrate yield nitric acid.

$$KNO_3(s) + H_2SO_4(aq) \rightarrow KHSO_4(aq) + HNO_3(aq)$$

The nitric acid is then reduced by some of the iron(II) sulphate to nitrogen oxide NO.

$$6FeSO_4(s) + 2HNO_3(aq) + 3H_2SO_4(aq) \rightarrow$$
$$3Fe_2(SO_4)_3(aq) + 4H_2O(l) + 2NO(g)$$
$$\text{iron(III) sulphate}$$

The nitrogen oxide reacts with more iron(II) sulphate to give the brown compound $FeSO_4.NO$, which appears as the brown ring.

$$FeSO_4(aq) + NO(g) \rightarrow FeSO_4.NO(aq)$$

The properties of some of the more important nitrates of the metallic elements are described below.

Potassium nitrate KNO₃

In the laboratory, this salt may be obtained by neutralization (page 274).

$$KOH(aq) + HNO_3(aq) \rightarrow KNO_3(aq) + H_2O(l)$$

On the industrial scale, it is prepared by double decomposition between potassium chloride (from the Stassfurt deposits) and sodium nitrate (from Chile).

$$KCl(aq) + NaNO_3(aq) \rightarrow KNO_3(aq) + NaCl(aq)$$

Boiling saturated solutions of potassium chloride and sodium nitrate are used and sodium chloride, being the least soluble of the four salts at this temperature (page 258) crystallizes and is filtered off. On cooling to ordinary temperature, potassium nitrate crystallizes as it is the least soluble of the four salts at this temperature.

When heated, potassium nitrate melts to a colourless liquid and decomposes slowly, liberating oxygen (test: glowing splint rekindled) and leaving, when cool, a pale yellow solid, potassium nitrite.

$$2KNO_3(s) \rightarrow 2KNO_2(s) + O_2(g)$$

Potassium nitrate is chiefly used for the making of fireworks and gunpowder, which usually contain about one part of charcoal, one part of sulphur, and six parts of potassium nitrate (by mass). When ignited, the mixture burns rapidly, producing nitrogen, oxides of carbon and sulphur, and other gases. These hot gases occupy a much greater volume than the original solids, and a very great pressure is set up which is used for propulsion or disruption.

Sodium nitrate (Chile saltpetre) $NaNO_3$

Large deposits of sodium nitrate, mixed with clay, occur in Chile, as 'caliche'. The material is broken up by blasting. The sodium nitrate is extracted by dissolving it out in water and evaporating the solutions by the heat of the sun. The area where the nitrates are found is practically rainless and water is supplied for the process by pipe-lines.

In the laboratory, sodium nitrate may be prepared by neutralizing caustic soda with nitric acid.

$$NaOH(aq) + HNO_3(aq) \rightarrow NaNO_3(aq) + H_2O(l)$$

When heated, sodium nitrate behaves exactly like potassium nitrate (see last section).

$$2NaNO_3(s) \rightarrow 2NaNO_2(s) + O_2(g)$$

The most important use of sodium nitrate is its application to the land as a fertilizer (see page 443). It is rapid in action and is applied in spring.

The Chilean deposits used to be the only source of nitrogenous 'chemical' fertilizer, but the Haber process now furnishes alternative supplies.

Large quantities of sodium nitrate are also used for the manufacture of nitric acid, potassium nitrate, and sodium nitrite. This substance is used extensively in the manufacture of aniline dyes, and is made by heating sodium nitrate with carbon or lead. The presence of these reducing agents hastens the conversion of of the nitrate to the nitrite, a process which is slow if heat alone is employed.

$$NaNO_3(s) + Pb(s) \rightarrow NaNO_2(s) + PbO(s)$$
$$2NaNO_3(s) + C(s) \rightarrow 2NaNO_2(s) + CO_2(g)$$

Calcium nitrate $Ca(NO_3)_2$

This compound may be made in the laboratory by the action of nitric acid upon slaked lime or chalk.

$$Ca(OH)_2(s) + 2HNO_3(aq) \rightarrow Ca(NO_3)_2(aq) + 2H_2O(l)$$
$$CaCO_3(s) + 2HNO_3(aq) \rightarrow Ca(NO_3)_2(aq) + H_2O(l) + CO_2(g)$$

It is usually met with as the tetrahydrate, $Ca(NO_3)_2.4H_2O$, which forms white deliquescent crystals.

In industry, some calcium nitrate (mixed with lime) is used as a fertilizer, 'air saltpetre'. It is prepared from the nitric acid obtained by oxidizing ammonia.

For the action of heat on calcium nitrate, see below.

Nitrates of magnesium, zinc, lead, copper

These compounds are all prepared, as explained in Chapter 24 by the action of nitric acid on the metal or on its oxide, hydroxide, or carbonate. Except lead(II) nitrate, which crystallizes anhydrous, they all form hydrated crystals. All the crystals are white in colour except those of copper(II) nitrate, which are blue.

The most important property of these nitrates is the reaction they undergo when heated, *e.g.*, for lead(II) nitrate (see Figure 186),

$$2Pb(NO_3)_2(s) \rightarrow 2PbO(s) + 4NO_2(g) + O_2(g)$$

This reaction is typical of the action of heat on the nitrates of common heavy metals.

The following equations express the reaction for other nitrates of heavy metals. In all cases the experimental observations are exactly as above except for the colours of the oxides, which are stated. With the nitrates given below there is no decrepitation.

$$2Ca(NO_3)_2(s) \rightarrow 2CaO(s) + 4NO_2(g) + O_2(g)$$

white   calcium
oxide
(white)

$$2Mg(NO_3)_2(s) \rightarrow 2MgO(s) + 4NO_2(g) + O_2(g)$$

white   magnesium
oxide
(white)

$$2Zn(NO_3)_2(s) \rightarrow 2ZnO(s) + 4NO_2(g) + O_2(g)$$

white   zinc oxide
(yellow when hot;
white when cold)

$$2Cu(NO_3)_2(s) \rightarrow 2CuO(s) + 4NO_2(g) + O_2(g)$$

blue   copper(II) oxide
(black)

Mercury(II) and silver nitrates

The oxides of mercury and silver are decomposed by heat, therefore the nitrates of these two metals leave the free metals when heated.

$$2AgNO_3(s) \rightarrow 2Ag(s) + 2NO_2(g) + O_2(g)$$

silver

$$Hg(NO_3)_2(s) \rightarrow Hg(l) + 2NO_2(g) + O_2(g)$$

mercury

Ammonium nitrate NH₄NO₃

This compound may be made in the laboratory by neutralization of ammonium hydroxide by nitric acid (page 274).

$$NH_4OH(aq) + HNO_3(aq) \rightarrow NH_4NO_3(aq) + H_2O(l)$$

It is colourless and very soluble in water. When it dissolves, heat is absorbed and, by dissolving a large quantity of the salt in water, a liquid of low temperature is rapidly obtained and may be used as a 'freezing-mixture'.

Ammonium nitrate is decomposed by heat into dinitrogen oxide and water (page 432).

$$NH_4NO_3(s) \rightarrow N_2O(g) + 2H_2O(l)$$

If the experiment is carried out in a test-tube, the usual test for dinitrogen oxide (rekindling of a glowing splint) will be masked by the steam which is also given off. If the heating is carried on to decompose all the ammonium nitrate, there will be no residue and the last traces of the salt will decompose with explosion.

A mixture of ammonium nitrate and aluminium powder is used as an explosive, 'ammonal'. When it is detonated, the following reaction occurs.

$$2Al(s) + 3NH_4NO_3(s) \rightarrow Al_2O_3(s) + 3N_2(g) + 6H_2O(l)$$

The gaseous nitrogen and steam, having a volume many times greater than that of the original solids, produce a very high pressure and hence an explosion.

Nitrogen cycle and the fixation of nitrogen

The fertility of soil depends in part on the presence in the soil of certain chemical elements. These elements are potassium, nitrogen, and phosphorus, together with traces of iron, sulphur, and others. We shall consider, for the present, nitrogen alone.

Every time a crop is taken from a given patch of soil, some of the nitrogen previously contained in the soil is removed in the form of complex organic compounds, which are part of the tissue of the plant. This nitrogen was absorbed from the soil as dissolved nitrates by the roots of the plant and this is the only manner in which the vast majority of plants can absorb and use nitrogen. It is obvious that unless nitrogen is continually supplied to the soil to balance the loss

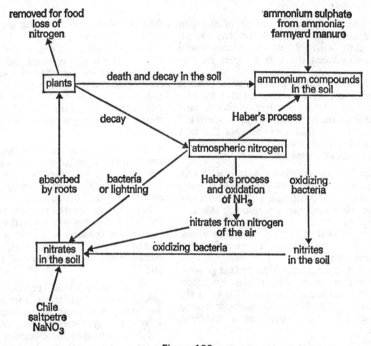

Figure 188

suffered by removal of crops, the fertility of the soil will decrease and its yield become meagre.

The soil receives some nitrogen by natural means. Certain plants, for example peas and beans, always have colonies of bacteria on their roots which are able to convert the nitrogen of the air into compounds which pass into the soil. Electrical discharges in the atmosphere, such as lightning, cause some slight combination of oxygen and nitrogen and this leads to the passage of nitrogen into the soil as nitrates, dissolved in rain water. This was a natural counterpart of Birkeland and Eyde's process, now obsolete. Nitrogen-fixing bacteria, living free in the soil, are another important agency supplying nitrogen to the soil from the air. Nitrogenous fertilizers are also used to make good the loss of nitrogen. These fertilizers fall into the following classes:

1. Ammonium sulphate derived from ammonia.
2. Sodium nitrate from the deposits of this substance in Chile – 'Chile salt-petre' $NaNO_3$.

3. Farmyard manure.

4. Fertilizers produced by manufacture, using the nitrogen of the air. These fertilizers include *ammonium sulphate*, for which the ammonia is prepared from nitrogen of the air by Haber's process (page 188), and *nitrates* also obtained from nitrogen of the air by oxidizing ammonia to nitric acid.

Though most plants can only use nitrogen as nitrates, it need not actually be supplied to the soil in this form. Bacteria in the soil will carry out the oxidation of any nitrogen compound to nitrates which the plants can then utilize.

The sum total of all these processes is called the '*Nitrogen cycle*'. A simplified form of it is given in Figure 188.

Questions

1. Outline briefly how a sample of nitrogen can be obtained from (*a*) the air, (*b*) an ammonium salt. What difference would there be between the two samples so prepared?

Describe in outline the industrial process by which nitrogen is converted into ammonia. The reaction involved in this process is reversible. Indicate *three* ways in which the manufacturer makes sure he gets the best possible yield of ammonia. State *two* industrial uses of ammonia. (S.)

2. Draw a fully labelled diagram and give an equation to show how you would prepare a dry sample of ammonia in the laboratory starting from a named ammonium salt and a named alkali.

Giving equations and reaction conditions, outline how nitric acid is manufactured from ammonia.

Nitrogen is necessary for all plant growth. (*a*) Name a plant which can assimilate nitrogen from the atmosphere. (*b*) Give the chemical name of a nitrogen-containing compound which is used as a fertilizer. (J.M.B.)

3. Write an equation for the process of nitrogen fixation which results in the formation of ammonia, and indicate whether heat is evolved or absorbed in the process. State and explain the conditions of temperature and pressure required to give a good yield of ammonia. Explain why a catalyst is also used. How is ammonia converted into nitric acid? (J.M.B.)

4. Describe the action of heat on ammonium chloride. By what reaction would you obtain (i) ammonia from ammonium chloride, (ii) nitrogen from ammonia? How can you account for the facts that a solution of ammonia in water will turn litmus blue and give a brown precipitate when mixed with a solution of iron(III) chloride?

Calculate the mass of ammonium sulphate obtainable from 100 g of ammonia. (O.)

5. Describe how ammonia is manufactured from its elements. *State* the source of each of these elements.

Describe *three* differences between a nitrogen/hydrogen mixture and ammonia.

How would you prepare dry ammonia gas from ammonium chloride? No diagram is required, but you should give the *names* of any other chemicals used, and indicate clearly how the gas could be collected. (O. and C.)

6. Explain how ammonia is converted into nitric acid on a large scale. Calculate the percentage of nitrogen in pure ammonium nitrate NH_4NO_3. Explain why the fixation of nitrogen is important.

Some ammonium nitrate was warmed with slaked lime. A gas was evolved which turned aqueous copper(II) sulphate a deep blue. Explain what has happened, writing an equation for the liberation of the gas. (O. and C.)

7. Give the essential conditions for the manufacture of ammonia from its elements. Say briefly how these elements are obtained. Why is this synthesis of great importance?

Given ammonium chloride, name *one* other chemical you would need to prepare ammonia and write the equation for the reaction.

Describe with equations, what happens when ammonia is passed (i) into dilute sulphuric acid, (ii) over heated copper(II) oxide. (O. and C.)

8. Ammonia is manufactured by the combination of nitrogen and hydrogen in the presence of finely divided iron as a catalyst at a temperature of about 500 °C and under a pressure of about 200 atm. The conversion of nitrogen and hydrogen to ammonia is not complete, and the ammonia is separated from the uncombined gases.

(*a*) Name the sources from which the nitrogen and hydrogen are obtained.

(*b*) In what ratio by volume would the nitrogen and hydrogen be mixed? Explain why this ratio is used.

(*c*) What is the purpose of the catalyst?

(*d*) Why is the catalyst in a finely divided form?

(*e*) Give one reason for the fact that the conversion of nitrogen and hydrogen to ammonia is not complete.

(*f*) The conversion to ammonia can be increased by operating the process at a pressure higher than 200 atm. Suggest a reason why a higher pressure increases the conversion to ammonia.

(*g*) How can the ammonia be separated from the uncombined nitrogen and hydrogen?

(*h*) What happens to the uncombined nitrogen and hydrogen after removal of the ammonia?

(*i*) State *two* important uses of ammonia.

(*j*) The composition of the gaseous mixture after passing over the catalyst was investigated by bubbling 2000 cm³ of the gas at room temperature and pressure through 100 cm³ of 0.3M hydrochloric acid. The final concentration of the hydrochloric acid was found to be 0.2M. Calculate the percentage of ammonia by volume in the mixture. (L.)

9. Describe briefly how ammonia is manufactured. Details of industrial plant are not required, but you should indicate the source of the raw materials and should state carefully the way in which the reaction conditions affect the formation of ammonia.

Give a brief account of (*a*) what happens when ammonia gas is burnt in oxygen (in the absence of a catalyst), and (*b*) how you would prepare a small specimen of a salt from ammonia gas.

Draw simple diagrams to show (*c*) the shape of the ammonia molecule, and (*d*) the arrangement of the electrons in it.

What similarity is there between the shapes of the ammonia and the methane molecules? (L.)

10. Describe in outline the manufacture of ammonia from nitrogen and hydrogen.

Calcium nitrate and sodium nitrate can both be used as nitrogenous fertilizers. Compare their nitrogen contents by calculating for *each* the mass of the anhydrous compound that contains 14 g of nitrogen.

What products are formed when dry ammonia is passed over heated copper(II) oxide (cupric oxide)? Draw a diagram of the apparatus you would use to carry out this reaction and to obtain samples of *each* of the products. (*No* further description is required and you are *not* required to show the production of the ammonia.) (C.)

11. Comment on or explain the reasons for the items in italic in the following account of the manufacture of ammonia.

In the manufacture of ammonia, hydrogen is obtained from water-gas and nitrogen from producer-gas. The gases are then mixed in the proportions by volume to give a mixture of *hydrogen* (3 *volumes*) *and nitrogen* (1 *volume*). The mixture is *heated, compressed to about* 200 *atmospheres* and then passed over *iron*. About 10 *per cent* of the mixture is converted into ammonia and the action is sufficiently *exothermic* to maintain the optimum working temperature, *viz.,* 500–600 °C. The gases leaving the reaction vessel are first *passed through a heat-interchanger* consisting of tubes through which the nitrogen–hydrogen mixture circulates before entering the reaction vessel. The gases then pass into another tube where they meet a *current of cold water*. The issuing gases pass through a tower packed with *soda-lime* and then *mix with the main nitrogen–hydrogen stream*.

Write equations to illustrate the action of heat on ammonium, sodium, and calcium nitrates. Name the products produced in each case. (S.)

12. In the laboratory, a steady stream of ammonia can be prepared by the reaction between ammonium sulphate and an alkali. (*a*) Name the alkali you would use. (*b*) State whether it is necessary to heat the mixture. (*c*) Write the equation for the reaction. (*d*) Name the reagent you would use to dry the gas and explain your choice. (*e*) State how you would collect a sample of dry ammonia. (No diagram is required.)

What change in volume would you expect if 100 cm³ of ammonia were completely decomposed into its elements? (Assume that all measurements are made at the same temperature and pressure.)

State and explain all that you would see when: (i) ammonium chloride is gently heated in a test-tube; (ii) a mixture of ammonia and air is passed over heated platinized asbestos, excess ammonia is removed, and the product is allowed to mix with more air. (A.E.B.)

13. Describe the process now used on an industrial scale to obtain nitric acid from ammonia. Describe *two* reactions in which nitric acid acts as an acid; give a different chemical property of nitric acid, and describe fully *one* experiment to demonstrate this property. (A.E.B.)

14. Outline the chemical reactions which are necessary for the industrial preparation of nitric acid, starting from ammonia. Give a careful description and explanation of a chemical test which would enable you to decide if a given colourless solution contained the nitrate ion. (S.)

15. Name the nitrogen-containing compound formed when (*a*) ammonium chloride is heated with calcium hydroxide. (*b*) Lead(II)

nitrate is heated strongly and the gaseous products are cooled in ice. (c) Lead(II) nitrate is heated strongly and the gaseous products are passed into water. (d) Ammonium nitrate is heated. (J.M.B.)

16. (a) Nitric acid can be prepared in the laboratory by heating solid sodium nitrate with concentrated sulphuric acid. Draw a labelled diagram of the apparatus you would use for this preparation and write an equation for the reaction.

Explain why sulphuric acid is used in this preparation rather than hydrochloric acid.

The concentrated nitric acid normally used in a laboratory contains about 70% by mass of HNO_3. State, with reasons, whether you would expect the acid prepared by the method above to be more or less concentrated than this.

(b) 1M nitric acid contains 1 mole (63 g) of HNO_3 per dm³. Calculate the masses of sodium nitrate formed when (i) 2 g of pure sodium hydroxide are added to 100 cm³ of 1M nitric acid; (ii) 4 g of pure sodium hydroxide are added to 50 cm³ of 1M nitric acid.

(c) How would you obtain a pure crystalline sample of sodium nitrate from a solution of sodium nitrate in water? (C.)

17. The following equation represents the thermal decomposition of lead(II) nitrate:

$$2Pb(NO_3)_2(s) \rightarrow$$
$$2PbO(s) + 4NO_2(g) + O_2(g)$$

(a) Nitrogen dioxide boils at 22 °C; oxygen boils at −183 °C. Sketch and label the apparatus you would use to separate the nitrogen dioxide and oxygen obtained in this reaction.

(b) Briefly state how you would obtain a sample of lead from lead(II) oxide.

(c) Calculate the volume of oxygen that would be obtained from the decomposition of 0.1 mole of lead(II) nitrate. (Assume that

1 mole of gas occupies 24 dm³ at the conditions of the experiment.)

(d) Name the products formed when potassium nitrate is decomposed by heat. Suggest why the nitrates of lead and potassium behave differently on heating.

Give a careful explanation of the following changes.

A little concentrated sulphuric acid was added to some potassium nitrate crystals in a test-tube and the mixture warmed. Colourless, oily drops were seen to condense on the walls of the tube. In addition, brown fumes were observed as the temperature was raised. (A.E.B.)

18. In each of the following preparations the product is impure. In each case name the main impurity, and explain why it is present. (a) Nitric acid from potassium nitrate and concentrated sulphuric acid. (b) Nitrogen from air and copper. (J.M.B.)

19. Nitric acid can be prepared by the action of sulphuric acid on sodium nitrate.

(a) Give the equation for the reaction, and state if the sulphuric acid is concentrated or dilute, and whether heat is needed.

(b) Give a diagram of the apparatus you would set up for the preparation and explain any points of interest concerning it.

(c) The acid when prepared is usually a yellow liquid. Why is this? How can the colour be removed?

(d) In what way does the action of dilute nitric acid on a metal like zinc differ from the action of dilute hydrochloric acid? How do you account for this difference? (A.E.B.)

20. Lead(II) nitrate crystals, $Pb(NO_3)_2$, were heated in the apparatus shown in Figure 189.

(a) Name the liquid X and the gas Y.

(b) If the lead(II) nitrate were heated until no more gas Y was evolved, what would remain in the test-tube?

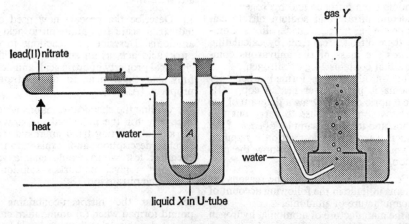

Figure 189

(c) Write the equation for the action of heat on lead(II) nitrate.

(d) If 0.01 mole (0.01 gramme-formula) of crystals were used, what would be the maximum volume of gas Y which could be collected at room temperature and pressure?

(e) If the cooling in beaker A was inefficient, some brown gas would pass through the U-tube unchanged and react with the water. Explain what happens in this reaction and indicate its industrial importance.

(f) A sample of liquid X is put in a sealed container, heated to 100 °C and then allowed to cool to room temperature again. What would you *see* happening? Represent the changes which occur as fully as possible by an equation or equations.

(g) Why is it not satisfactory to use copper(II) nitrate crystals, $Cu(NO_3)_2.3H_2O$, in this experiment to obtain a sample of the pure liquid X instead of lead(II) nitrate crystals?

(h) Name a metal nitrate that would give gas Y but no liquid X if heated in this apparatus. Give the equation for the thermal decomposition of this nitrate. (L.)

21. Draw a fully labelled diagram to illustrate the usual laboratory preparation and collection of nitrogen oxide.

Describe the conditions required for the industrial manufacture of nitric acid from ammonia. (A diagram is not required for this part of the question.)

By heating a coil of iron wire electrically in a known volume of nitrogen oxide it may be shown that the volume of nitrogen remaining is equal to half the original volume of nitrogen oxide. Show how the number of nitrogen atoms in one molecule of nitrogen oxide may be deduced from this information, assuming the formula of a molecule of nitrogen to be N_2. (J.M.B.)

22. When lead nitrate crystals are heated, a gas is evolved which, when cooled to 0 °C in dry apparatus, gives a yellow liquid, dinitrogen tetroxide. On heating this liquid, a gas is formed which quickly becomes a dark brown gas A. When A is strongly heated, the colour disappears leaving a colourless mixture B.

Identify A and B and write equations for *two* of the reactions. (J.M.B.)

23. Describe the preparation and collection of nitric acid in the laboratory. Starting from concentrated nitric acid, how would you prepare a sample of a *named* oxide of nitrogen?

Suggest a method by which you could obtain nitrogen from this oxide. How would your sample of nitrogen differ from nitrogen as usually obtained from the air? (No diagram is required in this section.) (C.)

24. Briefly describe and explain what happens when oxygen is bubbled into a jar of nitrogen oxide standing over water containing blue litmus. (S.)

25. Give an outline of the manufacture of nitric acid from ammonia.

If you were provided with concentrated nitric acid, copper, and water but *no other* chemical substances, describe briefly reactions by which nitrogen oxide (NO), nitrogen dioxide (NO_2), and oxygen could be produced. (Diagrams and descriptions of apparatus are *not* required.)

State briefly *three* differences in properties between the oxides NO and NO_2. (C.)

26. Give the name, formula, and colour of *three* gaseous oxides of nitrogen. Which of these gases is very soluble in water? Outline briefly the way in which samples of *two* of them can be prepared in the laboratory.

Which of the oxides of nitrogen could be confused with oxygen, and why is this? By what test can you distinguish between a jar of this gas and a jar of oxygen? (S.)

35 Phosphorus and its Compounds

Occurrence

Phosphorus is a very active non-metal and is not found uncombined in nature. It was first isolated by Brandt, a Hamburg chemist, in 1669, who obtained it by distilling concentrated urine with sand. It is extracted nowadays from bones and mineral phosphates. The residue left after burning away the organic matter from bones contains calcium phosphate $Ca_3(PO_4)_2$. This substance also occurs as the mineral, apatite.

Extraction

A charge of calcium phosphate, sand, and coke is fed continuously into an electric furnace (Figure 190). At the very high temperature obtained (the electric current

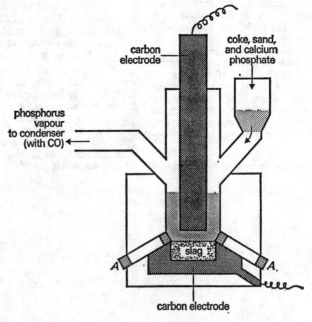

Figure 190

is merely to produce the high temperature; there is no electrolysis) the following reaction takes place:

$$2Ca_3(PO_4)_2(s) + 6SiO_2(s) + 10C(s) \rightarrow 6CaSiO_3(s) + 10CO(g) + P_4(l)$$
calcium silicate
slag

The phosphorus distils over with the carbon monoxide and the vapour is led below the surface of water, when the phosphorus solidifies to a white solid. The calcium silicate is formed as a molten slag which is run off from time to time at *A*.

Chemistry of the action

Once again (see preparation of hydrogen chloride and nitric acid, pages 380 and 437 respectively) an involatile acid (or acidic oxide, SiO_2) is displacing a more volatile one (P_4O_{10}). The carbon (a reducing agent) reduces the oxide of phosphorus to the element.

$$P_4O_{10}(s) + 10C(s) \rightarrow P_4(s) + 10CO(g)$$

On adding these equations together the main equation will be obtained.

Allotropes

Phosphorus exists in two chief allotropic forms, white or yellow phosphorus and red phosphorus. The latter is the stable form. (See page 93.)

Yellow phosphorus is a white solid which becomes pale yellow on exposure to light. It is of density 1.8 g cm^{-3} and is soluble in carbon disulphide. It smoulders in air owing to oxidation and this action causes it to glow in the dark.

$$P_4(s) + 3O_2(g) \rightarrow P_4O_6(s)$$

It is usually kept below the surface of water. **Great care should be taken when using it, for it gives off a very poisonous vapour and catches fire very readily.** It burns in air or oxygen, giving off white fumes of oxides of phosphorus. This variety of phosphorus is formed when the vapour of phosphorus is suddenly cooled.

Red phosphorus is the stable variety at all temperatures. It is not poisonous as is yellow phosphorus, and does not catch fire so readily. It can be made by heating yellow phosphorus in an inert atmosphere (usually with a little iodine to act as catalyst) to a temperature of about 250 °C. It is insoluble in carbon disulphide.

Experimental evidence that the two allotropes are chemically identical is the following. 1 g of white phosphorus heated to about 250–300 °C in an inert atmosphere yields exactly 1 g of red phosphorus and nothing else. Further, if equal masses of red and white phosphorus are converted fully to any given phosphorus compound, exactly the same masses of product are obtained. The following table compares the properties of the allotropes of phosphorus.

Red phosphorus	White (or yellow) phosphorus
Opaque red solid.	Colourless translucent solid (turns yellow).
Density 2.3 g cm^{-3}.	Density 1.8 g cm^{-3}.
Non-poisonous.	Very poisonous.
Sublimes at 400 °C.	Melting point 44 °C.
Insoluble in carbon disulphide.	Soluble in carbon disulphide.
Ignites in air at 260 °C.	Ignites in moist air at 30 °C.
No action with hot sodium hydroxide solution.	Forms phosphine with hot sodium hydroxide solution
Unoxidized at ordinary temperatures in air.	Rapidly oxidized at ordinary temperatures in air.

White phosphorus is the unstable variety. It has the higher vapour pressure and is very slowly reverting to red phosphorus in room conditions.

Phosphorus is used in the match industry, in rat poisons, and in making smoke bombs. About 80% of it is converted to phosphoric(V) acid H_3PO_4.

Match industry

In earlier days, matches contained white phosphorus and an oxidizing agent and the mixture was caused to burn by rubbing on sandpaper. The use of white phosphorus in industry caused hundreds of work-people to suffer from phosphorus poisoning, which took the form of the rotting of the bones of the face and jaw ('phossy-jaw'). Its use was then forbidden by law and matches nowadays consist of compounds of phosphorus (sulphides as a rule) and oxidizing agents. The friction of rubbing on a rough surface generates enough heat to start the combustion. In 'safety' matches the phosphorus (red variety) is on the side of the box and thus the match-head, which contains the oxidizing agent, is useless without the box.

Chemical properties of yellow phosphorus

Phosphorus will combine readily with oxygen, chlorine, and sulphur. It is a reducing agent and readily attacks oxidizing agents with the formation of oxides of phosphorus which are soluble in water, forming acids. It is attacked by alkalis forming phosphine. Phosphorus forms two series of compounds, exhibiting valencies of 3 and 5.

Preparation of phosphine PH₃

Fit up the apparatus as shown in Figure 191 in a fume-chamber. Place sodium hydroxide (caustic soda) solution and a few small pieces of yellow phosphorus in

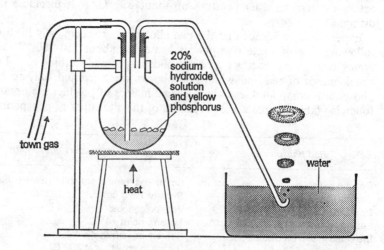

Figure 191

the flask, sweep out the air by means of town-gas, and warm the mixture. A gas is given off, phosphine, and on coming into contact with the air it ignites spontaneously (the ignition is due to an impurity) and forms white vortex rings of oxides of phosphorus.

$$3NaOH(aq) + P_4(s) + 3H_2O(l) \rightarrow 3NaH_2PO_2(aq) + PH_3(g)$$

sodium   phosphine
phosphinate

Phosphine is a colourless gas with a garlic-like odour and is extremely poisonous. It shows similarities to ammonia, forming salts, for example, PH_4Cl, phosphonium chloride (*cf.* ammonium chloride).

Phosphorus trichloride PCl₃

This is made by allowing dry chlorine to react with warm yellow phosphorus in an inert atmosphere.

$$P_4(s) + 6Cl_2(g) \rightarrow 4PCl_3(l)$$

Phosphorus trichloride is a liquid (boiling point 76 °C) which fumes in air owing to the action of moisture upon it. The liquid is attacked by water, forming phosphonic acid.

$$PCl_3(l) + 3H_2O(l) \rightarrow H_3PO_3(aq) + 3HCl(aq)$$

phosphonic
acid

Phosphorus pentachloride PCl₅

This is made from phosphorus trichloride by the action of chlorine upon it. A yellowish solid separates out. It sublimes when heated, decomposing into the trichloride and chlorine, but on cooling forms the original pentachloride.

$$PCl_5(s) \underset{cool}{\overset{heat}{\rightleftharpoons}} PCl_3(l) + Cl_2(g)$$

phosphorus   phosphorus   chlorine
pentachloride   trichloride

It also attacks water vigorously forming phosphoric(V) acid,

$$PCl_5(s) + 4H_2O(l) \rightarrow H_3PO_4(aq) + 5HCl(aq)$$

phosphoric(V)
acid

Phosphorus(III) oxide (phosphorus trioxide) P₄O₆

This is made by burning phosphorus in a limited supply of air and passing the mixture of trioxide and pentoxide through a tube surrounded by a water jacket at 50 °C, and containing a loose cotton-wool plug. The pentoxide remains solid and is retained by the plug. The trioxide vapour passes on and is condensed in a freezing-mixture.

$$P_4(s) + 3O_2(g) \rightarrow P_4O_6(s)$$

phosphorus(III)
oxide

$$P_4(s) + 5O_2(g) \rightarrow P_4O_{10}(s)$$

phosphorus(V)
oxide

Phosphorus(III) oxide is a white volatile solid which readily reacts with water to form phosphonic acid.

$$P_4O_6(s) + 6H_2O(l) \rightarrow 4H_3PO_3(aq)$$

phosphonic
acid

Phosphorus(V) oxide (phosphorus pentoxide) P_4O_{10}

This is made by igniting phosphorus in a plentiful supply of air. A small piece of phosphorus is placed in a crucible on a plate of glass under a bell-jar containing dry air. The phosphorus is ignited. It burns with a brilliant flame and a white solid finally settles on the plate. This is quickly scraped into a dry bottle.

$$P_4(s) + 5O_2(g) \rightarrow P_4O_{10}(s)$$

Phosphorus(V) oxide can be purified by heating in a current of dry oxygen. This converts any trioxide formed into pentoxide.

Phosphorus(V) oxide is a white solid which reacts vigorously with water, and is one of the best drying agents known. With hot water it forms phosphoric (V) acid.

$$P_4O_{10}(s) + 6H_2O(l) \rightarrow 4H_3PO_4(aq)$$
$$\text{phosphoric(V)}$$
$$\text{acid}$$

Phosphatic fertilizers

Phosphorus is an element essential to soil fertility. Plants absorb it from the soil and, if the crop is consumed by man, phosphorus passes into his bone structure and protoplasm. Much of it is lost to the soil in general by sewage disposal and by depositing human remains in cemeteries.

The loss must be made good. Bone-meal and basic slag are used, both of which contain calcium phosphate $Ca_3(PO_4)_2$. This compound is, however, almost insoluble in water and becomes available to the plants very slowly. A more soluble material, quicker in action, can be made by stirring calcium phosphate with an appropriate weight of 65% sulphuric acid. An acid calcium phosphate is formed.

$$Ca_3(PO_4)_2(s) + 2H_2SO_4(aq) \rightarrow Ca(H_2PO_4)_2(s) + 2CaSO_4(s)$$

The product is dried and sold as 'superphosphate'.

Questions

1. Describe a method (*a*) for the preparation of white phosphorus from calcium phosphate; (*b*) for the conversion of white phosphorus into the red form.
   Compare and contrast the properties of these two forms of phosphorus. What reaction, if any, has each form with chlorine, and with sodium hydroxide solution?

2. Compare the properties of the two forms of phosphorus. How may phosphorus be converted into *two* of the following: (*a*) phosphorus pentachloride; (*b*) phosphine; (*c*) phosphorus oxide? (O. and C.)

3. Describe with necessary detail how you would prepare from phosphorus reasonably pure specimens of (*a*) phosphoric(V) acid; (*b*) phosphorus trichloride. (O.)

4. Show by giving three chemical properties in each case that sodium is a metal and that phosphorus is a non-metal.
   Starting with metallic sodium and yellow

phosphorus, describe fully how you would prepare specimens of (*a*) phosphine, and (*b*) sodium phosphate(V). (L.)

5. How does phosphorus react with (*a*) oxygen; (*b*) chlorine; (*c*) sodium hydroxide?
   Describe how the products behave when brought into contact with water. (O. and C.)

6. Describe the preparation of (*a*) phosphorus trichloride, and (*b*) phosphine. Show how the latter is related to ammonia. (L.)

7. How is phosphorus obtained from bone ash? State how white phosphorus can be converted into red phosphorus, and name three respects in which they differ from one another. Write the formulae of two chlorides, one of hydride, and two oxides of phosphorus. How does the highest oxide of phosphorus react with water and for what purposes is this oxide used? (L.)

8. Contrast, tabulating your answer, *six* properties of two allotropes of phosphorus.

Starting with one of these forms, briefly describe the preparation of the other form from it. How is phosphine usually prepared in the laboratory? Give details. (L.)

9. Phosphorus is not found in nature as the free element. Explain very briefly why this is so. Name an important source of phosphorus and outline the chemistry of the extraction of the element from this source.

Read the following passage and then answer the questions which follow:

'Red phosphorus can be converted to phosphoric(V) acid by heating a small quantity with fairly concentrated nitric acid in a fume-cupboard. When the vigorous reaction has moderated, the reaction mixture is filtered (through glass-wool) if necessary, and then heated fairly gently to drive off any unused nitric acid. The reaction could be represented by the equation

$$P + 5HNO_3 = H_3PO_4 + 5NO_2 + H_2O$$

(a) Name the substance in the reaction which is being reduced.

(b) Suggest a reason for using red phosphorus rather than white phosphorus.

(c) Why is the reaction carried out in a fume-cupboard?

(d) Why may filtration be necessary?

(e) Give the formulae of the sodium salts that could be formed by phosphoric(V) acid (H_3PO_4).

(f) What is the maximum mass of phosphoric(V) acid that could be obtained from 6.2 g of red phosphorus? How many moles of NO_2 would be produced at the same time? (L.)

10. Phosphorus trichloride PCl_3, a colourless liquid, can be obtained by passing a stream of dry chlorine over phosphorus which is kept at 80 °C by suitable heating. Phosphorus trichloride vapour distils off and is collected in a cooled receiver.

(a) Draw a labelled sketch of the apparatus you would use to prepare phosphorus trichloride, including in this the apparatus for preparing the chlorine.

(b) Write an equation for the formation of phosphorus trichloride in this reaction.

(c) It is necessary to remove air from the apparatus before passing the chlorine. How could this be achieved, and why is it necessary?

(d) What information does the account of the experiment give about the boiling point of phosphorus trichloride?

(e) Suggest a possible shape for the molecule of phosphorus trichloride, and say why you have chosen this shape.

(f) When phosphorus trichloride is added to cold water, it reacts vigorously to produce a colourless solution of phosphonic acid H_3PO_3, and hydrochloric acid. Write an equation for this reaction.

(g) You have been provided with a colourless liquid which is either pure water or the solution formed when phosphorus trichloride is added to water. Outline *two* tests that you would carry out to decide which possibility was correct. (L.)

36 Metals; Extractions and Uses

Extraction and the reactivity series

In the table shown below the metals are arranged in order of the reactivity series (see page 474), and it indicates (in an approximate way only) the inverse order in which the elements were isolated. Thus metals low in the series such as gold, silver, and lead have been known since very early times. Metals high in the series proved very difficult to isolate and it was Davy's work on electrolysis which led to the isolation of potassium, sodium, calcium, magnesium, and aluminium over a period of years from 1807 (when Davy isolated potassium and sodium) to about 1850 (when aluminium was isolated).

K Na Ca Mg Al	Very reactive. Never found as free element. Extracted by electrolysis. All isolated after 1800.
Zn Fe Pb	Moderately reactive. Found as oxides, carbonates, or sulphides. Extracted by reaction with carbon or carbon monoxide.
Cu Hg Ag Au	Not very reactive. May be found in nature as the free element. Known for a very long time.

Aluminium was not obtained in the first place by electrolysis although its manufacture nowadays is entirely confined to that method, but it was isolated by the action of the very active element potassium on aluminium(III) chloride.

In view of what we have learnt about the electrochemical series, this order seems quite natural, for we should expect the compounds of the very active elements such as sodium and potassium to be very stable substances.

Metals low down in the series are frequently found as the free elements, although they may also be obtained from ores because the amounts found as the free metal are not sufficient for industrial purposes. Gold, however, the last element of the series, is found and mined almost entirely as the free element.

Extraction is a reduction

Metals are found as a rule as their oxides or salts, chiefly sulphides and carbonates, which are electrovalent compounds and contain the metal in ionized condition. During extraction, the metallic ion takes up the necessary number of electrons to convert it to the corresponding atom. This process of electron gain is a *reduction* and the electrons are supplied by the reducing agent concerned in the reaction. For example, in the ore, galena $Pb^{2+}.S^{2-}$, lead exists (as shown) in ionic form. After extraction, it exists as atoms of lead. Therefore, at some stage, the change:

$$Pb^{2+} + 2e^- \rightarrow Pb \quad \text{(a reduction)}$$

must have taken place. The electrons are supplied when the reducing agent, carbon, acts upon lead oxide, produced as an intermediate product in the extraction.

$$O^{2-} + C \rightarrow CO + 2e^-$$

The reduction in the case of the first members of the series is brought about by electrolysis, which can be regarded as a very powerful oxidation and reduction process which often results in the formation of the elements present in the compound electrolysed (see Electrolysis, page 147). As far as the metals are concerned the process is one of reduction, although no reducing agent (in the ordinary sense of the term) is used.

The cathode acts as a reducing region by supplying electrons. The usual reducing agents employed for the less electropositive metals are carbon in the form of coke, and carbon monoxide (made from the coke by passing a limited amount of air over the hot coke). You may ask why the usual laboratory reducing agents are not employed; for example, hydrogen. The reason is that, in an industrial process the chief concern is cost, and coke easily wins on that score.

Metals extracted by electrolysis

The first five metals in the reactivity series shown on page 454 are all extracted by electrolytic methods. These processes are usually expensive to install and maintain, and the cost of metals produced by electrolysis tends to be relatively high.

For each of these metals the names of the principal mineral sources are listed, the extraction process described in detail, and an indication of the uses to which the metal is commonly put is given.

Sodium and Potassium

Occurrence

These metals occur chiefly as the chloride, sodium as common salt in the huge salt deposits of Cheshire, England, and elsewhere, and potassium chloride as carnallite (together with magnesium chloride) in the deposits in Stassfurt, Germany (see page 386). Other sources of the elements are:

sodium carbonate	in Africa and in ash of sea plants.
sodium nitrate	Chile saltpetre in Chile.
sodium chloride	in sea-water and many salt lakes.
potassium nitrate	saltpetre found in India.
potassium carbonate	in the ash of land plants.

Extraction of sodium and potassium

Both metals are extracted by electrolysis of their fused chlorides, and since both metals are extracted in a similar manner, that for sodium will be described.

Sodium is extracted industrially by the Downs Process, in which common salt is electrolysed in the molten condition. As the melting point of the salt is high (about 800 °C), calcium chloride is added to lower the melting point; it becomes about 600 °C. The Downs cell (Figure 192) has an outer iron shell, lined with

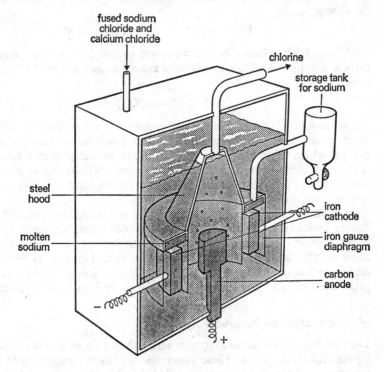

Figure 192

firebrick. A diaphragm of iron gauze screens the carbon anode from the ring-shaped iron cathode that surrounds it. Chlorine escapes via the hood. Sodium collects in the inverted trough, placed over the cathode, rises up the pipe, and is tapped off through the iron vessel. Sodium chloride produces ions Na^+ and Cl^-.

At the cathode	At the anode
$Na^+ + e^- \rightarrow Na$	$Cl^- - e^- \rightarrow Cl$
(a reduction)	(an oxidation)

$$\text{Then: } Cl + Cl \rightarrow Cl_2$$

Chlorine is a valuable by-product of the process.

Uses of sodium and potassium

Sodium is used in the manufacture of sodium cyanide $NaCN$, which is employed in the extraction of gold, and also in manufacturing sodamide $NaNH_2$, and sodium peroxide Na_2O_2. An alloy of sodium and lead has been employed in the manufacture of tetraethyl lead, an anti-knock additive used in petrol.

$$Pb(s) + 4Na(s) + 4C_2H_5Cl(g) \rightarrow 4NaCl(s) + Pb(C_2H_5)_4(l)$$

The alloy is heated in ethyl chloride vapour. Sodium vapour lamps are well known for the intensely yellow illumination they give. Sodium has also found use in the reduction of titanium chloride to the metal by heat. Titanium is used in high temperature steels, and the manufacture of heat-resistant alloys for use in rockets.

$$TiCl_4(l) + 4Na(s) \rightarrow Ti(s) + 4NaCl(s)$$

Potassium and sodium are used together in an alloy for the withdrawal of heat from nuclear reactors. Potassium is also used as a laboratory reducing agent.

Calcium and Magnesium

Occurrence

Calcium occurs abundantly and very widely distributed as *calcium(II) carbonate* $CaCO_3$, which is found as chalk, limestone, marble, calcite, Iceland spar, and aragonite.

It also occurs as *calcium sulphate* as gypsum, $CaSO_4.2H_2O$, and as anhydrite $CaSO_4$.

It occurs, less abundantly, as fluorspar, calcium fluoride, and as calcium phosphate.

Magnesium occurs as magnesium chloride in sea-water, which contains about 1 million tonnes of magnesium, as Mg^{2+}, per km^3.

Extraction of calcium and magnesium

Both metals are extracted by the electrolysis of their fused chlorides. (Notice that whereas potassium is obtained from the hydroxide, calcium cannot be obtained from the hydroxide by electrolysis. The reason is that if calcium hydroxide is heated strongly it becomes quicklime and, if that is heated strongly, it merely becomes white hot and does not melt. Quicklime is one of the most refractory substances known.)

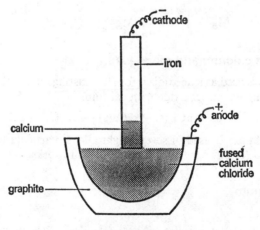

Figure 193

Calcium chloride, which is obtained as a by-product from the Solvay process (see page 318), is placed in a graphite (carbon) container which is the anode of the cell (Figure 193). The cathode is an iron rod which just dips below the surface of the calcium chloride. The calcium chloride is melted and the electrolysis is begun. As the calcium forms on the iron rod, the latter is withdrawn and an irregular stick of calcium is gradually formed. Chlorine is evolved at the anode. If the anode were made of metal it would be attacked by the chlorine, but chlorine has no effect on carbon, and the gas is a valuable by-product.

$$\underset{\text{(a reduction)}}{\underset{\text{At the cathode}}{Ca^{2+} + 2e^- \rightarrow Ca}} \qquad \underset{\text{(an oxidation)}}{\underset{\text{At the anode}}{2Cl^- - 2e^- \rightarrow Cl_2}}$$

In the case of magnesium, the melting point of the chloride is reduced by the addition of sodium chloride, and a lower working temperature for the cell is thus established. The cell is somewhat different from that of calcium (see Figure 194). Molten magnesium collects at the cathode and is protected from

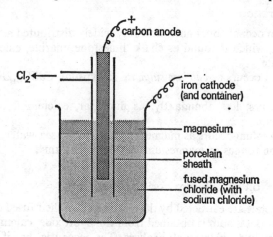

Figure 194

oxidation by a coating of electrolyte, and from anodic chlorine by a porcelain sheath.

$$\underset{\text{At the cathode}}{Mg^{2+} + 2e^- \rightarrow Mg} \qquad \underset{\text{At the anode}}{2Cl^- - 2e^- \rightarrow Cl_2}$$

Uses of calcium and magnesium

Calcium is used as a deoxidizer for steel castings and in the extraction of thorium by heating its tetrachloride with calcium.

$$ThCl_4(s) + 2Ca(s) \rightarrow Th(s) + 2CaCl_2(s)$$

Magnesium is used in several light alloys, particularly with aluminium; it is also used (in powder form) in flares and fireworks.

Aluminium

Occurrence

Compounds of this metal are quite abundant. Some of the better known are:

Mica, felspar $K_2Al_2Si_6O_{16}$
Kaolin (china clay) $Al_2Si_2O_7.2H_2O$ (used in making porcelain)
Corundum Al_2O_3
Cryolite Na_3AlF_6
Bauxite $Al_2O_3.2H_2O$

Extraction of aluminium

Purified bauxite (alumina Al_2O_3) is electrolysed (Figure 195) in solution in molten cryolite. If more purified bauxite is added as required, cryolite is

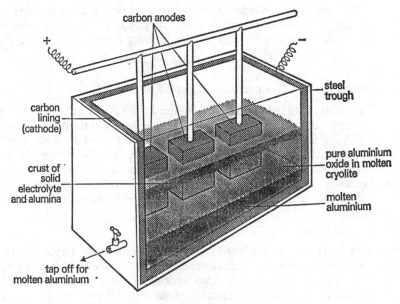

carbon anodes

steel trough

carbon lining (cathode)

pure aluminium oxide in molten cryolite

crust of solid electrolyte and·alumina

molten aluminium

tap off for molten aluminium

Figure 195

unchanged and can be used indefinitely. Pure aluminium is tapped from the cathode, but the carbon anode tends to be oxidized away by oxygen.

$$\textit{At the cathode} \qquad\qquad \textit{At the anode}$$
$$2Al^{3+} + 6e^- \rightarrow 2Al; \qquad 3O^{2-} - 6e^- \rightarrow 1\tfrac{1}{2}O_2$$

Cryolite is sodium aluminium fluoride.

Uses of aluminium

(1) *In alloys.* The metal is a constituent of several light alloys. They combine high tensile strength with lightness, and have been much used in aircraft construction. *Duralumin* (Al, Mg, Cu, Mn), *magnalium* (Mg, Al), and *aluminium bronze* (Al, Cu) are well known.

(2) *In cooking utensils.* Cheapness, low density, good appearance, good conductivity for heat, and resistance to attack by cooking solutions have combined to make aluminium very popular in the kitchen. Aluminium vessels must not be exposed to alkaline solutions (see above). The resistance to attack is due to a coherent oxide film (page 480) which would be destroyed rapidly by alkali since aluminium oxide is amphoteric. In addition, the oxide layer can be intensified by an electrolytic process known as *anodizing*, and dyes can be incorporated to give an attractive coloured finish.

(3) *In overhead electric cables.* The low density of aluminium is very favourable here. Thick cables of low resistance can be employed without undue weight.

(4) *In aluminium paint.* The powdered metal is used, with oils.

(5) *In thermit processes.* The reactions between aluminium powder and oxides of other metals are commonly very exothermic. If a mixture of iron(III) oxide

and aluminium powder, known as 'thermit', is 'fired' by burning a piece of magnesium ribbon stuck into it, a violent reaction will occur. Molten iron is produced with a slag of aluminium(III) oxide floating on it.

$$Fe_2O_3(s) + 2Al(s) \rightarrow 2Fe(l) + Al_2O_3(s)$$

This reaction was formerly used in welding steel parts *in situ* by means of the molten metal produced, and in incendiary bombs in the early part of World War II. Similar reactions are also used in isolating certain metals, *e.g.*, chromium.

$$Cr_2O_3(s) + 2Al(s) \rightarrow 2Cr(s) + Al_2O_3(s)$$

Extraction of the moderately reactive metals

The metals already described are obtained from their compounds by electrolysis. We now come to those metals which have been obtained for many years and which are used in far greater quantities than the more electropositive elements.

The metals zinc, iron, lead, and copper are found chiefly as the impure carbonates and sulphides, and iron is also found as the impure oxide. The following processes are amongst those commonly used in the extraction of metals from their ores *although all the processes are not used in connection with each individual metal.*

Concentration of ores

Very often, as ores are found contaminated with earthy impurities, methods are employed to pick out the richer ores, or those worth working up, and to reject the poorer grades. This may be done by hand, or the earthy matter may be washed away by means of a stream of water, leaving the heavier ores. In the case of copper ores, the latter drop through a magnetic separator where metallic ores are deflected into one pile, whilst the lower-grade ores and earthy impurities are not deflected and pass straight on. Crushed sulphide ores are in general now concentrated by a *froth-flotation* process in which the ore is powdered, mixed with oil and water, and air is blown through. The froth contains the ore and is skimmed off.

Roasting in air

Since many ores contain the sulphide or carbonate of the metal, a preliminary roasting in air will remove the sulphur as sulphur dioxide and drive off carbon dioxide from the carbonate. Thus:

$$2ZnS(s) + 3O_2(g) \rightarrow 2ZnO(s) + 2SO_2(g)$$
$$ZnCO_3(s) \rightarrow ZnO(s) + CO_2(g)$$

The oxides are usually easier to deal with than the sulphides or carbonates.

Reduction process

The roasted ore must now be reduced. The reduction in the case of zinc and iron is by means of carbon. In some cases (*e.g.*, copper), by a suitable adjustment of the roasting process, it is possible to oxidize some of the sulphide to oxide, and then by adding more of the ore to supply sufficient sulphide to react with the oxygen of the oxide, leaving the metal.

Purification

The product of the reduction process is seldom a pure specimen of the metal. Purification may be carried out electrolytically (as in the case of copper and zinc). By electrolysis a very pure product is usually obtainable. In other cases the impure metal is heated in a hearth open to the air, when impurities oxidize and rise to the surface as a scum and can be removed.

Zinc

Occurrence

Zinc occurs in various parts of the world, as

Zinc carbonate $ZnCO_3$, calamine, and
Zinc sulphide ZnS, zinc blende.

Extraction of zinc

The ores are first roasted in air when the oxide is formed whether the ore is calamine or zinc blende.

$$ZnCO_3(s) \rightarrow ZnO(s) + CO_2(g)$$
$$2ZnS(s) + 3O_2(g) \rightarrow 2ZnO(s) + 2SO_2(g)$$

The sulphur dioxide is frequently used for the manufacture of sulphuric acid.

The ore is now mixed with coke and placed in a fireclay retort to the end of which there is attached a fireclay condenser (Figure 196). On the end of this

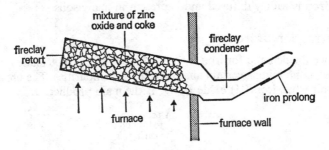

Figure 196

condenser is placed an iron 'prolong' which collects any zinc which escapes the condenser. The mixture is heated by means of producer-gas for about twenty-four hours. The zinc oxide is reduced to metallic zinc, the carbon becoming carbon monoxide, which burns at the mouth of the condenser.

$$ZnO(s) + C(s) \rightarrow Zn(s) + CO(g)$$

zinc   carbon   zinc   carbon
oxide   monoxide

The zinc distils out of the retort and the bulk of it condenses to molten zinc in the condenser, and is removed from time to time. Owing to the presence of air in the retort, some of the zinc burns to zinc oxide and condenses on the upper part of the condenser as 'zinc dust' (this is a mixture of zinc and zinc oxide). The impure zinc obtained in this way is purified by electrolysis. It is frequently 'granulated' by running the molten metal into water. (In this granulated form the zinc offers a larger area for action with, *e.g.*, dilute acids.)

Uses of zinc

Galvanizing. Small iron objects and iron wire or sheet are often coated with zinc (*i.e.*, *galvanized*) to delay rusting. This may be done by spraying, electrolytic deposition, or dipping into molten zinc. Zinc, in air, acquires a coherent, inert oxide layer by which rusting of the iron is prevented. Also, some protection is still given to the iron even if the zinc layer is broken. This is so because zinc is more electropositive than iron and the first stage of oxidation of zinc, $Zn \rightarrow Zn^{2+} + 2e^-$, occurs in preference to that of iron, $Fe \rightarrow Fe^{2+} + 2e^-$.

Zinc is also used in alloys, *e.g.*, brass (copper and zinc, 2 : 1) and in dry Leclanché batteries, as the negative pole.

Iron

Occurrence

The occurrence of iron, a metal of immense importance, has had a profound effect upon the development of the countries in which it is found, especially where iron ore and coal (which is necessary for the extraction of large quantities of the metal) have been found comparatively close together. The chief ores are the following:

Haematite, found in United States, Australia, and USSR, is impure iron(III) oxide Fe_2O_3.

Magnetite, or magnetic iron ore Fe_3O_4, occurs in Sweden and North America.

Spathic iron ore, iron(II) carbonate $FeCO_3$, is found in Great Britain.

Iron is widely diffused and is present in many soils.

Extraction of iron

Since the demand for iron is so great it has often to be made from poorer-grade ores, containing a certain amount of earthy impurities. The ores are first roasted in air when iron(III) oxide Fe_2O_3 is the main product.

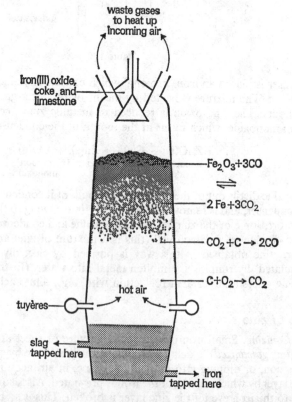

waste gases
to heat up
incoming air

Iron(III) oxide,
coke, and
limestone

$Fe_2O_3 + 3CO$

$\rightleftharpoons$

$2Fe + 3CO_2$

$CO_2 + C \rightarrow 2CO$

$C + O_2 \rightarrow CO_2$

tuyères

hot air

slag ← tapped here

→ Iron tapped here

Figure 197

This iron(III) oxide is mixed with coke and limestone and introduced into a blast furnace (Figure 197). The blast furnace is a tall structure about 30 m high and 9 m in diameter at the widest part. It contains a firebrick lining inside a steel shell, and a blast of hot air can be introduced low down in the furnace through several pipes known as tuyères. A well at the bottom of the furnace serves to hold the molten iron and slag until these can be run off. The mixture of ore, coke, and limestone is fed in continuously from the top, and a blast furnace, once started, is kept going for months at a time until repairs are necessary or work lacking.

Chemistry of the action

As the hot air comes into contact with the white-hot coke, the latter burns to form carbon dioxide.

$$C(s) + O_2(g) \rightarrow CO_2(g)$$

carbon   carbon
(coke)   dioxide

The above reaction liberates a very large quantity of heat, and it is this heat which keeps up the high temperature necessary for the reduction process.

As the gas is forced higher up the furnace the supply of oxygen (from the air) becomes less and the carbon dioxide coming into contact with white-hot coke is reduced to carbon monoxide.

$$CO_2(g) + C(s) \rightarrow 2CO(g)$$

This carbon monoxide at the high temperature (about 1000 °C) reduces the iron(III) oxide to metallic iron forming carbon dioxide.

$$Fe_2O_3(s) + 3CO(g) \rightleftharpoons 2Fe(s) + 3CO_2(g)$$

iron(III)   carbon   iron   carbon
oxide   monoxide   dioxide

The molten iron runs to the bottom of the furnace.

Action of the limestone. The limestone, which has been introduced together with the ore, is first decomposed at this high temperature to form calcium oxide.

$$CaCO_3(s) \rightarrow CaO(s) + CO_2(g)$$

calcium   carbon
oxide   dioxide

The earthy impurities contain a certain amount of silica (SiO_2), which is an acidic oxide, and this combines with the basic oxide, calcium oxide, to form calcium silicate.

$$SiO_2(s) + CaO(s) \rightarrow CaSiO_3(s)$$

silica   calcium silicate

$$(\text{compare } CO_2(g) + CaO(s) \rightarrow CaCO_3(s))$$

The earthy impurities and this calcium silicate form a molten slag which does not mix with iron but floats above it and can be run off separately. At one time this slag was a waste material, and the countryside has been defaced by the presence of huge slag heaps. The slag is being increasingly used at the present time for making roads.

The molten iron is run off into moulds, where it may be allowed to cool in long bars about 1 m long and 10 cm in diameter. It is known as 'cast iron' or 'pig iron'. In many works it is subjected to further treatment while still molten. This is obviously more economical in terms of energy used.

Cast iron

This is impure iron and contains varying amounts of impurities, such as carbon (4%), with smaller quantities of silicon, phosphorus, and sulphur. This impure iron melts at a lower temperature than pure iron and is brittle. It cannot be welded, and possesses little tensile strength. It is, however, used extensively for small castings, such as fire-grates, railings, hot-water pipes, Bunsen burner bases, and for many other purposes where little strain is imposed.

Wrought iron

This is the purest form of iron, and is obtained from cast iron by heating it with iron(III) oxide in a furnace by a process known as 'puddling'. The oxygen of the iron oxide oxidizes the impurities, carbon and sulphur, to the gaseous oxides which escape, and phosphorus to phosphates and silicon to silicates, and these form as a slag. The semi-molten mass is then hammered and rolled so that the slag is squeezed out and a mass of almost pure iron remains.

Wrought iron has a higher melting point than cast iron. It is malleable and can be forged, hammered, and welded when hot. It is tough and fibrous, and can withstand some strain, but is not elastic, and, if subjected to great strain, it will bend. It cannot be tempered. It is used to make iron nails, sheeting, ornamental work, horse-shoes, and agricultural implements. It has been replaced to a large extent in recent years by mild steel, which can be made more cheaply.

Steel

Ordinary steel is a material containing iron and a small proportion of carbon, the proportion being determined by the intended use of the steel. About 90% of the pig iron made is converted into steel.

Bessemer process. The usual Gilchrist-Thomas version of the Bessemer process is as follows. The *Bessemer converter* is made of steel plate lined with firebrick. It is roughly egg-shaped with an open top and a base perforated to take an air-blast. It has a *basic lining* of calcined dolomite (CaO and MgO). White-hot cast iron is run in and an air-blast is applied. The various impurities of the cast iron oxidize carbon to its gaseous oxides (CO and CO_2) which escape, manganese and silicon to oxides which form a slag, and phosphorus to phosphorus(V) oxide P_4O_{10}, which is absorbed as a phosphate by the basic lining. When the appearance of the flame indicates the end of these changes (about thirty minutes), the required content of carbon is added to the metal as anthracite. A short air-blast is applied for mixing and the steel can then be poured.

Recently, the Linz-Donawitz (L-D) modification has been introduced. The converter is similar but with a solid base. Instead of an air-blast from the base, an *oxygen* blast at 10 atm pressure is applied over the top of the white-hot cast iron, by a water-cooled copper tube. The rest of the process is essentially the same as before. Lime may be added to assist formation of slag. Advantages of the L-D process are greater speed and a less brittle steel, because the oxygen blow contains no nitrogen. Steel from this process is very suitable for pressing motor-car bodies.

Siemens-Martin open hearth process. The chief impurities of cast iron are non-metals, the oxides of which are acidic. The object of the Siemens process is to oxidize the impurities to their acidic oxides, and these combine with a basic lining supplied to the furnace. In this process pig iron, scrap iron, and iron(III) oxide (the last of these supplies the oxygen for oxidation of impurities) are melted in an open hearth which has been lined with a basic material (carbonates of calcium and magnesium are used, the oxides being formed at this high temperature). The amount of carbon is regulated at from 0.5% to 1%, and small quantities of various metals such as manganese, nickel, chromium, or tungsten, are added according to the quality of the steel and the use to which the steel will be put.

Properties of steel. Steel is hard, tough, and strong. If cooled gradually, steel can subsequently be hammered into shape or drilled, because it is fairly soft. By heating it and suddenly cooling it, the steel becomes very hard indeed, of very high tensile strength, and elastic. By reheating the steel to carefully regulated temperatures, steels of different degrees of hardness and brittleness can be obtained. This is called 'tempering'.

Uses of steel. Ordinary carbon steel, and alloy steels, have extensive and well-known uses, *e.g.*, as armour plate in warships and military tanks, as pressed sheet in automobile bodies, as girders and wire mesh in reinforced concrete building, as cutting and boring tools, crushing machinery, and stainless cutlery. The following are a few examples:

with chromium (1–4%)	armour plating, gears
with chromium (10–12%)	stainless steel
with cobalt (2–4%)	in electromagnets
with tungsten (5–18%)	drills and cutting tools.

Tin-plating. Thin iron sheet can also be *tin-plated* by passing the cleaned sheet through molten tin with a flux present, *e.g.*, zinc chloride. Tin-plate is used in canning fruit, meat, and fish (to which it imparts no taste). A continuous layer of tin protects iron from rusting (being unreactive with air). Tin, however, is less electropositive than iron, so, as soon as the tin layer is broken, oxidation of iron begins.

Lead

Occurrence

Lead occurs as galena, lead(II) sulphide PbS, and is distributed widely in the earth's crust, being found to some extent in most parts of the world. It has been known for a very long time – lead pipes were used by the Romans.

Extraction of lead

The galena is roasted with excess of air to form lead(II) oxide.

$$2PbS(s) + 3O_2(g) \rightarrow 2PbO(s) + 2SO_2(g)$$

The oxide is then reduced to lead by heating with carbon in a small blast furnace.

$$PbO(s) + C(s) \rightarrow Pb(s) + CO(g)$$

Some iron is added to reduce any remaining galena,

$$PbS(s) + Fe(s) \rightarrow Pb(s) + FeS(l)$$

and lime to combine with earthy impurities and form a molten slag. The molten iron(II) sulphide and slag are tapped off separately from the lead.

New process for joint extraction of zinc and lead

Substantial supplies of ores are known in which lead and zinc sulphides occur together. A new plant (in production in England in 1968) extracts both metals in a single process.

The ores (previously concentrated by flotation) are roasted in excess of air to form oxides.

$$2ZnS(s) + 3O_2(g) \rightarrow 2ZnO(s) + 2SO_2(g) \quad \text{(PbS similar)}$$

The oxides are dropped, with *coke* (reducing agent), into a blast furnace (page 426) and are reduced to the corresponding metals.

$$ZnO(s) + C(s) \rightarrow Zn(s) + CO(g); \quad PbO(s) + C(s) \rightarrow Pb(s) + CO(g)$$

Molten lead settles to the base of the furnace, with some slag, and is tapped off periodically. Zinc is vaporized and leaves the furnace by a pipe near the top, together with nitrogen and oxides of carbon. Zinc vapour is stripped from the gases by a spray of molten lead. The mixture of zinc and lead flows into shallow tanks where molten zinc separates above the much denser molten lead and can be tapped off separately. The lead is returned to the stripping circuit. The process is continuous and economical in both manpower and heating costs; a disadvantage is that 400 tonnes of molten lead must be pumped round per tonne of zinc extracted.

Uses of lead

(1) Lead was used in the manufacture of water and gas piping and as lead sheet for roofing. It was particularly valuable for piping as it is easily repaired, and joints are quickly made. It is also soft and bends easily at corners. It has been replaced for most purposes by copper or plastic piping. It is also used in the manufacture of lead shot.

(2) Lead is used in the manufacture of electrical accumulators and as a covering material for cables.

(3) Lead is used in making various alloys including typemetal, solder, and pewter.

(4) In paint, although white lead ($Pb(OH)_2.2PbCO_3$) is now little used because it is poisonous.

Copper

Occurrence

The principal ores of copper are copper pyrites ($CuFeS_2$), cuprite (Cu_2O), copper(I) sulphide (Cu_2S), and malachite ($CuCO_3.Cu(OH)_2$). It is mined as the free metal in parts of Canada and the U.S.A.

Extraction of copper

The ore is initially concentrated by a process of flotation, and then it is roasted in air to produce copper(I) sulphide.

$$2CuFeS_2(s) + 4O_2(g) \rightarrow Cu_2S(s) + 3SO_2(g) + 2FeO(s)$$

By adding silica, SiO_2, and heating in the absence of air, the iron(II) oxide is converted into a slag of iron(II) silicate, $FeSiO_3$, and poured away from the copper(I) sulphide. The copper(I) sulphide is reduced to the metal by heating in a regulated supply of air.

$$Cu_2S(s) + O_2(g) \rightarrow 2Cu(s) + SO_2(g)$$

The copper produced is too impure for many of the purposes for which it is to be used in industry, and it is therefore refined. This is achieved by making the impure copper the anode in an electrolytic cell, which also contains strips of pure copper as the cathode (Figure 198). The electrolyte is copper(II) sulphate

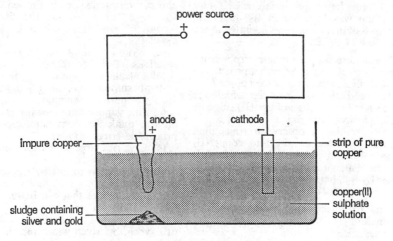

Figure 198

solution, and during the electrolysis copper is transferred from the impure anode to the pure cathode as follows.

<table>
<tr><td>*At the anode*</td><td>*At the cathode*</td></tr>
<tr><td>$Cu(s) - 2e^- \rightarrow Cu^{2+}(aq)$</td><td>$Cu^{2+}(aq) + 2e^- \rightarrow Cu(s)$</td></tr>
</table>

The copper produced is 99.98 per cent pure.

The metals silver and gold are not discharged in the process but collect in a sludge at the bottom of the cell, and they are subsequently recovered from it.

Uses of copper

(1). For conducting electric current. It must be very pure because impurities increase electrical resistance.

(2). For ornamental work, being little attacked by the air.

(3). In alloys, *e.g.*, brass (Cu and Zn), bronze (Cu and Sn), German silver (Cu, Zn, and Ni), and the copper coinage (Cu and Sn).

Questions

1. The metal iron is extracted from its ores in a blast furnace. Explain, with equations, the reactions which take place. Limestone is usually added to the mixture in the furnace. Explain why this is done.

Describe how you would prepare reasonably pure crystals of iron(II) sulphate, starting from the metal. (O. and C.)

2. Aluminium and iron form oxides with similar formulae. These oxides are used on the large scale as sources of the metals, but different methods of extraction have to be used. Give a brief account of both extraction processes, emphasizing the chemical principles involved. Technical details are not required.

Calculate the masses of aluminium and iron which could be obtained from 1000 tons of each oxide. Explain why these masses are different.

Choose one large-scale use of both aluminium and iron. Explain how each use is related to the properties of the metal. (L.)

3. Describe how aluminium is manufactured from pure aluminium oxide by an electrolytic process. Explain why (*a*) aluminium is often

used for overhead power cables, and (*b*) molten iron can be obtained by igniting a mixture of iron(III) oxide and aluminium powder.

The same quantity of electricity was passed through two appropriate cells. If 4.6 g of sodium was liberated in the first cell, what mass of aluminium was liberated in the second cell? (A.E.B.)

4. How is pure copper obtained from crude copper by electrolysis? What happens to the zinc and silver impurities in the copper?

Name and give the formulae of all the ions present in a solution of iron(II) sulphate in dilute hydrochloric acid.

What would you observe if (i) sodium hydroxide solution is added to copper(II) sulphate solution and the mixture is boiled; (ii) ammonia solution is gradually added to copper(II) sulphate solution until the ammonia is in excess? (C.)

5. Iron is manufactured by passing a blast of hot air through a mixture of iron ore, coke, and limestone.

(*a*) Explain carefully, with equations, the function of (i) the air, (ii) the coke, (iii) the limestone. (*b*) Give the reason why pig iron is brittle. (*c*) State *briefly* what has to be done to pig iron to convert it into steel. (*d*) If you were to build a new factory for manufacturing steel from imported iron ore, suggest two factors which would influence your choice of site. (J.M.B.)

6. Answer the following questions about the manufacture of iron and steel. (No diagrams are required.)

(i) Give the name and formula of *one* mineral from which iron is extracted.

(ii) Explain how carbon monoxide is formed in the blast furnace.

(iii) Write the equation for *one* reaction by which metallic iron is formed in the furnace.

(iv) Explain clearly why limestone (calcium carbonate) is used in the blast furnace and suggest what you think would happen if the limestone were not present.

(v) Name *three* impurities likely to be present in the 'pig iron' formed in the blast furnace. Give *one* effect of these impurities on the physical properties of the iron.

(vi) Explain how these impurities are removed during the conversion of pig iron into steel. (C.)

7. Zinc is extracted from zinc blende (impure zinc sulphide ZnS) by first heating the zinc blende in air to give zinc oxide and sulphur dioxide. The zinc oxide is then mixed with carbon and heated, when zinc vapour and carbon monoxide are formed. The zinc vapour is condensed and the zinc can be separated from its impurities by fractional distillation.

(*a*) Write equations for the *two* chemical reactions referred to above.

(*b*) Calculate the maximum mass of zinc and of sulphur dioxide obtainable from 1000 g of pure zinc sulphide.

(*c*) Zinc vapour burns readily in air. Why do you think this does not happen in the process described above?

(*d*) Give *two* large-scale uses each for zinc and sulphur dioxide.

(*e*) Say what is meant by the terms *condensed* and *fractional distillation*, and name *one* other mixture that can be separated by fractional distillation.

(*f*) Give a test, other than heating in air or oxygen, by which you could show that a given mineral is a metallic sulphide. (C.)

8. What raw materials are used in the blast furnace for the production of iron? State briefly the function of each. Give equations for *four* different reactions which take place in the furnace.

An ore of iron is known to contain 75 per cent of iron(III) (ferric) oxide and no other iron compound. What mass of iron would be obtained by the complete reduction of 100 kilogrammes of it?

Describe the reaction of steam on red-hot iron. Mention any one point of interest about this reaction. (S.)

9. Give the name and formula of a common ore of zinc and describe the chemistry involved in extracting zinc from it. Mention *two* industrial uses of zinc. How would you prepare a sample of zinc oxide from granulated zinc? What is meant by the statement *zinc oxide is amphoteric*? (S.)

10. Describe the electrolytic process for the refining of copper. Name four distinct uses of copper.

Starting with copper(II) oxide, how would you obtain (*a*) a dry crystalline specimen of copper(II) sulphate, (*b*) a sample of metallic copper?

37 Metals; General Properties

A great deal of the detailed chemistry of the compounds of the metals has already been covered, either in the previous chapter or in examining the reactions of the acid radicals (such as nitrates, chlorides, sulphates, carbonates, etc.). It is therefore the intention in this chapter only to give a broad summary of those reactions characteristic of metallic elements, and to indicate the major features of importance in connection with these reactions. Initially, therefore, it would be interesting to summarize those properties which are typical of metals as distinct from non-metals. Until recently physical differences between metals and non-metals have been regarded as quite significant. For example, gold and silver, with high density, lustre, malleability, and ductility, have been regarded as two of the most typical metals; sulphur, which has a much lower density and no lustre, and is very brittle, was regarded as a typical non-metal. This outlook has been largely replaced by the following.

Metals – chemical characteristics

A metal is now defined as **an element which can ionize by electron loss.** The number of electrons lost per atom is the valency of the metal and the ion carries an equal number of positive charges, as:

$$Na - e^- \rightarrow Na^+ \quad \text{(univalent)}$$
$$Mg - 2e^- \rightarrow Mg^{2+} \quad \text{(divalent)}$$
$$Al - 3e^- \rightarrow Al^{3+} \quad \text{(trivalent)}$$

The following are important chemical properties of metals which are derived from this ionization behaviour. There are, of course, exceptions to these properties, but they should be known as *typically metallic* properties.

1. *The oxide of a metal is basic, and, if soluble in water, gives an alkaline solution.*

This property follows from the fact that a typical metallic oxide is formed by oxygen accepting electrons from the metal as it ionizes, to form the ion, O^{2-}. For example,

$$\left. \begin{array}{r} Ca \rightarrow Ca^{2+} + 2e^- \\ \tfrac{1}{2}O_2 + 2e^- \rightarrow O^{2-} \end{array} \right\}$$

or adding these, $\quad Ca(s) + \tfrac{1}{2}O_2(g) \rightarrow Ca^{2+}O^{2-}(s)$

If the oxide dissolves in water, it gives the reaction:

$$O^{2-}(s) + H_2O(l) \rightleftharpoons 2OH^-(aq)$$

so providing the hydroxyl ion, which is characteristic of *alkalinity*. If insoluble in water, the oxide does not produce alkalinity but acts as a *basic oxide*.

This property arises from the behaviour of the O^{2-} ion or OH^- ion with hydrogen ion, H^+, which is characteristic of acidity. The behaviour is:

$$\left.\begin{array}{l} O^{2-}(s) + H_2O(l) \rightleftharpoons 2OH^-(aq) \\ 2OH^-(aq) + 2H^+(aq) \rightleftharpoons 2H_2O(l) \end{array}\right\}$$

The second of these reactions is the essential nature of neutralization, *i.e.*, the formation of water from its ions, H^+ and OH^-. The ions, Ca^{2+} and $2Cl^-$, derived from calcium oxide and hydrochloric acid, may remain as the *salt*, calcium chloride, and correspondingly for other oxides and acids. The complete reaction can be represented as:

$$Ca^{2+}O^{2-}(s) + H_2O(l) \rightleftharpoons Ca^{2+}(OH^-)_2(aq)$$
$$Ca^{2+}(OH^-)_2(aq) + 2(H^+Cl^-)(aq) \rightarrow Ca^{2+}(Cl^-)_2\,(aq) + 2H_2O(l)$$

It will be seen that the Ca^{2+} and Cl^- ions take no part in the reaction. Ions like this are often called *spectator ions*.

2. A metal can replace the H^+ ion in an acid and so produce a salt

This replacement may occur directly or indirectly. If directly, the metal ionizes, liberating electrons; the electrons are accepted by H^+ ions in the acid. These ions become hydrogen atoms, pair off as molecules, and are liberated as gas. The metallic ions pass into solution and, in conjunction with the negative ions of the acid, Cl^-, SO_4^{2-} etc., constitute a *salt*. For example:

$$\left.\begin{array}{l} Zn \rightarrow Zn^{2+} + 2e^- \\ 2H^+ + 2e^- \rightarrow H_2 \end{array}\right\}$$

For dilute sulphuric acid,

$$Zn(s) + (H^+)_2SO_4^{2-}(aq) \rightarrow Zn^{2+}SO_4^{2-}(aq) + H_2(g)$$

Indirect replacement occurs chiefly through the medium of the metallic oxide or hydroxide, as:

$$Cu(s) + \tfrac{1}{2}O_2(g) \rightarrow Cu^{2+}O^{2-}(s)$$
$$O^{2-}(s) + H_2O(l) \rightleftharpoons 2OH^-(aq)$$
$$2OH^-(aq) + 2H^+(aq) \rightleftharpoons 2H_2O(l)$$

If the acid is dilute sulphuric, Cu^{2+} ions are left in solution, with SO_4^{2-} ions, as the salt, *copper(II) sulphate*, the other product being water.

3. Metals form electrovalent chlorides

Metallic chlorides are typically electrovalent in type because, in their formation, a metal loses electrons, which are accepted by chlorine atoms, *e.g.*,

$$\left.\begin{array}{l} Ca \rightarrow Ca^{2+} + 2e^- \\ Cl_2 + 2e^- \rightleftharpoons 2Cl^- \end{array}\right\}$$

or adding these, $\quad Ca(s) + Cl_2(g) \rightarrow Ca^{2+}(Cl^-)_2(s)$

Such chlorides are electrovalent, electrolytes when molten or in aqueous solution, and solids of very low volatility.

4. Metals form few compounds with hydrogen

This situation arises because hydrogen forms compounds most readily by co-valency (*i.e.*, electron *sharing*) or, in acids, by electron *loss* to form the ion, H^+.

Neither of these modes of combination is well suited to metals, which tend to operate by electron loss.

A few very powerful metals (Na, K, Ca) can force hydrogen to accept electrons. The hydrides so formed are salt-like solids (compare the chlorides above) and liberate hydrogen with cold water, *e.g.*,

$$Na(s) + \tfrac{1}{2}H_2(g) \rightarrow Na^+H^-(s)$$
$$H^-(s) + H_2O(l) \rightarrow OH^-(aq) + H_2(g)$$

When molten, they act as electrolytes and liberate hydrogen at the *anode*.

$$H^- - e^- \rightarrow \tfrac{1}{2}H_2$$
(to anode)

In aqueous solutions, containing the usual ion, H^+, hydrogen appears at the *cathode* during electrolysis.

$$H^+ + e^- \rightarrow \tfrac{1}{2}H_2$$
(from cathode)

5. *Metals are reducing agents*

This follows from the modern definition of a reducing agent (page 174) as an electron donor, *e.g.*,

$$K \rightarrow K^+ + e^-$$
$$Zn \rightarrow Zn^{2+} + 2e^-$$

From the theoretical angle, the most characteristic physical property of a metal is its ability to conduct electricity, *i.e.*, pass a stream of electrons through itself if a potential difference is applied across it. This arises because metallic atoms part with their outermost electrons easily and a mass of metal always contains comparatively loose electrons. These move readily through the metal and are steadily replaced from the source of potential difference to maintain a flow of current. Conduction of heat, lustre, malleability, ductility, and high tensile strength are other physical properties possessed by some metals. They are very useful in practice but are not now accorded their former importance. Many metals have high density, but the most characteristic common metals, *i.e.*, those which ionize most readily by electron loss, are sodium and potassium, and they have densities below that of water. They are also soft enough to be cut with a pen-knife. The distinction must be clearly borne in mind between *strength as a metal* and *physical strength*. Sodium and potassium are the strongest metals: but no one would use them for building a bridge, even if they did not react with water.

Characteristics of non-metals

If a non-metal forms ions, it does so by electron gain, *i.e.*, the element is *electronegative*. The number of electrons gained per atom is the valency of the element for this purpose. (It may also exercise other kinds of valency.) The ion formed carries the corresponding number of negative charges but *they rarely exceed two*, and never exceed three.

$$\tfrac{1}{2}Cl_2 + e^- \rightarrow Cl^- \quad \text{(univalent)}$$
$$\tfrac{1}{2}Br_2 + e^- \rightarrow Br^- \quad \text{(univalent)}$$
$$\tfrac{1}{2}O_2 + 2e^- \rightarrow O^{2-} \quad \text{(divalent)}$$
$$S + 2e^- \rightarrow S^{2-} \quad \text{(divalent)}$$

The following are important chemical properties of non-metals which are connected with their tendency towards electron gain:

1. The oxide of a non-metal is never basic. Its characteristic oxide is acidic. Its other oxides are acidic or neutral

As explained on page 469, a basic oxide is formed when an oxygen atom accepts electrons from a metallic atom to give the ion, O^{2-}. Since a non-metal is an *acceptor* of electrons, it cannot form an oxide in this way. The characteristic oxide of a non-metal, *i.e.*, the oxide in which the non-metal exercises its *maximum* valency, is a *covalent* oxide, produced by formation of shared electron pairs.

When combined with water, it forms an acid, *e.g.*,

Non-metal	Maximum valency	Characteristic oxide and its action with water
Carbon	4	$CO_2(g) + H_2O(l) \rightleftharpoons H_2CO_3(aq) \rightleftharpoons 2H^+(aq) + CO_3^{2-}(aq)$
Nitrogen	5	$N_2O_5(g) + H_2O(l) \rightleftharpoons 2HNO_3(aq) \rightleftharpoons 2H^+(aq) + 2NO_3^-(aq)$
Sulphur	6	$SO_3(g) + H_2O(l) \rightleftharpoons H_2SO_4(aq) \rightleftharpoons 2H^+(aq) + SO_4^{2-}(aq)$
Chlorine	7	$Cl_2O_7(g) + H_2O(l) \rightleftharpoons 2HClO_4(aq) \rightleftharpoons 2H^+(aq) + 2ClO_4^-(aq)$

The H^+ ion, characteristic of acidity, is produced in all these cases.

When the non-metal exercises a reduced valency, the oxide is sometimes acidic, sometimes neutral, but never basic. Examples:

Non-metal	Oxide	Nature of oxide
Carbon	CO	neutral
Nitrogen	N_2O	neutral
	N_2O_4	acidic
Sulphur	SO_2	acidic
Chlorine	Cl_2O	acidic

2. A non-metal never replaces hydrogen in an acid to form a salt (as a metal does)

Replacement of hydrogen in an acid arises from the acceptance by H^+ of electrons supplied by a metallic atom:

$$2H^+ + 2e^- \rightarrow H_2$$

A non-metal is an electron acceptor and so cannot bring about the above replacement.

3. Non-metals form covalent chlorides

The behaviour of the non-metal, phosphorus, in forming its trichloride is typical:

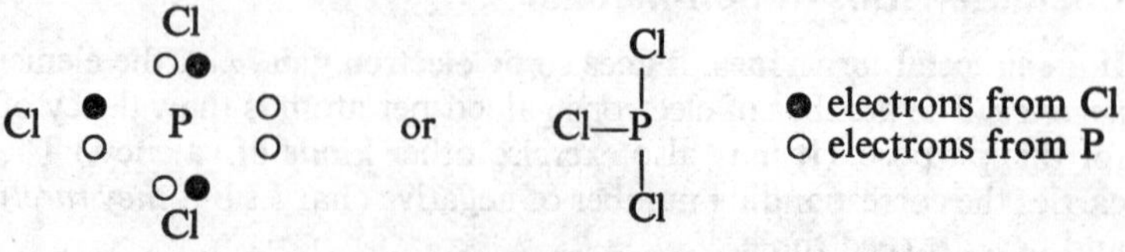

● electrons from Cl
○ electrons from P

Each Cl atom also has seven more electrons (not shown) in its outer electron layer. A covalent chloride of this kind is usually a volatile liquid, a non-electrolyte, and rapidly hydrolysed by water, as:

$$PCl_3(l) + 3H_2O(l) \rightarrow H_3PO_3(aq) + 3HCl(aq)$$

These properties are characteristic of non-metallic chlorides (though CCl_4 is not hydrolysed by water). Non-metals cannot supply electrons to form electrovalent chlorides. This is the behaviour of metals (see page 470).

4. *Non-metals combine with hydrogen to form many covalent hydrides*

These hydrides are formed by electron sharing, as:

(Each bond represents one shared pair of electrons)

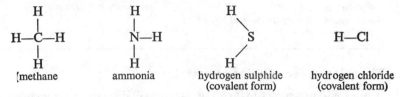

| methane | ammonia | hydrogen sulphide (covalent form) | hydrogen chloride (covalent form) |

Simple hydrides of this type have the properties of covalent compounds; they are gaseous, form no ions when anhydrous, and so are non-electrolytes. *Salt-like hydrides* are formed by metals (page 471) but non-metals cannot supply the electrons required for this.

5. *Non-metals are oxidizing agents*

This follows from the modern definition (page 174) of an oxidizing agent as an acceptor of electrons, *e.g.*,

$$\tfrac{1}{2}Cl_2 + e^- \rightarrow Cl^-$$
$$S + 2e^- \rightarrow S^{2-}$$

The most important physical property of non-metals is their very low electrical conductivity. Because of the tendency of a non-metal to attract electrons, a mass of it contains an absolute minimum of free electrons. This being so, no considerable electron flow can occur through the non-metal and it tends to be a very poor electrical conductor. (Graphite is unique as a non-metal, conducting electricity quite well.) In addition, non-metals are usually brittle and do not form sheets or wire. They have low tensile strength and no lustre, and cannot be polished.

Summary comparison of metals and non-metals

This may most conveniently be done in tabular form.

METALS	NON-METALS
Chemical properties	
Metallic oxides are *basic* and form *alkalis* if soluble in water.	Characteristic oxides of non-metals are *acidic*; other oxides are *acidic* or *neutral*.
Metals replace hydrogen in acids forming salts.	Non-metals do not form salts in this way.
Metallic chlorides are *electrovalent salts*, electrolytes containing Cl^- ions.	Non-metallic chlorides are covalent, non-electrolytes, and are generally hydrolysed by water.
Metals form few stable hydrides. Na, K, and Ca form salt-like hydrides, which contain ions, H^-.	Non-metals form many stable hydrides; they are covalent, the simple ones being gases (like NH_3 or CH_4) and non-electrolytes if water-free.

METALS	NON-METALS

Physical properties

Metals are:
  good conductors of electricity and heat;
  malleable (can be beaten into sheets);
  ductile (can be drawn into wire); lustrous,
  and can be polished
Some metals are very dense and have high
  tensile strength.

Non-metals are:
  generally bad conductors of electricity and
  heat (graphite conducts electricity well);
  generally brittle, not malleable or ductile;
  generally not lustrous.

The relative reactivities of metals

It is a well-known fact that certain metals will displace other metals from solutions of their salts in water. For example, iron will displace copper from copper(II) sulphate solution, and zinc will displace silver from silver nitrate solution.

$$Fe(s) + CuSO_4(aq) \rightarrow FeSO_4(aq) + Cu(s)$$
$$Fe(s) + Cu^{2+}(aq) \rightarrow Fe^{2+}(aq) + Cu(s)$$
$$Zn(s) + 2AgNO_3(aq) \rightarrow Zn(NO_3)_2(aq) + 2Ag(s)$$
$$Zn(s) + 2Ag^+(aq) \rightarrow Zn^{2+}(aq) + 2Ag(s)$$

We can arrange the metals in a series such that any metal higher up in the series will displace from its salts any metal below it. The greater the gap separating the metals in the series, the more readily does displacement take place.

Again, if a plate of zinc and a plate of copper are immersed in dilute sulphuric acid, a current will flow from the copper to the zinc outside the cell and from the zinc to the copper inside the cell, if the plates are connected by a wire. Thus the copper and the zinc must be at different potentials when in contact with dilute sulphuric acid. These potentials can be measured and by arranging the metals according to this potential difference the same series is obtained as by the displacement method. The list obtained is as follows, omitting the less common metals:

Metal	*Symbol*	
Potassium	K	Most electropositive metal.
Sodium	Na	
Calcium	Ca	
Magnesium	Mg	
Aluminium	Al	
Zinc	Zn	
Iron	Fe	
Lead	Pb	
(Hydrogen)	H	
Copper	Cu	
Mercury	Hg	
Silver	Ag	
Gold	Au	Least electropositive metal.

Hydrogen, although not a metal, is placed in the series to indicate the position it would occupy.

It will probably strike you at once that the metals occurring above hydrogen liberate that element from acids with an ease indicated by the interval separating the metal from hydrogen in the series. Thus magnesium and zinc liberate hydrogen readily (so would sodium and potassium, with such an 'ease' that the experiment would be dangerous), whereas copper, which is below hydrogen in the series, does not liberate hydrogen from acids at all.

Chemical activity of the more electropositive elements

Metals such as sodium and potassium, which occupy positions high up in the series, are said to be very electropositive. These metals are very active chemically, and metals lower in the series are less active. In modern terms, this means that the more electropositive metals ionize readily by loss of electrons, *e.g.*,

$$Na - e^- \rightarrow Na^+; \quad Ca - 2e^- \rightarrow Ca^{2+}$$

The less electropositive metals ionize much less readily. One result of this relation is that, if conditions are similar and neither metal is affected by water, a more electropositive metal will displace a less electropositive metal from its salt in solution, *e.g.*,

$$Zn(s) + CuSO_4(aq) \rightarrow ZnSO_4(aq) + Cu(s)$$

The essential feature of this relation is that the zinc atom (more electropositive) transfers two electrons to the copper *ion* in solution, which is converted to a copper *atom* and precipitated. The changes are:

$$Zn \rightarrow Zn^{2+} + 2e^-; \quad Cu^{2+} + 2e^- \rightarrow Cu$$

or, added together,

$$Zn(s) + Cu^{2+}(aq) \rightarrow Zn^{2+}(aq) + Cu(s)$$

The zinc ion is left in solution in association with the SO_4^{2-} ion, as the salt, zinc(II) sulphate.

The behaviour of metals in liberating hydrogen from water or dilute acid is a special case of this relation. All metals which are more electropositive than hydrogen displace it; the most electropositive metals, *e.g.*, Na, K, Ca, displace hydrogen from water. As electropositive nature decreases, metals require dilute acid, *e.g.*, Zn, Fe, while lead (very close to hydrogen in the electrochemical series) requires hot, concentrated hydrochloric acid. In all these cases, the metallic atom ionizes by supplying a number of electrons equal to its valency to hydrogen atoms and, by pairing, to molecules which are liberated. For example,

$$Na - e^- \rightarrow Na^+; \quad H^+ + e^- \rightarrow \tfrac{1}{2}H_2 \text{ (water)}$$
$$Zn - 2e^- \rightarrow Zn^{2+}; \quad 2H^+ + 2e^- \rightarrow H_2 \text{ (dilute HCl)}$$
$$Al - 3e^- \rightarrow Al^{3+}; \quad 3H^+ + 3e^- \rightarrow 1\tfrac{1}{2}H_2 \text{ (conc. HCl)}$$

Copper and silver are *less* electropositive than hydrogen. They cannot displace it from water or acid. They are attacked by no mineral acid except a strongly oxidizing compound such as nitric acid or hot, concentrated sulphuric acid.

In general the more electropositive metals (with the apparent exception of aluminium – see below) oxidize readily, while the least electropositive tend to be inactive; for example, copper and silver are not readily attacked by the oxygen of the atmosphere. They have been found as free metals in the earth's crust and used as coinage metals. The more electropositive metals do not occur free in nature but only as compounds, such as sodium chloride, zinc sulphide, and aluminium oxide.

Gradation in properties of compounds of metals according to position in the series

Not only does the series give us a good estimate of the chemical activity of the metal itself, but in many cases the properties of compounds of the metals are graded according to the position of the metal in the series. Thus the nitrates of sodium and potassium on heating decompose to the nitrite, the remainder of the metals of the series as far as copper (inclusive) form nitrates which decompose into the oxides on heating, liberating nitrogen dioxide and oxygen. The lowest members of the series, mercury, silver, and gold, form nitrates which decompose to the metal on heating. (See page 442.)

Table A, page 477, shows some of these properties of the metals and their compounds. It will be noticed that, as a rule, a change occurs in the region of calcium and another change in the region of copper. Be careful not to be too dogmatic about the properties as indicated in this way. Some metals show 'false' behaviour by acquiring, in air, a coherent coating of oxide which renders them inactive, *e.g.*, ordinary aluminium foil is inert towards air, water, or dilute sulphuric acid. If, however, it is dipped into mercury chloride solution, a layer of amalgam forms and an oxide layer cannot cohere on the foil. Then the metal attacks both water and acid, and rapidly forms flakes of oxide in air at ordinary temperature.

$$2Al(s) + 6H^+(aq) \rightarrow 2Al^{3+}(aq) + 3H_2(g); \ 4Al\ (s) + 3O_2(g) \rightarrow 2Al_2O_3(s)$$

The place of non-metals in the series

A list has been worked out to include common non-metals and is given below.

Electropositive

	Potassium	
	Sodium	
	Calcium	
	Magnesium	
	Aluminium	In this complete series, the
	Zinc	further apart two elements are,
METALS	Iron	the more likely they are to
	Lead	form a stable compound. Thus
	(Hydrogen)	oxygen combines very readily
	Copper	with sodium and potassium.
	Mercury	Elements close to one another
	Silver	either do not combine at all or
	Gold	form unstable compounds.
		Thus chlorine dioxide ClO_2
	Carbon	is an unstable explosive sub-
	Nitrogen	stance. Metals do not form
NON-	Phosphorus	stable compounds with each
METALS	Sulphur	other, but may form alloys.
	Oxygen	
	Chlorine	
	Fluorine	

Electronegative

Physical and chemical properties of the metals – a synopsis

For the purposes of this comparison the metals will be classified (arbitrarily) into three groups, based on the reactivity series, as follows:

the more reactive metals: potassium, sodium, calcium.

the reactive metals: aluminium, zinc, iron.

the less reactive metals: lead, copper.

	Combustion	Action on WATER	Action on ACIDS	Reduction of heated oxides by hydrogen	Action of heat on oxides	Action of water on oxides	Character of hydroxides	Character of carbonates	Action of heat on nitrates	Solubility of sulphides
K	Burn in air or oxygen readily	Decompose cold water	Attacked by dilute acids (Al is only attacked by dil.HCl)	Not reduced		Oxides react to form hydroxides	Soluble in water, therefore bases and alkalis	Soluble and not decomposed by heat	Nitrates decompose to nitrite	Sulphides soluble in water
Na										
Ca									(Ca acts as group below)	
Mg		Decompose steam at red heat			Stable when heated	Oxides do not react with water	Hydroxides insoluble bases only	Insoluble and decomposed by heat ($Al_2(CO_3)_3$ does not exist)	Nitrates decompose to oxide	Sulphides insoluble in water but soluble in dilute hydrochloric acid
Al										
Zn				Reduced						
Fe										
Pb	Oxidize when heated in air	Do not decompose water or steam at red heat	Attacked by oxidizing acids							Sulphides insoluble in water and dilute hydrochloric acid
Cu										
Hg					Decompose					
Ag	Unaffected by oxygen		Not attacked				Hydroxides not formed	Carbonates unstable	Nitrates decompose to metal	
Au										

Table A

(1) The more reactive metals; potassium, sodium, calcium

Physical properties

These are summarized in Table B.

PHYSICAL PROPERTY	POTASSIUM	SODIUM	CALCIUM
State	Solid	Solid	Solid
Appearance	White metal, possessing a lustre	White, silvery, shining metal. Rapidly tarnishes	Silvery, shining metal which rapidly tarnishes in air owing to formation of film of oxide
Density (g cm^{-3})	0.86	0.97	1.55
Malleability	Malleable and ductile	Very malleable. Can be cut with a knife	Malleable and ductile
Tensile strength	Does not possess tensile strength to any appreciable extent	Does not possess tensile strength to any appreciable extent	Possesses fair tensile strength
Melting point	63 °C	98 °C	850 °C
Conduction of heat and electricity	Good conductor of heat and electricity	Conducts both heat and electricity	Good conductor of heat and electricity

Table B

Chemical properties of potassium

In chemical properties potassium is very similar to sodium, but it is slightly more reactive. Thus when a small piece of potassium is placed on water it darts about, melts, and gives off hydrogen which at once ignites and burns with the oxygen of the air, giving a lilac-coloured flame.

$$2K(s) + 2H_2O(l) \rightarrow 2KOH(aq) + H_2(g)$$

Flame tests for potassium (see page 388). Compounds of potassium (especially the chloride) colour the Bunsen flame lilac and the colour is still visible through blue glass. Hence, although the lilac colour of potassium is easily masked by the presence even of traces of sodium, the potassium colour is visible when viewed through blue glass.

In general, the salts of potassium are less soluble in water than the corresponding salts of sodium (the principal exceptions are potassium hydroxide, potassium carbonate, and hydrogencarbonate).

Chemical properties of sodium

Action of sodium on exposure to air

Sodium is attacked by the oxygen of the air to form sodium oxide. The moisture present combines with some of the oxide to form the hydroxide and finally, after a time, the carbon dioxide of the air combines with the sodium hydroxide to form sodium carbonate, which crystallizes as the decahydrate. These crystals can effloresce to form the monohydrate.

$$4Na(s) + O_2(g) \rightarrow 2Na_2O(s)$$
$$Na_2O(s) + H_2O(l) \rightarrow 2NaOH(aq)$$
$$2NaOH(s) + CO_2(g) \rightarrow Na_2CO_3(s) + H_2O(l)$$

If heated in air or oxygen, sodium burns with a golden-yellow flame to form sodium peroxide.

$$2Na(s) + O_2(g) \rightarrow Na_2O_2(s)$$

Action of sodium on water

If a small piece of sodium (a 3 mm cube will be *ample*) is placed on the surface of water in a large dish or trough, the sodium will dart about and melt to a silvery ball of molten sodium, liberating hydrogen, and forming sodium hydroxide (see page 245).

$$2Na(s) + 2H_2O(l) \rightarrow 2NaOH(aq) + H_2(g)$$
$$\text{sodium}$$
$$\text{hydroxide}$$

If a light is applied the hydrogen given off will burn with a golden-yellow flame. Sodium and potassium are so readily attacked by the oxygen and water-vapour of the atmosphere that they are usually kept below the surface of petroleum oil.

Flame coloration

Sodium compounds impart a persistent golden-yellow coloration to the flame (see flame-test, page 388). This colour is invisible when viewed through blue glass. The above serves as a very definite and delicate test for the presence of sodium in a compound.

Chemical properties of calcium

Action of air on calcium

Calcium is not as reactive as sodium and potassium, and it is not necessary to keep it below the surface of petroleum.

A white film of oxide is formed on the surface on exposure to air.

Calcium will burn with a brick-red flame if heated in the air and forms quicklime, calcium oxide.

$$2Ca(s) + O_2(g) \rightarrow 2CaO(s)$$

Action of calcium on water (see page 245)

Calcium is attacked by water, liberating hydrogen and forming a suspension of calcium hydroxide. If this is filtered, a solution of calcium hydroxide (limewater) is obtained.

$$Ca(s) + 2H_2O(l) \rightarrow Ca(OH)_2(aq) + H_2(g)$$

The action is not so vigorous as that of sodium or potassium and a test-tube full of water can safely be placed over a piece of calcium in a dish containing water, and the hydrogen can be collected.

Flame coloration (see page 388)

Calcium compounds (especially the chloride) colour the flame brick red.

(2) The reactive metals; aluminium, zinc, iron

Physical properties

These are summarized in Table C.

PHYSICAL PROPERTY	ALUMINIUM	ZINC	IRON
State	Solid	Solid	Solid
Appearance	Silvery white	Bluish-white metal. Can be polished	Pure iron is a white metal and can be polished
Density (g cm^{-3})	2.69	7.1	7.9
Malleability	Can be rolled into foil	Malleable at temperatures between 100 and 150 °C	Extremely malleable
Tensile strength	Moderate (high in alloys)	High	High
Melting point	660 °C	419 °C	1535 °C
Conduction of heat and electricity	Good	Good	Good Iron can also be magnetized

Table C

Chemical properties of aluminium

Action of aluminium with air

The metal acquires a continuous, very thin coating of oxide and this resists further action. At 800 °C, it will burn in air, forming its oxide and nitride.

$$4Al(s) + 3O_2(g) \rightarrow 2Al_2O_3(s)$$
$$2Al(s) + N_2(g) \rightarrow 2AlN(s)$$

Action of aluminium with acids

The metal attacks dilute hydrochloric acid slowly and the concentrated acid rapidly, liberating hydrogen.

$$2Al(s) + 6HCl(aq) \rightarrow 2AlCl_3(aq) + 3H_2(g)$$

Aluminium has no action with dilute sulphuric acid, but the hot, concentrated acid is attacked by it with liberation of sulphur dioxide.

$$2Al(s) + 6H_2SO_4(aq) \rightarrow Al_2(SO_4)_3(aq) + 6H_2O(l) + 3SO_2(g)$$

Nitric acid does not react with aluminium at any concentration. This is probably because it produces on the metal a thin layer of insoluble oxide, which protects the metal from further attack. (See passive iron, page 482.)

Action of aluminium with caustic alkali solution

The metal, especially in powder form, reacts violently with bench sodium hydroxide solution, liberating hydrogen and leaving sodium aluminate in solution. (Potassium hydroxide similar.)

$$2Al(s) + 2OH^-(aq) + 6H_2O(l) \rightarrow 2Al(OH)_4^- + 3H_2(g)$$

Chemical properties of zinc

Action of zinc on exposure to air

Zinc is only very slightly attacked by air owing to the formation of a film of oxide which prevents further action. If heated in air, it will burn with a bluish-green flame forming the oxide.

$$2Zn(s) + O_2(g) \rightarrow 2ZnO(s)$$

Action of acids on zinc

Ordinary samples of zinc are attacked readily by the mineral acids. Very pure zinc is only slowly attacked, and in recent years the purity of even the commercial zinc is so high that very often the action of dilute sulphuric acid on zinc is slow at first. The common actions are expressed by the equations:

$$\underset{\text{zinc}}{Zn(s)} + \underset{\substack{\text{dilute} \\ \text{sulphuric} \\ \text{acid}}}{H_2SO_4(aq)} \rightarrow \underset{\substack{\text{zinc} \\ \text{sulphate}}}{ZnSO_4(aq)} + \underset{\text{hydrogen}}{H_2(g)}$$

With hot concentrated sulphuric acid, sulphur dioxide is formed.

$$Zn(s) + 2H_2SO_4(aq) \rightarrow ZnSO_4(aq) + 2H_2O(l) + SO_2(g)$$

$$\underset{\text{zinc}}{Zn(s)} + \underset{\substack{\text{dilute} \\ \text{hydrochloric} \\ \text{acid}}}{2HCl(aq)} \rightarrow \underset{\substack{\text{zinc} \\ \text{chloride}}}{ZnCl_2(aq)} + \underset{\text{hydrogen}}{H_2(g)}$$

$$3Zn(s) + 8HNO_3(aq) \rightarrow \underset{\substack{\text{zinc} \\ \text{nitrate}}}{3Zn(NO_3)_2(aq)} + 4H_2O(l) + 2NO(g)$$

In the last case other oxides of nitrogen and even ammonia, which combines to form ammonium nitrate, may be formed.

Action of alkalis on zinc

Zinc is attacked by a hot concentrated caustic alkali solution. (Zinc oxide is amphoteric, see page 287.) Hydrogen is evolved and sodium or potassium zincate solution is left.

$$Zn(s) + 2OH^-(aq) + 2H_2O(l) \rightarrow Zn(OH)_4{}^{2-}(aq) + H_2(g)$$

Action of water on zinc

Water does not attack zinc to any appreciable extent. Zinc at a red heat is attacked by steam with the formation of hydrogen.

$$Zn(s) + H_2O(l) \rightarrow ZnO(s) + H_2(g)$$

Chemical properties of iron

Action of iron on exposure to air

In the presence of air and moisture, iron readily rusts, forming a reddish-brown solid which consists mainly of hydrated iron(III) oxide ($Fe_2O_3.xH_2O$). If finely divided (for example iron filings), it will burn in air or oxygen to form the magnetic oxide of iron Fe_3O_4.

$$3Fe(s) + 2O_2(g) \rightarrow Fe_3O_4(s)$$

Action of steam on heated iron

Iron, at a red heat, is attacked by *excess* of steam, forming magnetic oxide of iron (tri-iron tetroxide) and hydrogen.

$$3Fe(s) + 4H_2O(g) \rightleftharpoons Fe_3O_4(s) + 4H_2(g)$$
tri-iron
tetroxide

The above action is reversible (see page 187).

Note that if air and water act together on iron, iron (III) oxide is formed, but if either of these substances act separately the product is tri-iron tetroxide.

Action of acids on iron

1. *Dilute sulphuric and hydrochloric acids.* Iron is attacked by these dilute acids in accordance with the following equations:

$$Fe(s) + H_2SO_4(aq) \rightarrow FeSO_4(aq) + H_2(g)$$
iron(II)
sulphate

$$Fe(s) + 2HCl(aq) \rightarrow FeCl_2(aq) + H_2(g)$$
iron(II)
chloride

The iron(II) salt is obtained because the hydrogen which is given off during the action is a reducing agent.

2. *Nitric acid.* Dilute nitric acid gives a series of complex reactions in which oxides of nitrogen and even ammonia are formed.

Passive state. If a piece of clean iron is dipped into concentrated nitric acid there is apparently no action, but the iron no longer behaves as a piece of ordinary iron; for example, it will not displace copper from copper(II) sulphate solution nor is it attacked by dilute nitric acid, which normally does attack it. If, however, the piece of iron is scratched while in contact with, say, dilute nitric acid, a vigorous reaction occurs. This 'passive state' is supposed to be due to a protective layer of oxide (Fe_3O_4) formed on the iron by the strong oxidizing agent, concentrated nitric acid.

Iron will readily combine with sulphur and chlorine when heated with them to form iron(II) sulphide (page 6) and iron(III) chloride (page 368) respectively.

Compounds of iron(II) and iron(III)

An iron atom possesses 26 electrons. The shell grouping normally shown for them is 2, 8, 14, 2. The 2 adjacent to the 14 are the valency electrons and the atom can part with them in electrovalent combination; it is then said to act in the *iron(II)* (formerly called *ferrous*) state and produces the ion, Fe^{2+}.

If, however, the iron(II) is exposed to suitable *oxidizing* conditions, it will utilize a futher electron for valency purposes. The resulting ion is then said to be in the *iron(III)* (formerly called *ferric*) state as Fe^{3+}.

$$Fe^{2+} \rightarrow Fe^{3+} + e^-$$

The following table shows the formulae (in molecular form) of the more important simple compounds of iron.

	Iron(II) ion Fe^{2+}	*Iron(III) ion Fe^{3+}*
Oxide	FeO	Fe_2O_3
Hydroxide	$Fe(OH)_2$	$Fe(OH)_3$
Chloride	$FeCl_2$	$FeCl_3$
Sulphate	$FeSO_4$	$Fe_2(SO_4)_3$
	Soluble iron(II) compounds give *green* solutions.	Soluble iron(III) compounds give *yellow* or *brown* solutions.

Solutions of pure iron(II) compounds are distinguished from those of pure iron(III) compounds by the colour differences just mentioned, though, in dilute solution the green colour of the iron(II) salts is very pale. Iron(III) hydroxide and iron(III) oxide are both brown, while iron(II) hydroxide, as usually precipitated, is green. Iron(II) oxide is so readily oxidized by oxygen of the air that it cannot be kept under ordinary laboratory conditions.

A simple test for an iron(II) salt. Dissolve a little iron(II) sulphate in water. To the solution, add sodium hydroxide (caustic soda) solution. A dirty-green gelatinous precipitate of iron(II) hydroxide is formed. This reaction is typical of an iron(II) salt.

$$Fe^{2+}(aq) + 2OH^-(aq) \rightarrow Fe(OH)_2(s)$$

Where it is exposed to the air, the precipitate will become brown because it is oxidized to iron(III) hydroxide.

$$2Fe(OH)_2(aq) + \tfrac{1}{2}O_2(g) + H_2O(l) \rightarrow 2Fe(OH)_3(s)$$

A simple test for an iron(III) salt. Using iron(III) chloride solution, repeat the test just given. In this case, the precipitate is reddish brown and is iron(III) hydroxide. This reaction is typical of an iron(III) salt.

$$Fe^{3+}(aq) + 3OH^-(aq) \rightarrow Fe(OH)_3(s)$$

Conversion of an iron(II) salt to an iron(III) salt. The conversion of an iron(II) salt to an iron(III) salt is an *oxidation* and is brought about by *oxidizing agents.*

$$Fe^{2+} - e^- \rightarrow Fe^{3+}$$

To a solution of iron(II) sulphate, which is green, add dilute sulphuric acid. Warm the mixture and add cautiously a few drops of concentrated nitric acid. (A dark brown coloration will probably appear. For an explanation of this see the 'brown ring' test, page 440.) Heat the mixture. Brown fumes of nitrogen dioxide are given off and a brown or yellow solution remains. It contains iron(III) sulphate (test as described above).

The nitric acid has oxidized the iron(II) sulphate to iron(III) sulphate and has itself been reduced to nitrogen oxide, which, on exposure to air, gives nitrogen dioxide.

$$6FeSO_4(aq) + 2HNO_3(aq) + 3H_2SO_4(aq) \rightarrow$$
$$3Fe_2(SO_4)_3(aq) + 4H_2O(l) + 2NO(g)$$

or $6Fe^{2+}(aq) + 8H^+(aq) + 2NO_3^-(aq) \rightarrow 6Fe^{3+}(aq) + 4H_2O(l) + 2NO(g)$

Iron(II) chloride is converted by the oxidizing agent, chlorine, to iron(III) chloride, in solution or when heated.

$$2FeCl_2(s) + Cl_2(g) \rightarrow 2FeCl_3(s)$$

or $\qquad 2Fe^{2+}(s) + Cl_2(g) \rightarrow 2Fe^{3+}(s) + 2Cl^-(s)$

Conversion of an iron(III) salt to an iron(II) salt. The conversion of an iron(III) salt to an iron(II) salt is a *reduction*, and is brought about by *reducing agents.*

$$Fe^{3+} + e^- \rightarrow Fe^{2+}$$

To a solution of yellow iron(III) chloride, add hydrochloric acid and zinc. There is vigorous effervescence with evolution of hydrogen. Leave the mixture for 20 to 30 minutes. The colour of the liquid is now green. It contains iron(II) chloride. (Test, after filtering, as described above.)

The explanation is that, in the presence of the acid, zinc supplies electrons which are taken up by the iron(III) ions, which are thereby reduced to iron(II) ions.

$$Zn \rightarrow Zn^{2+} + 2e^{-}$$
$$2Fe^{3+} + 2e^{-} \rightarrow 2Fe^{2+} \bigg\} \text{or } Zn + 2Fe^{3+} \rightarrow Zn^{2+} + 2Fe^{2+}$$

Other reducing agents will convert iron(III) salts to iron(II) salts. The action of two common ones is represented in the following equations.

$$\begin{cases} 2FeCl_3(aq) + H_2S(g) \rightarrow 2FeCl_2(aq) + 2HCl(aq) + S(s) \\ \text{or } 2Fe^{3+}(aq) + S^{2-}(g) \rightarrow 2Fe^{2+}(aq) + S(s) \end{cases}$$
$$\begin{cases} Fe_2(SO_4)_3(aq) + SO_2(g) + 2H_2O(l) \rightarrow 2FeSO_4(aq) + 2H_2SO_4(aq) \\ \text{or } 2Fe^{3+}(aq) + SO_2(g) + 2H_2O(l) \rightarrow 2Fe^{2+}(aq) + SO_4{}^{2-}(aq) + 4H^+(aq) \end{cases}$$

(3) The less reactive metals; lead and copper

Physical properties

These are summarized in Table D.

PHYSICAL PROPERTY	LEAD	COPPER
State	Solid	Solid
Appearance	Bluish white	A red-brown metal possessing a lustre. It can be polished
Density (g cm^{-3})	11.3	8.95
Malleability	Very malleable. Can be cut with a knife. Has a metallic lustre but speedily tarnishes	Very malleable and ductile
Tensile strength	Fair	Fairly high
Melting-point	327 °C	1080 °C
Conduction of heat and electricity	Good conductor of heat and electricity	It is an excellent conductor of both heat and electricity

Table D

Chemical properties of lead

Action of air and water on lead

Lead is attacked by air and water together, a white layer being formed on the lead which consists of a mixture of lead hydroxide and lead carbonate. Lead is used extensively as piping to carry water supplies and cases of poisoning have been traced to the removal of this layer of hydroxide and carbonate by the water passing through. If the water is slightly 'hard' (see page 248) a protective coat appears to be formed and none of the lead is removed. In many water supplies nowadays, the water is specially hardened by the addition of small quantities of lime to prevent any of these poisonous effects.

If lead is strongly heated in air it forms massicot (a yellow powder) which is a form of lead(II) oxide,

$$2Pb(s) + O_2(g) \rightarrow 2PbO(s)$$

but if heated to a carefully regulated temperature of about 450 °C red lead oxide is formed.

$$3Pb(s) + 2O_2(g) \rightarrow Pb_3O_4(s)$$
$$\text{red lead oxide}$$

Action of acids on lead

Dilute sulphuric acid and dilute hydrochloric acid have no action on lead. Hot concentrated sulphuric acid attacks lead (compare copper):

$$Pb(s) + 2H_2SO_4(aq) \rightarrow PbSO_4(s) + 2H_2O(l) + SO_2(g)$$
$$\text{lead(II)} \qquad\qquad\qquad \text{sulphur}$$
$$\text{sulphate} \qquad\qquad\qquad \text{dioxide}$$

Nitric acid attacks it forming nitrogen oxide (chiefly).

$$3Pb(s) + 8HNO_3(aq) \rightarrow 3Pb(NO_3)_2(aq) + 4H_2O(l) + 2NO(g)$$
$$\text{dilute nitric} \qquad \text{lead(II)} \qquad\quad \text{water} \qquad \text{nitrogen}$$
$$\text{acid} \qquad\qquad \text{nitrate} \qquad\qquad\qquad\quad \text{oxide}$$

The only satisfactory laboratory method of dissolving lead is by the action of dilute nitric acid to form lead(II) nitrate solution.

Chemical properties of copper

Action of air and water on copper

Copper is not attacked by pure air or water, but, when exposed to the atmosphere, it is slowly attacked on the surface with the formation of a green solid. The composition of the green solid depends on its location; at the seaside it corresponds to a basic chloride, but inland it corresponds to a basic sulphate.

When heated in the air, copper forms a layer of black copper(II) oxide on the surface.

$$2Cu(s) + O_2(g) \rightarrow 2CuO(s)$$

Action of acids on copper

Copper has no action on either dilute sulphuric acid or dilute hydrochloric acid.

With dilute nitric acid, oxides of nitrogen are liberated, chiefly nitrogen oxide, and a blue or bluish-green solution of copper(II) nitrate remains.

$$3Cu(s) + 8HNO_3(aq) \rightarrow 3Cu(NO_3)_2(aq) + 4H_2O(l) + 2NO(g)$$
$$\text{dilute nitric} \qquad \text{copper(II)} \qquad\quad \text{water} \qquad \text{nitrogen}$$
$$\text{acid} \qquad\qquad \text{nitrate} \qquad\qquad\qquad\quad \text{oxide}$$

With hot concentrated sulphuric acid sulphur dioxide is liberated and copper (II) sulphate is formed.

$$Cu(s) + 2H_2SO_4(aq) \rightarrow CuSO_4(aq) + 2H_2O(l) + SO_2(g)$$

Flame coloration

Copper salts colour the flame a characteristic bluish green.

Questions

1. Illustrate, by reference to two chemical and to four physical properties, the chief differences between the metals and the non-metals.

2. A piece of sodium which had been exposed to the air was found to be covered with a white powder. The whole specimen was dropped into 50 g of ethanol and 2400 cm³ of hydrogen measured at room temperature and pressure were set free. The unused ethanol was distilled off and a white solid remained.

(*a*) Write the equation for the reaction between sodium and ethanol and name the substance formed other than hydrogen.

(*b*) Calculate the mass of sodium which dissolved in the ethanol.

(*c*) What would be the mass of ethanol distilled off, assuming that there was no loss during the process?

(*d*) Suggest a safety precaution to observe when dealing with boiling ethanol.

(*e*) The ethanol distilled off at 80 °C, whilst the white solid remained unaffected at this temperature. Point out an essential difference in the structure of ethanol and the white solid.

(*f*) Name another liquid which produces hydrogen with sodium. What difference would you *observe* if identical pieces of sodium were dropped separately into small beakers containing ethanol and this other liquid respectively?

(*g*) What was the white powder coating the original piece of sodium? Explain in detail how it was formed.

(*h*) Describe one test by which you could identify the white powder which originally covered the sodium. (L.)

3. Two uncommon *metallic* elements have the following properties:

Element X

Low melting point
Forms only one chloride, XCl
Usually occurs as the compound with chlorine
The oxide reacts with water to form a soluble hydroxide

Element Y

High melting point
Forms two chlorides, YCl_2 and YCl_3
Usually occurs as the oxide, Y_2O_3
The oxide is not affected by water in any way

(*a*) Outline a method by which a sample of element X might be obtained from its compounds. Indicate why you have chosen this method.

(*b*) Outline a method by which a sample of element Y might be obtained from its compounds, and again indicate why you have chosen this method.

(*c*) What reaction, if any, would you expect to take place between these elements and cold water?

(*d*) What types of particle will be present in an aqueous solution of the chloride of X? How many of each type of particle will be present in 1 litre of molar XCl?

(*e*) 336 g of the chloride of X contain 71 g of chlorine. How many moles of chlorine are present in 71 g? Suggest a value for the relative atomic mass of X. (L.)

4. *A*, *B*, and *C* are three metals. From the following information write them down in order of reactivity, putting the most reactive first.

Carbon will reduce the hot oxides of *A* and *B* but not the oxide of *C*. *A* will remove the oxygen from the oxide of *B* on strong heating. Explain your reasoning. (S.)

5. Carbon at red heat will remove oxygen from the oxides of the metals *A*, *B*, and *C*, but not from the oxide of metal *D*. Metal *C* will remove oxygen from the oxide of *A*, but not from the oxide of *B*.

List the metals *A*, *B*, *C*, and *D* in order of decreasing activity. If metal *A* is divalent and metal *B* is trivalent, write (*a*) the formula for the nitrate of *A*, (*b*) the formula for the sulphate of *B*. (J.M.B.)

6. Name *four* metals, one for each reagent but a *different* one in each case, which react with the following specified reagents, also naming the gas formed in each reaction: cold dilute hydrochloric acid; hot concentrated sulphuric acid; cold dilute sulphuric acid; hot sodium hydroxide solution. (J.M.B.)

7. Name *one* ore of zinc. Outline the chemistry for the manufacture of zinc from this ore. Describe the changes which take place when: (*a*) zinc carbonate is strongly heated in air; (*b*) excess zinc is placed in copper(II) sulphate solution; (*c*) pure zinc is placed in dilute sulphuric acid and a few drops of copper(II) sulphate solution are added later. Give one use of zinc, explaining why it is used for the purpose mentioned. (S.)

8. Under what conditions, if any, do the following metals react with water: (*a*) sodium, (*b*) zinc, (*c*) iron, (*d*) calcium, (*e*) aluminium? State clearly the products formed. Select *two* of the products and outline how the original metal can be obtained. (A different method must be employed in each case.) State *one* use for any *two* of the metals. (S.)

9. Describe the reactions (if any) of the metals calcium, copper, iron, with (i) air, (ii) water, (iii) dilute hydrochloric acid. If there is a reaction, state the necessary conditions and give the *names and formulae* of the products. Using these reactions, and giving your reasons, arrange the three metals in order of *decreasing* chemical activity. (O. and C.)

10. Describe briefly *two* reactions which are characteristic of an element above magnesium in the electrochemical series. (S.)

11. If you were provided with iron in the form of filings, describe in detail how you would prepare from it: (*a*) iron(II) chloride crystals, (*b*) iron(III) chloride solution, (*c*) iron(III) oxide. (L.)

12. Describe briefly how you would prepare a soluble salt of each of the metals aluminium, lead, and zinc, starting with the metal in each instance.

Describe and explain the result of adding

(a) a little and (b) an excess of sodium hydroxide solution to the solutions of the salts of aluminium and zinc. (S.)

13. This question concerns the metals lead, iron, mercury, and aluminium.

(a) Place the four metals in order of reactivity, putting the most reactive first.

(b) State which of these metals would react with warm, dilute hydrochloric acid and give equations for the reactions which occur.

(c) Describe what you would see if a strip of lead were placed in a solution of silver nitrate ($AgNO_3$) and left for a few hours. Give an equation for the reaction occurring.

(d) One of the above metals is obtained from its oxide by electrolysis, another by heating its oxide, and the other two by heating their oxides with carbon monoxide. State which metals are obtained by each of the three processes.

(e) What is the connection between the way in which the metal is obtained from its oxide and your order of reactivity given in (a)? Give a reason for your answer. (J.M.B.)

14. Give the name and symbol of a metal (different in each instance) which fits the following information: (a) it must be kept out of contact with air or water; (b) it is prepared by electrolysis of its molten oxide; (c) limestone is one of its common compounds; (d) it has a green carbonate which, on heating, decomposes to give a black oxide. (S.)

15. Describe, in outline, one method in each case by which each of the changes numbered 1 to 4 below could be brought about. Give the essential conditions for the reactions involved and mention briefly how each product would be isolated in those cases where a solid is produced.

$$Cu(s) \underset{2}{\overset{1}{\rightleftarrows}} Cu^{2+}(aq) \overset{3}{\longrightarrow} CuCl(s)$$
$$\Big\downarrow 4$$
$$Cu(OH)_2(s)$$

Describe briefly *two* ways in which solid copper(I) chloride differs from solid copper(II) chloride. (L.)

38 Revision Notes

No two students will revise in quite the same way. For proper revision concentration is essential, and the following method is suggested, because it does ensure concentration.

(a) Self-expression

Suppose you are revising a topic such as 'Chlorine and Chlorides'. It does not matter what the subject is, the procedure is the same. *Begin with a pencil and a large sheet of paper*. Write down in note form as many important ways of making chlorine as you can. Write down briefly its properties, under headings if such classification is possible. Make a list of all the metallic chlorides you know, show how to prepare them, and state their principal properties. For the preparation of many substances a well-labelled diagram, apart from the equation, will be all that is required for this purpose. Draw the diagram freehand, or use a stencil, for all that is required of a diagram is to show clearly the apparatus you would use. This need not be an 'art' as well as a chemistry exercise, but a clear, neat diagram helps you.

Write down all you can, and do not give in too quickly, for there is a very great difference between a hasty revision and one in which you are determined to give in only when you have put down *all* the points you can possibly remember. Now turn to your book and you will find the correct answer to many points about which you were doubtful, and many points you did not know at all! The next time you revise that chapter, which should be some time after the first revision, you may content yourself with writing down only those points which you did not know before. The thing to remember is that **true revision must involve self-expression.** The following revision exercises are arranged with that point in mind.

(b) Approach the subject from a different angle

If you have followed the text of this book, you will have met most of the matter about the various salts classified according to the acid radical contained in them. Thus all metallic carbonates are discussed under 'carbonates', because they are similar in many ways, and all metal oxides are discussed under 'oxides' and so on.

Now attempt to revise in a different way by picking out one metal, for example copper, and writing down the preparation and properties of its various compounds.

(c) Read widely

You will find many excellent textbooks and 'popular' books on chemistry either at your school or in your public library. Every one of these will give you new ideas, presenting old material in a different form. Chemistry is a 'live' subject and of increasing importance to you in the civilized life you lead. Read about some of the applications and you will spend many interesting hours and widen and deepen your knowledge.

General statements

The following list of statements of a general type is given purposely without the common exceptions to them. **Some are perfectly general and others less so, but all are of wide application.** Learn them and attempt to write down (the symbols or formulae only will be quite sufficient to test yourself) any substances which are exceptions to the general statement. The exceptions are given on page 497.

1. Acids contain hydrogen which can be replaced by a metal.
2. Sodium, potassium, and ammonium salts are soluble in water.
3. Nitrates are soluble in water.
4. An ammonium salt heated with any alkali and water yields ammonia gas.
5. Heavy metal carbonates yield the oxide and carbon dioxide when heated.
6. Heavy metal nitrates decompose under the action of heat to yield the oxide, nitrogen dioxide, and oxygen.
7. Metals are attacked by nitric acid.
8. The action of an acid on a carbonate is to yield carbon dioxide.

Particular statements

The following list of questions summarizes many of the unusual facts which you are likely to overlook. Do not turn to the answers until you have tried them all.

1. What is the *only* common alkaline gas?
2. What is the *only* gas which turns brown on exposure to air?
3. Which common substance is almost insoluble in cold water but quite soluble in hot water?
4. What common substance increases its solubility in water very little for a large increase in temperature?
5. What substances are *less* soluble in hot water than in cold?
6. Which salts cannot be prepared, in solution, by the following method?

$$\left.\begin{array}{l}\text{base}\\\text{or alkali}\end{array}\right\} + \text{acid} \rightarrow \text{SALT} + \text{water}$$

7. What are the 2 common insoluble chlorides?
8. What are the common insoluble sulphates?
9. What are the 3 common soluble carbonates?
10. What are the 3 common soluble metallic hydroxides?
11. What are the common amphoteric oxides?

(Answers on page 497.)

Common gases

There are about a dozen common gases.

Consider the following types of apparatus (Figures 199, 200, and 201). Look at them for a few minutes and then close your book. Make a fair copy of each one on a separate page of your exercise book. Under each diagram make a table of gases which can be prepared using this type of apparatus. (Some gases may come under two headings, according to state of purity required.) Fill in the columns with the details required. The answers are given on page 496.

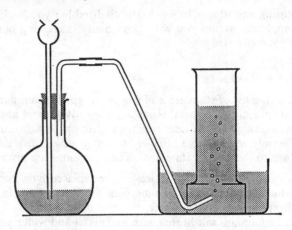

Figure 199

Example:

Gas	Materials under equation	Test

H_2	$Zn(s) + H_2SO_4(aq) \rightarrow ZnSO_4(aq) + H_2(g)$
	zinc dilute zinc hydrogen
	      sulphuric sulphate
	      acid
		Explodes with air when
		flame applied

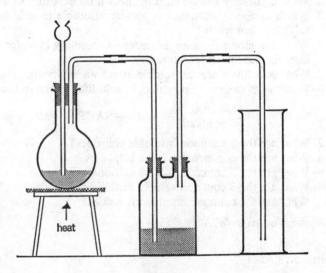

Figure 200

If you would use a second wash-bottle, state in third column the liquid you would put into it.

Gas	Materials under equation	Test

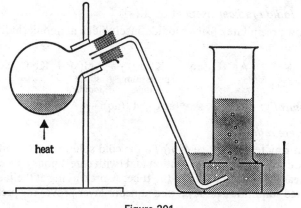

heat

Figure 201

Gas	Materials under equation	Test

Tests for acid radicals in solution (anions)

1. Sulphate radical

To the solution of the suspected sulphate in water, add dilute hydrochloric acid and then barium chloride solution. (Barium nitrate solution and dilute nitric acid can also be used.) A white precipitate of barium sulphate which is insoluble in excess acid indicates the presence of a sulphate in the original solution. For example,

$$BaCl_2(aq) + Na_2SO_4(aq) \rightarrow BaSO_4(s) + 2NaCl(aq)$$

barium   sodium   barium   sodium
chloride   sulphate   sulphate   chloride
(INSOLUBLE)

ionically: $\quad Ba^{2+}(aq) + SO_4^{2-}(aq) \rightarrow BaSO_4(s)$

2. (a) Chloride radical

To the solution of the suspected chloride in water, add dilute nitric acid, followed by silver nitrate solution. A white curdy precipitate of silver chloride (soluble in dilute ammonia) indicates the presence of the chloride radical in the original solution. For example,

$$AgNO_3(aq) + KCl(aq) \rightarrow AgCl(s) + KNO_3(aq)$$

silver   potassium   silver   potassium
nitrate   chloride   chloride   nitrate
(INSOLUBLE)

ionically: $\quad Ag^+(aq) + Cl^-(aq) \rightarrow AgCl(s)$

2. (b) Bromide radical. Repeat as 2. (a)

A pale yellow precipitate of silver bromide, sparingly soluble in concentrated ammonia, indicates the presence of a bromide. For example,

$$AgNO_3(aq) + NaBr(aq) \rightarrow AgBr(s) + NaNO_3(aq)$$

sodium   silver
bromide   bromide
(INSOLUBLE)

ionically: $\quad Ag^+(aq) + Br^-(aq) \rightarrow AgBr(s)$

2. (*c*) *Iodide radical*. Repeat as 2.(*a*)

A yellow precipitate of silver iodide, insoluble in ammonia, indicates the presence of an iodide.

$$AgNO_3(aq) + KI(aq) \rightarrow AgI(s) + KNO_3(aq)$$

potassium silver
iodide   iodide

ionically: $Ag^+(aq) + I^-(aq) \rightarrow AgI(s)$

3. *Nitrate radical*

(Brown ring test, see page 440.) To the cold solution of the nitrate in a boiling-tube add iron(II) sulphate solution and (with care!) pour concentrated sulphuric acid steadily down the side of the tube. A brown ring at the junction of the concentrated sulphuric acid layer and aqueous layer proves the presence of a nitrate.

4. *Carbonate radical*

Add dilute nitric acid to the substance in a test-tube (or to its solution in water). Effervescence is observed and carbon dioxide is evolved which, if passed into lime water, gives a white precipitate of chalk. For example,

$$CuCO_3(s) + 2HNO_3(aq) \rightarrow Cu(NO_3)_2(aq) + H_2O(l) + CO_2(g)$$

carbon
dioxide

ionically: $CuCO_3(s) + 2H^+(aq) \rightarrow Cu^{2+}(aq) + H_2O(l) + CO_2(g)$

For *sulphite* test, see page 408.

Tests for metal ions (cations)

The following are some simple tests for the cations present in single salts (not mixtures). The metallic ions are in combination, not free as elements. Thus the test for potassium will be given by any salt or compound of potassium.

Flame coloration (see page 388)

Lithium	Carmine (deep red).
Potassium	Lilac. Visible through blue glass.
Sodium	Persistent golden yellow.
Calcium	Brick-red
Copper	Green or bluish green

Action of sodium hydroxide solution on solution of soluble salt of metal

Zinc salt. White precipitate of zinc hydroxide. Soluble in excess of alkali. Precipitate ignited in crucible forms zinc oxide (yellow hot, white cold). For example:

$$ZnSO_4(aq) + 2NaOH(aq) \rightarrow Na_2SO_4(aq) + Zn(OH)_2(s)$$

zinc
hydroxide

ionically: $Zn^{2+}(aq) + 2OH^-(aq) \rightarrow Zn(OH)_2(s)$

$$Zn(OH)_2(s) \rightarrow ZnO(s) + H_2O(g)$$

zinc
oxide

Iron. Iron(II) salts. Green gelatinous precipitate of iron(II) hydroxide.
Iron(III) salts. Reddish-brown gelatinous precipitate of iron(III) hydroxide.

For example:

$$FeSO_4(aq) + 2NaOH(aq) \rightarrow Fe(OH)_2(s) + Na_2SO_4(aq)$$
iron(II)
hydroxide

ionically: $Fe^{2+}(aq) + 2OH^-(aq) \rightarrow Fe(OH)_2(s)$

$$FeCl_3(aq) + 3NaOH(aq) \rightarrow Fe(OH)_3(s) + 3NaCl(aq)$$
iron(III)
hydroxide

ionically: $Fe^{3+}(aq) + 3OH^-(aq) \rightarrow Fe(OH)_3(s)$

Lead(II) salts. White precipitate of lead(II) hydroxide soluble in excess of alkali. Precipitate ignited in crucible forms lead(II) oxide, a yellow powder. For example:

$$Pb(NO_3)_2(aq) + 2NaOH(aq) \rightarrow 2NaNO_3(aq) + Pb(OH)_2(s)$$

ionically: $Pb^{2+}(aq) + 2OH^-(aq) \rightarrow Pb(OH)_2(s)$
$$Pb(OH)_2(s) \rightarrow PbO(s) + H_2O(g)$$

Ammonium ion, NH_4^+. *Boil* with sodium hydroxide solution. Ammonia gas (turns red litmus blue) evolved. For example:

$$NH_4Cl(aq) + NaOH(aq) \rightarrow NH_3(g) + H_2O(l) + NaCl(aq)$$
ammonia

ionically: $NH_4^+(aq) + OH^-(aq) \rightarrow NH_3(g) + H_2O(l)$

$$(NH_4)_2SO_4(aq) + 2NaOH(aq) \rightarrow 2NH_3(g) + 2H_2O(l) + Na_2SO_4(aq)$$

It will be seen that the common metal ions (if present singly) can be detected either by

(*a*) Flame coloration or
(*b*) Action of sodium hydroxide solution.

Action of heat on common substances

Basic oxides of metals

No action except

$$2HgO(s) \rightarrow 2Hg(l) + O_2(g) \quad \text{(silver similarly)}$$
mercury(II)   mercury   oxygen
oxide

Peroxides and dioxides of metals

All decompose giving off oxygen (except sodium peroxide). For example:

$$2H_2O_2(aq) \rightarrow 2H_2O(l) + O_2(g)$$
hydrogen   water   oxygen
peroxide

$$2PbO_2(s) \rightarrow 2PbO(s) + O_2(g)$$
lead(IV)   lead(II)   oxygen
oxide   oxide

Hydroxides

These decompose under action of heat (except sodium hydroxide and potassium hydroxide) to give the oxide.

$$Cu(OH)_2(s) \rightarrow CuO(s) + H_2O(g)$$

Chlorides

No action except on ammonium chloride, which sublimes and dissociates.

$$NH_4Cl(s) \rightleftharpoons NH_3(g) + HCl(g)$$

Nitrates

(a) Sodium and potassium nitrates.

$$2KNO_3(s) \rightarrow 2KNO_2(s) + O_2(g)$$
potassium   oxygen
nitrite

(b) Heavy metal nitrates, *e.g.*,

$$2Pb(NO_3)_2(s) \rightarrow 2PbO(s) + 4NO_2(g) + O_2$$ (nitrates of silver and
lead(II)   nitrogen   oxygen   mercury decompose
oxide   dioxide   to metal)

(c) Ammonium nitrate.

$$NH_4NO_3(s) \rightarrow N_2O(g) + 2H_2O(g)$$
dinitrogen   water
oxide

Sulphates

$$2FeSO_4(s) \rightarrow Fe_2O_3(s) + SO_2(g) + SO_3(g)$$
iron(II)   iron(III)   sulphur   sulphur
sulphate   oxide   dioxide   trioxide

$$Fe_2(SO_4)_3(s) \rightarrow Fe_2O_3(s) + 3SO_3(g)$$
iron(III) sulphate

Carbonates

All decompose except sodium and potassium carbonate, *e.g.*,

$$CuCO_3(s) \rightarrow CuO(s) + CO_2(g)$$
copper(II)   copper(II)   carbon
carbonate   oxide   dioxide

Hydrogencarbonates (bicarbonates)

Decompose to give carbonate, water, and carbon dioxide, *e.g.*,

$$2NaHCO_3(s) \rightarrow Na_2CO_3(s) + H_2O(l) + CO_2(g)$$

Ammonium salts

These always decompose; sometimes sublime, *e.g.*, ammonium chloride.

$$NH_4Cl(s) \rightleftharpoons NH_3(g) + HCl(g)$$
ammonium
chloride

Ammonium nitrate. See Nitrates.

Metals

All metals oxidize if heated in air, except silver and gold, to form *basic oxides*, *e.g.*,

$$2Ca(s) + O_2(g) \rightarrow 2CaO(s)$$
calcium   calcium
oxide

$$2Mg(s) + O_2(g) \rightarrow 2MgO(s)$$
magnesium   magnesium
oxide

Non-metals

Carbon, phosphorus, and sulphur combine with the oxygen of the air to form *acidic oxides, e.g.,*

$$S(s) + O_2(g) \rightarrow SO_2(g)$$
$$\text{sulphur} \qquad\qquad \text{sulphur}$$
$$\text{dioxide}$$

Miscellaneous

Hydrates (that is, salts possessing water of crystallization) give off water vapour. For example:

$$Na_2CO_3.10H_2O(s) \rightarrow Na_2CO_3(s) + 10H_2O(g)$$
$$\text{sodium carbonate} \qquad \text{anhydrous} \qquad \text{water}$$
$$\text{crystals} \qquad\qquad \text{sodium}$$
$$\text{carbonate}$$

Potassium chlorate.

$$2KClO_3(s) \rightarrow 2KCl(s) + 3O_2(g)$$
$$\text{potassium} \qquad \text{potassium} \quad \text{oxygen}$$
$$\text{chlorate} \qquad\quad \text{chloride}$$

Red lead.

$$2Pb_3O_4(s) \rightarrow 6PbO(s) + O_2(g)$$
$$\text{red lead} \qquad \text{lead(II)} \quad \text{oxygen}$$
$$\text{oxide}$$

Answers

Gas preparations

The following remarks should be considered together with the diagrams on pages 490–1.

General notes on preparations of gases

1. Do not collect gases of approximately the same density as air by displacement of air. If the gases are required dry they must be dried by a suitable substance and collected in a syringe. This refers to O_2, N_2O, NO, N_2.

The following gases may be dried with concentrated sulphuric acid (symbol or other formula only given):

$$O_2 \quad H_2 \quad N_2 \quad N_2O \quad HCl \quad CO_2 \quad SO_2 \quad Cl_2 \quad CO$$

For the following, use special drying agents.

   Ammonia – quicklime.
   Hydrogen sulphide – anhydrous calcium chloride (quite satisfactory in practice).

2. **Never dry a gas and then collect it over water.**

3. Draw your diagrams with the following points in mind.

(*a*) The apparatus stands on the bench. Do not draw, for example, a washbottle in mid-air.

(*b*) Show a clear way through for the gas to pass. A diagram should be a section and not a pictorial illustration. Do not waste time on non-essentials.

(*c*) Label the diagram clearly. It makes a description much briefer and leaves no room for doubt.

(*d*) Indicate simply how the apparatus is supported, for example, tripod and gauze, clamps, etc.

4. If a steady supply of a dry gas is wanted (for example, hydrogen, or carbon monoxide for reduction of oxides) an anhydrous calcium chloride tube is to be preferred, as the drying apparatus, to a Woulff's bottle containing concentrated sulphuric acid. The latter gives a jerky supply of the gas, the former a steady supply.

5. Give the equation for the reaction concerned on or near the diagram. It is then clear to which action the equation refers, and you have then completed **two of the important steps in the description of any chemical process.**

Figure 199. Flask, thistle funnel – no heat – collect over water.

Gas	Materials under equation	Test
H_2	$Zn(s) + H_2SO_4(aq) \rightarrow ZnSO_4(aq) + H_2(g)$ zinc dilute zinc hydrogen sulphuric acid sulphate	Explodes with air when flame applied.
H_2S	$FeS(s) + 2HCl(aq) \rightarrow FeCl_2(aq) + H_2S(g)$ iron(II) fairly conc. iron(II) hydrogen sulphide hydrochloric chloride sulphide acid	Blackens lead(II) acetate paper.
CO_2	$CaCO_3(s) + 2HCl(aq) \rightarrow$ marble dilute hydrochloric acid $CaCl_2(aq) + H_2O(l) + CO_2(g)$ calcium carbon chloride dioxide	Turns lime water turbid.
NO	$3Cu(s) + 8HNO_3(aq) \rightarrow$ copper fairly conc. nitric acid $3Cu(NO_3)_2(aq) + 4H_2O(l) + 2NO(g)$ copper(II) nitrogen nitrate oxide	Forms brown fumes on exposure to air.
Cl_2	$2KMnO_4(aq) + 16HCl(aq) \rightarrow$ potassium conc. permanganate hydrochloric acid $2KCl(aq) + 2MnCl_2(aq) + 8H_2O(l) + 5Cl_2(g)$ potassium chlorine chloride	Bleaches damp litmus (greenish-yellow gas) (Collect over brine.)

Figure 200. Flask heated, wash-bottles, collect by displacement of air.

Gas	Materials under equation	Test
HCl	$NaCl(s) + H_2SO_4(aq) \rightarrow NaHSO_4(aq) + HCl(g)$ sodium conc. sodium hydrogen chloride sulphuric hydrogen- chloride acid sulphate (Heat not essential.)	Gives white precipitate of AgCl with silver nitrate in nitric acid solution. (Concentrated sulphuric acid in wash-bottle.)
SO_2	$Cu(s) + 2H_2SO_4(aq) \rightarrow$ copper hot conc. sulphuric acid $CuSO_4(aq) + 2H_2O(l) + SO_2(g)$ sulphur dioxide	Decolorizes potassium permanganate without precipitate of sulphur. (Concentrated sulphuric acid in wash-bottle.)

Gas	Materials under equation	Test

Cl_2	$MnO_2(s) + 4HCl(aq) \longrightarrow$
	man- conc.
	ganese(IV) hydrochloric
	oxide acid (or
	common salt
	and conc.
	sulphuric acid)
	      $MnCl_2(aq) + 2H_2O(l) + Cl_2(g)$
	    manganese(II) chlorine
	    chloride
		Bleaches damp litmus.
		(Water in first bottle and
		concentrated sulphuric
		acid in second.)
		Chlorine attacks mercury.

CO	$H_2C_2O_4(s) - H_2O(l) \longrightarrow CO(g) + CO_2(g)$ ethanedioic (oxalic) carbon carbon acid and conc. monoxide dioxide sulphuric acid to remove the elements of water	Burns with blue flame to carbon dioxide (pass through two bottles of caustic potash solution and collect over water).

Figure 201. Hard glass tube or small flask. Collect over water.

Gas	Materials under equation	Test
O_2	$2KClO_3(s) \xrightarrow{MnO_2} 2KCl(s) + 3O_2(g)$ potassium potassium oxygen chlorate chloride	Rekindles glowing splint – not soluble in water.
N_2	$2KNO_2/(NH_4)_2SO_4(aq) \longrightarrow 2N_2(g) + 4H_2O(l) + K_2SO_4(aq)$ ammonium nitrite nitrogen steam potassium solution sulphate	Inert gas. Gives negative test with splint and lime-water.
N_2O	$2KNO_3/(NH_4)_2SO_4(aq) \longrightarrow 2N_2O(g) + 4H_2O(l) + K_2SO_4(aq)$ ammonium nitrate dinitrogen steam potassium solution oxide sulphate (do not heat flask to dryness)	Rekindles glowing splint – soluble in cold water.

Special diagrams for ammonia and carbon monoxide. (See pages 426 and 315.)

Answers to general statements

(*N.B.* Common exceptions only given.) (See page 489.)

1–5. No exceptions.

6. Silver nitrate yields the metal. Mercury(II) nitrate similar.

$$2AgNO_3(s) \longrightarrow 2Ag(s) + 2NO_2(g) + O_2(g)$$

7. With concentrated nitric acid iron becomes 'passive' (see page 482).

8. Certain salts, being insoluble, form a protective layer round the carbonate, preventing the action of the acid. For example: calcium carbonate and dilute sulphuric acid, barium carbonate and dilute sulphuric acid.

Answers to particular statements

(*N.B.* Common exceptions only given.)

1. Ammonia.
2. Nitrogen oxide.
3. Lead(II) chloride (also lead(II) iodide).
4. Common salt.
5. All gases.
6. Lead(II) chloride, lead(II), calcium, and barium sulphates. (Carbonates are not usually prepared by this method.)

7. Silver chloride, lead(II) chloride.
8. Lead(II) sulphate, barium sulphate.
9. Sodium, potassium, and ammonium carbonates.
10. Sodium, potassium, and calcium hydroxides. (Calcium hydroxide is only slightly soluble.)
11. Zinc oxide, lead(II) oxide, aluminium oxide.

Relative Atomic Masses

(Scaled to the relative atomic mass $^{12}C = 12$ exactly)

Values quoted in the table, unless marked * or †, are reliable to at least ± 1 in the fourth significant figure. A number in parentheses denotes the atomic mass number of the isotope of longest known half-life.

Atomic Number	Name	Symbol	Relative Atomic Mass	Atomic Number	Name	Symbol	Relative Atomic Mass
1	Hydrogen	H	1.008	53	Iodine	I	126.9
2	Helium	He	4.003	54	Xenon	Xe	131.3
3	Lithium	Li	6.941*†	55	Caesium	Cs	132.9
4	Beryllium	Be	9.012	56	Barium	Ba	137.3
5	Boron	B	10.81†	57	Lanthanum	La	138.9
6	Carbon	C	12.01	58	Cerium	Ce	140.1
7	Nitrogen	N	14.01	59	Praseodymium	Pr	140.9
8	Oxygen	O	16.00	60	Neodymium	Nd	144.2
9	Fluorine	F	19.00	61	Promethium	Pm	(145)
10	Neon	Ne	20.18	62	Samarium	Sm	150.4
11	Sodium	Na	22.99	63	Europium	Eu	152.0
12	Magnesium	Mg	24.31	64	Gadolinium	Gd	157.3
13	Aluminium	Al	26.98	65	Terbium	Tb	158.9
14	Silicon	Si	28.09	66	Dysprosium	Dy	162.5
15	Phosphorus	P	30.97	67	Holmium	Ho	164.9
16	Sulphur	S	32.06†	68	Erbium	Er	167.3
17	Chlorine	Cl	35.45	69	Thulium	Tm	168.9
18	Argon	Ar	39.95	70	Ytterbium	Yb	173.0
19	Potassium	K	39.10	71	Lutetium	Lu	175.0
20	Calcium	Ca	40.08†	72	Hafnium	Hf	178.5
21	Scandium	Sc	44.96	73	Tantalum	Ta	180.9
22	Titanium	Ti	47.90*	74	Tungsten	W	183.9
23	Vanadium	V	50.94		(Wolfram)		
24	Chromium	Cr	52.00	75	Rhenium	Re	186.2
25	Manganese	Mn	54.94	76	Osmium	Os	190.2
26	Iron	Fe	55.85	77	Iridium	Ir	192.2
27	Cobalt	Co	58.93	78	Platinum	Pt	195.1
28	Nickel	Ni	58.70	79	Gold	Au	197.0
29	Copper	Cu	63.55	80	Mercury	Hg	200.6
30	Zinc	Zn	65.38	81	Thallium	Tl	204.4
31	Gallium	Ga	69.72	82	Lead	Pb	207.2†
32	Germanium	Ge	72.59*	83	Bismuth	Bi	209.0
33	Arsenic	As	74.92	84	Polonium	Po	(209)
34	Selenium	Se	78.96*	85	Astatine	At	(210)
35	Bromine	Br	79.90	86	Radon	Rn	(222)
36	Krypton	Kr	83.80	87	Francium	Fr	(223)
37	Rubidium	Rb	85.47	88	Radium	Ra	(226)
38	Strontium	Sr	87.62†	89	Actinium	Ac	(227)
39	Yttrium	Y	88.91	90	Thorium	Th	232.0
40	Zirconium	Zr	91.22	91	Protactinium	Pa	(231)
41	Niobium	Nb	92.91	92	Uranium	U	238.0†
42	Molybdenum	Mo	95.94*	93	Neptunium	Np	(237)
43	Technetium	Tc	(97)	94	Plutonium	Pu	(244)
44	Ruthenium	Ru	101.1	95	Americium	Am	(243)
45	Rhodium	Rh	102.9	96	Curium	Cm	(247)
46	Palladium	Pd	106.4	97	Berkelium	Bk	(247)
47	Silver	Ag	107.9	98	Californium	Cf	(251)
48	Cadmium	Cd	112.4	99	Einsteinium	Es	(254)
49	Indium	In	114.8	100	Fermium	Fm	(257)
50	Tin	Sn	118.7	101	Mendelevium	Md	(258)
51	Antimony	Sb	121.8	102	Nobelium	No	(259)
52	Tellurium	Te	127.6	103	Lawrencium	Lr	(260)

IUPAC 1975

Approximate Relative Atomic Masses

Aluminium	27	Helium	4	Nickel	59	Uranium	238
Argon	40	Hydrogen	1	Nitrogen	14	Xenon	131
Barium	137	Iodine	127	Oxygen	16	Zinc	65
Bismuth	209	Iron	56	Phosphorus	31		
Bromine	80	Krypton	84	Potassium	39		
Calcium	40	Lead	207	Silicon	28		
Carbon	12	Lithium	7	Silver	108		
Chlorine	35.5	Magnesium	24	Sodium	23		
Copper	63.5	Manganese	55	Sulphur	32		
Fluorine	19	Mercury	201	Tin	119		
Gold	197	Neon	20	Titanium	48		

Answers to Numerical Questions

Chapter 2, page 29
8. 3.17g; 1.59 g; X_2O
10. X_2O_3

Chapter 3, page 41
1. (a) 253.3 cm³; (b) 1638 cm³;
   (c) 1110 cm³; (d) 450 cm³; (e) 570 cm³;
   (f) 637 cm³; (g) 642 cm³; (h) 133 cm³;
   (i) 51.2 cm³; (j) 626 cm³; (k) 76.1 cm³;
   (l) 118 cm³; (m) 518 cm³
   (Answers to parts (g) to (m) are given to three significant figures.)
5. 800 mm Hg; 800 mm Hg
6. $\frac{273}{293}$ and $\frac{750}{760}$

Chapter 4, page 46
1.

Element	Mass /g	Number of moles	Number of particles
Sodium	9.2	0.4	2.4×10^{23}
Gold	0.394	2×10^{-3}	1.2×10^{21}
Iron	0.187	3.3×10^{-3}	2×10^{21}
Uranium	0.119	5×10^{-4}	3×10^{20}
Tin	1.98	0.0167	10^{22}
Silver	594	5.5	3.3×10^{24}
Copper	2.54	0.04	2.4×10^{21}
Helium	20	5	3×10^{24}
Carbon	0.72	6×10^{-2}	3.6×10^{22}
Lead	18.63	0.09	5.4×10^{22}
Potassium	26	0.67	4×10^{23}
Xenon	52.5	0.4	2.4×10^{23}
Mercury	0.167	8.33×10^{-4}	5×10^{20}
Hydrogen, H_2	0.2	0.1	6×10^{22}

2. 667 g

5. (a) 0.01X; (b) 0.8X; (c) 0.08X

Chapter 5, page 54

6. FeS_2

7. (i) 104 g in A, 208 g in B; (iii) XO_2

8. CuO; 16 g

12. A, PbO; B, PbO_2

Chapter 6, page 63

3. (b) 63.6 cm³; (c) 3508 s

4. (a) (i) 340 g; (ii) 216 g of Ag, 92 g of NO_2; (iii) 2 moles Ag, 2 moles NO_2, 1 mole O_2

5. 16.0 g; copper(II) oxide

6. (c) (i) 600 cm³, (ii) 20 cm³; (d) 25 cm³; (e) (i) 1 dm³, (ii) 24 dm³

Chapter 7, page 69

1. 15.875 tonnes

2. 1.55 g

3. (a) $ZnCl_2$; (b) NaCl; (c) $CuSO_4$; (d) PbN_2O_6

4. (a) Na, 27.38%; H, 1.19%; C, 14.29%; O, 57.14%
(b) Ca, 36.04%; Cl, 63.96%
(c) N, 21.21%; H, 6.06%; S, 24.24%; O, 48.48%
(d) Na, 29.11%; S, 40.51%; O, 30.38%

5. 63 g

6. 2.67 g

7. 20.04 g; 40.08 g

8. 3.085 g

9. 15.89 g; slaked lime by 5.42 g

10. 73 g

11. 97.5 %

12. $H_4N_2O_3$

13. 2.0 g

14. $x = 1$

15. PbO_2

16. $4X + 5O_2(g) \rightarrow 2X_2O_5$

17. 21.21%

18. 2 g; 1.4 dm³

19. (c) (i) 6.21 g; (ii) 0.96 g; (iii) 32 g; (iv) 2 moles; (v) PbO_2, lead(IV) oxide
(d) (i) 1 mole; (ii) PbO, lead(II) oxide

20. 2.65 g

Chapter 8, page 76

5. (a) 47; (b) 47; (c) 60; (d) 62. Ag = 108

Chapter 11, page 115

1. (a)17 g; (b) 34 g; (c) 28 g; (d) 71 g; (e) 44 g

2. (a) 32; (b) 64; (c) 30

3. C_2H_6

5. 71

6. (a) 23; 46

Chapter 12, page 120

1. 33.6 dm³

2. 1.083 dm³

3. 71

4. 1.55 dm³

5. 3.5 dm³; 2 dm³

6. 2.8 dm³

8. 4

9. 124

11. (a) 1.63 g; (b) 36.5

14. (b) 58, $x = 4$; (c) 0.70 g

Chapter 13, page 131

5. 0.515 g

7. (i) CO_2, 15 cm³; O_2, 50 cm³

8. (a)empirical formula H_2Se; (b) molecular formula H_2Se; (d)100 cm³

10. (b) 10 cm³, 20 cm³; (d) 10 cm³; (f) CH_4

Chapter 14, page 139

2. 0.0943M

6. 0.3M; 31.8 g dm⁻³

7. 30 cm³

8. 8g

9. (c) (i) 2.7×10^{-3}; (ii) 0.108M

10. (a) 5 dm³; (b) 31.25 cm³; (c) 50 cm³

Chapter 15, page 145

1. 0.25M

2. 6.27 g

3. (a) 0.0714M; (b) 4.00 g

4. 25.0 cm³

5. 18.75 cm³

6. 24.5 cm³

7. 0.217M; 7.92 g

8. 0.13M; 12.7 g

9. (a) 419 cm³; (b) 0.99M

10. 56

11. (i) 0.2M, (ii) 21.2 g dm⁻³

12. (i) 0.05M, (ii) 126

13. (a) 1.25M; 45.6 g dm⁻³; (b) 700 cm³

14. 20.8 cm³; 0.833M

15. 0.2M; (i) 112 cm³, (ii) 2.12 g

16. H_3PO_3; 2M

17. (b) (i) 4.32 g dm⁻³, (ii) 40 cm³

18. (a) 2.24 dm³; (b) 40 cm³

19. 197 cm³

Chapter 16, page 167

6. 22.4 dm³

7. 10.8 g; 560 cm³

8. (a) 3; (b) 1

11. 0.3175 g Cu; 0.08 g O

14. (d) 0.01 mole H_2; 0.005 mole O_2; 224 cm³ H_2, 112 cm³ O_2; (e) 965 s; (f) 0.01 mole H_2O

15. (a) 2 Faradays, 1 Faraday

16. (a) 2; (b) X^{2+}; (c) $X(OH)_2$

17. 6.35 g Cu; 2.24 dm³ H_2

18. (a) 31.75 g; (b) 9.0 g; (c) 11.2 dm³; (d) 5.6 dm³

25. 3.36 dm³

Chapter 17, page 178

13. 101.6 g

Chapter 19, page 199

5. 2.24 dm³
10. 1.12 dm³
14. 1.12 dm³; 2.24 dm³

Chapter 20, page 210

7. 1920 kJ
9. (*b*) methanol, 723 kJ mol⁻¹; ethanol, 1366 kJ mol⁻¹; propanol, 2004 kJ mol⁻¹
(*c*) approx. 2644 kJ mol⁻¹
10. (*a*) Mg, 4.17; Al, 3.70; Cu, 1.56; Mo, 1.04; Pt, 0.513
(*c*) about 2.2 moles in 100 g, hence Sc ≈ 45.5
(*d*) approx. 2.2 moles Sc, 6.65 moles Cl; ScCl₃
11. (*a*) −265.0 kJ; (*b*) −882.0 kJ mol⁻¹
12. (*a*) −57.3 kJ; (*b*) −114.6 kJ

Chapter 23, page 260

15. (*b*) (i) approx. 78 °C; (ii) approx. 75 g; (iii) approx. 57 °C
16. (*a*) 7 g/100 g water; (*g*) 23 g
17. (*c*) MgSO₄.7H₂O
18. 20 g/100 g water
22. (*a*) 25 g/100 g water at 70 °C; 14.9 g/100 g water at 40 °C
(*c*) (i) 2 g; (ii) 15 g
24. 0.212 g

Chapter 24, page 279

9. (*e*) 5.85 g
10. 90; 42.4%
11. 30 g ethanoic acid; 20 cm³ molar HCl; 224 cm³ CO₂
12. *n* = 2
13. (*f*) 287; (*g*) 57.4 g
18. 25 g

Chapter 25, page 298

1. (i) 211 cm³; (ii) 31.8
12. 150 cm³
14. 20 cm³ O₃; (*a*) 90 cm³; (*b*) 0.227 g
17. Fe₂O₃
19. Fe₂O₃
24. (*a*) 11.2 dm³; (*b*) 22.4 dm³

Chapter 26, page 306

1. (*b*) 0.65 g

6. (i) 22.4 dm³; (ii) 44.8 dm³; (iii) 44.8 dm³
7. 24 g

Chapter 27, page 324

1. (*c*) CH₃; C₂H₆
8. 11.2 dm³ CO; 11.2 dm³ H₂; 11.2 dm³ CO₂
11. (*a*) Yes; 560 cm³
16. (*d*) 100

Chapter 28, page 353

11. 1.5 dm³; 200 cm³

Chapter 30, page 377

10. (*c*) *A* IO₂(I₂O₄); *B* I₂O₅.
10 IO₂(s) → 4I₂O₅(s) + I₂(s)

Chapter 33, page 419

5. 10%
21. Mo = 96
22. 400 cm³ of M NaOH
23. 42.5%
25. (*e*) 2.8 dm³
34. (*f*) 0.2 moles

Chapter 34, page 444

4. 388 g
6. 35%
8. (*j*) 11.2%
10. 82 g Ca(NO₃)₂; 85 g NaNO₃
12. Volume increases by 100 cm³
16. (*b*) (i) 4.25 g; (ii) 4.25 g
17. (*c*) 1.2 dm³
20. (*d*) 120 cm³

Chapter 35, page 452

9. (*f*) 19.6 g; 1 mole

Chapter 36, page 467

2. 529 tons Al; 700 tons Fe
3. 1.8 g
7. (*b*) 667 g Zn; 6.67 g SO₂
8. 52.5 kg

Chapter 37, page 485

2. (*b*) 4.6 g Na; 40.8 g ethanol
3. (*e*) 2 moles Cl atoms; 132.5

General Questions

1. 1.60 g of the oxide of a metal M gave 1.28 g of the metal (M = 64) when reduced in hydrogen. Deduce from the data the formula of the oxide and write the equation for the reduction. (O = 16.)

2. 3.60 g of magnesium were burnt in air. After addition of water (to eliminate nitride), drying, and heating to constant mass, 6.00 g of oxide remained. From the data, deduce the simplest formula of magnesium oxide. (Mg = 24; O = 16.) Write the corresponding formula for the nitride and give the equation for its decomposition by water (to ammonia and magnesium hydroxide).

3. 0.888 g of a metal M (M = 64) were converted to its oxide, which weighed exactly 1 g. Show from the data that the simplest formula of the oxide is M_2O and write corresponding formulae for its chloride and sulphate. (O = 16.)

4. 0.225 g of a metal M (of relative atomic mass 27) liberated 303 cm³ of hydrogen (measured dry at 17 °C and 740 mmHg) from excess of dilute hydrochloric acid. Deduce from the data the corresponding equation and write formulae for the oxide and sulphate of M. (22.4 dm³ of hydrogen at s.t.p. weigh 2.016 g.)

5. When 1.95 g of metal M (M = 65) reacted with excess of hot copper(II) sulphate solution, the mass of copper deposited (when purified and dried) was 1.92 g (Cu = 64). Show that these figures correspond to the equation:

$$CuSO_4 + M \rightarrow Cu + MSO_4.$$

6. 1.30 g of zinc containing carbon as impurity was dissolved in nitric acid. After filtration and washing of the residue, the entire liquid was evaporated to dryness and the solid ignited. The mass of zinc oxide left was 1.59 g. Calculate the percentage of pure zinc in the sample. (Zn = 65; O = 16.)

7. 2.33 g of a metal M (M = 116) were converted to its anhydrous chloride and, by heating with concentrated sulphuric acid, to 5.22 g of the anhydrous sulphate of M. From the data, deduce the equation for the conversion of the chloride to sulphate and write the formulae of the corresponding oxide an nitrate of M. (SO_4 = 96.)

8. Two oxides of a metal contained respectively 92.59% and 96.15% of the metal. Show these facts to agree with the law of multiple proportions.

9. One oxide of phosphorus contains 43.6% of oxygen, whilst another contains 43.6% of phosphorus. Show this is in accordance with the law of multiple proportions.

10. Iron(II) chloride contains 56.0% of chlorine by mass and 1 g of it was converted into iron(III) chloride by the action of chlorine, 97 cm³ of which were absorbed (measured at 17 °C and 735 mmHg). Show the above to agree with the law of multiple proportions.

11. A metal, M, formed a chloride which, on analysis, was found to contain 55.6% by mass of chlorine. When heated in dry hydrogen chloride, 0.800 g of M yield 1.814 g of a chloride; when heated in dry chlorine, 0.700 g of M yielded 2.031 g of a chloride. In each of the three cases, calculate the mass of chlorine which combines with one gramme of M. Are these figures in accord with one or more chemical laws?

12. Four samples of oxides of a certain metal were reduced to the metal with results quoted below. In each case, calculate the mass of metal which combines with one gramme of oxygen. Are the figures in accord with one or more chemical laws?

	Mass of oxide (g)	Mass of metal (g)
(a)	2.973	2.760
(b)	3.585	3.105
(c)	2.230	2.070
(d)	9.133	8.280

13. What mass of metallic mercury will be obtained by completely decomposing 1.08 g of mercury(II) oxide by heating?

14. 4.6 g of metallic sodium are converted into common salt by heating sodium in a current of chlorine. What mass of common salt is obtained?

15. 0.6 g of magnesium is heated in a current of oxygen. What mass of magnesium oxide will remain?

16. What is the percentage composition by mass of anhydrous zinc sulphate?

17. What mass of dilute nitric acid containing 20% of the pure acid will be needed to dissolve 10 g of calcium carbonate?

18. 17.4 g of pyrolusite, containing 80% by mass of manganese(IV) oxide and the rest inactive matter, were warmed with excess of concentrated hydrochloric acid. What is the volume of chlorine given off at s.t.p.?

19. What mass of ammonia would be evolved by heating together 100 g of ammonium chloride and 70 g of calcium hydroxide? Which of the reagents is in excess and by what mass?

20. 2.10 g of a mixture of potassium chlorate and potassium chloride were heated until the evolution of oxygen was complete. This residue weighed 1.62 g. What is the percentage by mass of potassium chloride in the mixture?

21. 4.6 g of sodium are added slowly to 10 g of water in a small dish. What is the total mass of the solution left when the action is complete?

22. What mass of water would be required to react with the calcium oxide which could be obtained by strongly heating 100 g of calcium carbonate?

23. What mass of copper would be required to react with 100 g of nitric acid, and what mass of copper(II) nitrate would be formed? Use the equation

$$3Cu + 8HNO_3 \rightarrow$$
$$3Cu(NO_3)_2 + 4H_2O + 2NO.$$

24. 0.920 g of a mixture of metallic copper and copper(II) oxide was heated in a stream of hydrogen until reduction was complete. The residue weighed 0.752 g. What was the percentage of metallic copper in the original mixture?

25. 1.12 g of a mixture of sodium chloride and ammonium chloride was dissolved in water and an excess of silver nitrate solution was added to precipitate the silver chloride. This precipitate was found to weigh 2.87 g. What were the masses of sodium chloride and ammonium chloride in the original mixture?

26. On exposure to air, washing soda, $Na_2CO_3.10H_2O$, effloresces to leave sodium carbonate monohydrate. Calculate the *loss* of mass (*a*) per 7.15 g of washing soda, (*b*) per metric ton of washing soda if this process of efflorescence is completed.

27. Some zinc sulphate crystals were heated to constant mass with the following results:

Mass of crucible	20.00 g
Mass of crucible and crystals	25.74 g
Mass of crucible and residue	23.22 g

From the data, calculate x in the formula, $ZnSO_4.xH_2O$.

28. A powder contains sodium sulphate anhydrous and decahydrate. On heating to constant mass to produce the pure anhydrous salt, 2.50 g of the powder left 1.60 g of residue. Calculate the percentage of each form of sodium sulphate in the powder.

Assume in questions 29 to 52 that the molar volume of any gas at s.t.p. is 22.4 dm³. Assume the pressure is 1 atm and the temperature is that of a room, i.e. about 15 °C, unless otherwise stated.

29. What volume of oxygen at s.t.p. will be obtained by decomposing 4.32 g of mercury(II) oxide by the action of heat?

30. 50 cm³ of oxygen are mixed with 50 cm³ of hydrogen at ordinary room temperature and exploded. What is the volume of the residual gas on cooling to the original temperature?

31. What volume of carbon dioxide measured at s.t.p. can be obtained by heating 75 g of calcium carbonate to a high temperature?

32. What volume of carbon dioxide at s.t.p. could be obtained by dissolving 50 g of pure marble (calcium carbonate) in dilute hydrochloric acid?

33. 50 cm³ of ammonia are sparked continuously until there is no further volume change. If 98% by volume of the ammonia is decomposed into its elements, what is the final volume of gas?

34. What volume of hydrogen at s.t.p. could be obtained by the action of 2.32 g of magnesium on dilute hydrochloric acid?

35. 1000 cm³ of chlorine are mixed with 1500 cm³ of hydrogen sulphide, both gases at s.t.p. What is (*a*) the mass of sulphur deposited, (*b*) the volume and name of the gas left at s.t.p. if the reaction is complete?

36. 100 dm³ of sulphur dioxide measured at 0 °C and 740 mmHg are mixed with an equal volume of hydrogen sulphide at the same temperature and pressure. What is the mass of sulphur deposited if the reaction between the two gases is complete?

37. What volume of steam measured at 150 °C and 760 mmHg could be obtained by exploding 1 g of hydrogen with excess oxygen?

38. What volume of hydrogen measured at 20 °C and 760 mmHg would be obtained by dissolving 1.30 g of zinc in dilute sulphuric acid?

39. 50 cm³ of hydrogen are mixed with 10 cm³ of oxygen, both gases at 120 °C and 750 mmHg, and exploded. What is the total volume of gas left after cooling to the original

temperature and pressure? What percentage of this gas by volume is steam?

40. What mass of sodium nitrite would be necessary to react completely with 1.07 g of ammonium chloride? What volume of nitrogen measured at 25 °C and 730 mmHg would be given off if the mixture was dissolved in water and heated?

41. 200 cm³ of hydrogen were sparked with 75 cm³ oxygen, both gases measured at 14 °C. What was the total volume of residual gas (*a*) at 127 °C, (*b*) at 14 °C? Pressure constant throughout at 760 mmHg.

42. 73 g of ammonia are mixed with an equal mass of hydrogen chloride. What is the mass of ammonium chloride formed? What is the volume of gas remaining measured at 14 °C and 750 mmHg? What gas is it?

43. 5 g of zinc were added to dilute hydrochloric acid containing 10 g of the pure acid. Calculate (*a*) the volume of hydrogen given off at 15 °C and 770 mmHg, (*b*) the mass of pure acid remaining at the end of the action.

44. 2 g of a mixture of potassium chlorate and potassium chloride yielded, on heating, 336 cm³ of oxygen measured dry at 15 °C and 747 mmHg. What is the percentage by mass of potassium chlorate in the mixture?

45. To 40 cm³ of a mixture of hydrogen and nitrogen were added 50 cm³ oxygen. After explosion and cooling, the residual gases occupied 45 cm³. What is the percentage by volume of hydrogen in the original mixture? (All measurements at 15 °C and 760 mmHg.)

46. 16.32 dm³ of ammonia measured at 27 °C and 700 mmHg were passed slowly over heated copper(II) oxide. What volume of nitrogen measured at 35 °C and 700 mmHg pressure was obtained?

47. 20 g of benzene, C_6H_6, were completely burnt to carbon dioxide and water. What volume would be occupied at 127 °C and 740 mmHg by the residual gas?

48. 25 cm³ of air were mixed with 50 cm³ of hydrogen and the mixture was exploded. The volume of gas left was 60 cm³ after cooling and at the same conditions as before. What is the percentage by volume of oxygen in the air?

49. When dehydrated by hot, concentrated sulphuric acid, ethanedioic (oxalic) acid yields equal volumes of carbon monoxide and carbon dioxide. If 100 cm³ of this mixture is added to 50 cm³ of oxygen and exploded, what volume of gas remains? What volume of gas remains after absorption by excess of potassium hydroxide solution? (Measurements at constant temperature and pressure.)

50. 50 cm³ of a sample of water-gas containing carbon monoxide and hydrogen were mixed with 100 cm³ of oxygen and exploded. After cooling to the previous conditions, 100 cm³ of gas were left. After absorption with excess of potassium hydroxide solution, 75 cm³ of oxygen remained. What is the volume composition of the water-gas? Show that it satisfies all the measurements.

51. 100 cm³ of ozonized oxygen were taken under ordinary conditions and then kept at about 400 °C for some time. On cooling to the original conditions, the volume was 107 cm³. What would be the effect of exposing 25 cm³ of the original ozonized oxygen to turpentine for some time?

52. 50 cm³ of a mixture of hydrogen and hydrogen chloride under room conditions were exposed to sodium amalgam for some time. The volume of gas (under similar conditions) was then 42.5 cm³. If 100 cm³ of the original mixture was added to 50 cm³ of ammonia, what would be the volume of the final gaseous mixture? What would be the effect of exposing it to excess of dilute sulphuric acid in similar conditions? Explain.

53. What is the molarity of the following solutions? (*a*) 0.53 g of sodium carbonate in 100 cm³ of solution. (*b*) 2.45 g of sulphuric acid in 500 cm³ of solution. (*c*) 1 g of sodium hydroxide in 1 dm³ of solution. (*d*) 15.75 g of nitric acid in 250 cm³ of solution. (*e*) 5.6 g of potassium hydroxide in 750 cm³ of solution. (*f*) 13.8 g of potassium carbonate in 200 cm³ of solution. (*g*) 18.25 g of hydrogen chloride in 500 cm³ of solution.

54. 12.5 cm³ of 0.5M sulphuric acid neutralize 50 cm³ of a given solution of sodium hydroxide. What is the molarity of the alkali?

55. 25 cm³ of a solution of sulphuric acid required 32 cm³ of 0.1M sodium hydroxide for neutralization. Calculate the concentration of the acid in g dm⁻³.

56. 36 cm³ of a solution of sodium hydroxide required 25 cm³ of 0.5M sulphuric acid to neutralize it. Calculate the concentration of the alkali in g dm⁻³.

57. 25 cm³ of a sulphuric acid solution are neutralized by 27.0 cm³ of 0.1M sodium hydroxide solution. What is the concentration of the acid solution in terms of (*a*) molarity, (*b*) g dm⁻³?

58. 70 cm³ of a sodium carbonate solution required 50 cm³ of M sulphuric acid for complete neutralization. What is the concentration of the solution of sodium carbonate in g dm⁻³?

59. 25 cm³ of a solution of potassium carbonate required 26.8 cm³ of 0.1M hydro-

chloric acid for neutralization. If 50 cm³ of the potassium carbonate solution are mixed with 20 cm³ of 0.5M hydrochloric acid, how many cubic centimetres of 0.25M sodium hydroxide solution must be added to make the solution neutral?

60. 1 g of sodium hydroxide is added to 30 cm³ of M hydrochloric acid. How many cubic centimetres of 0.1M potassium hydroxide solution will be needed to neutralize the excess acid?

61. 10 g of calcium carbonate were dissolved in 250 cm³ of M hydrochloric acid, and the solution was then boiled. What volume of 2M potassium hydroxide solution would be required to neutralize the excess of acid?

62. What volume of 0.1M sulphuric acid would be required to neutralize a mixture of 1 g of anhydrous sodium carbonate and 1 g of sodium hydroxide?

63. 10 g of a mixture of sodium chloride and anhydrous sodium carbonate were made up to 1 dm³ of aqueous solution. 25 cm³ of this solution required 20 cm³ of 0.2M hydrochloric acid for neutralization. What was the mass of sodium chloride in the mixture?

64. 5.0 g of a mixture of sodium chloride and sodium carbonate (anhydrous) were made up to 250 cm³ of aqueous solution. 25 cm³ of this solution required 40 cm³ of 0.1M hydrochloric acid for neutralization. What is the percentage by mass of anhydrous sodium carbonate in the mixture?

65. 3.5 g of a mixture of potassium carbonate and potassium sulphate (both anhydrous) were made up to 250 cm³ of aqueous solution. 25 cm³ of this solution required 24.6 cm³ of 0.110M hydrochloric acid to neutralize. What is the percentage by mass of potassium carbonate in the mixture?

66. 50 cm³ of a solution of ammonium chloride containing 2 g dm⁻³ were boiled with 50 cm³ of 0.1M sodium hydroxide solution. What was the mass of sodium hydroxide left in excess?

67. 100 cm³ of solution containing 10 g of pure sulphuric acid were mixed with an equal volume of solution containing 15 g of hydrogen chloride. The mixture was stirred to make it uniform. If there was no volume change during the mixing, how many cubic centimetres of M sodium hydroxide solution would be needed to neutralize 25 cm³ of the mixture?

68. 100 cm³ of 0.05M sulphuric acid were placed in a flask and a small quantity of anhydrous sodium carbonate was added. The mixture was boiled to expel carbon dioxide, cooled and the volume restored by addition of distilled water to 100 cm³. 25 cm³ of the solution now required 18 cm³ of 0.1M sodium hydroxide solution to neutralize it. What was the mass of sodium carbonate added?

69. Calculate the number of molecules of water of crystallization in ethanedioic (oxalic) acid crystals, $H_2C_2O_4.xH_2O$, from the following data: 5 g of the crystals were made up to 250 cm³ of aqueous solution and 25 cm³ of this solution required 15.9 cm³ of 0.5M sodium hydroxide solution to neutralize it.

70. Equal volumes of oxygen and hydrogen are weighed at constant temperature and pressure. The masses are oxygen, x g, and hydrogen y g. Select the correct statement from the following: (i) $x = y$, (ii) $x = 2y$, (iii) $x = 8y$, (iv) $x = 16y$, (v) $x = \frac{1}{2}y$.

71. A certain atom has atomic number, Z, and mass number A. Select the correct statements from the following.

(a) The number of *neutrons* in the atom is (i) $Z - A$, (ii) $A - Z$, (iii) $A + Z$.

(b) The number of *protons* in the atom is (i) A, (ii) Z, (iii) $\frac{1}{2}(A + Z)$.

(c) The number of *electrons* in the atom is (i) Z, (ii) $Z + A$, (iii) A.

72. The ammonia molecule forms a valency bond with hydrogen ion. Which one of the following statements is true: the bond (i) results from an electron pair equally shared, (ii) results from a pair of electrons donated by the ammonia molecule, (iii) is an electrovalent bond, (iv) results from a pair of electrons donated by the hydrogen ion?

73. Elements X and Y are in Group II and in Periods 2 and 3 of the Periodic Table. Which of the following statements is true: (i) X and Y are both electronegative elements, (ii) X and Y can both ionize by losing two electrons per atom, (iii) X and Y are equally electropositive, (iv) X and Y have equal electrical conductivities?

74. Which of the following statements is correct: sulphur is classified as a *non-metal* because (i) it has a low m.p., (ii) it is an electronegative element, (iii) it is much less dense than most metals, (iv) it forms a stable hydride?

75. Which of the following statements is correct: iron is classified as a metal beause (i) it is an electropositive element, (ii) it is much denser than any non-metal, (iii) it forms a solid chloride, (iv) it shows two states of oxidation?

76. Atoms A and B are isotopes of the same element. Which of the following statements are true: A and B have (i) the same number of electrons, (ii) the same mass number, (iii) the same number of protons, (iv) the same number of neutrons, (v) the same number of electrons in the highest energy level?

77. Which of the following statements is true: magnesium is a good electrical conductor because (i) it can be formed into filaments, (ii) its atoms contain positively charged nuclei, (iii) some of its electrons are free to move through the metal, (iv) it is extracted from its chloride by electrolysis?

78. Which of the following statements is true: sulphur is a very poor electrical conductor because (i) it is less dense than most metals, (ii) its atoms contain positively charged nuclei, (iii) its electrons are tightly bound to atomic nuclei, (iv) it cannot be formed into filaments at room temperature?

79. Which of the following statements is true: the Avogadro number is (i) the number of electrons in a mole of any solid element, (ii) the number of atoms in a mole of any gas at s.t.p., (iii) the number of atoms in a mole of any element, (iv) the number of electrons needed to liberate one gramme of a univalent metal in electrolysis, (v) the number of protons in a mole of any element?

80. Which of the following statements is true: an aqueous solution of pH = 6 is (i) neutral, (ii) slightly acidic, (iii) strongly alkaline, (iv) strongly acidic?

81. Which of the following statements is true: an aqueous liquid of pH = 8 is (i) slightly alkaline, (ii) neutral, (iii) strongly acidic, (iv) weakly acidic?

82. Which of the following aqueous solutions is most likely to liberate hydrogen by the action of magnesium powder, a solution of (i) pH = 7, (ii) pH = 2, (iii) pH = 12, (iv) pH = 6.5?

83. Which of the following aqueous solutions is most likely to liberate hydrogen by the action of aluminium powder, a solution of (i) pH = 13, (ii) pH = 7, (iii) pH = 7.4?

84. Two gaseous compounds, A and B, have relative molecular masses of 2 and 50 respectively. In the same conditions, how many times faster does A diffuse than B: (i) 25 times, (ii) $12\frac{1}{2}$ times, (iii) 5 times, (iv) 100 times?

85. Equal volumes of M hydrochloric acid and M sulphuric acid are neutralized by dilute sodium hydroxide solution and x kilojoules and y kilojoules of heat are liberated respectively. Which of the following is true: (i) $x = y$, (ii) $x = \frac{1}{2}y$, (iii) $x = 2y$?

86. Equal volumes of M hydrochloric acid and M sulphuric acid are neutralized separately by dilute solutions of sodium hydroxide and potassium hydroxide. The heat evolved in each case is:

M–HCl—NaOH a kJ
M–HCl—KOH b kJ
M–H$_2$SO$_4$—NaOH c kJ
M–H$_2$SO$_4$—KOH d kJ

Which of the following statements are true: (i) $a = b$, (ii) $a = c$, (iii) $c = d$, (iv) $b = d$, (v) $a = d$?

87. Consider the reaction:

$$\tfrac{1}{2}Cl_2 + I^- \rightarrow Cl^- + \tfrac{1}{2}I_2$$

Which of the following statements are true with respect to it: (i) chloride ion is a reducing agent, (ii) chlorine is an oxidizing agent, (iii) iodide ion is oxidized, (iv) iodine is an ozidizing agent?

88. A and B are atoms of elements in the same early period of the Periodic Table; A is in Group II and B in Group III. Which of the following statements are true: (i) B has one more proton in its nucleus than A, (ii) the relative atomic mass of B must be one unit greater than the relative atomic mass of A, (iii) A has one electron fewer than B in the highest energy level, (iv) B must have one neutron more than A in its nucleus, (v) A contains one electron fewer than B?

89. A and B are atoms of elements in the same Group of the Periodic Table. A is in Period 2 and B in Period 3. Which of the following statements is true: (i) A and B have equal numbers of protons in their nuclei, (ii) A and B have equal numbers of electrons in their highest energy levels, (iii) A and B must have equal numbers of neutrons in their nuclei, (iv) A has two electrons, and B three electrons, in its highest energy level?

90. Explain the following facts: (*a*) aqueous hydrogen chloride is strongly acidic while a solution of the gas in pure toluene is non-acidic; (*b*) methane is gaseous while sodium chloride is a solid at room temperature and pressure; (*c*) diamond is hard and very resistant to attack by chemicals, while graphite is soft and less resistant to chemical attack; (*d*) copper conducts electricity readily, sulphur hardly at all.

91. The halogen elements, chlorine and bromine, both show: (*a*) univalency; (*b*) oxidizing action. Explain this in electronic terms. Give one example of the oxidizing action of chlorine and one (different) example of the same for bromine, mentioning experimental conditions. Give evidence to show that chlorine is the stronger oxidizing agent. How does the silver salt of bromine behave on exposure to light?

92. The atom of an element, X, has electron shells 2,8,8,2. Without identifying the element, discuss, in electronic terms: (*a*) the valency of the element; (*b*) whether it is likely to have oxidizing or reducing properties; (*c*) the likely nature and formula of its chloride; (*d*) the likely nature, formula and solubility in water of its hydroxide. Suggest *with reasons* a likely method of extraction of the element on the large scale.

93. Illustrate the idea of *periodicity* as applied to chemical elements by reference to the elements sodium, nitrogen, chlorine, potassium, fluorine and phosphorus. Quote the electronic structures of their atoms and give at least two examples of periodicity of properties among them.

94. Two metals, A and B, are both divalent and without action on water. A is much the more electropositive. Explain, in electronic terms, what you would expect to happen if A is placed in a solution of a salt of B. What arrangement could be made, involving A and B, to obtain electric current? Explain how the arrangement works and why it would probably be unsuitable for giving current over a long period. Suggest, with reasons, likely forms in which A and B would occur in nature and likely means of extracting the metals.

95. Give (a) *two* examples of *photocatalysis*, one from organic and one from inorganic chemistry; (b) *one* example each of an exothermic reaction and an endothermic reaction; (c) *two* reasons for considering air to be a mixture; (d) *one* example of an isotopic element, explaining the occurrence of the isotopes. Why, in spite of isotopy, does such an element usually have only very slight variations of relative atomic mass?

96. Outline the manufacture of calcium carbide. What is meant by calling this an *electrothermal* reaction? In contrast, describe the electrolytic extraction of calcium and explain the essential differences between the two methods. Contrast the behaviour of cold water with calcium and calcium carbide.

97. The essential reaction for the Haber process is:

$$N_2 + 3H_2 \rightleftharpoons 2NH_3; \quad A \text{ kJ evolved}$$

Deduce the conditions required to produce the greatest degree of efficiency in the conduct of the process and indicate briefly how they are realized in practice. How is the ammonia extracted after production? How is it converted to nitric acid on the large scale?

98. Give a concise account of the chemistry of the chlorides, oxides and hydrides of nitrogen and phosphorus (excluding preparations) to show that these two elements are properly included in the same group. Relate your answer as much as possible to electronic considerations.

Answers to General Questions

1. $MO + H_2 \rightarrow M + H_2O$
2. MgO
   $Mg_3N_2 + 6H_2O \rightarrow 3Mg(OH)_2 + 2NH_3$
3. M_2SO_4; MCl
4. $2M + 6HCl \rightarrow 2MCl_3 + 3H_2$
   M_2O_3; $M_2(SO_4)_3$
6. 98.2
7. $2MCl_3 + 3H_2SO_4 \rightarrow M_2(SO_4)_3 + 6HCl$
   M_2O_3; $M(NO_3)_3$
8. 2 : 1
9. 3 : 5
10. 2 : 3
13. 1 g
14. 11.7 g
15. 1.0 g
16. Zn, 40.4%; S, 19.9%; O, 39.7%
17. 63 g
18. 3.55 dm³
19. 31.8 g; $Ca(OH)_2$ by 0.84 g
20. 41.7%
21. 14.4 g
22. 18.0 g
23. 37.5 g Cu; 111.3 g nitrate
24. 9.84%
25. 0.581 g; 0.539 g
26. 4.05 g; 576 kg
27. 7
28. 64.4%; 35.6%
29. 0.224 dm³
30. 25 cm³ oxygen
31. 16.8 dm³
32. 11.2 dm³
33. 99.0 cm³
34. 2.165 dm³
35. 1.43 g; 500 cm³ H_2S
36. 209 g
37. 17.4 dm³
38. 0.481 dm³
39. 50 cm³; 40%
40. 1.38 g; 516 cm³
41. 279 cm³; 50 cm³
42. 107 g; 54.7 dm³; NH_3
43. (a) 1.794 dm³; (b) 4.39 g
44. 57.1%
45. 75%
46. 8.38 dm³
47. 77.8 dm³
48. 20%
49. 125 cm³; 25 cm³
50. 50% each
51. 3.5 cm³ fall in volume
52. 90 cm³ (20 cm³ fall)
53. (a) 0.05; (b) 0.05; (c) 0.025; (d) 1;
    (e) 0.133; (f) 1; (g) 1
54. 0.25
55. 6.27 g
56. 27.8 g
57. 0.054; 5.29 g
58. 75.7 g
59. 18.6 cm³
60. 50 cm³
61. 25 cm³
62. 219 cm³
63. 1.52 g
64. 42.4%
65. 53.3%
66. 0.125 g
67. 76.9 cm³
68. 0.148 g
69. 2
70. (iv)
71. (a) (ii)
    (b) (ii)
    (c) (i)
72. (ii)
73. (ii)
74. (ii)
75. (i)
76. (i) (iii) (v)
77. (iii)
78. (iii)
79. (iii)
80. (ii)
81. (i)
82. (ii)
83. (i)
84. (iii)
85. (ii)
86. (i) (iii)
87. (ii) (iii)
88. (i) (iii) (v)
89. (ii)

Index

Periodic Table

Periodic Table of

Period	Group 1	Group 2

Relative Atom

1							1 **H** Hydrogen 1		

Atomic Number

2	7 **Li** Lithium 3	9 **Be** Beryllium 4
3	23 **Na** Sodium 11	24 **Mg** Magnesium 12

Period	Group 1	Group 2								
4	39 **K** Potassium 19	40 **Ca** Calcium 20	45 **Sc** Scandium 21	48 **Ti** Titanium 22	51 **V** Vanadium 23	52 **Cr** Chromium 24	55 **Mn** Manganese 25	56 **Fe** Iron 26	**Co** Cobalt 27
5	85.5 **Rb** Rubidium 37	88 **Sr** Strontium 38	89 **Y** Yttrium 39	91 **Zr** Zirconium 40	93 **Nb** Niobium 41	96 **Mo** Molybdenum 42	99 **Tc** Technetium 43	101 **Ru** Ruthenium 44	**Rh** Rhodium 45
6	133 **Cs** Cæsium 55	137 **Ba** Barium 56	* 139 **La** Lanthanum 57	178.5 **Hf** Hafnium 72	181 **Ta** Tantalum 73	184 **W** Tungsten 74	186 **Re** Rhenium 75	190 **Os** Osmium 76	**Ir** Iridium 77
7	223 **Fr** Francium 87	226 **Ra** Radium 88	† 227 **Ac** Actinium 89						

Lanthanide Series	* 139 **La** Lanthanum 57	140 **Ce** Cerium 58	141 **Pr** Praseodymium 59	144 **Nd** Neodymium 60	147 **Pm** Promethium 61	150 **Sm** Samarium 62	**Eu** Europi 63
Actinide Series	† 227 **Ac** Actinium 89	232 **Th** Thorium 90	231 **Pa** Protactinium 91	238 **U** Uranium 92	237 **Np** Neptunium 93	242 **Pu** Plutonium 94	**Am** Americ 95